Preface

As you will have noticed, this seventh edition of *Building Regulations in Brief* is in an exciting new colour format. We have continued to provide a simplified explanation of the numerous regulations which relate to buildings that our readers have told us they value. We are proud of this book and believe that it provides a comprehensive guide to a complicated area of legislation. Indeed, there is no other book on the market that provides the detail we do in such an easy to use format.

We recognize that it is only 12 months since the last edition of this book was published, but much has changed with the source legislation, and therefore we have found it necessary to produce a new edition. Particular changes this time include the adoption of the new Building Regulations 2010 (SI 2010/2214), which consolidate the Building Regulations (SI 2000/2531), and which came into force in October 2010. All the amendments made since the 2000 Regulations came into effect are incorporated in this instrument. As part of the government's planned changes there have also been amendments to Parts F (Ventilation) and J (Combustion Appliances and Fuel Storage Systems), and the wholesale amendment of all four sections of Part L (Conservation of Fuel and Power).

Although the Regulations themselves are comparatively short, they rely on their technical detail being available in a series of Approved Documents and a vast number of British, European and International standards, codes of practice, drafts for development, published documents and other non-statuary guidance documents. The main problem, from the point of view of the average builder and DIY enthusiast, is that the Building Regulations are far too professional for their purposes. Because they cover every aspect of building, they are often far too detailed and contain far too many options. All the builder or DIY person really requires is sufficient information to enable them to comply with the regulations in the simplest and most cost-effective manner possible.

Building inspectors, acting on behalf of local authorities, are primarily concerned with whether a building complies with the requirements of the Building Regulations, and to assess this, they need to 'see the calculations'. But how does the DIY enthusiast and/or builder obtain these calculations? Where can they find, for instance, the policy and requirements for the load-bearing elements of a structure?

Builders, through experience, are normally aware of the overall requirements for building projects such as laying drains, building walls, or installing central heating or glazing. However, they still need a reminder when they come across a different situation for the first time (e.g. how deep will the foundations have to be if they are going to construct a building on soft soil?).

On the other hand, the DIY enthusiast, keen on building his own extension, conservatory, garage or workshop usually has no past experience, and needs the relevant information – but in a form that he can easily understand without having had the advantage of many years of experience. In fact, what he really needs is a rule-of-thumb guide to the basic requirements.

We know from a number of surveys that the majority of builders and virtually all DIY enthusiasts are self-taught, and most of their knowledge is gained through experience. When they hit a problem, it is usually discussed over a drink with friends in the building trade, as opposed to seeking professional help. What they really need is a reference book to enable them to understand (or remind themselves of) the official requirements.

The aim of this book, therefore, is to provide the reader with an in-brief guide that can act as an *aide-mémoire* to the current requirements of the Building Regulations. Intended readers are primarily builders and the DIY fraternity (who need to know the Regulations but do not require the detail), but the book, with its ready reference and no-nonsense approach, will be equally useful to students, architects, designers, building surveyors and inspectors as well.

The structure of this book

In essence, this book is in two parts. The first part (Chapters 1 to 5) provides a user-friendly introduction to the Building Act 1984 and its associated Building Regulations. It explains the meaning of the Building Regulations, their current status, requirements, associated documentation and how local authorities and councils view their importance. We also describe the content of the guidance documents (such as the overview of the key elements of each Approved Document contained in Chapter 3), and provide details of how to get planning permission and how much it will cost. This first part concludes with a series of 'What if?' scenarios, providing answers to the most common questions that DIY enthusiasts and builders might ask concerning building projects (for ease of reference this list is alphabetical). The second part is the bit of the book of which we are most proud, and the part about which we get the most comments. It contains Chapter 6 – Meeting the Requirements of the Building Regulations – and its supporting appendices. In essence, this part provides a foundations-up approach, detailing all the regulations relating to specific projects (e.g. building walls). We have developed Chapter 6 extensively over the 11 years of writing previous versions of this book.

In summary, the chapters in this book are:

Chapter 1 – The Building Act 1984
Chapter 2 – The Building Regulations 2010

Chapter 3 – The requirements of the Building Regulations
Chapter 4 – Planning permission
Chapter 5 – Requirements for planning permission and Building Regulations approval
Chapter 6 – Meeting the requirements of the Building Regulations

Chapter 6 is further supported by appendices which cover specific areas in greater detail as entities in their own right, rather than how they apply to a particular project. All the appendices are available at http://www.routledge.com/books/details/9780415809696 and they are also available in the printed version:

Appendix A – Access and facilities for disabled people
Appendix B – Conservation of fuel and power
Appendix C – Sound insulation
Appendix D – Guidance to the requirements of Part P – electrical safety
Appendix E – Fire resistance
Appendix F – Means of escape
Appendix G – Entrance and access

The book concludes with a list of acronyms, a bibliography, useful names and addresses, and a full index.

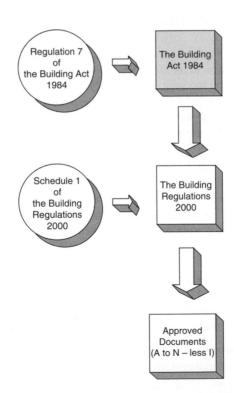

The following symbols will help you get the most out of this book: In the margins you will find:

An important requirement or point

A good idea or suggestion

and within the text:

Notes: these are used to provide further amplification or information.

Blue italic text, which indicates a direct quote from the relevant Act, Regulation or Approved Document

Shaded boxes are used in Chapter 6 to show either the full text of the Building Regulation's *legal requirements* or a paraphrased version of these requirements.

The current legislation is the Building Regulations 2010. It is made by the Secretary of State for the Environment under powers delegated by Parliament under the Building Act 1984. These Regulations have replaced **all** former regulations and Statutory Instruments.

The Building Act 1984

The Secretary of State ensures that the health, welfare and convenience of persons living in or working in (or nearby) buildings is secured through the Building Act 1984. One of the Act's prime purposes is to assist in the conservation of fuel and power, prevent waste, undue consumption and the misuse and contamination of water.

The Act imposes on owners and occupiers of buildings a set of requirements concerning the design and construction of buildings, and the provision of services, fittings and equipment used in (or in connection with) buildings.

The Building Act 1984 does not apply to Scotland or to Northern Ireland.

The Building Act 1984 has five parts:

Part 1 – The Building Regulations
Part 2 – Supervision of Building Work etc. other than by a local authority
Part 3 – Other provisions about buildings
Part 4 – General
Part 5 – Supplementary

Part 5 then contains seven schedules, the prime function of which is to list the principal areas requiring regulation and to show how the Building Regulations are to be controlled by local authorities. These schedules are:

Schedule 1 – Building Regulations
Schedule 2 – Relaxation of Building Regulations

Schedule 3 – Inner London
Schedule 4 – Provisions consequential upon public body's notice
Schedule 5 – Transitional provisions
Schedule 6 – Consequential amendments
Schedule 7 – Repeals

Schedule 1 is the most important (from the point of view of builders), as it shows, in general terms, how the Building Regulations are to be administered by local authorities, the approved methods of construction and the approved types of materials that are to be used in (or in connection with) buildings.

Regulation 7 of the Building Act 1984 covers materials and workmanship and states that building work shall be carried out:

(a) *with adequate and proper materials which –*
  (i) are appropriate for the circumstances in which they are used,
  (ii) are adequately mixed or prepared, and
  (iii) are applied, used or fixed so as adequately to perform the functions for which they are designed; and
(b) *in a workmanlike manner.*

The Building Regulations

Building Regulations 2010 (SI 2010/2214) form Schedule 1 of the Building Act and are a set of minimum requirements and basic performance standards designed to secure the health, safety and welfare of people in and around buildings and to conserve fuel and energy in England and Wales.

The Building Regulations describe the mandatory requirements for completing **all** building work, including:

- accommodation for specific purposes (e.g. for disabled persons);
- air pressure plants;
- cesspools (and other methods for treating and disposing of foul matter);
- dimensions of rooms and other spaces (inside buildings);
- drainage (including waste disposal units);
- emission of smoke, gases, fumes, grit or dust (or other noxious or offensive substances);
- fire precautions (services, fittings and equipment, means of escape);
- lifts (escalators, hoists, conveyors and moving footways);
- materials and components (suitability, durability and use);
- means of access to and egress from buildings;
- natural lighting and ventilation of buildings;
- open spaces around buildings;
- prevention of infestation;
- provision of power outlets;
- resistance to moisture and decay;
- site preparation;

- solid fuel, oil, gas, electricity installations (including appliances, storage tanks, heat exchangers, ducts, fans and other equipment);
- standards of heating, artificial lighting, mechanical ventilation and air-conditioning;
- structural strength and stability (overloading, impact and explosion, underpinning, safeguarding of adjacent buildings);
- telecommunications services (wiring installations for telephones, radio and television);
- third-party liability (danger and obstruction to persons working or passing by building work);
- transmission of heat;
- transmission of sound;
- waste (storage, treatment and removal);
- water services, fittings and fixed equipment (including wells and boreholes for supplying water); and
- matters connected with (or ancillary to) any of the foregoing matters.

They are legal requirements laid down by Parliament and are based on the Building Act 1984. The Building Regulations:

- are approved by Parliament;
- are designed to ensure structural stability;
- contribute to meeting the needs of disabled people;
- deal with the minimum standards of design and building work for the construction of domestic, commercial and industrial buildings;
- ensure the health and safety of people in and around buildings (by providing functional requirements for building design and construction);
- promote energy efficiency in buildings;
- promote the use of suitable materials to provide adequate durability, fire and weather resistance, and the prevention of damp;
- set out the procedure for ensuring that building work meets the standards laid down;
- stipulate the minimum amount of ventilation and natural light to be provided for habitable rooms.

The level of safety and standards acceptable are set out as guidance in **Approved Documents**, which are discussed below. Compliance with the detailed guidance of the Approved Documents is usually considered as evidence that the Building Regulations themselves have been complied with.

Approved Documents

The Building Regulations are supported by separate documents which correspond to the different areas covered by the Regulations. These are called *Approved Documents* and they contain practical and technical guidance on ways in which the requirements of Schedule 1 and Regulation 7 of the Building Act 1984 can be met.

Each Approved Document reproduces the actual *requirements* contained in the Building Regulations relevant to the subject area. These are then followed by *practical and technical guidance* (together with examples) showing how the requirements can be met in some of the more common building situations. There may, however, be alternative ways of complying with the requirements to those shown in the Approved Documents, and you are, therefore, under no obligation to adopt any particular solution contained in an Approved Document if you prefer to meet the requirement(s) in some other way.

 If you intend to carry out building work you should always check with the building control body, either the local authority or an approved inspector, that your proposals comply with Building Regulations.

The current set of Approved Documents are in 14 parts, A to P (less I and O) and consist of:

A   Structure
B   Fire safety
C   Site preparation and resistance to contaminants and moisture
D   Toxic substances
E   Resistance to the passage of sound
F   Ventilation
G   Sanitation, hot water safety and water efficiency
H   Drainage and waste disposal
J   Combustion appliances and fuel storage systems
K   Protection from falling, collision and impact
L   Conservation of fuel and power
M   Access to and use of buildings
N   Glazing – safety in relation to impact, opening and cleaning
P   Electrical safety – dwellings

 The Planning Portal actually shows these as different titles.

In accordance with Regulation 8 of the Building Regulations, the requirements in Parts A to D, F to K and N (except for paragraphs G2 and J6) of Schedule 1 to the Building Regulations do not require anything to be done except for the purpose of securing reasonable standards of health and safety for persons in or about buildings (and any others who may be affected by buildings or matters connected with buildings).

 Note:

(1) Paragraphs H2 and J6 are excluded from Regulation 8 because they deal directly with the prevention of contamination of water.
(2) Parts E and M (which deal, respectively, with resistance to the passage of sound, and access to and use of buildings) are excluded from Regulation 8 because they address the welfare and convenience of building users.
(3) Part L is excluded from Regulation 8 because it addresses the conservation of fuel and power.

Planning permission

Planning permission is the single biggest hurdle for anyone who has acquired land on which to build a house, or wants to extend or carry out other building work on a property. There is never a guarantee that permission will be given, and without that permission no project can start. The planning system itself is not at all user-friendly. There are a bewildering array of formalities to go through and ever more stringent requirements to satisfy. Planning permission has never been more difficult to get, nor so sought after. Every year over half a million applications are made, and the number is rising.

The purpose of the planning system is to protect the environment as well as public amenities and facilities. It is **not** designed to protect the interests of one person over another. Within the framework of legislation approved by Parliament, councils are tasked to ensure that development is allowed where it is needed, while ensuring that the character and amenity of the area are not adversely affected by new buildings or changes in the use of existing buildings and/or land.

Provided that the work you are completing does not affect the external appearance of the building, you are allowed to make certain changes to your home without having to apply to the local council for permission. These are called *Permitted Development Rights*. However, the majority of building work, that you are likely to undertake will probably require you to have planning permission – so be warned!

The actual details of planning requirements are complex but, for most domestic developments, the planning authority is only really concerned with construction work such as an extension to a house (e.g. a conservatory) or the provision of a new garage or new outbuildings. Structures such as walls, fences and decking also need to be considered because their height or siting might well infringe the rights of neighbours and other members of the community. The planning authority will also want to approve any change of use, such as converting a house into flats or running a business from premises previously occupied as a dwelling only.

The voluntary Code for Sustainable Homes was launched as part of a package of measures towards zero-carbon development in 2006. Full technical guidance on how to comply with the code was published in April 2007. Further details of the code can be found in Chapter 4.

Main changes in this edition of *Building Regulations in Brief*

For this new edition we have incorporated the updated Building Regulations 2010, which consolidate the Building Regulations (SI 2000/2531) and subsequent amending regulations, and came into force in October 2010. This consolidation has brought about changes to the numbering of the Building

Regulations, as outlined in Appendix 2A at the end of Chapter 2. As part of planned changes there have also been amendments to Parts F (Ventilation) and J (Combustion Appliances and Fuel Storage Systems), and further large-scale amendments of all four sections of Part L (Conservation of Fuel and Power).

The main changes to the legal requirements and the supporting guidance since the previous issues of Approved Documents F, J and L are as set out below.

Legal requirements

Conservation of fuel and power

In new dwellings and buildings:

- *extensions consisting of a conservatory or porch are only exempted from the energy-efficiency provisions of Parts L where the existing walls, windows or doors are retained, or replaced provided that the heating system of the building does not extended into the conservatory or porch;*
- *a CO_2 emission rate needs to be carried out and given to the building control body before the start of building work on the erection of a new building.*

In existing dwellings and buildings:

- *the installation of thermal insulation in a roof space or loft space need not be notified to the building control body if this is the only work being carried out and the work is not carried out so as to comply with any requirement in the Building Regulations.*

Ventilation

In new and existing dwellings and buildings:

- *all fixed mechanical ventilation systems shall be commissioned and a commissioning notice given to the building control body; and*
- *the owner shall be given sufficient information about the ventilation system and its maintenance requirements.*

In new dwellings:

- *the air flow rates of mechanical ventilation systems shall be measured on site and a notice given to the building control body.*

Combustion appliances and fuel storage systems:

- *a new requirement exists for the provision of carbon monoxide alarms where solid fuel appliances are installed.*

Changes in the technical guidance

Conservation of fuel and power

For a new dwelling:

- the annual CO_2 emission rate of a new dwelling must be 25 per cent better than the previous 2006 requirement;
- secondary heating is now counted as part of the annual CO_2 emission rate;
- credit is allowed if low-energy lighting is installed;
- some performance specifications for building fabrics and services have been strengthened;
- revised guidance is provided for avoiding thermal bridging at construction joints; and
- new guidance on how to limit heat loss from a swimming pool basin has been included.

For a new building:

- revised guidance is provided:
  - on shell and core developments and first fit-out work;
  - on avoiding thermal bridging at construction joints;
  - to limit heat loss from a swimming pool basin where this is constructed as part of a new building;
- a revised procedure is provided for demonstrating that reasonable provision has been made to limit the effects of solar gain in summer.

In existing dwellings and buildings:

- guidance is generally based on an elemental approach to demonstrating compliance;
- the main technical changes comprise a general strengthening of energy-efficiency standards for work on thermal elements, controlled fittings and controlled services in existing dwellings;
- amended guidance is given:
  - for historic and traditional buildings which may have an exemption from the energy-efficiency requirements (or where special considerations apply);
  - where an extension is a conservatory or porch that is not exempt from the energy-efficiency requirements;
  - for the renovation of a thermal element through the provision of a new layer or the replacement of an existing layer has been expanded;
  - for swimming pool basins (walls and floor) in existing dwellings and buildings other than dwellings.

Ventilation

- Ventilation provisions have been increased for dwellings with a design air permeability tighter than or equal to $5\,m^3/h\,m^2$) at 50 Pa.

- For passive stack ventilators, the stack diameter has been increased to 125 mm for all room types, and use of passive stack ventilation in inner wet rooms has been clarified.
- The guidance for ventilation when a kitchen or bathroom in an existing dwelling is refurbished has been clarified.

Combustion appliances and fuel storage systems

Clarification and guidance has been updated and included for:

- the provision of hearths and wall clearances for solid fuel appliances;
- flue outlet clearances adjacent to pitched roofs;
- the visual inspection of concealed flues;
- the need for permanent ventilation openings for open-flued appliances in very airtight houses (particularly older houses);
- the use of liquid biofuel and blends of mineral oil and liquid biofuel with combustion installations designed to burn oil; and
- identifying where secondary containment for oil tanks is necessary

 Note: If any reader has any thoughts about the contents of this book (such as areas where perhaps they feel we have not given sufficient coverage, omissions and/or mistakes, etc.) then please let me know by emailing us at raysam@ herne.org.uk, and we will make suitable amendments in the next edition of this book.

Foreword

In my role, quick reference to the Approved Documents is sometimes essential on-site, and with so much useful information contained in one clear volume, I find this book invaluable.

This simple approach to the regulations, without delving too deeply into detail, gives a perfect guide for the DIY person, student, site operative and architect, as well as other professionals across the board; the key being that it helps you to locate the vital information. It is also extremely cost-effective compared with the full documents, which are not easy to read or as accessible due to the multiple volumes.

A wide range of angles is covered, including Statutory Appeals and what you can do if you do not agree with the process. There is a guide to what requirements you need for planning, which could well save the reader a great deal of time and cost. The book clearly identifies a range of common issues, and this can be reassuring at times for the individual reading it, where comparisons can be made when making final decisions. For example, it advises that there may be variations in the planning requirements and to some extent the Building Regulations from one area of the country to another, and that this can sometimes be misunderstood.

Even the more experienced personnel refer back to the Approved Documents, but in this edition the references are contained in one handy volume, enabling you to easily keep up-to-date with the latest regulations as well.

This book helps you to understand what to expect from a legal point of view, the hurdles you might encounter along the way, and the interface of the governing bodies. With this complexity in mind, this book is essential for understanding the Building Regulations and the requirements of the Building Act.

Appendix B Conservation of Fuel and Power typifies the writers' simplicity in explaining the additions to Part L, whereas the actual Approved Documents take some reading to understand the sections completely.

In each of the sections, the book gives you guidelines for meeting the requirements for Building Regulations and the Act, assisting everyone from the DIY enthusiast to the professional, and providing an excellent platform for the legal requirements and where to begin in achieving compliance.

This is an essential reference book for understanding the Building Regulations, and it is cleverly designed to put the building regulations in brief by giving you all the information needed to gain compliance, but at the same time simplifying the requirements. The book is a very cost-effective option rather than buying the full volumes and, in my opinion, is excellent reading.

A must for all people involved in Building Regulations and construction.

Mark Heggs MBEng, MIHEEM, MICWCI
University of Leicester

About the authors

Ray Tricker (MSc, IEng, FIET, FCMI, FCQI, FIRSE) is the Senior Consultant (Management Systems) of Herne European Consultancy Ltd (a company specializing in business enhancement and outsourced back office functions). He is also an established author (over 27 titles published). He served with the Royal Corps of Signals (for a total of 37 years), during which time he held various senior managerial posts, culminating in being appointed as the Chief Engineer of NATO's Communication Security Agency (ACE COMSEC).

Most of Ray's work since joining Herne has centred on the European railways. He has held a number of posts with the Union International des Chemins de Fer (UIC) (e.g. Quality Manager of the European Train Control System (ETCS)) and with the European Union (EU) Commission (e.g. T500 Review Team Leader, European Rail Traffic Management System (ERTMS) Users Group Project Co-ordinator, HEROE Project Co-ordinator), and currently (as well as writing books on diverse subjects such as optoelectronics, medical devices, ISO 9001:2008, building, wiring and water regulations for Routledge) he is busy assisting small businesses from around the world (usually on a no-cost basis) to produce their own auditable quality and/or integrated management systems to meet the requirements of ISO 9001:2000, ISO 14001 and OHSAS, etc. He is also a UKAS Assessor for the assessment of certification bodies for the harmonization of the trans-European high-speed rail network.

Recently he was appointed as the Quality and Safety Manager for the Project Management Consultant overseeing the multi-billion dollar Trinidad Rapid Rail System, and is currently the Quality Director for the Independent Safety Authority for a multi-billion dollar Abu Dhabi rail project (one day he might retire!).

For this edition of *Building Regulations in Brief*, Ray is joined by **Samantha Alford** (MSc, MCIPS). Samantha served with the Royal Air Force for 17 years, where she held various logistics and managerial posts, the majority of which were in the operations and planning areas. Samantha also has a wealth of experience in the catering and fuels fields. She holds an MSc in Defence Logistics Management, which was completed in 2006; she is an expert in integrated (i.e. quality, environmental and safety)

management and is a qualified Internal, External and Third Party Quality Auditor, an experienced instructor and a published author.

On leaving the RAF, Samantha started her own company specializing in the provision of supplemental author and co-author services to a number of publishers, and has worked on projects as diverse as marketing and quantitative techniques. She also provides *ad hoc* administrative services to businesses in her local area, as well as running a successful events management company.

1

The Building Act 1984

The Building Act 1984 applies in England and Wales and is the United Kingdom statute under which the Building Regulations have been made.

1.1 What is the Building Act 1984? (Building Act 1984 Section 1)

The Building Act 1984 is the mechanism by which the Secretary of State ensures that the health, welfare and convenience of persons living in or working in or near buildings is secured.

The primary purpose of the Building Act 1984 is to assist in the conservation of fuel and power, and to prevent waste, undue consumption, misuse and contamination of water. The Building Act 1984 imposes a set of requirements on owners and occupiers of buildings which cover the design and construction of buildings, and the provision of services, fittings and equipment used in (or in connection with) buildings. These involve and cover:

- a method of controlling (inspecting and reporting) buildings;
- how services, fittings and equipment may be used;
- the inspection and maintenance of any service, fitting or equipment used.

1.1.1 What about the rest of the United Kingdom?

The Building Act 1984 only applies to England and Wales. Separate acts and regulations apply in Scotland or Northern Ireland; these are shown in Table 1.1.

Scotland

In Scotland, the requirements for buildings are controlled by the *Building (Scotland) Regulations 2003* as amended by the *Building (Scotland) Amendment Regulations 2011*. The methods for implementing these requirements are similar to those for England and Wales, except that the guidance documents for achieving compliance are contained in two *Technical Handbooks* (2010), one for domestic work and one for non-domestic work, and other associated guidance.

The main procedural difference between the Scottish system and the others is that a building warrant is **still** required before work can start in Scotland.

Table 1.1 Building legislation within the United Kingdom

	Act	Regulations	Implementation
England and Wales	Building Act 1984	Building Regulations 2010	Approved Documents
Scotland	Building (Scotland) Act 2004	Building (Scotland) Regulations 2004	Technical Handbooks
Northern Ireland	Building Regulations (Northern Ireland) Order 1979	Building Regulations (Northern Ireland) 2000 (as amended)	Technical Booklets

Northern Ireland

On the other hand, in Northern Ireland the primary legislation for buildings are the *Building Regulations (Northern Ireland) Order 1979* under which *Building Regulations (Northern Ireland) 2000* (as amended) are made. These regulations revoked and replaced with amendments the previous 1994 Building Regulations, they have since been amended by *The Building (Amendment) Regulations (Northern Ireland) 2010*. The Principal Regulations comprise 15 Parts and are supported by *Technical Booklets* which are then used to ensure that the requirements are implemented.

Table 1.2 provides a summary of the titles of the sections of Building Regulations that are apply in the United Kingdom.

1.2 What does the Building Act 1984 contain?

Table 1.2 Building Regulations for the United Kingdom

England and Wales	Scotland		Northern Ireland
Part A Structure	Section 1 – Structure	Technical Handbooks 2010	Part D – Structure
Part B Fire safety	Section 2 – Fire	Technical Handbooks 2010	Part E – Fire safety
Part C Site preparation and resistance to contaminants and water	Section 3 – Environment	Technical Handbooks 2010	Part C – Preparation of site and resistance to moisture
Part D Toxic substances	Section 3 – Environment	Technical Handbooks 2010	Part B – Materials and workmanship
Part E Resistance to the passage of sound	Section 5 – Noise	Technical Handbooks 2010	Part G – Sound insulation of dwellings

Table 1.2 (Continued)

England and Wales	Scotland		Northern Ireland
Part F Ventilation	Section 3 – Environment	Technical Handbooks 2010	Part K – Ventilation
Part G Sanitation, hot water safety and water efficiency	Section 3 – Environment	Technical Handbooks 2010	Part P – Sanitary appliances and unvented hot water storage systems
Part H Drainage and waste disposal	Section 3 – Environment	Technical Handbooks 2010	Part N – Drainage
Part J Combustion appliances and fuel storage	Section 3 – Environment Section 4 – Safety	Technical Handbooks 2010	Part L – Heat-producing appliances and liquefied petroleum gas installations
Part K Protection from falling, collision and impact	Section 4 – Safety	Technical Handbooks 2010	Part H – Stairs, ramps and protection from impact
Part L Conservation of fuel and power	Section 6 – Energy	Technical Handbooks 2010	Part F – Conservation of fuel and power
Part M Access and facilities for disabled people	Section 4 – Safety	Technical Handbooks 2010	Part R – Access and facilities for disabled people
Part N Glazing	Section 6 – Energy	Technical Handbooks 2010	Part V – Glazing
Part P Electrical safety	Section 4 – Safety	Technical Handbooks 2010	

The Building Act 1984 is made up of five parts:

Part 1	The Building Regulations
Part 2	Supervision of building work, etc. other than by a local authority
Part 3	Other provisions about buildings
Part 4	General
Part 5	Supplementary

These parts are then broken down into a number of sections and sub-sections as shown in Appendix 1A.

1.3 What are the Supplementary Regulations?

The Supplementary Regulations make up Part 5 of the Building Act. They comprise seven schedules whose function is to list the principal areas requiring regulation and to show how the Building Regulations are to be controlled by local authorities. These schedules are:

Schedule 1	Building Regulations
Schedule 2	Relaxation of Building Regulations for existing work
Schedule 3	Inner London
Schedule 4	Provisions consequential upon public body's notice
Schedule 5	Transitional provisions
Schedule 6	Consequential amendments
Schedule 7	Repeals

The details of what is contained within each of these schedules are shown in alphabetical order in Appendix 1B. We also discuss each area on the following pages.

1.4 What are 'Approved Documents'?
(Building Act 1984 Section 6)

The Secretary of State makes available a series of documents (called '**Approved Documents**') which are intended to provide practical guidance with respect to the requirements of the Building Regulations. More details about Approved Documents are contained in Chapter 3 and on the Planning Portal (http://www.planningportal.gov.uk).

1.5 How are buildings classified? (Building Act 1984 Section 35)

For the purpose of the Building Act, the six normal classifications for buildings are:

- by reference to size;
- by description;
- by design;
- by purpose;
- by location;
- or '*any other suitable characteristic*'!

1.6 Who polices the Building Act?

Under the terms of the Building Act 1984, **local authorities** are responsible for ensuring that any building work that takes place in their area conforms to the requirements of the associated Building Regulations. They have the authority to:

*   make you take down and remove or rebuild anything that contravenes a regulation;
*   make you complete alterations so that your work complies with the Building Regulations;
*   employ a third party (and then send you the bill!) to take down and rebuild non-conforming buildings or parts of buildings.

 You can be prosecuted or ordered to carry out remedial work on a property whether you are the owner or merely the occupier.

They can, in certain circumstances, even take you to court and have you fined – especially if you fail to complete the removal or rebuilding of the non-conforming work.

1.7 How is my building work evaluated for conformance with the Building Regulations? (Building Act 1984 Section 33)

Part of the local authority's duty is to make regular checks that all building work being completed is in conformance with the approved plan and the Building Regulations. These checks would normally be completed at certain stages of the work (e.g. the excavation of foundations), and tests will include:

*   tests of the soil or sub-soil of the site of the building;
*   tests of any material, component or combination of components that has been, is being or is proposed to be used in the construction of a building;
*   tests of any service, fitting or equipment that has been, is being or is proposed to be provided in or in connection with a building;
*   the cost of carrying out these tests will normally be charged to the owner or occupier of the building.

 The local authority has the power to ask the person responsible for the building work to complete some of these tests on their behalf.

1.8 What are the duties of the local authority?
(Building Act 1984 Section 91)

It is the duty of local authorities to ensure that the requirements of the Building Act 1984 are carried out and that the appropriate associated Building Regulations are enforced, subject to:

- the provisions of Part I of the Public Health Act 1936 (relating to united districts and joint boards);
- Section 151 of the Local Government, Planning and Land Act 1980 (relating to urban development areas);
- Section 1(3) of the Public Health (Control of Disease) Act 1984 (relating to port health authorities).

Limitations are placed on a local authority's ability to enforce building regulations for work on buildings which are owned by exempt bodies, work that is subject to an initial notice or work relating to a plans certificate which was issued or accepted before an initial notice ceases to be in force.

1.8.1 What document controls must local authorities have in place? (Building Act 1984 Sections 92 and 93)

The Secretary of State mandates a written format for all notices, applications, orders, consents, demands and other documents, that are required by this Act or by a local authority (or an officer of a local authority.

All the documents that a local authority provides under the Building Act 1984 must be correctly authenticated and signed by:

- the proper officer for the authority;
- the district surveyor (for documents relating to matters within his specialization);
- an officer authorized by the authority to sign documents (of a particular kind).

 If a document bears the signature of an officer of the local authority it is deemed to have been issued by that authority.

1.8.2 How do local authorities 'serve' notices and documents? (Building Act 1984 Section 94)

Notices, orders, consents, demands or other documents which are authorized or required by the Building Act 1984 may be given or served to a person in the following ways:

- by delivering it to the person concerned;
- by leaving it, or sending it in a pre-paid letter addressed to him, at his usual or last known residence.

Notices for an officer of the local authority are considered to be 'served' if they are left or sent in a prepaid letter addressed to the officer at his office.

If it is not possible to ascertain the name and address of the person to or on whom it should be given or served (or if the premises are unoccupied) then the notice, order, consent, demand or other document can be addressed to the 'owner' or 'occupier' of the premises (naming them) and delivering it to 'some person on the premises'. Or, if there is not anyone at the premises to whom it can be delivered, a copy of the document can be fixed to a conspicuous part of the premises.

1.9 What are the powers of the local authority? (Building Act 1984 Sections 97–101)

The local authority has the following powers under the Building Act 1984 and its associated Building Regulations:

*   overall responsibility for the construction and maintenance of sewers and drains and the laying and maintenance of water mains and pipes;
*   the authority to make the owner or occupier of any premises complete essential and remedial work in connection with the Building Act 1984 (particularly with respect to the construction, laying, alteration or repair of a sewer or drain);
*   the authority to complete remedial and essential work themselves (on repayment of expenses) if the owner or occupier refuses to do this work himself;
*   the ability to sell any materials that have been removed, by them, from any premises when executing works under this Act (paying all proceeds, less expenses, from this sale to the owner or occupier).

 This does not apply to any refuse that is, or has been, removed by the local authority.

1.9.1 Has the local authority any power to enter premises?
(Building Act 1984 Section 95)

An authorized officer of a local authority has a right to enter any premises, at all 'reasonable hours', for the following reasons:

*   to check if there is (or has been) a contravention of the Building Act or of Building Regulations that the local authority has a duty to enforce;
*   to see if there are any circumstances that would require local authority action or for them to carry out any work;
*   to carry out work or take any action that is authorized or required by the Building Act, or by Building Regulations;
*   to carry out their function as a local authority.

If the premises is a factory or workplace the local authority must give 24 hours' notice of their intention to enter the property.

If the local authority is refused admission to any premises (or the premises are unoccupied) then the local authority can apply to a Justice of the Peace for a warrant authorizing entry.

1.10 Who are approved inspectors? (Building Act 1984 Section 49)

An approved inspector is a person who is approved by the Secretary of State (or a body such as a local authority or county council designated by the Secretary of State) to inspect, supervise and authorize building work. Lists of approved inspectors are available from all local authorities and can be found on the Construction Industry Council's website (http://www.cic.org.uk/services/AIregister.shtml).

If an approved inspector gives a notice or certificate that falsely claims to comply with the Building Regulations and/or the Building Act then he is liable to prosecution.

1.10.1 What is an initial notice? (Building Act 1984 Section 47)

An approved inspector and the person intending to carry out the work will have to present an initial notice and plan of work to the local authority. Once accepted, the approved inspector is authorized to inspect and supervise all work being completed and to provide certificates and notices. Acceptance of an initial notice by a local authority is treated as 'depositing plans of work'.

Under Section 47 of the Building Act, the local authority is required to accept all certificates and notices, unless the initial notice and plans contravene a local ruling. Whilst the initial notice continues to be in force the local authority is not allowed to give a notice in relation to any of the work being carried out or take any action for a contravention of Building Regulations.

If the local authority rejects the initial notice for any reason, then the approved inspector can appeal to a magistrates' court for a ruling. If still dissatisfied, he can appeal to the Crown Court.

Cancellation of the initial notice (Building Act 1984 Sections 52 and 53)

If an approved inspector is unable to carry out or complete his functions, or is of the opinion that there is a contravention of the Building Regulations, then he can cancel the initial notice lodged with the local authority.

Equally, if the person carrying out the work has good reason to consider that the approved inspector is unable (or unwilling) to carry out his functions, then that person can cancel the initial notice given to the local authority.

 The fact that the initial notice has ceased to be in force does not affect the right of an approved inspector to give a new initial notice relating to any of the work that was previously specified in the original notice.

1.10.2 What are plans certificates? (Building Act 1984 Section 50)

When an approved inspector has inspected and is satisfied that the plans of work specified in the initial notice do not contravene the Building Regulations in any way, he will provide a certificate (referred to as a 'plans certificate') to the local authority. This plans certificate:

• can relate to the whole or part of the work specified in the initial notice;
• does not have any effect unless the local authority accepts it;
• may only be rejected by the local authority 'on prescribed grounds'.

1.10.3 What are final certificates? (Building Act 1984 Section 51)

Once the approved inspector is satisfied that all work has been completed in accordance with the work specified in the initial notice, he will provide a certificate (referred to as a 'final certificate') to the local authority and the person who carried out the work. This certificate will detail his acceptance of the work and, once acknowledged by the local authority, the approved inspector's job will have been completed and (from the point of view of local authority) he will have been considered 'to have discharged his duties'.

 If work has not commenced within three years the local authority can cancel the initial notice.

1.10.4 Who retains all these records? (Building Act 1984 Section 56)

Local authorities are required to keep a register of all initial notices and certificates given by approved inspectors and to retain all relevant and associated documents concerning those notices and certificates. The local authority is further required to make this register available for public inspection during normal working hours.

1.10.5 Can public bodies supervise their own work?
(Building Act 1984 Sections 54 and 55)

If a public body (e.g. local authority or county council) is of the opinion that building work that is to be completed on one of its own buildings can be adequately supervised by one of its employees and/or agents, then they can provide the local authority with a notice (referred to as a 'public body's notice') together with their plan of work.

Once accepted by the local authority, the public body is authorized to inspect and supervise all work being completed and to provide certificates and notices. Acceptance by a local authority of a public body's notice is treated as 'depositing plans of work'.

If the local authority rejects the public body's notice for any reason, then the public body can appeal to a magistrates' court for a ruling. If still dissatisfied, they can appeal to the Crown Court.

1.11 Can I appeal against a local authority's ruling? (Building Act 1984 Sections 40, 41, 55, 81, 83, 86, 102 and 103)

If you are aggrieved by the local authority's rejection of any of the following:

* initial notice;
* amendment notice;
* public body's notice;
* plans certificate;
* final certificate;
* public body's certificate;
* public body's final certificate;

you may appeal to the magistrates' court (within 21 days of the notice of the local authority's requirement) in the area where the work is to be carried out. If your appeal is unsuccessful at magistrates' court you may appeal to the Crown Court.

 Where the Secretary of State has given a ruling, this ruling shall be considered as being final.

If you have grounds for disagreeing with the local authority's notice requiring works, ordering demolition or a ruling to remove or renew 'offending work', then you are entitled to appeal to the local magistrates' court. They will rule whether the local authority was correct and entitled to give you this ruling, or whether it should withdraw the notice. If you then disagree with the magistrates' ruling, you have the right to appeal to the Crown Court.

1.11.1 What about compensation? (Building Act 1984 Sections 106, 107 and 108)

If an owner or occupier considers that a ruling obtained from the local authority is incorrect, he can appeal (in the first case) to the local magistrates' court. If, on appeal, the magistrates rule against the local authority, then the owner/occupier of the building concerned is entitled to compensation from the local authority. If, on the other hand, the magistrates rule in favour of the local authority, then the local authority is entitled to recover any expenses that it has incurred.

Be sure of your facts before you ask a magistrates' court for a ruling!

1.12 Are there any exemptions from Building Regulations? (Building Act 1984 Sections 3, 4 and 5)

The following are exempt from the Building Regulations:

* prescribed classes of buildings, services, fittings or equipment;
* buildings or classes of buildings at a particular location that have been declared exempt by the Secretary of State;
* a 'public body' (i.e. local authorities, county councils and any other body 'that acts under an enactment for public purposes and not for its own profit') – this can be rather a grey area and it is best to seek advice if you think that you come under this category;
* educational buildings and buildings belonging to 'statutory undertakers', the UK Atomic Energy Authority or the Civil Aviation Authority.

1.12.1 What about Crown buildings? (Building Act 1984 Sections 44 and 87)

Although the majority of the requirements of the Building Regulations are applicable to Crown buildings (i.e. a building in which there is a Crown or Duchy of Lancaster or Duchy of Cornwall interest) or Government buildings (held in trust for Her Majesty) there are occasional deviations, and therefore you should seek the advice of the Treasury before submitting plans for work on a Crown building.

1.12.2 What about buildings in inner London? (Building Act 1984 Sections 44, 46 and 88)

You will find that the majority of the requirements found in the Building Regulations are also applicable to buildings in inner London boroughs (i.e. Inner Temple and Middle Temple). There are, however, some important deviations, which are contained in Schedule 3 of Part 5 and reproduced in Appendix

1B. You should seek the advice of the local authority concerned before submitting plans.

1.12.3 What about the United Kingdom Atomic Energy Authority? (Building Act 1984 Section 45)

The Building Regulations do **not** apply to buildings belonging to or occupied by the United Kingdom Atomic Energy Authority (UKAEA), unless they are dwelling houses or offices.

1.13 Can I apply for a relaxation in certain circumstances? (Building Act 1984 Sections 8–11 and 39)

The Building Act allows the local authority to dispense with, or relax, a Building Regulation if they believe that that requirement is unreasonable in relation to a particular type of work being carried out.

The local authority may charge a fee for reviewing and deciding on these matters.

In the majority of cases, applications to dispense with or relax Building Regulations can be settled locally. In more complicated cases the local authority can seek guidance from the Secretary of State who will give a direction as to whether the requirement may be relaxed or dispensed with (either unconditionally or subject to certain conditions).

If a question arises between the local authority and the person who has executed (or proposes to execute any) work regarding:

*   the application of Building Regulations;
*   whether the plans are in conformity with the Building Regulations;
*   whether the work has been executed in conformance with these plans;

then the question can be referred to the Secretary of State for determination. In these cases, the Secretary of State's decision will be deemed final.

The proposal to relax building regulations must be published in a local newspaper not less than 21 days before work commences.

Note: Schedule 2 of the Building Act 1984 provides guidance and rules for the application of Building Regulations to work that has been carried out prior to the local authority (under the Building Act 1984 Section 36) dispensing with, or relaxing, some of the requirements contained in the Building Regulations. This schedule is quite difficult to understand and, if it affects you, I would strongly advise that you discuss it with the local authority before proceeding any further.

1.14 What is 'type approval'? (Building Act 1984 Sections 12 and 13)

Type approval is where the Secretary of State is empowered to approve a particular type of building matter as complying, either generally or specifically, with a particular requirement of the Building Regulations. This power of approval is normally delegated by the Secretary of State to the local council or other nominated public body.

1.15 What causes some plans for building work to be rejected? (Building Act 1984 Sections 16 and 17)

If a plan for proposed building work is accompanied by a certificate from a person(s) approved by the Secretary of State then only in extreme circumstances can the local authority reject the plans.

The local authority will reject all plans for building work that:

* are defective;
* contravene any of the Building Regulations.

In all cases the local authority will advise the person putting forward the plans why they have been rejected (giving details of the relevant regulation or section) and, where possible, indicate what amendments and/or modifications will have to be made in order to get them approved. The person who initially put forward the plans is then responsible for making amendments/alterations and resubmitting them for approval.

1.16 Must I complete the approved work in a certain time? (Building Act 1984 Section 32)

Once a building plan has been passed by the local authority, then '**work must commence**' within three years from the date that it was approved. Failure to do so could result in the local authority cancelling the approved plans and you will have to resubmit them if you want to carry on with your project.

Note: The term 'work must commence' can vary but normally it means physically laying foundations of the building (in some cases it could mean more work has to be completed). It is always best to check with the local authority for clarification when your plans are first approved.

1.17 What happens if I contravene any of these requirements? (Building Act 1984 Sections 2, 7, 35, 36 and 112)

If you contravene the Building Regulations or wilfully obstruct a person acting 'in the execution of the Building Act 1984 or of its associated Building Regulations', then on summary conviction you could be liable to a fine or, in exceptional circumstances, even a short holiday in one of HM Prisons!

1.18 What about civil liability? (Building Act 1984 Section 38)

It is an aim of the Building Act 1984 that all building work is completed safely and without risk to people employed on the site or visiting the site. Any contravention of the Building Regulations that causes injury (or death) to any person is liable to prosecution in the normal way.

1.19 What is the 'Building Regulations Advisory Committee'? (Building Act 1984 Section 14)

The Building Act allows the Secretary of State to appoint a committee (known as the Building Regulations Advisory Committee) to review, amend, improve and produce new Building Regulations and associated documentation (e.g. such as Approved Documents – see earlier).

1.20 Does the Fire Authority have any say in Building Regulations? (Building Act 1984 Section 15)

When a requirement 'encroaches' on something that is normally handled by the Fire Authority, such as the provision of means of escape and structural fire precautions, the local authority **must** consult the fire authority before making any decision.

1.21 Can I change a plan of work once it has been approved? (Building Act 1984 Section 31)

If the person intending to carry out building work has had their plan (or plans) passed by the local authority, but then wants to change them, that person will have to submit (to the local authority) a set of revised plans showing precisely how they want to deviate from the approved plan and ask for their approval. If the deviation or change is a small one this can usually be achieved by talking to the local planning officer, but if it is a major change it could result in the resubmission of a complete plan of the revised building work.

1.22 What about dangerous buildings? (Building Act 1984 Sections 77 and 78)

The local authority can make an order restricting the use of a dangerous building until such time as a magistrates' court is satisfied that all necessary works have been completed.

If a building, or part of a building or structure, is in such a dangerous condition (or is used to carry loads which would make it dangerous) then the local authority may apply to a magistrates' court to make an order requiring the owner:

*   to carry out work to avert the danger;
*   to demolish the building or structure, or any dangerous part of it, and remove any rubbish resulting from the demolition.

Before the authority exercises its powers the owner must be informed.

1.22.1 Emergency measures

In emergencies, the local authority can make the owner take immediate action to remove the danger or they can complete the necessary action themselves. In these cases, the local authority is entitled to recover from the owner such expenses reasonably incurred by them. For example:

*   fencing off the building or structure;
*   arranging for the building/structure to be monitored.

1.22.2 Can I demolish a dangerous building? (Building Act 1984 Section 80)

Be careful, penalties can be very severe for demolishing something illegally!

You must have good reasons for knocking down a building, such as making way for rebuilding or improvement (which in most cases would be incorporated in the same planning application).

You are not allowed to begin any demolition work (even on a dangerous building) unless you have given the local authority notice of your intention **and** this has either been acknowledged by the local authority or the relevant notification period has expired. In this notice you will have to:

- specify the building to be demolished;
- state the reason(s) for wanting to demolish it;
- show how you intend to demolish it.

Copies of this notice will have to be sent to:

- the local authority;
- the occupier of any building adjacent to the building in question;
- any public gas supplier in whose area the building is;
- any public electricity supplier in whose area the building is;

Note: This regulation does not apply to the demolition of an internal part of an occupied building, or a greenhouse, conservatory, shed or pre-fabricated garage (that forms part of that building) or an agricultural building.

1.22.3 Can I be made to demolish a dangerous building?
(Building Act 1984 Sections 81, 82 and 83)

If the local authority considers that a building is so dangerous that it should be demolished, it is also entitled to issue a notice to the owner requiring them:

- to shore up any building adjacent to the building to which the notice relates;
- to weatherproof any surfaces of an adjacent building that are exposed by the demolition;
- to repair and make good any damage to an adjacent building caused by the demolition or by the negligent act or omission of any person engaged in it;
- to remove material or rubbish resulting from the demolition and clear the site;
- to disconnect, seal and remove any sewer or drain in or under the building;
- to make good the surface of the ground that has been disturbed in connection with this removal of drains;
- in accordance with the Housing Act 1985 to arrange with the relevant statutory undertakers for the disconnection of gas, electricity and water supplies to the building;
- to leave the site in a satisfactory condition following completion of all demolition work.

Before complying with this notice, the owner must give the local authority 48 hours' notice of commencement.

Note: In certain circumstances, the owner of an adjacent building may be liable to assist in the cost of shoring up their part of the building and waterproofing the surfaces. It could be worthwhile checking this point with the local authority!

1.23 What about defective buildings? (Building Act 1984 Sections 76, 79 and 80)

If a building or structure is, because of its ruinous or dilapidated condition, liable to cause damage to (or be a nuisance to) the amenities of the neighbourhood, then the local authority can require the owner:

*   to carry out necessary repairs and/or restoration; or
*   to demolish the building or structure (or any part of it) and to remove all the rubbish or other material resulting from this demolition.

If, however, the building or structure is in a defective state and remedial action (envisaged under Section 93of the Environmental Protection Act 1990) would cause an unreasonable delay, then the local authority can serve an abatement notice stating that within nine days **it** intends to complete such works as it deems necessary to remedy the defective state and recover the 'expenses reasonably incurred in so doing' from the person on whom the notice was sent.

If appropriate, the owner can (within seven days), after the local authority's notice has been served, serve a counter-notice stating that he intends to remedy the defects specified in the first-mentioned notice himself.

 Note: A local authority is not entitled to serve a notice, or commence any work in accordance with a notice that they have served, if the execution of the works would (to their knowledge) be in contravention of a building preservation order that has been made under Section 29 of the Town and Country Planning Act.

1.24 What are the rights of the owner or occupier of the premises? (Building Act 1984 Sections 102–107)

When a person has been given a notice by a local authority to complete work, he has the right to appeal to a magistrates' court on any of the following grounds:

*   that the notice or requirement is not justified by the terms of the provision under which it purports to have been given;
*   that there has been some informality, defect or error in (or in connection with) the notice;
*   that the authority has refused (unreasonably) to approve completion of alternative works, or that the works required by the notice to be executed are unreasonable or unnecessary;

- that the time limit set to complete the work is insufficient;
- that the notice should lawfully have been served on the occupier of the premises in question instead of on the owner (or vice versa);
- that some other person (who is likely to benefit from completion of the work) should share in the expense of the works.

Appendix 1A Contents of the Building Act 1984

Sub-section title	Part	Sub-section	Section title
Advertisement of proposal for relaxation of building regulations	1	10	Relaxation of building regulations
Appeal against notice requiring works	4	102	Execution of works
Appeal against notice under Section 81	3	83	Defective premises, demolition, etc.
Appeal against refusal, etc. to relax building regulations	1	39	Appeals in certain cases
Appeal against Section 36 notice	1	40	Appeals in certain cases
Appeal and statement of case to High Court in certain cases	1	42	Appeals in certain cases
Appeal to Crown Court	1	41	Appeals in certain cases
Appeal to Crown Court	3	86	Appeal to Crown Court
Appeals	2	55	Supplementary
Application for relaxation	1	9	Relaxation of building regulations
Application of provisions to Crown property	3	87	Application of provisions to Crown property
Application to Crown	1	44	Application of building regulations to Crown, etc.
Application to United Kingdom Atomic Energy Authority	1	45	Application of building regulations to Crown, etc.
Approval of documents for purposes of building regulations	1	6	Approved Documents
Approval of persons to give certificates, etc.	1	17	Passing of plans
Approved inspectors	2	49	Supervision of plans and work by approved inspectors

Sub-section title	Part	Sub-section	Section title
Arbitration	4	111	Compensation and recovery of sums
Authentication of documents	4	93	Documents
Breaking open of streets	4	101	Execution of works
Building over sewer	1	18	Passing of plans
Building regulations	5	1	Schedule 1
Cancellation of initial notice	2	52	Supervision of plans and work by approved inspectors
Cellars and rooms below sub-soil water level	3	74	Buildings
Change of person intending to carry out work	2	51c	Supervision of plans and work by approved inspectors
Civil liability	1	38	Breach of building regulations
Classification of buildings	1	34	Classification of buildings
Commencement	5	134	Supplementary
Compensation for damage	4	106	Compensation and recovery of sums
Compliance or non-compliance with Approved Documents	1	7	Approved Documents
Consents under Section 74	3	75	Buildings
Consequential amendments and repeals	5	133	Supplementary
Consequential amendments	5	6	Schedule 6
Construction and availability of sewers	4	125	Interpretation
Construction of certain references concerning temples	4	127	Interpretation
Construction of Part 11	2	58	Supplementary
Consultation with Building Regulations Advisory Committee and other bodies	1	14	Consultation
Consultation with fire authority	1	15	Consultation
Content and enforcement of notice requiring works	4	99	Execution of works
Continuing offences	4	114	Prosecutions

Appendix 1A (Continued)

Sub-section title	Part	Sub-section	Section title
Continuing requirements	1	2	Power to make building regulations
Dangerous building – emergency measures	3	78	Defective premises, demolition, etc.
Dangerous building	3	77	Defective premises, demolition, etc.
Default powers of Secretary of State	4	116	Default powers
Defective premises	3	76	Defective premises, demolition, etc.
Delegation of power to approve	1	13	Type approval of building matter
Determination of questions	1	30	Determination of questions
Disconnection of drain	3	62	Drainage
Drainage of building	3	59	Drainage
Drainage of buildings in combination	1	22	Passing of plans
Duties of local authorities	4	91	Duties of local authorities
Effect of amendment notice	2	51b	Supervision of plans and work by approved inspectors
Effect of initial notice ceasing to be in force	2	53	Supervision of plans and work by approved inspectors
Effect of initial notice	2	48	Supervision of plans and work by approved inspectors
Entrances, exits, etc. to be required in certain cases	3	71	Buildings
Erection of public conveniences	3	68	Provision of sanitary conveniences
Exemption of educational buildings and buildings of statutory undertakers	1	4	Exemption from building regulations
Exemption of particular classes of buildings, etc.	1	3	Exemption from building regulations
Exemption of public bodies from procedural requirements of building regulations	1	5	Exemption from building regulations
Expenses of Secretary of State	4	117	Default powers
Facilities for inspecting local acts	3	90	Miscellaneous

Appendix 1A (Continued)

Sub-section title	Part	Sub-section	Section title
Notices under Section 81	3	82	Defective premises, demolition, etc.
Obstruction	4	112	Obstruction
Obtaining of report where Section 36 notice given	1	37	Breach of building regulations
Offences	2	57	Supplementary
Orders	4	120	Orders
Passing or rejection of plans	1	16	Passing of plans
Paving and drainage of yards and passages	3	84	Yards and passages
Payments by instalments	4	108	Compensation and recovery of sums
Penalty for contravening building regulations	1	35	Breach of building regulations
Plans certificates	2	50	Supervision of plans and work by approved inspectors
Power of Secretary of State to approve type of matter building matter	1	12	Type approval of building matter
Power to enter premises	4	95	Entry on premises
Power to execute work	4	97	Execution of works
Power to make building regulations	1	1	Power to make building regulations
Power to require occupier to permit work	4	98	Execution of works
Procedure on appeal or application to magistrates' court	4	103	General provisions about appeals and applications
Procedure on appeal to Secretary of State on certain matters	1	43	Appeals in certain cases
Proposed departure from plans	1	31	Proposed departure from plans
Prosecution of offences	4	113	Prosecutions
Protection for dock and railway undertakings	4	128	Savings
Protection of members, etc. of authorities	4	115	Protection of members, etc. of authorities

Sub-section title	Part	Sub-section	Section title
Provision of closets in building	3	64	Provision of sanitary conveniences
Provision of drainage	1	21	Passing of plans
Provision of exits, etc.	1	24	Passing of plans
Provision of facilities for refuse	1	23	Passing of plans
Provision of food storage accommodation in house	3	70	Buildings
Provision of sanitary conveniences in workplace	3	65	Provision of sanitary conveniences
Provision of water supply in occupied house	3	69	Buildings
Provision of water supply	1	25	Passing of plans
Provisions consequential upon public body's notice	5		Schedule 4
Raising of chimney	3	73	Buildings
Recording and furnishing of information	2	56	Supplementary
Recovery of expenses, etc.	4	107	Compensation and recovery of sums
References in acts to building by-laws	3	89	Miscellaneous
Relaxation of building regulations	1	8	Relaxation of building regulations
Relaxation of building regulations	5		Schedule 2
Removal or alteration of offending work	1	36	Breach of building regulations
Repair, etc. of drain	3	61	Drainage
Repeals	5		Schedule 7
Replacement of earth closets, etc.	3	66	Provision of sanitary conveniences
Restriction of application of Part IV to Schedule 3	4	131	Savings
Ruinous and dilapidated buildings and neglected sites	3	79	Defective premises, demolition, etc.
Sale of materials	4	100	Execution of works
Saving for Local Land Charges Act 1975	4	129	Savings

Appendix 1A (Continued)

Sub-section title	Part	Sub-section	Section title
Saving for other laws	4	130	Savings
Service of documents	4	94	Documents
Short title and extent	5	135	Supplementary
Supplementary provisions as to entry	4	96	Entry on premises
Tests for conformity with building regulations	1	33	Tests for conformity with building regulations
Transitional provisions	5	132	Supplementary
Transitional provisions	5		Schedule 5
Type relaxation of building regulations	1	11	Relaxation of building regulations
Use and ventilation of soil pipes	3	60	Drainage
Use of materials unsuitable for permanent building	1	20	Passing of plans
Use of short-lived materials	1	19	Passing of plans
Variation of work to which initial notice relates	2	51a	Supervision of plans and work by approved inspectors
Variation or revocation of order transferring powers	4	118	Default powers

Appendix 1B Contents of the Schedules forming Part 5 of the Building Act 1984

What is Schedule 1 of Part 5 of the Building Act 1984?

Building Regulations also apply to alterations and extensions being completed on buildings erected before the date on which the regulations came into force.

Schedule 1 of Part 5 of the Building Act 1984 contains Building Regulations gives details of how the generic requirements of the Building Act are to be met. Compliance with Building Regulations is required for **all**:

*   alterations and extensions of buildings (including services, fixtures and fittings);
*   provision of new services, fittings or equipment;

unless (in most circumstances) **the increased area of the alteration or extension is less than 30 m^2** (35.9 yrds^2) in which case the Building Regulations provide the generic and specific requirements for this work.

Schedule 1 of the Building Act 1984 shows, in general terms, how the Building Regulations are to be administered by local authorities, the approved methods of construction and the approved types of materials that are to be used in (or in connection with) buildings.

How are the Building Regulations controlled?

To assist local authorities, Section 1 shows:

- how notices are given;
- how plans of proposed work (or work already executed) are deposited;
- how copies of deposited plans are administered and retained;
- how documents are to be controlled;
- how work is inspected and tested;
- how samples are taken;
- how local authorities can seek external expertise to assist them in their duties;
- how certificates signifying compliance with the Building Regulations are to be issued;
- how local authorities can accept certificates from a person (or persons) nominated to act on their behalf;
- how proposed work can be prohibited;
- when a dispute arises, how local authorities can refer the matter to the Secretary of State;
- what fees (and what level of fees) local authorities can charge.

What are the requirements of the Building Regulations?

Schedule 1 describes the mandatory requirements for completing **all** building work. These include:

- accommodation for specific purposes (e.g. for disabled persons);
- air pressure plants;
- cesspools (and other methods for reception, treatment and disposal of foul matter);
- emission of smoke, gases, fumes, grit or dust (or other noxious and/or offensive substances);
- dimensions of rooms and other spaces (inside buildings);
- drainage (including waste disposal units);
- electrical safety;
- fire precautions (services, fittings and equipment, means of escape);
- lifts (escalators, hoists, conveyors and moving footways);
- materials and components (suitability, durability and use);
- means of access to and egress from;

- natural lighting and ventilation of buildings;
- open spaces around buildings;
- prevention of infestation;
- provision of power outlets;
- resistance to moisture and decay;
- site preparation;
- solid fuel, oil, gas and electricity installations (including appliances, storage tanks, heat exchangers, ducts, fans and other equipment);
- standards of heating, artificial lighting, mechanical ventilation and air-conditioning;
- structural strength and stability (overloading, impact and explosion, underpinning, safeguarding of adjacent buildings);
- telecommunications services (including telephones and radio and television wiring installations).
- third party liability (danger and obstruction to persons working or passing by building work);
- transmission of heat;
- transmission of sound;
- waste (storage, treatment and removal);
- water services, fittings and fixed equipment (including wells and boreholes for supplying water);

and matters connected with (or ancillary to) any of the foregoing matters.

What is Schedule 2 of Part 5 of the Building Act 1984?

This Schedule provides guidance in connection with work that has been carried out prior to a local authority (under the Building Act 1984 Section 36) dispensing with or relaxing some of the requirements contained in the Building Regulations.

Schedule 2 is quite difficult to understand and if it affects you, then I would strongly advise that you discuss it with the local authority before proceeding any further.

What is Schedule 3 of Part 5 of the Building Act 1984?

Schedule 3 applies to how Building Regulations are to be used in inner London. As well as ruling which sections of the Act may be omitted this schedule also provides details of how by-laws concerning the demolition of buildings in inner London may be made.

What sections of the Building Act 1984 are not applied to inner London?

In inner London, because of its existing and changed circumstances (compared to other cities in England and Wales), certain sections of the Building Act are inappropriate (these are listed in Tables 1.3 and 1.4). In this case additional requirements, which are applicable to inner London **only,** have been approved instead. These primarily cover the demolition of buildings.

What about inner London's by-laws?

By authority of the Building Act 1984, the council of any inner London borough may make by-laws in relation to the demolition of buildings in the borough. Requiring:

* the fixing of floor level fans on buildings undergoing demolition;
* the hoarding up of windows in a building where all the sashes and glass have been removed;
* the demolition of internal parts of buildings before any external walls are taken down;
* the placing of screens or mats, the use of water or the taking of other precautions to prevent nuisances arising from dust;
* regulating the hours during which ceilings may be broken down and mortar may be shot, or be allowed to fall, into any lower floor;

Table 1.3 Sections inapplicable to inner London

Section	Sub-section
Buildings	• Provision of exits, etc. (except fire escape) • Provision of water supply • Entrances, exits, etc. to be required in certain cases • Means of escape from fire • Raising of chimney • Cellars and rooms below sub-soil water level • Consents under Section 74
Defective premises, demolition, etc.	• Dangerous building • Dangerous building – emergency measures • Ruinous and dilapidated buildings and neglected sites • Notice to local authority of intended demolition • Local authority's power to serve notice about demolition • Notices under Section 81 • Appeal against notice under Section 81

Table 1.4 Sections inapplicable to Temples

Section	Sub-section
Drainage	• Drainage of building • Use and ventilation of soil pipes • Repair, etc. of drains

- requiring any person proposing to demolish a building to give to the borough council such notice of his intention to do so as may be specified in the by-laws.

What is Schedule 4 of Part 5 of the Building Act 1984?

Schedule 4 of the Building Act 1984 concerns the authority and ruling of public bodies' notices and certificates.

What is a public body's plans certificate?

When a public body is satisfied that the work specified in their (as well as another) public bodies' notices has been completed as detailed (and in full accordance with the Building Regulations) then that public body will give that local authority a certificate of completion.

This certificate is called a 'public body's plans certificate' and can relate either to the whole or to part of, the work specified in the public body's notice. Acceptance by the local authority signifies satisfactory completion of the planned work and the public body's notice ceases to apply to that work.

What is a public body's final certificate?

When a public body is satisfied that all work specified in their (or another's) public body's notice has been completed in compliance with the Building Regulations, then that public body will give the local authority a certificate of completion. This is referred to as a 'Final Certificate'.

How long is the duration of a public body's notice?

A public body's notice comes into force when it is accepted by the local authority and continues in force until the expiry of an agreed period of time.

Local authorities are authorized by the Building Regulations to extend the notice in certain circumstances.

What is Schedule 5 of Part 5 of the Building Act 1984?

Schedule 5 of Part 5 lists the transitional effect of the Building Act 1984 on the following acts:

- The Public Health Act 1936;
- The Public Health Act 1961;
- The London Government Act 1963;
- The Local Government Act 1972;
- The Health and Safety at Work, etc. Act 1974;
- The Local Government (Miscellaneous Provisions) Act 1982.

What is Schedule 6 of Part 5 of the Building Act 1984?

Schedule 6 lists the consequential amendments that will have to be made to existing Acts of Parliament owing to the acceptance of the Building Act 1984. These amendments concern:

- The Public Health Act 1936;
- The Atomic Energy Authority Act 1954;
- The Clean Air Act 1956;
- The Housing Act 1957;
- The Radioactive Substances Act 1960;
- The Public Health Act 1961;
- The London Government Act 1963;
- The Offices, Shops and Railway Premises Act 1963;
- The Faculty Jurisdiction Measure 1964;
- The Fire Precautions Act 1971;
- The Local Government Act 1972;
- The Safety of Sports Grounds Act 1975;
- The Local Land Charges Act 1975;
- The Development of Rural Wales Act 1976;
- The Local Government (Miscellaneous Provisions) 1976;
- The Interpretation Act 1978;
- The Highways Act 1980;
- New Towns Act 1981;
- The Public Health (Control of Disease) Act 1984.

What is Schedule 7 of Part 5 of the Building Act 1984?

Schedule 7 lists the cancellation (repeal) of some sections of existing Acts of Parliament, owing to acceptance of the Building Act 1984. These cancellations concern:

- The Public Health Act 1936;
- The Education Act 1944;
- The Water Act 1945;
- The Atomic Energy Authority Act 1954;
- The Radioactive Substances Act 1960;
- The Public Health Act 1961;
- The London Government Act 1963;
- The Greater London Council (General Powers) Act 1967;
- The Fire Precautions Act 1971;
- The Local Government Act 1972;
- The Water Act 1973;
- The Health and Safety at Work, etc. Act 1974;
- The Control of Pollution Act 1974;
- The Airports Authority Act 1975;
- The Local Government (Miscellaneous Provisions) Act 1976;

- The Criminal Law Act 1977;
- The City of London (Various Powers) Act 1977;
- The Education Act 1980;
- The Highways Act 1980;
- The Water Act 1981;
- The Civil Aviation Act 1982;
- The Local Government (Miscellaneous Provisions) Act 1982.
- The Housing and Building Control Act 1984.

2

The Building Regulations 2010

 Building Regulations approval is **separate** from planning permission. Receiving planning permission is not the same as taking action to ensure that the building works comply with the Building Regulations.

The Building Regulations apply to building works in England and Wales and set standards for the design and construction of buildings to ensure the safety and health for people in or about those buildings. They also include requirements to ensure that fuel and power are conserved and that facilities are provided for people, including those with disabilities, to get into and move around inside buildings.

The Building Regulations contain both procedural regulations and technical requirements. The former state what work needs Building Regulations approval and how to get it, while the latter sets the standards for the work being undertaken.

The 2010 edition takes into account all amending regulations to the 2000 Building Regulations and, although there have been no major changes to the legislation, there have been a number of changes to Regulation numbers, and Requirements F, J and L have been revised. Details of the changes to regulation numbers can be found in Appendix 2A.

Even when planning permission is not required, most building works, including alterations to existing structures, are subject to minimum standards of construction to safeguard public health and safety.

2.1 What is the purpose of the Building Regulations?

The Building Regulations are legal requirements laid down by Parliament, based on the Building Act 1984. They are approved by Parliament and deal with the **minimum** standards of design and building work for the construction of domestic, commercial and industrial buildings.

 Building standards, including matters concerning drainage or sanitary installations, are enforced by your local building control officer.

Building Regulations ensure that new developments or alterations and/or extensions to buildings are all carried out to an agreed standard that protects the health and safety of people in and around the building.

Builders and developers are required by law to obtain building control approval, which is an independent check that the Building Regulations have been complied with. There are two types of building control providers – the local authority and approved inspectors.

2.2 Why do we need the Building Regulations?

The Great Fire of London in 1666 was the single most significant event to have shaped today's legislation. The rapid growth of the fire through adjoining timber buildings highlighted the need for builders to consider the possible spread of fire between properties when rebuilding work commenced. This resulted in the first building construction legislation that required all buildings to have some form of fire resistance.

 Alternative ways of achieving the same level of safety, or accessibility, are also acceptable.

During the Industrial Revolution (200 years after the Great Fire) poor living and working conditions in ever expanding, densely populated urban areas caused outbreaks of cholera and other serious diseases. Poor sanitation, damp conditions and lack of ventilation forced the government to take action and building control took on the greater role of health and safety through the first Public Health Act of 1875. This act had two major revisions in 1936 and 1961, and led to the first set of national building standards – the Building Regulations 1965.

The current legislation is the Building Regulations 2010 (Statutory Instrument No. 2214) which is made by the Secretary of State for the Environment under powers delegated by Parliament under the Building Act of 1984.

The Building Regulations are basic performance standards and the level of safety and acceptable standards are set out as guidance in the Approved Documents (which are quite frequently referred to as 'Parts' of the Building Regulations). Compliance with the detailed guidance of the Approved Documents is usually considered as evidence that the Regulations themselves have been complied with.

2.3 What building work is covered by the Building Regulations?

 Some building work may also be subject to other statutory requirements such as planning permission, fire precautions, water regulations or licensing/ registration.

The Building Regulations cover all new building work. This means that if you want to put up a new building, extend or alter an existing one, or provide new and/or additional fittings in a building such as drains or heat-producing appliances, washing and sanitary facilities and hot water storage (particularly unvented hot water systems), the Building Regulations will probably apply.

It should be remembered that although it may appear that the Regulations do not apply to some of the work you wish to undertake, the end result of doing that work could well lead to you contravening some of the regulations. You should also recognize that some work – whether it is controlled or not – could have implications for an adjacent property. In such cases it would be advisable to take professional advice and consult the local authority or an approved inspector. Some examples are:

- removing a buttressed support to a party wall;
- underpinning part of a building;
- removing a tree close to a wall of an adjoining property;
- building parapets which may increase snow accumulation and lead to an excessive increase in loading on a shared roof.

2.4 What are the requirements associated with the Building Regulations?

The Building Regulations contain a list of requirements (referred to as 'Schedule 1') which are designed to ensure the health and safety of people in and around buildings; to promote energy conservation and to provide access and facilities for disabled people. In total there are 14 parts (A–P less I and O) to these requirements and these cover subjects such as structure, fire and electrical safety, ventilation, drainage, etc.

The requirements are expressed in broad, functional terms in order to give designers and builders the maximum flexibility in preparing their plans.

2.5 What are the Approved Documents?

If guidance in an Approved Document is followed, there will be a presumption of compliance with the requirement(s) covered by the guidance. However, this presumption is not conclusive, so simply following guidance does not actually guarantee you compliance in an individual case!

Approved Documents contain practical and technical guidance on ways in which the requirements of each part of the Building Regulations can be met. Each Approved Document reproduces the *requirements* contained in the Building Regulations relevant to the subject area. This is then followed by *practical and technical guidance*, with examples, on how the requirements can be met in some of the more common building situations.

There may be alternative ways of complying with the requirements to those shown in the Approved Documents and you are, therefore, under no obligation to adopt any particular solution in an Approved Document if you prefer to meet the relevant requirement(s) in some other way.

If you are intending to carry out building work, it is best to always check with your Building Control Body (BCB) or one of their local authorities' approved inspectors first, to ensure that their proposals comply with Building Regulations!

In accordance with Regulation 8 of the Building Regulations, the requirements in Parts A–D, F–K and N and P (except for paragraphs G2, H2 and J7) of Schedule 1 to the Building Regulations do not require anything specific to be done except those things which are needed to secure reasonable standards of health and safety for persons in or about buildings (and any others who may be affected by buildings or matters connected with buildings).

 Note: The Building Regulations are constantly reviewed to meet the growing demand for better, safer and more accessible buildings as well as the need to reflect emerging harmonized European Standards. Where there are any issues common to one or more parts (such as the guidance on air tightness in Part L corresponding to the requirements for ventilation in Part F) these have been taken into consideration. Any changes necessary are brought into operation after consultation with all interested parties.

The Approved Documents are in 14 parts (A–P less I and O) and are listed in Table 2.1.

You can download pdf copies of the Building Act and Statutory Instruments from the UK government website (http://www.legislation.gov.uk) and Approved Documents from the planning portal (http://www.planningportal. gov.uk/buildingregulations/approveddocuments).

Alternatively you can buy a copy of the Approved Documents (and the Building Act 1984 if you wish) from The Stationery Office (TSO) or some book shops. Occasionally they are available from libraries.

2.5.1 Part A Structure

The requirements cover the three main structural areas of loading, ground movement and collapse, to ensure that:

(1)   all structural elements of a building can safely carry the loads expected to be placed on them;

(2)   foundations are adequate for any movement of the ground (e.g. caused by landslip or subsidence);

(3)   large buildings are strong enough to withstand an accident without collapsing.

Table 2.1 List of Approved Documents

Section	Title	Edition	Latest amendment
A	Structure	2004	
B	Fire safety	2006	2007
C	Site preparation and resistance to moisture	2004	
D	Toxic substances	1992	2002
E	Resistance to the passage of sound	2003	2004 and 2010
F	Ventilation	2010	
G	Sanitation, hot water safety and water efficiency	2010	
H	Drainage and waste disposal	2002	
J	Combustion appliances and fuel storage systems	2010	
K	Protection from falling, collision and impact	1998	2000
L	Conservation of fuel and power	2010	
M	Access to and use of buildings	2004	
N	Glazing – safety in relation to impact, opening and cleaning	1998	2000
P	Electrical safety	2006	
	Approved Document to support		
	Regulation 7 – Materials and workmanship	1999	2000

2.5.2 Part B Fire safety

The Regulations consider nine aspects of fire safety in the construction of buildings: both dwelling houses and other buildings. These are:

(1) the buildings shall have means of early warning and escape in the event of a fire;

(2) that the internal spread of fire should be inhibited within the building by ensuring linings adequately resist the spread of flame over their surface and have a rate of heat release, or fire growth that is reasonable;

(3) that the building shall be designed and constructed so that, in the event of fire, its stability will be maintained for a reasonable period;

(4) that walls common to two or more dwellings are designed and constructed to adequately resist the spread of fire between those buildings;

(5) that where a building is sub-divided, fire spread shall be inhibited through the use of fire resisting materials or fire suppression systems;

(6) that the building is constructed and designed so that the unseen spread of fire and smoke within concealed spaces in its structure is inhibited;

(7) that external walls of a building are able to resist the spread of fire from one building to another and that roofs are able to resist the spread of fire from one room to another;

(8)   that the roof of the building can resist the spread of fire over the roof and from one building to another;

(9)   that buildings are designed and constructed so as to provide reasonable assistance to fire-fighters in the protection of life and to enable fire appliances to gain access to the building.

2.5.3 Part C Site preparation and resistance to contaminants and moisture

There are four requirements to this part:

(1)   that before any building works commence, all vegetation and topsoil are removed;

(2)   that any contaminated ground is either treated, neutralized or removed before a building is erected;

(3)   that sub-soil drainage is provided to waterlogged sites;

(4)   that **all** floors, walls and roof of a building should not be adversely affected by ground moisture, precipitation, interstitial and surface condensation, and the spillage of water from sanitary fixings.

2.5.4 Part D Toxic substances

This part requires walls to be constructed in such a way that reasonable precautions have been taken to prevent the permeation of toxic fumes emanating from any insulating material that is inserted into cavities (including cavity walls).

2.5.5 Part E Resistance to passage of sound

This part has four main requirements:

(1)   that dwellings shall provide reasonable resistance to sound from other parts of the same building and/or from adjoining buildings;

(2)   that internal walls and floors of dwellings shall provide reasonable resistance to sound;

(3)   that the construction of common internal parts of buildings (containing flats or rooms for residential purposes) shall prevent unreasonable reverberation;

(4)   that school rooms shall be acoustically insulated against noise.

2.5.6 Part F Ventilation

This part requires that people in the building are provided with an adequate means of ventilation and that mechanical ventilation systems are suitably commissioned.

2.5.7 Part G Hygiene

There are ten elements in this part:

(1) wholesome water (see Chapter 6) has to be supplied to any place where drinking water is drawn off and to any sink provided in any area where food is prepared;

(2) wholesome water or softened wholesome water shall be supplied to any washbasin or bidet, fixed bath or shower in a bathroom;

(3) for the prevention of undue consumption of water, all fittings and fixed appliances must use water efficiently;

(4) the potential consumption of wholesome water by persons occupying a dwelling must not exceed 125 litres per person per day;

(5) hot water storage vessels shall prevent the temperature of the water stored from exceeding 100° C;

(6) discharges from a safety device shall be safely conveyed to where it is visible – without causing danger to persons in or about the building;

(7) the hot water supply to a bath shall not exceed 48° C;

(8) buildings are required to have satisfactory sanitary conveniences and washing facilities;

(9) all dwellings are required to have a fixed bath or shower with hot and cold water;

(10) unvented hot water systems over a certain size are required to have safety provisions to prevent explosion.

2.5.8 Part H Drainage and waste disposal

 H1 does not apply to the diversion of water which has been used for personal washing or for the washing of clothes, linen or other articles to collection systems for reuse. Requirement H3 does not apply to the gathering of rainwater for reuse.

There are six aspects of this part:

(1) that adequate drains are provided to take foul water from within buildings into a public or private sewer, septic tank or cesspool;

(2) where no public sewer is available, a suitable wastewater treatment system should be made available which is affixed with a durable notice containing details of continuing maintenance required;

(3) that adequate provision is made to take rainwater from roofs of buildings and paved areas to an appropriate soakaway, watercourse or sewer;

(4) that building work should not be detrimental to the continued maintenance of a drain, sewer or disposal main;

(5) that a system for discharging water to a sewer as described in paragraph 3 above shall be separate from that provided to carry foul water from the building;

(6)   that adequate provision is made for the storage of solid waste and access provided for building occupants to the place of storage and from the place of storage to a refuse collection point.

2.5.9 Part J Combustion appliances and fuel storage systems

 Requirements J1, J2 and J4 only apply to fixed combustion appliances (including incinerators). Requirement J3 only applies to fixed combustion appliances in dwellings. Requirement J5 applies only to fixed oil storage tanks with a capacity greater than 90 litres, LPG storage tanks with a capacity greater than 150 litres and their associated pipes.

There are seven main aspects to this part:

(1)   that combustion appliances are provided with a supply of fresh air to prevent overheating and for efficient working of any flue;
(2)   that adequate provision is made to discharge the products of combustion from a combustion appliance to the outside air;
(3)   that there should be a means of warning of the release of carbon monoxide from a combustion appliance;
(4)   that combustion appliances, flue pipes, chimneys and fireplaces are constructed in such a manner as to reduce the risk of people suffering burns or the building catching fire;
(5)   that where a hearth, fireplace, flue or chimney is provided or extended, a durable notice should be placed at a suitable place in the building providing information on the performance capabilities of the facility;
(6)   that liquid fuel storage systems and the pipes connecting them to combustion appliances are separated from buildings and the boundary of the premises to reduce the risk of the fuel igniting in the event of a fire;
(7)   that oil storage tanks and pipes are constructed in a manner to minimize the risk of oil escaping and provide a notice giving details of how to respond in the event of an oil escape.

2.5.10 Part K Protection from falling, collision and impact

There are five main aspects to this part:

(1)   that stairs, ladders and ramps which form part of the building are designed in a manner that allows the safe passage of people between levels in or about the building;
(2)   to avoid persons falling off stairwells, balconies, floors, some roofs, light wells and basement areas (or similar sunken areas) connected to a building they need to be suitably guarded according to the building's use;
(3)   to avoid vehicles falling off buildings, car park floors, ramps and other raised areas they need to be provided with vehicle barriers;
(4)   to avoid danger to people from colliding with an open window, skylight or ventilator, some form of guarding may be needed;

(5)   that measures are taken to avoid injury to people by the falling onto them of a door or gate that opens or slides upwards; or that opening powered doors and gates do not trap people and are capable of being opened in the event of a power failure.

2.5.11 Part L Conservation of fuel and power

This part is split into four separate parts which cover the conservation of fuel and power in:

(1)   new dwellings (L1A);
(2)   existing dwellings (L1B);
(3)   new buildings other than dwellings (L2A);
(4)   existing buildings other than dwellings (L2B).

The requirements of parts L1A, L1B, L2A and L2B have been amended completely since the last version of this book.

L1A

There are three main aspects to this part covering target CO_2 emissions, energy performance certificates and the conservation of fuel and power in new dwellings. This part requires;

(1)   that the building does not exceed the CO_2 limits set for it;
(2)   that the person carrying out work in a new dwelling that affects the heating, hot water, air conditioning or ventilation provides suitable energy performance certificates to the owner and local authority.
(3)   that fuel and power is conserved by limiting heat gains and losses;
(4)   that appropriately commissioned, energy-efficient fixed building services are provided with effective controls;
(5)   that the owner is provided with sufficient information about the building, the fixed building services and their maintenance requirements so that the building can be operated in an energy-efficient manner.

L1B

There are four main aspects to this part, covering the renovation of thermal elements, consequential improvements to thermal elements in buildings over $1000\,m^2$, energy performance certificates and the conservation of fuel and power in existing dwellings. This part requires:

(1)   that renovations to or replacement of thermal elements comply with schedule 1;
(2)   that the person carrying out work in a new dwelling that affects the heating, hot water, air conditioning or ventilation provides suitable energy performance certificates to the owner and local authority.
(3)   that fuel and power is conserved by limiting heat gains and losses;
(4)   that appropriately commissioned, energy efficient fixed building services are provided with effective controls;

(5) that the owner is provided with sufficient information about the building, the fixed building services and their maintenance requirements so that the building can be operated in an energy-efficient manner.

L2A

There are three main aspects to this part, covering target CO_2 emissions, energy performance certificates and the conservation of fuel and power in new buildings other than dwellings. This part requires:

(1) the building does not exceed the CO_2 limits set for it;
(2) the person carrying out work in a new dwelling that affects the heating, hot water, air conditioning or ventilation provides suitable energy performance certificates to the owner and local authority;
(3) that the word building refers to the whole of the building as well as any parts that have been designed to be used separately;
(4) that fuel and power is conserved by limiting heat gains and losses;
(5) that appropriately commissioned, energy efficient fixed building services are provided with effective controls;
(6) the owner is provided with sufficient information about the building, the fixed building services and their maintenance requirements so that the building can be operated in an energy-efficient manner.

L2B

There are five main aspects to this part, covering the renovation of thermal elements, consequential improvements to thermal elements in buildings over $1000\,m^2$, interpretation of the word building, the use of energy performance certificates and the conservation of fuel and power in new existing buildings other than dwellings. This part requires:

(1) that renovations to or replacement of thermal elements comply with schedule 1;
(2) that the person carrying out work in a new dwelling that affects the heating, hot water, air conditioning or ventilation provides suitable energy performance certificates to the owner and local authority;
(3) that fuel and power is conserved by limiting heat gains and losses;
(4) that appropriately commissioned, energy-efficient fixed building services are provided with effective controls;
(5) the owner is provided with sufficient information about the building, the fixed building services and their maintenance requirements so that the building can be operated in an energy-efficient manner.

2.5.12 Part M Access to and use of buildings

There are four main aspects to this part:

(1) that reasonable provision is made for people to gain access to and use a building and its facilities;

(2)   that suitable independent access is provided to an extension in a building other than a dwelling;
(3)   that if sanitary conveniences are provided in any building that is to be extended, reasonable provision is made within the extension for sanitary conveniences;
(4)   that reasonable provision is made in the entrance storey for sanitary conveniences or, where this is not possible, reasonable provision is made for sanitary conveniences in the entrance storey or principal storey.

2.5.13 Part N Glazing – safety in relation to impact, opening and cleaning

There are four main aspects in this part:

(1)   that glazing with which people are likely to come into contact whilst moving in or about, shall if broken, break in a way which is unlikely to cause injury;
(2)   that in locations where people might collide with the glass, this glass should either be robust enough not to break, or be constructed of safety glass, or be provided with suitable guarding;
(3)   that windows within the building should be capable of being operated safely by the building's occupants;
(4)   that there should be suitable access should be provide for the safe cleaning of windows, skylights, or any transparent or translucent walls ceilings or roofs.

2.5.14 Part P Electrical safety

This part requires that reasonable provision should be made in the design and installation of electrical installations in order to protect those who are operating, maintaining or altering the installations from fire or injury.

The government document *new Rules for Electrical Safety in the Home* is principally associated with the guidance relating to electrical safety and is available from the planning portal (http://www.planningportal.gov.uk/uploads/br/electrical_safety.pdf).

2.5.15 Future Approved Documents and guidance

Following the change of government in the United Kingdom, there is currently an ongoing review of all future Approved Documents and guidance. Therefore, at this stage, it is not appropriate to comment further until more details are available. Details of any forthcoming changes will be made available via the authors' website (http://www.herne.org.uk) and the Routledge website.

2.5.16 What happens if I do not comply with an Approved Document?

Not actually complying with an Approved Document (which is, after all, only meant as a guidance document) does not mean that you are liable to any civil or criminal prosecution. If, however, you have contravened a Building Regulation, then not having complied with the recommendations contained in the Approved Documents may be held against you.

2.6 Health and safety responsibilities

Clients, contractors and designers may also have duties under health and safety legislation and may need to notify the Health and Safety Executive (HSE). Although a domestic client does not have duties under Construction (Design and Management) Regulations 2007 (CDM 2007), those who work for them on construction projects will (see the CDN 2007 at http://www.hse.gov.uk/construction/cdm.htm).

2.7 Are there any exemptions?

The Building Regulations do not apply to:

- A 'public body' (i.e. local authorities, county councils and any other body 'that acts under an enactment for public purposes and not for its own profit'). This can be rather a grey area and it is best to seek advice if you think that you come under this category.
- Buildings belonging to 'statutory undertakers' (e.g. a water board).
- Crown buildings. Although the majority of the requirements of the Building Regulations are applicable to Crown buildings (i.e. a building in which there is a Crown or Duchy of Lancaster or Duchy of Cornwall interest) or government buildings (held in trust for Her Majesty) there are occasional deviations, and before submitting plans for work on a Crown building you should seek the advice of the Treasury.
- Buildings in inner London. The majority of the requirements found in the Building Regulations are also applicable to buildings in inner London boroughs (i.e. Inner Temple and Middle Temple). There are some important deviations (see Appendix 1B), and before submitting plans you should seek the advice of the local authority concerned.
- United Kingdom Atomic Energy Authority (UKAEA). Building Regulations do not apply to buildings belonging to or occupied by the United Kingdom Atomic Energy Authority unless they are dwelling houses and offices.
- Any building in which explosives are manufactured or stored under a licence granted under the Manufacture and Storage of Explosives Regulations 2005.

- Any building (other than a building containing a dwelling or a building used for office or canteen accommodation) erected on a site in respect of which a licence under the Nuclear Installations Act 1965 is, for the time being, in force.
- A building included in the schedule of monuments maintained under Section 1 of the Ancient Monuments and Archaeological Areas Act 1979.
- Certain detached buildings that are not frequented by people.
- Greenhouses and agricultural buildings.
- Temporary buildings that are not intended to remain where they are erected for more than 28 days.
- A building on a site, being a building which is intended to be used only in connection with the disposal of buildings or building plots on that site.
- A building on the site of construction or civil engineering works, which is intended to be used only during the course of those works and contains no sleeping accommodation.
- A building, other than a building containing a dwelling or used as an office or showroom, erected for use on the site of and in connection with a mine or quarry.
- A detached single-storey building, having a floor area that does not exceed $30\,m^2$ and which contains no sleeping accommodation.
- A detached building designed and intended to shelter people from the effects of nuclear, chemical or conventional weapons, and not used for any other purpose.
- A detached building, having a floor area that does not exceed $15\,m^2$ and which contains no sleeping accommodation.
- The extension of a building by the addition at ground level of a conservatory, porch, covered yard or covered way or a carport open on at least two sides where the floor area of that extension does not exceed $30\,m^2$, as long as in the case of a glazed conservatory or porch that the glazing satisfies the requirements of Part N of Schedule 1.
- The Metropolitan Police Authority (except Regulation 29).

Purpose-built student living accommodation (including flats) should be treated as hotel/motel accommodation in respect of space requirements and internal facilities.

2.8 Energy- and water-efficiency requirements

Building regulations includes a number of energy-efficiency requirements which apply equally to the erection of buildings, extensions and the carrying out of work within buildings and extensions. These regulations apply to all buildings with roofs, walls and that use energy to control the internal temperature with the following exceptions:

- places of worship;
- listed buildings;
- within a Conservation Area;
- a scheduled monument;
- a temporary building if it is expected to be used for only two years or less;
- stand-alone buildings with a floor area of less than $50\,m^3$.

Any changes to the energy status or the thermal elements within the building must comply with Part L of Schedule 1. New buildings are required to conform to minimum energy-efficiency performance and CO_2 emission rates. Certificates must be produced by an appropriately qualified and registered energy assessor, and these must detail any changes to the energy-efficiency of an existing building and provide information on the energy-efficiency characteristics of a new building.

All new dwellings are subject to water-efficiency requirements, and this will include the need to carry out a wholesome water consumption calculation for the residents of the building and provide details of this calculation to the local authority within five days of work being complete.

2.9 Do I need Building Regulations approval?

If you are in any doubt about whether you need to apply for permission, you should contact your Local Authority Building Control (LABC) department before commencing any work (in all cases, you may require planning permission).

If you are considering carrying out building work to your property, then you may need to apply to your local authority for *Building Regulations approval*.

For most types of building work (e.g. extensions, alterations, conversions and drainage works), you will be required to submit a Building Regulations application prior to commencing any work. Certain types of extensions and small detached buildings are exempt from Building Regulations control – particularly if the increased area of the alteration or extension is less than $30\,m^2$. However, you may **still** be required to apply for planning permission.

2.9.1 Building work needing formal approval

The Building Regulations apply to any building that involves:

- erecting or extending a building;
- the installation or extension of a service or fitting which is controlled under the regulations;
- an alteration project involving work which will temporarily or permanently affect the ongoing compliance of the building, service or fitting with the requirements relating to structure, fire or access to and use of buildings;

- work that is required under regulation 6 (material change of use)
- the insertion of insulation into a cavity wall;
- the underpinning of the foundations of a building;
- work affecting the thermal elements, energy status or energy performance of a building.

2.9.2 Typical examples of work needing approval

- altered openings for new windows in roofs or walls;
- cellars (particularly in London);
- electrical installations;
- erection of new buildings that are not exempt;
- home extensions such as for a kitchen, bedroom, lounge, etc.;
- installation of baths, showers or water closets (WCs) which involve new drainage or waste plumbing;
- installation of cavity insulation;
- installation of new heating appliances;
- internal structural alterations, such as the removal of a load-bearing wall or partition;
- loft conversions;
- new chimneys or flues;
- replacing roof coverings (unless exactly like-for-like repair);
- underpinning of foundations.

2.9.3 Self-certification

There are certain types of work that are exempt from control as long as it is carried out by a person authorized under a self-certification scheme. The installations listed in Table 2.2 are exempt from having to give building notice or deposit full plans, **provided** that the person carrying out the work is as indicated in the third column.

Table 2.2 Self-certification schemes providing exemption from giving building notice or depositing full plans

Area	Type of work	Person authorized to self-certify
Cavity walls	Insertion of insulating material into the cavity walls of an existing building	A person registered under the Cavity Wall Insulation Self Certification Scheme by Cavity Insulation Guarantee Agency Ltd in respect of that type of work
Electrical installations	Installation of a lighting system or electric heating system, or associated electrical controls	A person registered by Ascertiva Group Ltd, Building Engineering Services Competence Accreditation Ltd, ECA Certification Ltd, NAPIT Registration Ltd or Stroma Certification Ltd in respect of that type of work

Table 2.2 (Continued)

Area	Type of work	Person authorized to self-certify
	Installation of fixed low or extra-low voltage electrical installations	A person registered by Ascertiva Group Ltd, Benchmark Certification Ltd, British Standards Institution, Building Engineering Services Competence Accreditation Ltd, ECA Certification Ltd or NAPIT Registration Ltd in respect of that type of work
	Installation of fixed low or extra-low voltage electrical installations as a necessary adjunct to or arising out of other work being carried out by the registered person	A person registered by Ascertiva Group Ltd, Association of Plumbing and Heating Contractors (Certification) Ltd, Benchmark Certification Ltd, Building Engineering Services Competence Accreditation Ltd, ECA Certification Ltd, NAPIT Registration Ltd or Oil Firing Technical Association Ltd in respect of that type of electrical work
Energy production	Installation in a building of a system to produce electricity, heat or cooling: (a) by microgeneration, or (b) from renewable sources (as defined in European Parliament and Council Directive 2009/28/EC of 23 April 2009 on the promotion of the use of energy from renewable sources)	A person registered by Ascertiva Group Ltd, Association of Plumbing and Heating Contractors (Certification) Ltd, Benchmark Certification Ltd, British Standards Institution, Building Engineering Services Competence Accreditation Ltd, ECA Certification Ltd, HETAS Ltd, NAPIT Registration Ltd, Oil Firing Technical Association Ltd or Stroma Certification Ltd in respect of that type of work
Gas	Installation of a heat-producing gas appliance	A person, or an employee of a person, who is a member of a class of persons approved in accordance with regulation 3 of the Gas Safety (Installation and Use) Regulations 1998
Hot water systems	Installation of a heating or hot water system connected to an oil-fired combustion appliance or its associated controls	A person registered by Ascertiva Group Ltd, Association of Plumbing and Heating Contractors (Certification) Ltd, Benchmark Certification Ltd, Building Engineering Services Competence Accreditation Ltd, ECA Certification Ltd, HETAS Ltd, NAPIT Registration Ltd, Oil Firing Technical Association Ltd or Stroma Certification Ltd in respect of that type of work
	Installation of a heating or hot water system connected to a heat-producing gas appliance, or associated controls	A person registered by Ascertiva Group Ltd, Association of Plumbing and Heating Contractors (Certification) Ltd, Benchmark Certification Ltd, Building Engineering Services Competence Accreditation Ltd, Capita Gas Registration and Ancillary Services Ltd, ECA Certification Ltd, HETAS Ltd, NAPIT Registration Ltd, Oil Firing Technical Association Ltd or Stroma Certification Ltd in respect of that type of work

Area	Type of work	Person authorized to self-certify
	Installation of a heating or hot water system connected to a solid-fuel burning combustion appliance or its associated controls	A person registered by Ascertiva Group Ltd, Association of Plumbing and Heating Contractors (Certification) Ltd, Benchmark Certification Ltd, Building Engineering Services Competence Accreditation Ltd, ECA Certification Ltd, HETAS Ltd, NAPIT Registration Ltd, Oil Firing Technical Association Ltd or Stroma Certification Ltd in respect of that type of work
	Installation of a heating or hot water system connected to an electric heat source or its associated controls	A person registered by Ascertiva Group Ltd, Benchmark Certification Ltd, Building Engineering Services Competence Accreditation Ltd, ECA Certification Ltd, HETAS Ltd, NAPIT Registration Ltd, Oil Firing Technical Association Ltd or Stroma Certification Ltd in respect of that type of work
Lighting	Installation of a lighting system or electric heating system, or associated electrical controls	A person registered by Ascertiva Group Ltd, Building Engineering Services Competence Accreditation Ltd, ECA Certification Ltd, NAPIT Registration Ltd or Stroma Certification Ltd in respect of that type of work
Oil-fired appliances	Installation of: (a) an oil-fired combustion appliance; or (b) oil storage tanks and the pipes connecting them to combustion appliances	A person registered by Ascertiva Group Ltd, Association of Plumbing and Heating Contractors (Certification) Ltd, Benchmark Certification Ltd, Building Engineering Services Competence Accreditation Ltd, ECA Certification Ltd, HETAS Ltd, NAPIT Registration Ltd or Oil Firing Technical Association Ltd in respect of that type of work
Plumbing	Installation of a sanitary convenience, sink, washbasin, bidet, fixed bath, shower or bathroom in a dwelling which does not involve work on shared or underground drainage	A person registered by Ascertiva Group Ltd, Association of Plumbing and Heating Contractors (Certification) Ltd, Benchmark Certification Ltd, Building Engineering Services Competence Accreditation Ltd or NAPIT Registration Ltd in respect of that type of work
	Installation of a wholesome cold water supply or a softened wholesome cold water supply	A person registered by Ascertiva Group Ltd, Association of Plumbing and Heating Contractors (Certification) Ltd, Benchmark Certification Ltd, Building Engineering Services Competence Accreditation Ltd or NAPIT Registration Ltd in respect of that type of work
	Installation of a supply of non-wholesome water to a sanitary convenience fitted with a flushing device which does not involve work on shared or underground drainage	A person registered by Ascertiva Group Ltd, Association of Plumbing and Heating Contractors (Certification) Ltd, Benchmark Certification Ltd, Building Engineering Services Competence Accreditation Ltd or NAPIT Registration Ltd in respect of that type of work

Area	Type of work	Person authorized to self-certify
Roof coverings	Installation, as a replacement, of the covering of a pitched or flat roof and work carried out by the registered person as a necessary adjunct to that installation. This paragraph does not apply to the installation of solar panels	A person registered by National Federation of Roofing Contractors Ltd in respect of that type of work
Solid-fuel appliances	Installation of a solid-fuel burning combustion appliance	A person registered by Ascertiva Group Ltd, Association of Plumbing and Heating Contractors (Certification) Ltd, Benchmark Certification Ltd, Building Engineering Services Competence Accreditation Ltd, ECA Certification Ltd, HETAS Ltd or NAPIT Registration Ltd in respect of that type of work
Ventilation systems	Installation of a mechanical ventilation or air conditioning system or associated controls which does not involve work on a system shared with parts of the building occupied separately, in a building other than a dwelling	A person registered by Ascertiva Group Ltd, Benchmark Certification Ltd or Building Engineering Services Competence Accreditation Ltd in respect of that type of work
	Installation of an air conditioning or ventilation system in a dwelling which does not involve work on systems shared with other dwellings	A person registered by Ascertiva Group Ltd, Benchmark Certification Ltd, Building Engineering Services Competence Accreditation Ltd or NAPIT Registration Ltd in respect of that type of work
Windows	Installation, as a replacement, of a window, rooflight, roof window or door in an existing dwelling	A person registered under the Fenestration Self-Assessment Scheme by Fensa Ltd, or a person registered by BM Trada Certification Ltd, the British Standards Institution, CERTASS Ltd or Network VEKA Ltd in respect of that type of work

Where this regulation applies the person authorized under the self-certification scheme must provide the local authority with a certificate stating what work has been carried out within 30 days of completion of the work. This does not apply for work carried out under schedule 4 (descriptions of work where no building notice or deposit of full plans is required).

2.9.4 Where can I obtain assistance in understanding the requirements?

Local councils can provide assistance with:

*   advice about how to incorporate the most efficient energy safety measures into your scheme;
*   advice about the use of materials;
*   advice on electrical safety;
*   advice on fire safety measures (including safe evacuation of buildings in the event of an emergency);
*   at what stages local councils need to inspect your work;
*   deciding what type of application is most appropriate for your proposal;
*   how to apply for Building Regulations approval;
*   how to prepare your application (and what information is required);
*   how to provide adequate access for disabled people;
*   what your Building Regulation Completion Certificate means to you.

2.10 How do I obtain Building Regulations approval?

You, as the owner or builder, are required to fill in an application form and return it, along with basic drawings and relevant information, to the building control office at least two days before work commences. Alternatively, you may submit full detailed plans for approval. Whatever method you adopt, it may save time and trouble if you make an appointment to discuss your scheme with the building control officer well before you intend carrying out any work.

The building control officer will be happy to discuss your intentions, including proposed structural details and dimensions together with any lists of the materials you intend to use, so that he can point out any obvious contravention of the Building Regulations before you make an official application for approval. At the same time he can suggest whether it is necessary to approach other authorities to discuss planning, sanitation, fire escapes and so on.

 When the building is finished you must notify the local authority.

The building control officer will ask you to inform the office when crucial stages of the work are ready for inspection (by a surveyor) in order to make sure the work is carried out according to your original specification. Should the surveyor be dissatisfied with any aspect of the work, he may suggest ways to remedy the situation.

It would be to your advantage to ask for written confirmation that the work was satisfactory as this will help to reassure a prospective buyer when you come to sell the property.

2.11 What are building control bodies?

Your local authority has a general duty to see that all building work complies with the Building Regulations. To ensure that your particular building work complies with the Building Regulations you must use one of the two services available to check and approve plans and to inspect your work as appropriate. The two services are the LABC service or the service provided by the private sector in the form of an approved inspector. Both building control bodies will charge for their services. Both may offer advice before work is started. Certain types of building work can be self-certificated as compliant with building regulations by a member of a Competent Person Scheme without the need to notify a BCB.

2.11.1 What will the local authority do?

 The total fee is the same whichever method is chosen.

This rather depends on whether you are submitting:

(1)   full plans application submission; or
(2)   building notice application.

In both cases the building control office will carry out site inspections at various stages. The LABC is an organization that co-ordinates local authority services regionally and nationally.

Full plans

If you use the full plans procedure you must submit detailed drawings to the local authority, who will check your plans and consult any appropriate authorities (such as fire and water authorities). If your plans comply with the Building Regulations the local authority will then issue an approval notice before work starts. This process can take between five and eight weeks, depending on the project. The LABC has developed an online service for creating and submitting building control applications. You can apply to any local authority in England and Wales using the service (http://www.submit-a-plan.com) or make an online application on the planning portal (http://www.planningportal.gov.uk/england/public/planning/applications).

If the local authority is not satisfied with your plans, you may be asked to make amendments or provide more details. Alternatively, a conditional approval may be issued, which will either specify modifications that must be made to the plans, or will specify further plans that must be deposited. A local authority may only apply conditions if you have either requested them to do so or have consented to them doing so. A request or consent must be made in writing. If your plans are rejected, the reasons will be stated in the notice.

Note: You must contact your local authority to obtain these forms, as each authority produces its own. The LABC does not produce these forms but you can enter your postcode into the 'Find Your Local Authority' section on their website to obtain your local authority contact details.

Building notice

If you use the building notice procedure, as with full plans applications, the work will normally be inspected as it proceeds; but you will not receive any notice indicating whether your proposal has been passed or rejected. Work can then commence and regular site inspections will be made at agreed stages after which you will be advised where the work itself is found (by the building control officer) not to comply with the Regulations. A building notice cannot be used for commercial developments.

 Note: Building notice approval requires the person carrying out building work or making a material change of use to provide plans showing how they intend to conform to the requirements of Building Regulations. The local authority may also require further information such as structural design calculations of plans.

2.11.2 What will the approved inspector do?

Approved inspectors are companies or individuals who have been authorized under the Building Act 1984 to carry out building control work in England and Wales. The Construction Industry Council (CIC) is responsible for deciding on all applications for approved inspector status. A list of approved inspectors can be viewed at the Association of Corporate Approved Inspectors (ACAI) website or the CIC website.

If you use an approved inspector they will give you advice, check plans, issue a plans certificate, inspect the work, etc. as agreed between you. You and the inspector will jointly notify the local authority on what is termed an 'initial notice'. Once that has been accepted by the local authority, the approved inspector will then be responsible for the supervision of building work. Although the local authority will have no further involvement, you may still have to supply it with limited information to enable it to be satisfied about certain aspects linked to Building Regulations (e.g. about the point of connection to an existing sewer).

If the approved inspector is not satisfied with your proposals you may alter your plans according to his advice; or you may seek a ruling from the Secretary of State regarding any disagreement between you. The approved inspector might also suggest an alternative form of construction, and, provided that the work has not been started, you can apply to the local authority for a relaxation or a dispensation from one (or more) of the Regulations' requirements and, in the event of a refusal by the local authority, appeal to the Secretary of State.

If, however, you do not exercise these options and you do not do what the approved inspector has advised to achieve compliance, the inspector will not be able to issue a final certificate. The inspector will also be obliged to notify the local authority so that they can consider whether to use their powers of enforcement.

2.11.3 What is the difference between a full plans application and the building notice procedure?

A person who intends carrying out any building work or making a material change of use to a building, shall either:

*   provide the local authority with a building notice; or
*   deposit full plans with the local authority.

For a full plans application, plans need to be produced showing all constructional details, preferably well in advance of your intended commencement on site. For the building notice procedure, less detailed plans are required. In both cases, your application or notice should be submitted to the local authority and should be accompanied by any relevant calculations, to demonstrate compliance with safety requirements concerning the structure of the building.

A 'full plans' submission is required in the following circumstances:

*   if the building use falls under Section 1 of the Fire Precautions Act 1971;
*   if use falls under Part II of the Fire Precautions (Workplace) Regulations 1997;
*   if work will be close to or over the top of drains shown on the 'map of sewers';
*   where a new building will front onto a private street.

Approved plans are valid for at least three years.

2.12 How do I apply for building control?

 Take advantage of the free advice that local authorities offer, and discuss your ideas well in advance.

If your prospective work will involve any form of structure, you could need building control approval. Some types of work may need both planning permission and building control approval; others may need only one or the other. The process of assessing a proposed building project is carried out through an evaluation of submitted information and plans and the inspection of work as the building progresses.

2.12.1 What applications do not require submission plans?

The following building works do **not** require the submission of plans:

* in respect of any work specified in an initial notice, an amendment notice or a public body's notice, which is in force;
* where a person intends to have electrical installation work completed by a competent firm registered under the NICIEC Approved Contractor scheme;
* where a person intends to have installed (by a person, or an employee of a person approved in accordance with Regulation 3 of the Gas Safety (Installation and Use) Regulations 1998) a heat-producing gas appliance;
* where Regulation 19 of the Building (Approved Inspectors, etc.) Regulations 2010 (Local authority powers in relation to partly completed work) applies.

2.12.2 Other considerations

Depending on the type of work involved, you may need to get approval from several sources before starting. The list below provides a few examples:

* there may be legal objections to alterations being made to your property;
* a solicitor might need to be consulted to see if any covenants or other forms of restriction are listed in the title deeds to your property and if any other person or party needs to be consulted before you carry out your work;
* you may need planning permission for a particular type of development work;
* if a building is listed or is within a Conservation Area or an Area of Outstanding Natural Beauty, special rules apply.

2.13 Full plans application

This type of application can be used for any type of building work, but it **must** be used where the proposed premises are to be used as a factory, office, shop, hotel, and boarding house or railway premises.

 The full plans application may be accompanied by a request (from the person carrying out such building work) that on completion of the work, he wishes the local authority to issue a completion certificate.

A full plans application requires the submission of fully detailed plans, specifications, calculations and other supporting details to enable the building control officer to ascertain compliance with the Building Regulations. The amount of detail depends on the size and type of building works proposed, but as a minimum will have to consist of:

* a description of the proposed building work or material change of use;
* plan(s) showing what work will be completed; and
* a location plan showing where the building is located relative to neighbouring streets.

Two copies of the full plans application need to be sent to the local authority, except in cases where the proposed building work relates to the erection, extension or material alteration of a building (other than a dwelling-house or flat) and where fire safety imposes an additional requirement, in which case five copies are required.

A full plans application will be thoroughly checked by the local authority that is required to pass or reject your plans within a certain time limit (usually eight weeks); or they may add conditions to an approval (with your written agreement). If they are satisfied that the work shown on the plans complies with the Regulations, you will be issued with an approval notice (within a period of five weeks or up to two months) showing that your plans were approved as complying with the Building Regulations.

If your plans are rejected, and you do not consider it is necessary to alter them, you will have two options available to you:

(1)   You may seek a 'determination' from the Secretary of State if you believe your work complies with the Regulations (but you must apply before work starts).
(2)   If you acknowledge that your proposals do not necessarily comply with a particular requirement in the Regulations and feel that it is too onerous in your particular circumstances, you may apply for a relaxation or dispensation of that particular requirement from the local authority. You can make this sort of application at any time you like but it is obviously sensible to do so as soon as possible and preferably before work starts. If the local authority refuses your application, you may then appeal to the Secretary of State within a month of the date of receipt of the rejection notice.

2.13.1 Consultation with sewerage undertaker

Where applicable, the local authority shall consult the sewerage undertaker as soon as practicable after the plans have been deposited, and before issuing any completion certificate in relation to the building work.

2.13.2 Advantages of submitting full plans application

The advantages of the full plans method are that:

*   a (free) completion certificate will be issued on satisfactory completion of the work;
*   a formal notice of approval or rejection will be issued within five weeks (unless the applicant agrees to extend this to two months);
*   only when work starts on site (and the building control officer has completed his initial visit) is the remaining part of the fee invoiced;
*   the plans can be examined and approved in advance (for an advance payment of (typically) 25% of the total fee).

2.14 Building notice procedure

 The submission of a marked-up sketch showing the location of the building, although not mandatory, is recommended.

Under the building notice procedure no approval notice is given. There is also no procedure to seek a determination from the Secretary of State if there is a disagreement between you and the local authority – unless plans are subsequently deposited. However, the advantage of the building notice procedure is that it will allow you to carry out **minor works** without the need to prepare full plans. You must, however, feel confident that the work will comply with the Regulations or you risk having to correct any work you carry out at the request of the local authority.

A building notice is particularly suited to minor works (e.g. a householder wishing to install another WC). For such building work, detailed plans are unnecessary and most matters can be agreed when the building control officer visits your property. You do not need to have detailed plans prepared, but in some cases you may be asked to supply extra information.

As no formal approval is given, good liaison between the builder and the building control officer is essential to ensure that work does not have to be re-done.

This type of application may be used for all types of building work, so long as no part of the premises is used for any of the purposes mentioned earlier under the full plans application.

2.14.1 What do I have to include in a building notice?

A building notice shall:

- state the name and address of the person intending to carry out the work;
- be signed by that person or on that person's behalf;
- contain, or be accompanied by:

    - a description of the proposed building work or material change of use;
    - particulars of the location of the building;
    - the use or intended use of that building.

 Note: Councils have differing requirements for details of what should be included on building notices. The following lists are examples of what some councils require, other areas (even those adjacent) may have less stringent requirements. In all cases you are recommended to approach your local council for further details.

New buildings and extensions

When planning a building extension, the building notice needs to be accompanied by:

- a plan to a scale of not less than 1:1250 showing:

    - the size and position of the building, or the building as extended, and its relationship to adjoining boundaries (commonly known as a block plan);

- particulars of:

    - the provisions to be made for the drainage of the building or extension.

Insertion of insulating material into the cavity walls of a building

For cavity wall insulations, the building notice needs to be accompanied by a statement which specifies:

- the name and type of insulating material to be used;
- the name of any European Technical Approval issuing body that has approved the insulating material;
- the requirements of Schedule 1 in relation to which the issuing body has approved the insulating material;
- any European Economic Area (EEA) national standard with which the insulating material conforms;
- the name of any body that has issued any current approval to the installer of the insulating material.

Although if the insulating material has been installed by an approved installer, they will usually submit at Type Approval Notice on your behalf.

Provision of a hot water storage system

A building notice in respect of a proposed hot water system shall be accompanied by a statement which specifies:

- the name, make, model and type of hot water storage system to be installed;
- the name of the body (if any) that has approved or certified the system;
- the name of the body (if any) that has issued any current registered operative identity card to the installer or proposed installer of the system.

Electrical installations

All proposals to carry out electrical installation work **must** be notified to the local authority's BCB before work begins, **unless** the proposed installation work is undertaken by a competent person registered with an electrical self-certification scheme and does not include the provision of a new circuit.

Non-notifiable work (such as replacing a socket outlet or other fixed electrical equipment) can be completed by a DIY enthusiast (family member or friends), but needs to be installed in accordance with manufacturers' instructions and done in such a way that they do not present a safety hazard. This work does **not** need to be notified to a local authority BCB (unless it is installed in an area of high risk such as a kitchen or a bathroom, etc.), **but** all DIY electrical work (unless completed by a qualified professional – who is responsible for issuing a Minor Electrical Installation Certificate) will still need to be checked, certified and tested by a competent electrician.

Any work that involves adding a new circuit to a dwelling will need to be either notified to the BCB (who will then inspect the work) or need to be carried out by a competent person who is registered under a Government Approved Part P Self-Certification Scheme.

Work involving any of the following will also have to be notified:

*   consumer unit replacements;
*   electric floor or ceiling heating systems;
*   extra-low-voltage lighting installations, other than pre-assembled, CE-marked lighting sets;
*   garden lighting or power installations;
*   installation of a socket outlet on an external wall;
*   installation of outdoor lighting and/or power installations in the garden or that involves crossing the garden;
*   installation of new central heating control wiring;
*   outdoor lighting and power installations;
*   solar Photovoltaic (PV) power supply systems;
*   small-scale generators such as micro-CHP units.

Note: Where a person who is **not** registered to self-certify, intends to carry out the electrical installation, then a Building Regulation (i.e. a building notice or full plans) application will need to be submitted together with the appropriate fee, based on the estimated cost of the electrical installation. The BCB will then arrange to have the electrical installation inspected at first fix stage and tested upon completion.

2.15 How long is a building notice valid?

The approved plans may be used (i.e. built to) for at least three years, **even if the Building Regulations change during this time**.

A building notice shall cease to have effect three years from the date when that notice was given to the local authority, unless, before the expiry of that period:

*   the building work to which the notice related has commenced; or
*   the material change of use described in the notice was made.

2.16 What can I do if my plans are rejected?

In the first two cases below, the address to write to is the Department for Communities and Local Government (DCLG). In Wales, you should refer the matter to the Secretary of State for Wales.

If your plans were initially rejected, you can start work **provided** you give the necessary notice of commencement required under Regulation 14 of the Building Regulations and are satisfied that the building work itself now complies with the Regulations. However, it would **not** be advisable to follow this course if you are in any doubt and have not taken professional advice. Instead:

- you should resubmit your full plans application with amendments to ensure that they comply with Building Regulations; or
- if you think your plans comply (and that the decision to reject is, therefore, unjustified) you can refer the matter to the Secretary of State for the Environment, Transport and the Regions, or the Secretary of State for Wales (as appropriate) for their determination, but usually only before the work has started; or
- you could (in particular cases) ask the local authority to relax or dispense with their rejection. If the local authority refuses your application you could then appeal to the appropriate Secretary of State within one month of the refusal.

A fee is payable for determinations but not for appeals. The fee is half the plan fee (excluding VAT) subject to a minimum of £50 and a maximum of £500. The Department of the Environment, Transport and the Regions (DETR) or the Welsh Office will then seek comments from the local authority on your application (or appeal), which will be copied to you. You will then have a further opportunity to comment before a decision is issued by the Secretary of State.

2.16.1 Do my neighbours have the right to object to what is proposed in my Building Regulations application?

A free explanatory booklet on the Party Wall, etc. Act 1996 is available to download at http://www.communities.gov.uk.

Basically – **no**! However, while there is no requirement in the Building Regulations to consult neighbours, it would be prudent to do so. In any event, you should be careful that the work does not encroach on their property, as this could well lead to bad feeling and possibly an application for an injunction for the removal of the work.

Objections may be raised under other legislation, particularly if your proposal is subject to approval under the Town and Country Planning legislation or the Party Wall, etc. Act of 1996.

The Party Wall, etc. Act 1996 came into force on 1 July 1997 and is largely based on Part VI of the London Building Acts (Amendment) Act 1939 – which started life as a Private Members Bill sponsored by the Earl of Lytton.

In a nutshell, this act says that if you intend to carry out building work which involves:

- work on an existing wall shared with another property;
- building on the boundary with a neighbouring property;
- excavating near an adjoining building;

you **must** find out whether that work falls within the scope of the Act. If it does, then you must serve the statutory notice on all those defined by the Act as 'adjoining owners'. You should, however, remember that reaching agreement with adjoining owners on a project that falls within the scope of the Act does **not** remove the possible need for planning permission or Building Regulations approval.

Note: If you are not sure whether the Act applies to the work that you are planning, you should seek professional advice (see the useful contact names and addresses in Appendix I at the end of this book).

2.17 What happens if I wish to seek a determination but the work in question has started?

If you start work before you receive a decision on your full plans application, you will prejudice your ability to seek a determination from the Secretary of State if there is a dispute.

You will only need to seek a determination if you believe the proposals in your full plans application comply with the Regulations but the local authority disagrees. You may apply for a determination either before or after the local authority has formally rejected your full plans application. The legal procedure is intended to deal with compliance of 'proposed' work only and, in general, applications relating to work which is substantially completed cannot be accepted. Exceptionally, however, applications for 'late' determinations may be accepted – but it is in your best interest to always ensure that you apply for a determination well before you start work.

2.18 When can I start work?

Again, it depends on whether you are using the local authority or the approved inspector.

2.18.1 Using the local authority

Once you have given a building notice or submitted a full plans application, you can start work at any time. However, you must give the local authority a commencement notice at least two clear days (not including the day on which you give notice and any Saturday, Sunday, bank or public holiday) before you start.

2.18.2 Using an approved inspector

If you use an approved inspector you may, subject to any arrangements you may have agreed with the inspector, start work as soon as the initial notice is accepted by the local authority (or is deemed to have been accepted if nothing is heard from the local authority within five working days of the notice being given). Work may not start if the initial notice is rejected, however.

2.19 Planning officers

Before construction begins, planning officers determine whether the plans for the building or other structure comply with the Building Regulations and if they are suited to the engineering and environmental demands of the building site. Building inspectors are then responsible for inspecting the structural quality and general safety of buildings.

2.20 Building inspectors

Building inspectors examine the construction, alteration, or repair of buildings, highways and streets, sewer and water systems, dams, bridges, and other structures to ensure compliance with building codes and ordinances, zoning regulations and contract specifications.

Building codes and standards are the primary means by which building construction is regulated in the United Kingdom to assure the health and safety of the general public. Inspectors make an initial inspection during the first phase of construction and then complete follow-up inspections throughout the construction project in order to monitor compliance with regulations.

The inspectors will visit the worksite before the foundation is poured to inspect the soil condition and positioning and depth of the footings. Later, they return to the site to inspect the foundation after it has been completed. The size and type of structure, as well as the rate of completion, determine the number of other site visits they must make. Upon completion of the project, they make a final comprehensive inspection.

2.21 Notice of commencement and completion of certain stages of work

A person who proposes carrying out building work shall not start work unless:

*   he has given the local authority notice that he intends to commence work; and
*   at least two days have elapsed since the end of the day on which he gave the notice.

2.21.1 Notice of completion of certain stages of work

 This requirement does not apply in respect of any work specified in an initial notice, an amendment notice or a public body's notice that is in force.

The person responsible for completing the building work is also responsible for notifying the local authority a minimum of five days **prior** to commencing any work involving excavations for foundations, foundations themselves, any damp-proof course any concrete or other material to be laid over a site and drains or sewers.

Upon completion of this work (especially work that will eventually be covered up by later work) the person responsible for the building work shall give five days' notice of intention to backfill.

A person who has laid, haunched or covered any drain or sewer shall (not more than five days after that work has been completed) give the local authority notice to that effect.

Where a building is being erected and that building (or any part of it) is to be occupied before completion, the person carrying out that work shall give the local authority at least five days' notice before the building, or any part of it is, occupied.

The person carrying out the building work shall **not**:

*   cover up any foundation (or excavation for a foundation), any damp-proof course or any concrete or other material laid over a site; or
*   cover up (in any way) any drains or sewers **unless** he has given the local authority notice that he intends to commence that work and at least one day has elapsed since the end of the day on which he gave the notice.

 Note: Where a person fails to comply with the above, then the local authority can insist that he shall cut into, lay open or pull down 'so much of the work as to enable the authority to ascertain whether these Regulations have been complied with, or not'. If the local authority then notifies the owner/builder that certain work contravenes the requirements in these Regulations, then the owner/builder shall, after completing the remedial work, notify the local authority of its completion.

2.21.2 What kind of tests are local authorities likely to make?

To establish whether building work has been carried out in conformance with the Building Regulations, local authorities will test to ensure that all work has been carried out:

- in a workmanlike manner;
- with adequate and proper materials which:

    - are appropriate for the circumstances in which they are used;
    - are adequately mixed or prepared; and
    - are applied, used or fixed so as to adequately perform the functions for which they are designed;

- in compliance with the requirements of Part H of Schedule 1 (drainage and waste disposal);
- so as to enable them to ascertain whether the materials used comply with the provisions of these Regulations.

2.22 What are the requirements relating to building work?

In all cases, building work shall be carried out so that it:

(a)   it complies with the applicable requirements contained in Schedule 1; and

(b)   in complying with any such requirement there is no failure to comply with any other such requirement.

Building work shall be carried out so that, **after** it has been completed:

(a)   any building which is extended or to which a material alteration is made; or

(b)   any building in, or in connection with which, a controlled service or fitting is provided, extended or materially altered; or

(c)   any controlled service or fitting, complies with the applicable requirements of Schedule 1 or, where it did not comply with any such requirement, is no more unsatisfactory in relation to that requirement than before the work was carried out.

2.23 Do I need to employ a professional builder?

Unless you have a reasonable working knowledge of building construction it would be advisable before you start work to get some professional advice (e.g. from an architect, or a structural engineer or a building surveyor) and/or choose a recognized builder to carry out the work. It is also advisable to consult the LABC officer or an approved inspector in advance.

2.24 Unauthorized building work

If, for any reason, building work has been done without a building notice or full plans of the work being deposited with the local authority, or a notice of commencement of work being given, then the applicant may apply, in writing, to the local authority for a regularization certificate.

This application will need to include:

* a description of the unauthorized work;
* a plan of the unauthorized work; and
* a plan showing any additional work that is required for compliance with the requirements relating to building work in the Building Regulations.

Local authorities may then 'require the applicant to take such reasonable steps, including laying open the unauthorized work for inspection by the authority, making tests and taking samples, as the authority think appropriate to ascertain what work, if any, is required to secure that the relevant requirements are met'.

When the applicant has taken any such steps required by the local authority, the local authority will notify the applicant:

* if no work is required to secure compliance with the relevant requirements;
* of the work that is required to comply with the relevant requirements;
* of the requirements that can be dispensed with or relaxed.

2.24.1 What happens if I do work without approval?

The local authority has a general duty to see that all building work complies with the Regulations – except where it is formally under the control of an approved inspector. Where a local authority is controlling the work and finds after its completion that it does not comply, then the local authority may require you to alter or remove it. If you fail to do this, the local authority may serve a notice requiring you to do so and you will be liable for the costs.

2.24.2 What are the penalties for contravening the Building Regulations?

If you contravene the Building Regulations by building without notifying the local authority or by carrying out work which does not comply, the local authority can prosecute. If you are convicted, you are liable to a penalty not exceeding £5000 (at the date of publication of this book) plus £50 for each day on which each individual contravention is not put right after you have been convicted. If you do not put the work right when asked to do so, the local authority has power to do the work itself and recover costs from you.

2.25 Why do I need a completion certificate?

A completion certificate is a valuable document that should be kept in a safe place!

A completion certificate certifies that the local authority are satisfied that the work complies with the relevant requirements of Schedule 1 of the Building Regulations, 'in so far as they have been able to ascertain after taking all reasonable steps'.

2.26 How do I get a completion certificate when the work is finished?

The local authority shall give a completion certificate only when they have received the completion notice and have been able to ascertain that the relevant requirements of Schedule 1 (specified in the certificate) have been satisfied.

Where full plans are submitted for work that is also subject to the Fire Precautions Act 1971, the local authority must issue you with a completion certificate concerning compliance with the fire safety requirements of the Building Regulations once work has finished. In other circumstances, you may ask to be given one when the work is finished, but you must make your request **when you first submit your plans**.

If you use an approved inspector, they must issue a final certificate to the local authority when the work is completed.

2.27 Where can I find out more?

You can find out more from:

- the local authority's building control department;
- an approved inspector; or
- other sources.

2.27.1 Local authority

Each local authority in England and Wales (i.e. unitary, district and London boroughs in England and county borough councils in Wales) has a building control section whose general duty is to see that work complies with the Building Regulations – except where it is formally under the control of an approved inspector. Most local authorities have their own website and these usually contain a wealth of useful information, the majority of which is downloadable as read-only pdf files.

Individual local authorities co-ordinate their services regionally and nationally (and provide a range of national approval schemes) via LABC Services. You can find out more about LABC services through its website (http://www.labc.uk.com) but your LABC department will be pleased to give you information and advice. They may offer to let you see their copies of the Building Act 1984, the Building Regulations 2010 and their associated Approved Documents that provide additional guidance.

The *Fire and Building Regulations Procedural Guide* which deals with procedures for building work to which the Fire Precautions Act 1971 applies, and the DETR leaflet on safety of garden walls, are amongst the documentation and advice that is available, free of charge, from your local authority.

The DETR's (and the Welsh Office's) separate booklets on planning permission for small businesses and householders are also available free of charge from your local authority.

2.27.2 Approved inspectors

Approved inspectors are companies or individuals authorized under the Building Act 1984 to carry out building control work in England and Wales.

The Construction Industry Council (CIC) is responsible for deciding all applications for approved inspector status. You can find out more about the CIC's role (including how to apply to become an approved inspector) through its website (http://www.cic.org.uk).

A list of approved inspectors can be viewed at the Association of Consultant Approved Inspectors (ACAI) website (http://approvedinspectors.org.uk/2011/02/122).

2.27.3 Other sources

Most of the documents can be purchased from The Stationery Office or from any main bookshop. Copies should also be available in public reference libraries.

Appendix 2A

The table below shows changes which have been made in the 2010 edition to the numbering of the sections within the Building Regulations. All other sections remain the same.

New Regulation Number in Building Regulations 2010	Original Regulation Number in Building Regulations 2000	New Regulation Number in Building Regulations 2010	Original Regulation Number in Building Regulations 2000	New Regulation Number in Building Regulations 2010	Original Regulation Number in Building Regulations 2000
2(3)	2(2A)	18(5)	21(6)	44(2)	20C(1)
2(4)	2(2B)	18(6)	21(7)	44(3)	2OC(2)
2(5)	2(2C)	18(7)	21(8)	44(4}	20C(3)
3(1)(g)	3(1)(h)	18(8)	21(2)	45	18
3(1)(h)	3(i)(g)	19	20	46	19
4(2)	4(1A)	20	16A	47	22
4(3)	4(2)	21 (1)	9(3)	48(1)(a)	22B(1)(a)
6(1)(d)	6(1)(cc)	21(2)	9(4)	48(1)(b)	22B(1)(b)
6(1)(e)	6(1)(d)	21(3)	9(5)	48(1)(c)	22B(1)(c)
6(1)(f)	6(1)(e)	21(4)	9(5A)	48(1)(d)	22B(1)(e)
6(1)(g)	6(1)(f)	21(5)	9(6)	48(1)(e)	22B(1)(l)
6(1)(h)	6(1)(ff)	22	4B(1)	48(1)(f)	22B(1)(m)
6(1)(i)	6(1)(g)	23	4A	48(1)(g)	22B(1)(d)
9(2)	9(1 A)	24	17A	48(1)(h)	22B(1)(k)
9(3)	9(2)	25	17B	48(1)(i)	22B(1)(1)
12(1)	12(2)	26	17C	48(1)(j)	22B(1)(g)
12(2)	12(2A)	27	20D	48(1)(k)	22B(1)(ka)
12(5)	12(4A)	28	17D	48(1)(l)	22B(1)(h)
12(6)	12(5)	29(4)	17E(5)	48(1)(m)	22B(1)(ha)
12(7)	12(6)	29(5)	17E(4)	48(1)(n)	22B(1)(i)
12(8)	12(7)	30	17F	48(1)(o)	22B(1)(j)
13(2)(ii)	13(2)(c)(iii)	31	17G	48(2)	22B(2)
13(3)	13(5)	32	17H	J3	J2A
13(4)	13(6)	33	17I	J4	J3
13(5)	13(7)	35	17J	J5	J4
14(3)(b)	14(3)(aa)	36	17K	J6	J5
14(3)(c)	14(3)(b)	37	20E	J7	J6
15	14A	38	16B	Regulation 40	L1(c)
16	15	39	16C	Schedule 3	Schedule 2A
18(1)	21(1)	41	20A	Schedule 4	Schedule 2B
18(2)	21(3)	42	20AA	deleted	2(3)
18(3)	21(4)	43	20B	deleted	4B(2)
18(4)	21(5)	44(1)	20C(A1)	deleted	13(3)

3

The requirements of the Building Regulations

3.1 Introduction

This chapter provides a breakdown of the key elements of each Approved Document, showing not only the regulation, but also an overview of the actual requirement 'in a nutshell'. This acts as a good starting point to understand the regulations, which are then expanded upon in much more detail by specific building elements in Chapter 6. The current editions of the Approved Documents are shown in Table 3.1.

If you intend to carry out work that involves any of these areas you are recommended to view the appropriate instrument on line or purchase a copy for reference.

All statutory instruments can be accessed at on the UK government legislation website (http://www.legislation.gov.uk) and can be purchased from The Stationery Office.

Table 3.1 Current editions of approved documents

Section	Title	Edition	Latest amendment
A	Structure	2004	
B	Fire safety	2006	2007
C	Site preparation and resistance to moisture	2004	
D	Toxic substances	1992	2002
E	Resistance to the passage of sound	2003	2004 and 2010
F	Ventilation	2010	
G	Sanitation, hot water safety and water efficiency	2010	
H	Drainage and waste disposal	2002	
J	Combustion appliances and fuel storage systems	2010	
K	Protection from falling, collision and impact	1998	2000
L	Conservation of fuel and power	2010	
M	Access to and use of buildings	2004	
N	Glazing – safety in relation to impact, opening and cleaning	1998	2000
P	Electrical safety	2006	
	Approved Document to support		
	Regulation 7 – Materials and workmanship	1999	2000

Table 3.2 Part A – Structure

Number	Title	Regulation	Requirement (in a nutshell)
A1	Loading	(1) *The building shall be constructed so that the combined dead, imposed and wind loads are sustained and transmitted by it to the ground:* (a) *safely; and* (b) *without causing such deflection or deformation of any part of the building, or such movement of the ground, as will impair the stability of any part of another building.* (2) *In assessing whether a building complies with sub-paragraph (1) regard shall be had to the imposed and wind loads to which it is likely to be subjected in the ordinary course of its use for the purpose for which it is intended.*	The safety of a structure depends on: • the loading (see BS 6399: Parts 1 and 3); • properties of materials; • design analysis; • details of construction; • safety factors; • workmanship.
A2	Ground movement	*The building shall be constructed so that ground movement caused by:* (a) *swelling, shrinkage or freezing of the subsoil; or* (b) *land-slip or subsidence (other than subsidence arising from shrinkage, in so far as the risk can be reasonably foreseen), will not impair the stability of any part of the building.*	• Horizontal and vertical ties should be provided.
A3	Disproportionate collapse	*The building shall be constructed so that in the event of an accident the building will not suffer collapse to an extent disproportionate to the cause.*	

Table 3.3 Part B – Fire safety

Number	Title	Regulation	Requirement (in a nutshell)
B1	Means of warning and escape	*The building shall be designed and constructed so that there are appropriate provisions for the early warning of fire, and appropriate means of escape in case of fire from the building to a place of safety outside the building capable of being safely and effectively used at all material times.* B1 does not apply to any prison provided under Section 33 of the Prisons Act 1952.	• There shall be an early warning fire alarm system for persons in the building. • There shall be sufficient escape routes that are suitably located to enable persons to evacuate the building in the event of a fire. • Safety routes shall be protected from the effects of fire. • In an emergency, the occupants of any part of the building shall be able to escape without any external assistance.
B2	Internal fire spread (linings)	*(1) To inhibit the spread of fire within the building, the internal linings shall:* *(a) adequately resist the spread of flame over their surfaces; and* *(b) have, if ignited, a rate of heat release or a rate of fire growth which is reasonable in the circumstances.* *(2) In this paragraph 'internal linings' means the materials or products used in lining any partition, wall, ceiling or other internal structure.*	• The spread of flame over the internal linings of the building shall be restricted. • The heat released from the internal linings shall be restricted.
B3	Internal fire spread (structure)	*(1) The building shall be designed and constructed so that, in the event of fire, its stability will be maintained for a reasonable period.* *(2) A wall common to two or more buildings shall be designed and constructed so that it adequately resists the spread of fire between those buildings.*	Dependent on the use of the building, its size and the location of the element of construction: • load-bearing elements of a building structure shall be capable of withstanding the effects of fire for an appropriate period without loss of stability;

Table 3.3 Part B – Fire safety (Continued)

Number	Title	Regulation	Requirement (in a nutshell)

		For the purposes of this sub-paragraph a house in a terrace and a semi-detached house are each to be treated as a separate building.
		
		(3) Where reasonably necessary to inhibit the spread of fire within the building, measure shall be taken, to an extent appropriate to the size and intended use of the building, comprising either or both of the following:
		
		  (a) sub-division of the building with fire-resisting construction;
		
		  (b) installation of suitable automatic fire suppression systems.
		
		(4) The building shall be designed and constructed so that the unseen spread of fire and smoke within concealed spaces in its structure and fabric is inhibited.
		
		B3(3) does not apply to material alterations to any prison provided under Section 33 of the Prisons Act 1952.
			• the building shall be sub-divided by elements of sub-divided by elements of fire-resisting construction into compartments;
			• all openings in fire-separating elements shall be suitably protected in order to maintain the integrity of the element (i.e. the continuity of the fire separation);
			• any hidden voids in the construction shall be sealed and sub-divided to inhibit the unseen spread of fire and products of combustion.

B4	External fire spread	*(1) The external walls of the building shall adequately resist the spread of fire over the walls and from one building to another, having regard to the height, use and position of the building.* *(2) The roof of the building shall adequately resist the spread of fire over the roof and from one building to another, having regard to the use and position of the building.*	• External walls shall be constructed so as to have a low rate of heat release and thereby be capable of reducing the risk of ignition from an external source and the spread of fire over their surfaces. • The amount of unprotected area in the side of the building shall be restricted so as to limit the amount of thermal radiation that can pass through the wall. • The roof shall be constructed so that the risk of spread of

Table 3.3 Part B – Fire safety

Number	Title	Regulation	Requirement (in a nutshell)
B5	Access and facilities for the fire service	*(1) The building shall be designed and constructed so as to provide reasonable facilities to assist fire fighters in the protection of life.* *(2) Reasonable provision shall be made within the site of the building to enable fire appliances to gain access to the building.*	flame and/or fire penetration from an external fire source is restricted. • There shall be sufficient means of external access to enable fire appliances to be brought near to the building for effective use. • There shall be sufficient means of access into and within the building for fire-fighting personnel to affect search and rescue and fight fire. • The building shall be provided with sufficient internal fire mains and other facilities to assist fire-fighters in their tasks. • The building shall be provided with adequate means for venting heat and smoke from a fire in a basement. The extent to which these arrangements and facilities apply vary depending on the size and use of building.

Table 3.4 Part C – Site preparation and resistance to contaminants and moisture

Number	Title	Regulation	Requirement (in a nutshell)
C1	Site preparation and resistance to contaminants and moisture	(1) *The ground to be covered by the building shall be reasonably free from any material that might damage the building or affect its stability, including vegetable matter, topsoil and pre-existing foundations.* (2) *Reasonable precautions shall be taken to avoid danger to health and safety caused by contaminants on or in the ground covered, or to be covered by the building and any land associated with the building.* (3) *Adequate subsoil drainage shall be provided if it is needed to avoid:* (a) *the passage of the ground moisture to the interior of the building;* (b) *damage to the building, including damage through the transport of water-borne contaminants to the foundations of the building.* (4) *For the purpose of this requirement, 'contaminant' means: any substance that is or may become harmful to persons or buildings including substances, which are corrosive, explosive, flammable, radioactive or toxic.*	Buildings should be safeguarded from the adverse effects of: • vegetable matter; • contaminants on or in the ground to be covered by the building; • ground water.

Table 3.4 Part C – Site preparation and resistance to contaminants and moisture

Number	Title	Regulation	Requirement (in a nutshell)
C2	Resistance to moisture	*The floors, walls and roof of the building shall adequately protect the building and people who use the building from harmful effects caused by:* *(a) ground moisture;* *(b) precipitation and wind-driven spray;* *(c) interstitial and surface condensation; and* *(d) spillage of water from or associated with sanitary fittings or fixed appliances.*	• A solid or suspended floor shall be built next to the ground to prevent undue moisture from reaching the upper surface of the floor. • A wall shall be erected to prevent undue moisture from the ground reaching the inside of the building, and (if it is an outside wall) adequately resisting the penetration of rain and snow to the inside of the building. • The roof of the building shall be resistant to the penetration of moisture from rain or snow to the inside of the building. • All floors next to the ground, walls and roof shall not be damaged by moisture from the ground, rain or snow and shall not carry that moisture to any part of the building which it would damage.

Table 3.5 Part D – Toxic substances

Number	Title	Regulation	Requirement (in a nutshell)
D1	Cavity insulation	*If insulating material is inserted into a cavity in a cavity wall, reasonable precautions shall be taken to prevent the subsequent permeation of any toxic fumes from that material into any part of the building occupied by people.*	Fumes given off by insulating materials such as by urea formaldehyde (UF) foams should not be allowed to penetrate occupied parts of buildings to an extent where they could become a health risk to persons in the building by reaching an irritant concentration.

Table 3.6 Part E – Resistance to the passage of sound

Number	Title	Regulation	Requirement (in a nutshell)
E1	Protection against sound from other parts of the building and adjoining buildings	*Dwelling-houses, flats and rooms for residential purposes shall be designed and constructed in such a way that they provide reasonable resistance to sound from other parts of the same building and from adjoining buildings.*	Dwellings shall be designed so that the noise from domestic activity in an adjoining dwelling (or other parts of the building) is kept to a level that: • does not affect the health of the occupants of the dwelling; • will allow them to sleep, rest and engage in their normal activities in satisfactory conditions. Dwellings shall be designed so that any domestic noise that is generated internally does not interfere with the occupants' ability to sleep, rest and engage in their normal activities in satisfactory conditions.
E2	Protection against sound within a dwelling-house etc.	*Dwelling-houses, flats and rooms for residential purposes shall be designed and constructed in such a way that:* *(a) internal walls between a bedroom or a room containing a water closet, and other rooms; and* *(b) internal floors provide reasonable resistance to sound.* E2 does not apply to: (a) an internal wall which contains a door; (b) an internal wall which separates an en suite toilet from the associated bedroom; (c) existing walls and floors in a building which is subject to a material change of use.	

Table 3.6 Part E – Resistance to the passage of sound

Number	Title	Regulation	Requirement (in a nutshell)
E3	Reverberation in the common internal parts of buildings containing flats or rooms for residential purposes	*The common internal parts of buildings which contain flats or rooms for residential purposes shall be designed and constructed in such a way as to prevent more reverberation around the common parts than is reasonable.* E3 only applies to corridors, stairwells, hallways and entrance halls which give access to the flat or rooms for residential purposes.	Suitable sound-absorbing material shall be used in domestic buildings so as to restrict the transmission of echoes.
E4	Acoustic conditions in schools	*(1) Each room or other space in a school building shall be designed and constructed in such a way that it has the acoustic conditions and the insulation against disturbance by noise appropriate to its intended use.* *(2) For the purposes of this Part – 'school' has the same meaning as in Section 4 of the Education Act 1996 [4]; and 'school building' means any building forming a school or part of a school.*	Suitable sound-insulation materials shall be used within a school building so as to reduce the level of ambient noise (particularly echoing in corridors etc.).

Table 3.7 Part F – Ventilation

Number	Title	Regulation	Requirement (in a nutshell)
F1	Means of ventilation	*(1) There shall be adequate means of ventilation provided for people in the building.* *(2) Fixed systems for mechanical ventilation and any associated controls must be commissioned by testing and adjusting as necessary to secure that the objective referred to in sub-paragraph (1) is met.* F1 does not apply to a building or space within a building *(a) into which people do not normally go; or* *(b) which is used solely for storage; or* *(c) which is a garage used solely in connection with a single dwelling.*	Ventilation (mechanical and/or air-conditioning systems designed for domestic buildings) shall be capable of restricting the accumulation of moisture and pollutants originating within a building.

Requirements in the Building Regulations 2010

16C	Information about ventilation	*(1) This regulation applies where Part F(1) of Schedule 1 imposes a requirement in relation to building work.* *(2) The person carrying out the work shall not later than five days after the work has been completed give sufficient information to the owner about the building's ventilation system and its maintenance requirements so that the ventilation system can be operated in such a manner as to provide adequate means of ventilation.*	
20AA	Mechanical ventilation air flow rate testing	*(1) This regulation applies where paragraph F1(1) of Schedule 1 imposes a requirement in relation to the creation of a new dwelling by building work.* *(2) The person carrying out the work shall, for the purpose of ensuring compliance with paragraph F1(1) of Schedule 1 –* *(a) Ensure that testing of the mechanical ventilation air flow rate is carried out in accordance with a procedure approved by the Secretary of State; and* *(b) Give notice of the results of the testing to the local authority.*	

Table 3.7 Part F – Ventilation

Number	Title	Regulation	Requirement (in a nutshell)
		(3) The notice referred to in paragraph 2(b) shall –	
		(a) Record the results and the date upon which they are based in a manner approved by the Secretary of State; and	
		(b) Be given to the local authority not later than five days after the final test is carried out.	
20C	Commissioning	(A1) This regulation applies to building work in relation to which paragraph F1(2) of Schedule 1 imposes a requirement, but does not apply to the provision or extension of any fixed system for mechanical ventilation or any associated controls where testing and adjustment is not possible.	
		(1) This regulation applies to building work in relation to which paragraph L1(b) of Schedule 1 imposes a requirement, but does not apply to the provision or extension of any fixed building service where testing and adjustment is not possible or would not affect the energy efficiency of that fixed building service.	
		(2) Where this regulation applies the person carrying out the work shall, for the purpose of complying with F1(2) or L1(b) of Schedule 1, give to the local authority a notice confirming that the fixed building services have been commissioned in accordance with a procedure approved by the Secretary of State.	
		(3) The notice shall be given to the local authority –	
		(a) not later than the date on which the notice required by regulation 15(4) is required to be given; or	
		(b) where the regulation does not apply, not more than 30 days after the completion of the work.	

Table 3.7 Part F – Ventilation (Continued)

Number	Title	Regulation	Requirement (in a nutshell)
		Requirements in the Building (Approved Inspectors etc.) Regulations 2010	

12AA	Mechanical ventilation air flow rate testing	(1) *This regulation applies where paragraph F1(1) of Schedule 1 imposes a requirement in relation to the creation of a new dwelling by building work.*
		(2) *The person carrying out the work shall, for the purpose of compliance with paragraph F1(1) of Schedule 1 –*
		   (a) *Ensure that testing of the mechanical ventilation air flow rate is carried out in accordance with a procedure approved by the Secretary of State; and*
		   (b) *Give notice of the results of the testing to the building control body.*	

12C	Commissioning	(A1) *This regulation applies to building work which is the subject of and initial notice, and in relation to which paragraph F1(2) of Schedule 1 imposes a requirement, but does not apply to the provision or extension of any fixed system for mechanical ventilation or any associated controls where testing and adjustment is not possible.*
		(1) *This regulation applies to building work which is the subject of an initial notice, and in relation to which paragraph L1(b) of Schedule 1 imposes a requirement, but does not apply to the provision or extension of any fixed building service where testing and adjustment is not possible or would not affect the energy efficiency of that fixed building service.*
		(2) *Where this regulation applies the person carrying out the work shall, for the purpose of complying with F1(2) or L1(b) of Schedule 1, give to the approved inspector a notice confirming that the fixed building services have been commissioned in accordance with a procedure approved by the Secretary of State.*
		(3) *The notice shall be given to the approved inspector –*
		   (a) *not later than the date on which the notice required by regulation 15(4) is required to be given; or*
		   (b) *where the regulation does not apply, not more than 30 days after the completion of the work.*	

Table 3.8 Part G – Hygiene

Number	Title	Regulation	Requirement (in a nutshell)
G1	Cold water supply	(1) *There must be a suitable installation for the provision of:* (a) *wholesome water to any place where drinking water is drawn off;* (b) *wholesome water or softened wholesome water to any washbasin or bidet provided in or adjacent to a room containing a sanitary convenience;* (c) *wholesome water or softened wholesome water to any washbasin, bidet, fixed bath or shower in a bathroom; and* (d) *wholesome water to any sink provided in any area where food is prepared.* *There must be a suitable installation for the provision of water of suitable quality to any sanitary convenience fitted with a flushing device.*	• The cold water supply must be reliable; • the water supplied must be wholesome; • the pressure and flow rate must be sufficient for the operation of all appliances and locations planned in the building; and • the installation must convey wholesome water or softened wholesome water without waste, misuse, undue consumption or contamination of water. For sanitary conveniences: • the water supplied may be either wholesome, softened wholesome or of a suitable quality.
G2	Water efficiency	*Reasonable provision must be made by the installation of fittings and fixed appliances that use water efficiently for the prevention of undue consumption of water.*	For new dwellings: • the estimated consumption of wholesome water for both cold and hot water systems shall not be greater than 125 litres per person per day of

Table 3.8 Part G – Hygiene (Continued)

Number	Title	Regulation	Requirement (in a nutshell)
		G2 applies only when a dwelling is: (a) erected; or (b) formed by a material change of use of a building.	wholesome water; sanitary appliances and white goods shall be installed in accordance with the requirements of this Approved Document; • a record of all installed sanitary appliances and relevant white goods (washing machines and dishwashers) shall be provided; • a record of the alternative sources of water supplied to the dwelling shall be provided.

Water efficiency of new dwellings

17K		(1) *The potential consumption of wholesome water by persons occupying a dwelling to which this regulation applies must not exceed 125 litres per person per day, calculated in accordance with the methodology set out in the document 'The Water Efficiency Calculator for New Dwellings'.* (2) *This regulation applies to a dwelling which is:* (a) *erected; or* (b) *formed by a material change of use of a building within the meaning of regulation 5(a) or (b).*	

Wholesome water consumption calculation

20E		(1) *Where regulation 17K applies, the person carrying out the work must give the local authority a notice which specifies the potential consumption of*	

Table 3.8 Part G – Hygiene

Number	Title	Regulation	Requirement (in a nutshell)
		wholesome water per person per day calculated in accordance with the methodology referred to in that regulation in relation to the completed dwelling. *(2) The notice shall be given to the local authority not later than five days after the work has been completed.*	
12E		*(1) Where regulation 17K of the Principal Regulations applies to work which is the subject of an initial notice, the person carrying out the work must give the approved inspector a notice which specifies the potential consumption of wholesome water per person per day calculated in accordance with the methodology referred to in that regulation in relation to the completed dwelling.* *(2) The notice shall be given to the approved inspector not later than:* *(a) five days after the work has been completed; or* *(b) the date on which, in accordance with regulation 18, the initial notice ceases to be in force, whichever is the earlier.*	

G3	Hot water supply and systems	*(1) There must be a suitable installation for the provision of heated wholesome water or heated softened wholesome water to:*
		
		(a) any washbasin or bidet provided in or adjacent to a room containing a sanitary convenience;
		(b) any washbasin, bidet, fixed bath or shower in a bathroom; and
		(c) any sink provided in any area where food is prepared.
		
		(2) A hot water system, including any cistern or other vessel that supplies water to or receives expansion water from a hot water system, shall be designed, constructed and installed so as to resist the temperature and pressure that may occur either in normal use, or has the event of such malfunctions as may reasonably be anticipated, and must be adequately supported.
			• The water supplied shall be heated wholesome water or heated softened water;
			• the installation shall convey hot water without waste, misuse or undue consumption of water;
			• hot water systems (including cisterns) shall safely contain the hot water:
			during normal operation of
			  ○ the hot water system;
			  ○ following failure of any

Table 3.8 Part G – Hygiene (Continued)

Number	Title	Regulation	Requirement (in a nutshell)
(3)		*A hot water system that has a hot water storage vessel shall incorporate precautions to:*	thermostat used to control temperature; and
		(a) prevent the temperature of the water stored in the vessel at any time exceeding 100°C; and	○ during operation of any of the safety devices.
		(b) ensure that any discharge from safety devices is safely conveyed to where it is visible but will not cause a danger to persons in or about the building.	• storage vessels shall have a suitable vent pipe connecting the top of the vessel to a point open to the atmosphere above the level of the water in the cold water storage cistern;
		G3(3) does not apply to a system which heats or stores water only for the purposes of an industrial process.	• in addition to any thermostat, either the heat source, or the storage vessel shall be fitted with a device to prevent the temperature of the stored water exceeding 100° C;
(4)		*The hot water supply to any fixed bath must be so designed and installed as to incorporate measures to ensure that the temperature of the water that can be delivered to that bath does not exceed 48° C.*	• the hot water system shall have pipework that allows hot water from the safety devices to discharge into a place open to the atmosphere w here it will cause no danger to persons in or about the building;
		G3(4) applies only when a dwelling is:	• storage vessels shall have at least two independent safety devices that release pressure
		(a) erected; or	
		(b) formed by a material change of use.	

Table 3.8 Part G – Hygiene

Number	Title	Regulation	Requirement (in a nutshell)
			and in doing so, prevent the temperature of the stored water at any time exceeding 100° C. hot water outlet temperature devices used to limit the maximum temperature that is supplied at the outlet shall not be easily altered by building users.
G4	Sanitary conveniences and washing machines	(1) Adequate and suitable sanitary conveniences must be provided in rooms provided to accommodate them or in bathrooms. (2) Adequate stand washing facilities must be provided in: (a) rooms containing sanitary conveniences; or (b) rooms or spaces adjacent to rooms containing sanitary conveniences. (3) Any room containing a sanitary convenience, a bidet, or any facility for washing hands provided in accordance with paragraph (2)(b) must be separated from any kitchen or any area where food is prepared.	
G5	Bathrooms	A bathroom must be provided containing a wash basin and either a fixed bath or a shower. G5 applies only to dwellings and to buildings containing one or more rooms for residential purposes.	
G6	Food preparation areas	A suitable sink must be provided in any area where food is prepared.	

Table 3.9 Part H – Drainage and waste disposal

Number	Title	Regulation	Requirement (in a nutshell)
H1	Foul water	(1) An adequate system of drainage shall be provided to carry foul water from appliances within the building to one of the following, listed in order of priority: (a) a public sewer; or, where that is not reasonably practicable; (b) a private sewer communicating with a public sewer; or, where that is not reasonably practicable; (c) either a septic tank which has an appropriate form of secondary treatment or another wastewater treatment system; or, where that is not reasonably practicable; (d) a cesspool. (2) In this Part 'foul water' means wastewater which comprises or includes: (a) waste from a sanitary convenience, bidet or appliance used for washing receptacles for foul waste; or (b) water which has been used for food preparation, cooking or washing. H1 does not apply to the diversion of water which has been used for personal washing or for the washing of clothes, linen or other articles to collection systems for reuse.	The foul water drainage system shall: • convey the flow of foul water to a foul water outfall (i.e. sewer, cesspool, septic tank or settlement (i.e. holding) tank); • minimize the risk of blockage or leakage; • prevent foul air from the drainage system from entering the building under working conditions; • be ventilated; • be accessible for clearing blockages; • not increase the vulnerability of the building to flooding.
H2	Wastewater treatment systems and cesspools	(1) Any septic tank and its form of secondary treatment, other wastewater treatment system or cesspool, shall be so sited and constructed that: (a) it is not prejudicial to the health of any person; (b) it will not contaminate any watercourse, underground water or water supply;	Wastewater treatment systems shall: • have sufficient capacity to enable breakdown and settlement of solid matter

Table 3.9 Part H – Drainage and waste disposal

Number	Title	Regulation	Requirement (in a nutshell)
		(c) there are adequate means of access for emptying and maintenance; and	in the wastewater from the buildings;
		(d) where relevant, it will function to a sufficient standard for the protection of health in the event of a power failure.	• be sited and constructed so as to prevent overloading of the receiving water.
		(2) Any septic tank, holding tank which is part of a wastewater treatment system or cesspool shall be:	Cesspools shall have sufficient capacity to store the foul water from the building until they are emptied.
		(a) of adequate capacity; (b) so constructed that it is impermeable to liquids; and (c) adequately ventilated.	Wastewater treatment systems and cesspools shall be sited and constructed so as not to:
		(3) Where a foul water drainage system from a building discharges to a septic tank, wastewater treatment system or cesspool, a durable notice shall be affixed in a suitable place in the building containing information on any continuing maintenance requirement to avoid risks to health.	• be prejudicial to health or a nuisance; • adversely affect water sources or resources; • pollute controlled waters; and • be in an area where there is a risk of flooding.
			Septic tanks and wastewater treatment systems and cesspools shall be constructed and sited so as to:

Table 3.9 Part H – Drainage and waste disposal (Continued)

Number	Title	Regulation	Requirement (in a nutshell)
			• have adequate ventilation; • prevent leakage of the contents and ingress of subsoil water; and • having regard to water table levels at any time of the year and rising groundwater levels. Drainage fields shall be sited and constructed so as to: • avoid overloading of the soakage capacity; and • provide adequately for the availability of an aerated layer in the soil at all times.
H3	Rainwater drainage	*(1) Adequate provision shall be made for rainwater to be carried from the roof of the building.* *(2) Paved areas around the building shall be so constructed as to be adequately drained.* *(3) Rainwater from a system provided pursuant to sub-paragraphs (1) or (2) shall discharge to one of the following, listed in order of priority:* *(a) an adequate soakaway or some other adequate infiltration system; or, where that is not reasonably practicable,* *(b) a watercourse; or, where that is not reasonably practicable,* *(c) a sewer.*	Rainwater drainage systems shall: • minimize the risk of blockage or leakage; • be accessible for clearing • ensure that rainwater soaking into the ground is distributed sufficiently so that it does not damage foundations of the proposed building or any adjacent structure;

Table 3.9 Part H – Drainage and waste disposal

Number	Title	Regulation	Requirement (in a nutshell)
			• ensure that rainwater from roofs and paved areas is carried away from the surface either by a drainage system or by other means;
			• ensure that the rainwater drainage system carries the flow of rainwater from the roof to an outfall (e.g. a soakaway, a watercourse, a surface water or combined sewer).
		H3(2) applies only to paved areas:	
		(a) which provide access to the building pursuant to paragraph M2 of Schedule 1 (access for disabled people);	
		(b) which provide access to or from a place of storage pursuant to paragraph H6(2) of Schedule 1 (solid waste storage); or	
		(c) in any passage giving access to the building where this is intended to be used in common by the occupiers of one or more other buildings.	
		H3(3) does not apply to the gathering of rainwater for reuse.	
H4	Building over sewers	(1) *The erection or extension of a building or work involving the underpinning of a building shall be carried out in a way that is not detrimental to the building or building extension or to the continued maintenance of the drain, sewer or disposal main.*	Building or extension or work involving underpinning shall:
			• be constructed or carried out in a manner which will not overload or otherwise cause damage to the drain, sewer or disposal main either during or after the construction;
		(2) *In this paragraph 'disposal main' means any pipe, tunnel or conduit used for the conveyance of effluent to or from a sewage disposal works, which is not a public sewer.*	
		(3) *In this paragraph and paragraph H5 'map of sewers' means any records*	

Table 3.9 Part H – Drainage and waste disposal (Continued)

Number	Title	Regulation	Requirement (in a nutshell)
		kept by a sewerage undertaker under Section 199 of the Water Industry Act 1991.	• not obstruct reasonable access to any manhole or inspection chamber on the drain, sewer or disposal main; • in the event of the drain, sewer or disposal main requiring replacement, not unduly obstruct work to replace the drain, sewer or disposal main, on its present alignment; • reduce the risk of damage to the building as a result of failure of the drain, sewer or disposal main.
		H4 applies only to work carried out: (a) over a drain, sewer or disposal main which is shown on any map of sewers; or (b) on any site or in such a manner as may result in interference with the use of, or obstruction of the access of any person to, any drain, sewer or disposal main which is shown on any map of sewers.	
H5	Separate systems of drainage	*Any system for discharging water to a sewer which is provided pursuant to paragraph H3 shall be separate from that provided for the conveyance of foul water from the building.* H5 applies only to a system provided in connection with the erection or extension of a building where it is reasonably practicable for the system	Separate systems of drains and sewers shall be provided for foul water and rainwater where: • the rainwater is not contaminated; and

Table 3.9 Part H – Drainage and waste disposal

Number	Title	Regulation	Requirement (in a nutshell)
		to discharge directly or indirectly to a sewer for the separate conveyance of surface water which is: (a) shown on a map of sewers; or (b) under construction either by the sewerage undertaker or by some other person (where the sewer is the subject of an agreement to make a declaration of vesting pursuant to Section 104 of the Water Industry Act 1991).	• the drainage is to be connected either directly or indirectly to the public sewer system and either: ◦ the public sewer system in the area comprises separate systems for foul water and surface water; or ◦ a system of sewers which provides for the separate conveyance of surface water is under construction either by the sewerage undertaker or by some other person (where the sewer is the subject of an agreement to make a declaration of vesting pursuant to Section 104 of the Water Industry Act 1991).
H6	Solid waste storage	(1) *Adequate provision shall be made for storage of solid waste.* (2) *Adequate means of access shall be provided:* (a) *for people in the building to the place of storage; and* (b) *from the place of storage to a collection point (where one has*	Solid waste storage shall be: • designed and sited so as not to be prejudicial to health

Table 3.9 Part H – Drainage and waste disposal (Continued)

Number	Title	Regulation	Requirement (in a nutshell)
		been specified by the waste collection authority under Section 46 (household waste) or Section 47 (commercial waste) of the Environmental Protection Act 1990) or to a street (where no collection point has been specified).	• of sufficient capacity having regard to the quantity of solid waste to be removed and the frequency of removal; • sited so as to be accessible for use by people in the building and of ready access from a street for emptying and removal.

Table 3.10 Part J – Combustion appliances and fuel storage systems

Number	Title	Regulation	Requirement (in a nutshell)
J1	Air supply	*Combustion appliances shall be so installed that there is an adequate supply of air to them for combustion, to prevent over-heating and for the efficient working of any flue.* J1 only applies to fixed combustion appliances (including incinerators).	The building shall: • enable the admission of sufficient air for: ○ the proper combustion of fuel and the operation of flues; and ○ the cooling of appliances where necessary; • enable normal operation of appliances without the products of combustion becoming a hazard to health; • enable normal operation of appliances without their causing danger through

Table 3.10 Part J – Combustion appliances and fuel storage systems

Number	Title	Regulation	Requirement (in a nutshell)
			damage by heat or fire to the fabric of the building; • have been inspected and tested to establish suitability for the purpose intended; • have been labelled to indicate performance capabilities.
J2	Discharge of products of combustion	*Combustion appliances shall have adequate provision for the discharge of products of combustion to the outside air.* J2 only applies to fixed combustion appliances (including incinerators).	
J2A	Warning of release of carbon monoxide	*Where a fixed combustion appliance is provided, appropriate provision shall be made to detect and give warning of the release of carbon monoxide.* J2A only applies to fixed combustion appliances located in dwellings.	
J3	Protection of building	*Combustion appliances and flue-pipes shall be so installed, and fireplaces and chimneys shall be so constructed and installed, as to reduce to a reasonable level the risk of people suffering burns or the building catching fire in consequence of their use.* J3 only applies to fixed combustion appliances (including incinerators).	
J4	Provision of information	*Where a hearth, fireplace, flue or chimney is provided or extended, a durable notice containing information on the performance capabilities of the hearth, fireplace, flue or chimney shall be affixed in a suitable place in the building for the purpose of enabling combustion appliances to be safely installed.*	

Table 3.10 Part J – Combustion appliances and fuel storage systems (Continued)

Number	Title	Regulation	Requirement (in a nutshell)
J5	Protection of liquid fuel storage systems	*Liquid fuel storage systems and the pipes connecting them to combustion appliances shall be so constructed and separated from buildings and the boundary of the premises as to reduce to a reasonable level the risk of the fuel igniting in the event of fire in adjacent buildings or premises.* J5 only applies to: (a) fixed oil storage tanks with capacities greater than 90 litres and connecting pipes; and (b) fixed liquefied petroleum gas storage installations with capacities greater than 150 litres which are located outside the building and which serve fixed combustion appliances (including incinerators) in the building.	Oil and Liquid Petroleum Gas (LPG) fuel storage installations shall be located and constructed so that they are reasonably protected from fires that may occur in buildings or beyond boundaries. Oil storage tanks used wholly or mainly for private dwellings shall be: • reasonably resistant to physical damage and corrosion; • designed and installed so as to minimize the risk of oil escaping during the filling or maintenance of the tank; • incorporate secondary containment when there is a significant risk of pollution; • be labelled with information on how to respond to a leak.
J6	Protection against pollution	*Oil storage tanks and the pipes connecting them to combustion appliances shall:*	

Table 3.10 Part J – Combustion appliances and fuel storage systems

Number	Title	Regulation	Requirement (in a nutshell)
		(a) be so constructed and protected as to reduce to a reasonable level the risk of the oil escaping and causing pollution; and *(b) have affixed in a prominent position a durable notice containing information on how to respond to an oil escape so as to reduce to a reasonable level the risk of pollution.*	
		J6 applies only to fixed oil storage tanks with capacities of 3500 litres or less, and connecting pipes, which are: (a) located outside the building; and (b) serve fixed combustion appliances (including incinerators) in a building used wholly or mainly as a private dwelling, but does not apply to buried systems.	

Table 3.11 Part K – Protection from falling, collision and impact

Number	Title	Regulation	Requirement (in a nutshell)
K1	Stairs, ladders and ramps	*Stairs, ladders and ramps shall be so designed, constructed and installed as to be safe for people moving between different levels in or about the building.* K1 applies only to stairs, ladders and ramps which form part of the building.	All stairs, steps and ladders shall provide reasonable safety between levels in a building. Stairs etc. in public buildings may need to be of a higher standard.
K2	Protection from falling	*(a) Any stairs, ramps, floors and balconies and any roof to which people have access; and* *(b) any light well, basement area or similar sunken area connected to a building, shall be provided with barriers where it is necessary to protect people in or about the building from falling.* K2 (a) applies only to stairs and ramps which form part of the building.	Pedestrian guarding should be provided for any part of a floor, gallery, balcony, roof, or any other place to which people have access and any light well, basement area or similar sunken area next to a building.
K3	Vehicle barriers and loading bays	*(1) Vehicle ramps and any levels in a building to which vehicles have access shall be provided with barriers where it is necessary to protect people in or about the building.* *(2) Vehicle loading bays shall be constructed in such a way, or be provided with such features, as may be necessary to protect people in them from collision with vehicles.*	Vehicle barriers should be provided that are capable of resisting or deflecting the impact of vehicles. Loading bays shall be provided with an adequate number of exits (or refuges) to enable people to avoid being crushed by vehicles.
K4	Protection from collision with open windows etc.	*Provision shall be made to prevent people moving in or about the building from colliding with open windows, skylights or ventilators.* K4 does not apply to dwellings.	All windows, skylights, and ventilators shall be capable of being left open without danger of people colliding with them.

Table 3.11 Part K – Protection from falling, collision and impact

Number	Title	Regulation	Requirement (in a nutshell)
K5	Protection against impact from and trapping by doors	*(1) Provision shall be made to prevent any door or gate:* *(a) which slides or opens upwards, from falling onto any person; and* *(b) which is powered, from trapping any person.* *(2) Provision shall be made for powered doors and gates to be opened in the event of a power failure.* *(3) Provision shall be made to ensure a clear view of the space on either side of a swing door or gate.* K5 does not apply to: (a) dwellings; or (b) any door or gate that is part of a lift.	

Table 3.12 Part L – Conservation of fuel and power

Number	Title	Regulation	Requirement (in a nutshell)
L1A			
17c	New buildings	Where a building is erected, it shall not exceed the target CO_2 emission rate for the building that has been approved pursuant to regulation 17B.	
17E	Energy performance certificates	(1) This regulation applies where –	
		(a) A building is erected; or	
		(b) A building is modified so that it has a greater of fewer number of parts designed or altered for separate us than it previously had, where the modification includes the provision or extension of any of the fixed services for heating, hot water, air conditioning or mechanical ventilation.	
		(2) The person carrying out the work shall –	
		(a) Give an energy performance certificate for the building to the owner of the building; and	
		(b) Give to the local authority a notice to that effect, including the reference number under which the energy performance certificate has been registered in accordance with regulation 17F(4).	
		(3) The energy performance certificate and notice shall be given no later than five days after the work has been completed.	
		(4) The energy performance certificate must be accompanied by a recommendation report containing recommendations for the improvement to the energy performance of the building, issued by the energy assessor who issued the energy performance certificate.	
		(5) The energy performance certificate must –	
		(a) Express the asset rating of the building in a way approved by the Secretary of State under regulation 17A;	
		(b) Include a reference value such as a current legal standard or benchmark;	
		(c) Be issued by an energy assessor who is accredited to produce energy performance certificates for that category of building; and	
		(d) Include the following information	

Table 3.12 Part L – Conservation of fuel and power

Number	Title	Regulation	Requirement (in a nutshell)
		i. The reference number under which the certificate has been registered in accordance with regulation 17F(4);	
		ii. The address of the building;	
		iii. An estimate of the total useful floor area of the building;	
		iv. The name of the energy assessor who issued it;	
		v. The name and address of the energy assessors employer, or, if he is self-employed, the name under which he trades and his address;	
		vi. The date on which it was issued; and	
		vii. The name of the approved accreditation scheme of which the energy assessor is a member.	
		(6) Certification for apartments or units designed or altered for separate use in blocks may be based –	
		(a) Except in the case of a dwelling, on a common certification of the whole building for blocks with a common heating system; or	
		(b) On the assessment of another representative apartment or unit in the same block.	
		(7) Where –	
		(a) A block with a common heating system is divided into parts designed or altered for separate use; and	
		(b) One or more, but not all, of the parts are dwellings; certification for those parts which are not dwellings may be based on a common certification of all the parts which are not dwellings.	
L1	Conservation of fuel and power (new dwellings)	Reasonable provision shall be made for the conservation of fuel and power in buildings by:	New regulations that introduce new energy efficiency
		(a) limiting heat gains and losses:	
		i. through thermal elements and other parts of the building fabric; and	
		ii. from pipes, ducts and vessels used for space heating, space cooling and hot water services;	
		(b) providing and commissioning energy efficient fixed building services with effective controls; and	

Table 3.12 Part L – Conservation of fuel and power (Continued)

Number	Title	Regulation	Requirement (in a nutshell)
		(c) providing to the owner sufficient information about the building, the fixed building services and their maintenance requirements so that the building can be operated in such a manner as to use no more fuel and power than is reasonable in the circumstances.	requirements and other relevant changes to the existing regulations for new dwellings.
L1B			
4A	Requirements relating to thermal elements	(1) Where a person intends to renovate a thermal element, such work shall be carried out as is necessary to ensure that the whole thermal element complies with the requirements of paragraph L1(a)(i) of Schedule 1.	
		(2) Where a thermal element is replaced, the new thermal element shall comply with the requirements of paragraph L1(a)(i) of Schedule 1 requirements of paragraph L1(a)(i) of Schedule 1 requirements of paragraph L1(a)(i) of Schedule 1.	
17D	Consequential improvements to energy regulation 17D	(1) Paragraph (2) applies to an existing building with a total useful floor area over 1000 m² where the proposed building work consists of or includes –	
		(a) An extension;	
		(b) The initial provision of any fixed building services; or	
		(c) An increase to the installed capacity of any fixed building services.	
		(2) Subject to paragraph (3), where this paragraph applies, such work, if any, shall be carried out as is necessary to ensure that the building complies with the requirements of Part L of Schedule 1.	
		(3) Nothing in paragraph (2) requires work to be carried out if it is not technically, functionally or economically feasible.	

Table 3.12 Part L – Conservation of fuel and power

Number	Title	Regulation	Requirement (in a nutshell)
17E	Energy performance certificates	(1) *This regulation applies where –*	
		(a) *A building is erected; or*	
		(b) *A building is modified so that it has a greater of fewer number of parts designed or altered for separate us than it previously had, where the modification includes the provision or extension of any of the fixed services for heating, hot water, air conditioning or mechanical ventilation.*	
		(2) *The person carrying out the work shall –*	
		(a) *Give an energy performance certificate for the building to the owner of the building; and*	
		(b) *Give to the local authority a notice to that effect, including the reference number under which the energy performance certificate has been registered in accordance with regulation 17F(4).*	
		(3) *The energy performance certificate and notice shall be given no later than five days after the work has been completed.*	
		(4) *The energy performance certificate must be accompanied by a recommendation report containing recommendations for the improvement to the energy performance of the building, issued by the energy assessor who issued the energy performance certificate.*	
		(5) *The energy performance certificate must –*	
		(a) *Express the asset rating of the building in a way approved by the Secretary of State under regulation 17A;*	
		(b) *Include a reference value such as a current legal standard or benchmark;*	
		(c) *Be issued by an energy assessor who is accredited to produce energy performance certificates for that category of building; and*	
		(d) *Include the following information*	
		i. *The reference number under which the certificate has been registered in accordance with regulation 17F(4);*	
		ii. *The address of the building;*	
		iii. *An estimate of the total useful floor area of the building;*	
		iv. *The name of the energy assessor who issued it;*	

Table 3.12 Part L – Conservation of fuel and power (Continued)

Number	Title	Regulation	Requirement (in a nutshell)
		v. The name and address of the energy assessors employer, or, if he is self-employed, the name under which he trades and his address;	
		vi. The date on which it was issued; and	
		vii. The name of the approved accreditation scheme of which the energy assessor is a member.	
		(6) Certification for apartments or units designed or altered for separate use in blocks may be based –	
		(a) Except in the case of a dwelling, on a common certification of the whole building for blocks with a common heating system; or	
		(b) On the assessment of another representative apartment or unit in the same block.	
		(7) Where –	
		(a) A block with a common heating system is divided into parts designed or altered for separate use; and	
		(b) One or more, but not all, of the parts are dwellings; certification for those parts which are not dwellings may be based on a common certification of all the parts which are not dwellings.	
L1B	Conservation of fuel and power (existing dwellings)	Reasonable provision shall be made for the conservation of fuel and power in buildings by: (a) limiting heat gains and losses: i. through thermal elements and other parts of the building fabric; and ii. from pipes, ducts and vessels used for space heating, space cooling and hot water services; (b) providing and commissioning energy efficient fixed building services with effective controls; and (c) providing to the owner sufficient information about the building, the fixed building services and their maintenance requirements so that the building can be operated in such a manner as to use no more fuel and power than is reasonable in the circumstances.	New regulations that introduce new energy efficiency requirements and other relevant changes to the existing regulations for existing dwellings.

Table 3.12 Part L – Conservation of fuel and power

Number	Title	Regulation	Requirement (in a nutshell)
L2A			
17c	New buildings	Where a building is erected, it shall not exceed the target CO_2 emission rate for the building that has been approved pursuant to regulation 17B.	
17E	Energy performance certificates	(1) This regulation applies where – (a) A building is erected; or (b) A building is modified so that it has a greater of fewer number of parts designed or altered for separate us than it previously had, where the modification includes the provision or extension of any of the fixed services for heating, hot water, air conditioning or mechanical ventilation. (2) The person carrying out the work shall – (c) Give an energy performance certificate for the building to the owner of the building; and (d) Give to the local authority a notice to that effect, including the reference number under which the energy performance certificate has been registered in accordance with regulation 17F(4). (3) The energy performance certificate and notice shall be given no later than five days after the work has been completed. (4) The energy performance certificate must be accompanied by a recommendation report containing recommendations for the improvement to the energy performance of the building, issued by the energy assessor who issued the energy performance certificate. (5) The energy performance certificate must – (a) Express the asset rating of the building in a way approved by the Secretary of State under regulation 17A; (b) Include a reference value such as a current legal standard or benchmark; (c) Be issued by an energy assessor who is accredited to produce energy performance certificates for that category of building; and (d) Include the following information i. The reference number under which the certificate has been registered in accordance with regulation 17F(4);	

Table 3.12 Part L – Conservation of fuel and power (Continued)

Number	Title	Regulation	Requirement (in a nutshell)
		ii. The address of the building;	
		iii. An estimate of the total useful floor area of the building;	
		iv. The name of the energy assessor who issued it;	
		v. The name and address of the energy assessors employer, or, if he is self-employed, the name under which he trades and his address;	
		vi. The date on which it was issued; and	
		vii. The name of the approved accreditation scheme of which the energy assessor is a member.	
		(6) *Certification for apartments or units designed or altered for separate use in blocks may be based –*	
		(a) *Except in the case of a dwelling, on a common certification of the whole building for blocks with a common heating system; or*	
		(b) *On the assessment of another representative apartment or unit in the same block.*	
		(7) *Where –*	
		(a) *A block with a common heating system is divided into parts designed or altered for separate use; and*	
		(b) *One or more, but not all, of the parts are dwellings; certification for those parts which are not dwellings may be based on a common certification of all the parts which are not dwellings.*	
L2A	Conservation of fuel and power (new buildings other than dwellings)	*Reasonable provision shall be made for the conservation of fuel and power in buildings by:*	New regulations that introduce new energy efficiency requirements and other relevant changes to the existing regulations for new buildings other than dwellings.
		(a) *limiting heat gains and losses:*	
		i. through thermal elements and other parts of the building fabric; and	
		ii. from pipes, ducts and vessels used for space heating, space cooling and hot water services;	
		(b) *providing and commissioning energy efficient fixed building services with effective controls; and*	
		(c) *providing to the owner sufficient information about the building, the fixed building services and their maintenance requirements so that the building can be operated in such a manner as to use no more fuel and power than is reasonable in the circumstances.*	

Table 3.12 Part L – Conservation of fuel and power

Number	Title	Regulation	Requirement (in a nutshell)
L2B			
4A	Requirements relating to thermal elements	(1) Where a person intends to renovate a thermal element, such work shall be carried out as is necessary to ensure that the whole thermal element complies with the requirements of paragraph L1(a)(i) of Schedule 1. (2) Where a thermal element is replaced, the new thermal element shall comply with the requirements of paragraph L1(a)(i) of Schedule 1 requirements of paragraph L1(a)(i) of Schedule 1 requirements of paragraph L1(a)(i) of Schedule 1.	
17D	Consequential improvements to energy regulation 17D	(1) Paragraph (2) applies to an existing building with a total useful floor area over 1000m2 where the proposed building work consists of or includes – (a) An extension; (b) The initial provision of any fixed building services; or (c) An increase to the installed capacity of any fixed building services. (2) Subject to paragraph (3), where this paragraph applies, such work, if any, shall be carried out as is necessary to ensure that the building complies with the requirements of Part L of Schedule 1. (3) Nothing in paragraph (2) requires work to be carried out if it is not technically, functionally or economically feasible.	
17J	Interpretation	In this part – 'building' means the building as a whole or parts of it that have been designed or altered to be used separately.	
17E	Energy performance certificates	(1) This regulation applies where – (a) A building is erected; or (b) A building is modified so that it has a greater of fewer number of parts designed or altered for separate us than it previously had, where the modification includes the provision or extension of any of the fixed services for heating, hot water, air conditioning or mechanical ventilation. (2) The person carrying out the work shall – (a) Give an energy performance certificate for the building to the owner of the building; and	

Table 3.12 Part L – Conservation of fuel and power (Continued)

Number	Title	Regulation	Requirement (in a nutshell)
		(b) Give to the local authority a notice to that effect, including the reference number under which the energy performance certificate has been registered in accordance with regulation 17F(4).	
(3)		The energy performance certificate and notice shall be given no later than five days after the work has been completed.	
(4)		The energy performance certificate must be accompanied by a recommendation report containing recommendations for the improvement to the energy performance of the building, issued by the energy assessor who issued the energy performance certificate.	
(5)		The energy performance certificate must –	
		(a) Express the asset rating of the building in a way approved by the Secretary of State under regulation 17A;	
		(b) Include a reference value such as a current legal standard or benchmark;	
		(c) Be issued by an energy assessor who is accredited to produce energy performance certificates for that category of building; and	
		(d) Include the following information	
		i. The reference number under which the certificate has been registered in accordance with regulation 17F(4);	
		ii. The address of the building;	
		iii. An estimate of the total useful floor area of the building;	
		iv. The name of the energy assessor who issued it;	
		v. The name and address of the energy assessors employer, or, if he is self-employed, the name under which he trades and his address;	
		vi. The date on which it was issued; and	
		vii. The name of the approved accreditation scheme of which the energy assessor is a member.	
(6)		Certification for apartments or units designed or altered for separate use in blocks may be based –	
		(a) Except in the case of a dwelling, on a common certification of the whole building for blocks with a common heating system; or	
		(b) On the assessment of another representative apartment or unit in the same block.	

Table 3.12 Part L – Conservation of fuel and power

Number	Title	Regulation	Requirement (in a nutshell)
		(7) Where –	
		(a) A block with a common heating system is divided into parts designed or altered for separate use; and	
		(b) One or more, but not all, of the parts are dwellings; certification for those parts which are not dwellings may be based on a common certification of all the parts which are not dwellings.	

L2B	Conservation of fuel and power (existing buildings other than dwellings)	Reasonable provision shall be made for the conservation of fuel and power in buildings by:
		
		(a) limiting heat gains and losses:
		
		  i. through thermal elements and other parts of the building fabric;; and
		  ii. from pipes, ducts and vessels used for space heating, space cooling and hot water services;
		
		(b) providing and commissioning energy efficient fixed building services with effective controls; and
		(c) providing to the owner sufficient information about the building, the fixed building services and their maintenance requirements so that the building can be operated in such a manner as to use no more fuel and power than is reasonable in the circumstances.
			New regulations that introduce new energy efficiency requirements and other relevant changes to the existing regulations for existing buildings other than dwellings.

Table 3.13 Part M – Access to and use of buildings (Continued)

Number	Title	Regulation	Requirement (in a nutshell)
M1	Access and use	*Reasonable provision shall be made for people to:* (a) *gain access to; and* (b) *use;* *the buildings and its facilities.* M1 does not apply to: (a) an extension of, or material alteration of, a dwelling; or (b) any part of a building which is used solely to enable the building to be inspected, repaired or maintained.	In addition to the requirements of the Disability Discrimination Act 1995, Approved Document M also requires: (1) Precautions are taken to ensure that: • new non-domestic buildings and/or dwellings (e.g. houses and flats used for student living accommodation etc.); • extensions to existing non-domestic buildings; • non-domestic buildings that have been subject to a material change of use (e.g. so that they become a hotel, boarding house, institution, public building or shop);
M2	Access to extensions of buildings other than dwellings	*Suitable independent access shall be provided to the extension where reasonably practical.* M2 does not apply where suitable access to the extension is provided through the building that is extended.	(2) Are capable of allowing people, regardless of their disability, age or gender to: • gain access to buildings; • gain access within buildings; • be able to use the facilities of the buildings (both as visitors and as people who live or work in them);

Table 3.13 Part M – Access to and use of buildings

Number	Title	Regulation	Requirement (in a nutshell)
M3	Sanitary conveniences in extensions to buildings other than dwellings	*If sanitary conveniences are provided in any building that is to be extended, reasonable provision shall be made within the extension for sanitary conveniences.* M3 does not apply where there is reasonable provision for sanitary conveniences elsewhere in the building, such that people occupied, or otherwise having occasion to enter the extension, can gain access to and use those sanitary conveniences.	(3) Use sanitary conveniences in the principal storey of any new dwelling.
M4	Sanitary conveniences in dwellings	*(1) Reasonable provision shall be made in the entrance storey for sanitary conveniences, or where the entrance contains no habitable rooms, reasonable provision for sanitary convenience shall be made in either the entrance storey or principal storey.* *(2) In this paragraph 'entrance storey' means the storey which contains the principal entrance and 'principal storey' means the storey nearest to the entrance storey which contains a habitable room, or if there are two stories equally near, either such storey.*	

Table 3.14 Part N – Glazing – safety in relation to impact, opening and cleaning

Number	Title	Regulation	Requirement (in a nutshell)
N1	Protection against impact	*Glazing with which people are likely to come into contact whilst moving in or about the building shall:* *(a) if broken on impact, break in a way which is unlikely to cause injury; or* *(b) resist impact without breaking; or* *(c) be shielded or protected from impact.*	All glazing installed in buildings shall be: • sufficiently robust to withstand impact from a falling or passing person; or • protected from a falling or passing person.
N2	Manifestation of glazing	*Transparent glazing with which people are likely to come into contact while moving in or about the building, shall incorporate features which make it apparent.* N2 does not apply to dwellings.	
N3	Safe opening and closing of windows etc.	*Windows, skylights and ventilators which can be opened by people in or about the building shall be so constructed or equipped that they may be opened, closed or adjusted safely.* N3 does not apply to dwellings.	
N4	Safe access for cleaning windows etc.	*Provision shall be made for any windows, skylights, or any transparent or translucent walls, ceilings or roofs to be safely accessible for cleaning.* N4 does not apply to: (a) dwellings; or (b) any transparent or translucent elements whose surfaces are not intended to be cleaned.	

Table 3.14 Part P – Electrical safety

Number	Title	Regulation	Requirement (in a nutshell)
P1	Design and installation	*Reasonable provision shall be made in the design and installation of electrical installations in order to protect persons operating, maintaining or altering the installations from fire or injury.*	These requirements only apply to electrical installations that are intended to operate at low or extra-low voltage: • in (or attached to) a dwelling; • in common parts of a building serving one or more dwellings (but excluding power supplies to lifts); • in a building that receives its electricity from a source located within (or shared with) a dwelling; • in a garden or in or on land associated with a building where the electricity is from a source located within or shared with a dwelling.

4

Planning permission

Before undertaking any building project, you must first obtain the approval of local government authorities. Many people (particularly householders) are initially reluctant to approach local authorities because, according to local gossip, they are 'likely to be obstructive'. In fact the reality is quite the reverse, as it is their purpose to protect all of us from irresponsible builders and developers, and they are normally most sympathetic and helpful to any builder and/or DIY person who wants to comply with the statutory requirements and has asked for their advice.

There are two main controls that districts rely on to ensure that adherence to the local plan is ensured, namely planning permission and Building Regulation approval. There is quite a lot of confusion with these terms. While both are associated with gaining planning permission, actually receiving planning permission does not automatically confer Building Regulation approval, and vice versa. Additionally, you **may** require **both** sets of approval before you can proceed with your building project. Indeed, there may be a variation in the planning requirements (and to some extent the Building Regulations) from one area of the country to another. Consequently, the information given in the following pages should be considered as a guide only and not as an authoritative statement of the law.

 You do not require planning permission to carry out any internal alterations to your home, house, flat or maisonette, provided that it does not affect the external appearance of the building.

You are allowed to make certain changes to your home without having to apply to the local council for permission **provided** that it does not affect the external appearance of the building. These are called 'permitted development rights'. The majority of building work that you are likely to complete will, however, probably require you to have planning permission, and the nation's planning system plays an important role in today's society by helping to protect the environment in our towns, cities and the countryside.

You might even require consent or be required to follow certain procedures if carrying out work to trees. If you are thinking about carrying out work on a listed building or work that requires the pruning or felling of a tree protected by a Tree Preservation Order, you will need to contact your local authority planning department before carrying out any work.

Note: The Localism Bill is currently passing through Parliament, and if it is adopted it will introduce a number of reforms to the planning system.

4.1 Planning controls

Councils use planning controls to protect the character of an area through the regulation of the use, siting and appearance of buildings and other constructions within the area. What might seem to be a minor development in itself could have far-reaching implications that you had not previously considered (e.g. erecting a structure that would ultimately obscure vision at a busy junction and thereby constitute a danger to traffic). Equally, the local authority might refuse permission on the grounds that the planned scheme would not blend sympathetically with its surroundings. Your property could also be affected by legal restrictions such as a right of way, which could prejudice planning permission.

The actual details of planning requirements are complex, but in respect of domestic developments the planning authority is concerned primarily with the construction work such as an extension to a house or the provision of a new garage or new outbuilding that is being carried out. Structures such as walls and fences also need to be considered because their height or siting might well infringe the rights of neighbours and other members of the community. The planning authority will also want to approve any change of use, such as converting a house into flats or running a business from premises previously occupied as a dwelling only.

4.1.1 Why are planning controls needed?

If you live in a listed building of historical or architectural interest or your house is in a Conservation Area, you should seek advice before considering any alterations.

The purpose of the planning system is to protect the environment as well as public amenities and facilities. It is not designed to protect the interests of one person over another. Within the framework of legislation approved by Parliament, councils are tasked to ensure that development is allowed where it is needed, while ensuring that the character and amenity of the area are not adversely affected by new buildings or changes in the use of existing buildings or land.

Some people think the planning system should be used to prevent any change in their local environment, while others may think that planning controls are an unnecessary interference in their individual rights. The present position is that **all** major works need planning permission from the council but many minor works do not. Parliament thinks this is the right balance as it enables councils to protect the character and amenity of their area, while individuals have a reasonable degree of freedom to alter their property.

4.2 Who requires planning permission?

 If you are a leaseholder, you may first need to get permission from your landlord or management company.

Although the rules and requirements vary according to whether you actually own a house or a flat/maisonette, generally speaking the principles and procedures for making planning applications are exactly the same for owners of houses and for freeholders (or leaseholders) of flats and maisonettes. Planning regulations, however, have to cover many different situations, and so even the provisions that affect the average householder are quite detailed.

4.3 Who controls planning permission?

The planning system is made up of a cascade of documents. Currently, under the provisions of the Building Act 1984, national policy is set out in Planning Policy Statements (PPSs). Regions set out regional policy through Regional Spatial Strategy (RSS). Local Development Frameworks set out broad planning policies at county, local, district and unitary level (Figure 4.1).

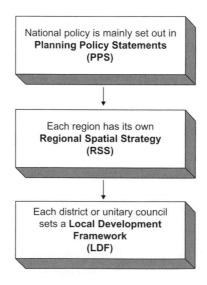

Figure 4.1 Planning responsibilities.

Note: The Localism Bill is currently passing through Parliament and if it is adopted it will introduce a number of reforms to planning system including the removal of Regional Spatial Strategies.

4.3.1 Planning and Compulsory Purchase Act 2004

The Planning and Compulsory Purchase Act 2004 made significant changes to the UK's town planning and compulsory purchase framework. The act created a hierarchy of policies and abolished Local Plans and Structure Plans and replaced them with Local Development Frameworks.

4.3.2 Local Development Framework

 Local Development Plans must take account of national and regional planning policy.

 More details about planning law can be found in your local library. Advice on a legal problem involving planning can be obtained from your local Citizens Advice Bureau or from a legal professional.

The way that planning is managed in your area is set out in the Local Development Framework. The framework is a collection of documents prepared by district councils that outline the planning strategy for the local area. The published framework is used as a guide to the location of development over a ten-year period. For example, they:

- will identify where new homes, jobs and other types of development may be built;
- may require related development to be provided, such as children's play areas, parking facilities and road improvements;
- will outline restrictions where certain types of development are unacceptable.

In preparing the Local Development Framework the council is responsible for consulting local people and for ensuring that their views are taken into account, thereby giving them a chance to influence the way in which their area is affected.

 You can find details about a particular planning case through your local planning department or on their website.

There must be a core strategy and proposals map within the Local Development Framework and there also may be Area Action Plans and Other Development Plan Documents within the folder. Together, these elements of the plan:

- allow local people to clearly see if their homes, businesses or other property would be affected by what is proposed;
- give guidance to anyone who wants to build on a piece of land or change the use of a building in the area;
- provide a basis for decisions on planning applications.

The district council is responsible for keeping their plan under constant review and for making it available to everybody (usually via their council website).

 You do not need planning permission to let one or two of your rooms to lodgers.

Planning permission is required for most building works, engineering works and use of land, and the following are some common examples of when you would need to apply for planning permission:

- you want to make additions or extensions to a flat or maisonette (including those converted from houses);
- you want to divide off part of your house for use as a separate home (e.g. a self-contained flat or bed-sit);
- you want to use a caravan in your garden as a separate residence for some-one else;
- you want to build a separate house in your garden;
- you want to divide off part of your home for business or commercial use (e.g. a workshop);
- you want to build a parking place for a commercial vehicle;
- you want to build something which goes against the terms of the original planning permission for your house (e.g. your house may have been built with a restriction to stop people putting up fences in front gardens because it is on an 'open plan' estate);
- the work you want to complete might obstruct the view of road users;
- the work would involve a new or wider access to a trunk or classified road.

 Note: A Department for Communities and Local Government DCLG book-let (*Planning Permission, A Guide for Business*) giving advice about work-ing from home and whether planning permission is likely to be required is available from councils or can be downloaded from http://www.communities. gov.uk.

4.4 What is planning permission?

The planning control process is administered by your local authority and the system 'exists to control the development and use of land and buildings for the best interests of the community'.

The process is intended to make the environment better for everyone and acts as a service to manage the types of constructions, modifications of premises, uses of land, and ensures the right mix of premises in any one vicinity (that individuals may plan to make) is maintained. The key feature of the process is to allow a party to propose a plan and for other parties to object if they wish, or are qualified, to do so.

4.5 What types of planning permission are available?

There are three types of planning permission available: outline, reserved and full.

4.5.1 Outline

This is an application for a development 'in principle' without giving too much detail on the actual building or construction. It basically lets you know, in advance, whether the development is likely to be approved. Assuming permission is granted under these circumstances you will then have to submit a further application in greater detail. In the main, this applies to large-scale developments only and you will probably be better off making a full application in the first place.

4.5.2 Reserved matters

This is the follow-up stage to an outline application to give more substance and more detail.

4.5.3 Full planning permission

This is most widely used and is for erection or alteration of buildings or changes of use. There are no preliminary or outline stages, and when consent is granted it is for a specific period of time. If this is due to lapse, a renewal of limited permission can be applied for.

4.6 How do I apply for planning permission?

There are fees to pay for each application for planning permission, and your local planning office can provide you with the relevant details. You must make sure you have paid the correct fee – as permission can be refused if there is a discrepancy on fees paid.

When your forms and plans are ready, they should be submitted to the planning office. The planning office will arrange for them to be listed in the local newspaper under 'latest planning applications' and will write to each neighbouring property and (normally) give them 21 days in which to raise any objections.

There is an appeal procedure, which your local authority planning officer can advise you about.

At the planning office, officials will produce a file after the 21 days have expired, with any objections or supporting information, and will make recommendation on the application, ready for presenting it at the next planning committee or sub-committee meeting. At this meeting, they will discuss the case, reject it, ask for modifications or accept it. Whichever the decision, the planning officer will feed back the decision to the applicant.

The Planning Portal is the UK government's planning resource. It provides extensive details about the planning system, applying for planning permission, finding out about development near you and appeals against a planning decision, and researches the latest government policy.

The portal is split into three sections:

- **general public** – a guide to applying for planning permission and accessing local information;
- **planning professionals** – the complete resource for researching and submitting planning applications;
- **government users** – a dedicated knowledge base for all levels of government.

For more details go to http://www.planningportal.gov.uk.

4.7 Do I really need planning permission?

 It is best to employ a skilled designer when preparing plans for extensions and alterations. However, the authority's planning officers are able to offer general design guidance.

Most alterations and extensions to property and changes of use of land need to have some form of planning permission, which is achieved by submitting a planning application to the local authority. The purpose of this control is to protect and enhance our surroundings, to preserve important buildings and natural areas, and strengthen the local economy.

However, not all extensions and alterations to dwelling houses require planning permission. Certain types of development are permitted without the need to make an official request, and it is always wise to contact the local authority before commencing any work.

Whether or not planning permission is required, good design is always important. Extensions and alterations should be in scale and in harmony with the remainder of the house. The builder should ensure that details such as window openings and matching materials are taken into account. The Planning Portal also features an interactive house which you can use to identify what permissions you will require for your specific project (http://www.planningportal.gov.uk/permission/house).

Chapter 5 provides an indication of the requirements for planning permission and Building Regulation approval.

4.8 How should I set about gaining planning permission?

 New work requires details of materials used, dimensions and all related installations, similar to that required for Building Regulations.

If you are in the planning stages for your work and you know planning permission will be required, it is wise to get the plans passed before you go to any
expense or make any decisions that you may find hard to reverse – such as
signing a contract for work. If your plans are rejected, you will still have to pay
your architect or whoever prepared your plans for submission, but you will not
have to pay any penalty clauses to the building contractor.

An architect (surveyor or general contractor) can be asked to prepare and
submit your plans on your behalf if you like, but as the owner and person
requiring the development it will be your name that goes on the application,
even if all the correspondence goes between your architect and the planning
department.

 It is always best to submit an application in the early stages – if you try to be
clever by submitting plans at the last minute (in the hope that neighbours will
not have time to react) then you could be in for an expensive mistake! It is
much better to do things properly and up front.

You don't have to own the land to make a planning application, but you will
need to disclose your interest in the property. This might happen if you plan
to buy land, with the intention of developing it, subject to planning approval.
It would, therefore, be in your best interest to obtain the consent before the
purchase proceeds.

To submit your application you will need to use the official forms, available
from the local authority planning department. It is a good idea to collect these
personally, as you may get the opportunity to talk through your ideas with a
planning officer, and in doing so probably get some useful feedback. You will
also need to include detailed plans of the present and proposed layout as well
as the property's position in relation to other properties and roads or other
features.

4.9 What sort of plans will I have to submit?

There are three types of plans (namely site, block and building) that can accompany your application and, as indicated above, the choice will depend on the
work proposed.

4.9.1 Site plan

You do not have to own the land to make a planning application, but you will need to disclose your interest in the property (i.e. if you plan to buy land with the intention of developing it, subject to planning approval).

A site plan indicates the development location and relationship to neighbouring property and roads etc. The minimum scale is 1:2500 (or 1:1250 in a built-up area). The land to which the application refers is outlined in red ink. Adjacent land, if owned by the applicant, is outlined in blue ink.

Block plan

A block plan is a detailed plan of a construction or structural alteration that shows the existing and proposed building, all trees, waterways, ways of access, pipes and drainage and any other important features. Minimum scale is 1:1500.

Building plans

Building plans are the detailed drawings of the proposed building works and would show plans, elevations and cross-sections that accurately describe every feature of the proposal. These plans are normally very thorough and include types of material, colour and texture, the layers of foundations, floor constructions and roof constructions etc.

4.10 What is meant by 'building works'?

In the context of the Building Regulations, 'building works' means:

- the erection or extension of a building;
- the provision or extension of a controlled service or fitting;
- the material alteration of a building, or a controlled service or fitting;
- work required by Regulation 6 (requirements relating to material change of use);
- the insertion of insulating material into the cavity wall of a building;
- work involving the underpinning of a building.

4.11 What important areas should I take into consideration?

The following are some of the most important areas that should be considered before you submit a planning application.

4.11.1 Advertisement applications

If your proposal is to display an advertisement, you will need to make a separate application on a special set of forms, as this is subject to separate legislation. Three copies of the forms and the relevant drawings must be supplied. These must include a location plan and sufficient detail to show the size, materials and colour of the sign and its position. No certificate of ownership is needed, but it is illegal to display signs on the property without the consent of the owner.

4.11.2 Conservation Area consent

If you live in a Conservation Area, you will need Conservation Area consent to do the following:

*   demolish a building within a Conservation Area;
*   demolish a gate, fence, wall or railing over 1 m high if it is next to a highway (including a public footpath or bridleway) or public open space; or over 2 m high elsewhere.

4.11.3 Listed building consent

 It is a criminal offence to carry out work which needs listed building consent without obtaining it beforehand.

You will need to apply for listed building consent if either of the following cases apply:

*   you want to demolish a listed building;
*   you want to alter or extend a listed building in a manner which would affect its character as a building of special architectural or historic interest.

 You may also need listed building consent for any works to separate buildings within the grounds of a listed building. Check the position carefully with the council.

4.11.4 Trees

 Ask the council for a copy of the department's free leaflet *Protected Trees: A Guide to Tree Preservation Procedures*.

Many trees are protected by a Tree Preservation Order (TPO), which means that, in general, you need the council's consent to prune or fell them. In addition, there are controls over many other trees which have a trunk diameter greater than 75 mm in Conservation Areas. You have to give your local planning authority six weeks' notice before carrying out work on trees located in a Conservation Area, even if they are not yet subject to a TPO.

4.11.5 Lawful Development Certificate

In certain circumstances where no previous planning permission has been applied for, and if no action has been taken following the development of a property, an application can be submitted for a Lawful Development Certificate.

4.12 What are the government's restrictions on planning applications?

All applications for planning permission will have to take into account the following acts and regulations.

4.12.1 Planning (Listed Buildings and Conservation Areas) Act 1990

Under the terms of the Planning (Listed Buildings and Conservation Areas) Act 1990, local councils must maintain a list of buildings within their boroughs that have been classified as being of special architectural or historic interest. Councils are also required to keep maps showing which properties are within Conservation Areas.

4.12.2 Town and Country Planning (Control of Advertisements) (England) Regulations 2007

In accordance with the Town and Country Planning (Control of Advertisements) (England) Regulations 2007, councils need to maintain a publicly available register of applications and decisions for consent to display advertisements.

4.12.3 The Local Government (Access to Information) (Variation) Order 2006

The Local Government (Access to Information) (Variation) Order 2006 ensures that information relating to proposed development by councils cannot be treated as exempt when the planning decision is made.

4.12.4 Town and Country Planning (General Permitted Development) (Amendment) (England) Order 2010

Every council must keep the following registers available for public inspection in accordance with the Town and Country Planning (General Permitted Development) (Amendment) (England) Order 2010:

- planning applications, including accompanying plans and drawings;
- applications for a certificate of lawfulness of existing or proposed use or development;
- Enforcement Notices and any related stop notices.

All applications for planning permission must receive publicity.

4.12.5 Other areas

As well as the legal requirement to make the planning register available for public inspection, councils will also allow the public to have access to all other relevant information, such as letters of objection/support for an application or correspondence about considerations. Three clear days before any committee meeting, the file will normally be made available for public inspection and this file will remain available (i.e. for further public inspection) after the committee meeting. Although commercial confidentiality could well be a valid consideration, the council will not use it so as to prevent important information about materials and facilities also being available.

4.13 How do I apply for planning permission?

 Under normal circumstances you will have to pay a fee in order to seek planning permission, but there are exceptions. The planning department will advise you.

Once you have established that planning permission is required, you will need to submit a planning application. Remember, it may take up to eight weeks, or even longer, to get planning permission, so apply early. You will have to prepare a plan showing the position of the site in question (i.e. the site plan) so that the authority can determine exactly where the building is located. You must also submit another, larger scale, plan to show the relationship of the building to other premises and highways (i.e. the block plan). In addition, it would help the council if you also supplied drawings to give a clear idea of what the new proposal will look like, together with details of both the colour and the kind of materials you intend using. You may prepare the drawings yourself, provided you are able to make them accurate.

4.13.1 Application forms and plans

It is important to make sure that you make your planning application correctly. The following checklist may help:

- Obtain the application forms from the planning department or from the local council's website or apply on line using the Planning Portal.
- Read the 'Notes for Applicants' carefully – again available from the planning department, or the local council's website.

- Fill in the relevant parts of the forms and remember to sign and date them.
- Submit the correct number and type of supporting plans. Each application should be accompanied by a site plan of not less than 1:2500 scale, and detailed plans, sections and elevations, where relevant.
- Fill in and sign the relevant certificate relating to land ownership.

It is in your own interest to provide plans of good quality and clarity, and so it is probably advisable to get help from an architect, surveyor or similarly qualified person to prepare the plans and carry out the necessary technical work for you. You can obtain the necessary application form from the planning department of your local council, and you will find that this is laid out simply, with guidance notes to help you fill it in. Alternatively, you can ask a builder or architect to make the application on your behalf. This is sensible if the development you are planning is in any way complicated, because you will have to include measured drawings with the application form.

Applying online via the Planning Portal

You can apply for planning permission online via the Planning Portal (http://www.planningportal.gov.uk).

The Portal's service will also let you:

- create a site location plan (compulsory for all applications);
- attach supporting documents (such as photographs);
- pay the application fee online (where enabled).

Local authorities working with the Planning Portal offer two different ways of applying for permission:

- If your local authority has integrated its systems with the portal you can complete the whole process online and pay for the application electronically (where enabled).
- If the council systems have not been integrated you can still use the portal's service to complete the application forms electronically then print and post them to your local authority. Some local authorities request up to five copies of the forms, so completing them electronically can save time and make sure there are no discrepancies.

The portal provides a detailed map and list of all authorities who currently allow applications on line.

4.14 What is the planning permission process?

 Under the Local Government Act 1972 (as amended), the public have the right to inspect and copy the following:

- the agenda for a council committee or sub-committee meeting;
- reports for the public part of the meeting;
- the minutes of such meetings and any background papers, including planning applications, used in preparing reports.

These documents can be inspected and copied from three clear days before a meeting. There is no charge to inspect a document but councils will charge for making photocopies. If you think you might need to apply for planning permission, then this is the process to follow. Figure 4.2 is a detailed flow chart of this process.

4.14.1 How do I make an application for planning permission?

Step 1

Contact your local council planning department to discuss what you want to do and ask for their advice.

Step 2

If your local council planning department thinks you need to apply for planning permission or Building Regulations approval, ask for an application form and discuss how many copies of the form you will need to send back and how much the application fee will be. Also check if they foresee any difficulties which could be overcome by amending your proposal. It can save time or trouble later if the proposals you want to carry out also reflect what the council would like to see.

Step 3

Decide on the type of application you need to make. In most cases this will be a full application, but there are a few circumstances when you may want to make an outline application – for example, if you want to see what the council thinks of the building work you intend to carry out before you go to the trouble of making detailed drawings (but you will still need to submit details at a later stage).

Step 4

Send the completed application forms and supporting documents to your council, together with the correct fee. Each form must be accompanied by a plan of the site and a copy of the drawings showing the work you propose to carry out (the council will advise you on what drawings are needed). Extracts from Ordnance Survey maps can be supplied for planning applications submitted by private individuals and for school/college use. There is usually a charge for this service.

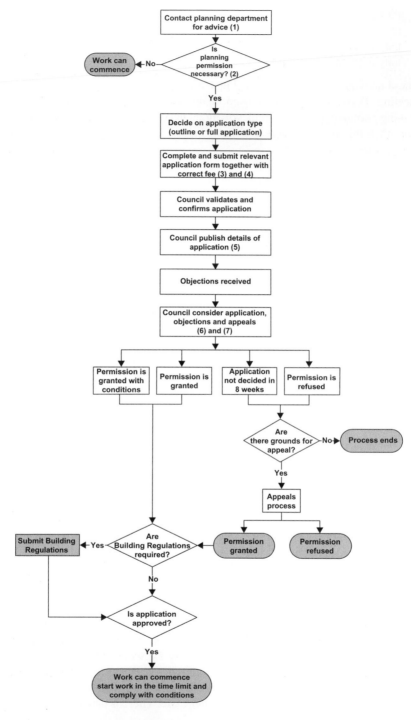

Figure 4.2 Flow chart of the planning permission process.

Step 5

The planning department will acknowledge receipt of your application, and publicly announce it – via letters to the neighbourhood parish council and anyone directly affected by the proposal, by publishing details of the application in the local press, notifying your neighbours and/or putting up a notice on or near the site. The council may also consult other organizations, such as the highway authority or the parish council (or community council in Wales). A copy of the application will also be placed on the planning register at the council offices so that it can be inspected by any interested member of the public. Anyone can object to the proposal, but there is a limited period of time in which to do this and they must specify the grounds for objection.

Step 6

You are entitled to see and have a copy of any report submitted to a local government committee. You are also entitled to see certain background papers (comments of consultees, objections and supporters) used in the preparation of reports.

The planning department may prepare a report for the planning committee, which is made up of elected councillors. Or the council may give a senior officer in the planning department the responsibility for deciding your application on its behalf. If a report has been made, then this will be presented to a meeting of the council committee, with recommendations on the decision to be made, based on the implications and objections received.

Step 7

The councillors or council officers who decide your application must consider whether there are any good planning reasons for refusing planning permission or for granting permission subject to conditions. The council cannot reject a proposal simply because many people oppose it. It will also look at whether your proposal is consistent with the development plan for the area.

The committee will consider the merits of a proposal; ensure the proposed work meets all the conditions of any local plan or requirements for a district, and that the process has been followed properly. The kinds of planning issue it can also consider include potential traffic problems, the effect on amenity and the impact the proposal may have on the appearance of the surrounding area. Moral issues, the personal circumstances of the applicant or the effect the development might have on nearby property prices are not relevant to planning and will not normally be taken into account by the council. The committee will arrive at its decision and the result will be communicated back to the applicants via the planning department.

4.14.2 How long will the council take?

Discuss the proposal with a representative of the planning department before you submit your application. They will do their best to help you meet the requirements.

You can expect to receive a decision from the planning department within eight weeks and, **once granted, planning permission is valid for five years**. If the work is not begun within that time, you will have to apply for planning permission again.

If the council cannot make a decision within eight weeks then it must obtain your written consent to extend the period. If it has not done so, you can appeal to the Secretary of State for the Environment, Transport and the Regions, or, in Wales, to the National Assembly for Wales (see below). But appeals can take several months to decide and it may be quicker to reach agreement with the council.

4.14.3 What can I do if my application is refused?

A second application is normally exempt from a fee.

If the council refuses permission or imposes conditions, it must give reasons. If you are unhappy or unclear about the reasons for refusal or the conditions imposed, talk to staff at the planning department. Ask them if changing your plans might make a difference. If your application has been refused, you may be able to submit another application with modified plans free of charge within 12 months of the decision on your first application.

The planning department will always grant planning permission unless there are very sound reasons for refusal, in which case the department must explain the decision to you so that you can amend your plans accordingly and resubmit them for further consideration. The following are some of the main objection areas that your application may meet.

The property is a listed building

Listed buildings are protected for their special architectural or historical value. A *Listed Building Consent* may be needed for alterations but grants could be available towards repair and restoration! If a building is listed it probably has some historic importance and will have been listed by the Department of the Environment. This could apply to houses, factories, warehouses and even walls or gateways. Most alterations which affect the external appearance or design will require Listed Building Consent in addition to other planning consents.

The property is in a Conservation Area

This is an area defined by the local authority, which is subject to special restrictions in order to maintain the character and appearance of that area. Again, other planning consents may be needed for areas designated as Green Belt, an Area of Outstanding Natural Beauty, a National Park or a Site of Specific Scientific Interest.

The application does not comply with the Local Development Plan

Local authorities often publish a development plan, which sets out policies and aims for future development in certain areas. These are to maintain specific environmental standards, and can include very detailed requirements such as minimum or maximum dimensions of plot sizes, number of dwellings per acre, height and style of dwellings etc. It is important to check if a plan exists for your area, as proposals can meet with some fierce objections from residents protecting their environment.

The property is subject to a covenant

This is an agreement between the original owners of the land and the persons who acquired it for development. Covenants were implemented to safeguard residential standards and can include things such as the size of outbuildings, banning the use of front gardens for parking cars, or even just the colours of exterior paintwork.

There is existing planning permission

A previous resident or owner may have applied for planning permission, which may not have expired yet. This could save time and expense if a new application can be avoided.

The proposal infringes a right of way

If your proposed development would obstruct a public path that crosses your property, you should discuss the proposals with the council at an early stage. The granting of planning permission will **not** give you the right to interfere with, obstruct or move the path. A path cannot be legally diverted or closed unless the council has made an order to divert or close it to allow the development to go ahead. The order must be advertised and anyone may object. You must not obstruct the path until any objections have been considered and the order has been confirmed. You should bear in mind that confirmation is not automatic; for example, an alternative line for the path may be proposed, but not accepted.

If you are considering a planning application, you should consider the above questions. Normally your retained expert – architect, surveyor or builder – can advise and help you to get an application passed. Most information can be collected from your local planning department; if you need to find out about covenants, look for the appropriate land registry entry.

4.14.4 What matters cannot be taken into account?

- Competition;
- disturbance from construction work;
- loss of property value;
- loss of view;
- matters controlled under other legislation such as Building Regulations (e.g. structural stability, drainage, fire precautions etc.);
- moral issues;
- need for development;
- private issues between neighbours (e.g. land and boundary disputes, damage to property, private rights of way, deeds, covenants etc.);
- Sunday trading;
- the identity or personal characteristics of the applicant.

4.14.5 What are the most common stumbling blocks?

In no particular order of priority, these are:

- adequacy of parking;
- archaeology;
- design, appearance and materials;
- effect on a listed building or Conservation Area;
- government advice;
- ground contamination;
- hazardous materials;
- landscaping;
- light pollution;
- local planning policies;
- nature conservation;
- noise and disturbance from the use (but not from construction work);
- overlooking and loss of privacy;
- previous planning decisions;
- previous appeal decisions;
- road access;
- size, layout and density of buildings;
- the effect on the street or area (but not loss of private view);
- traffic generation and overall highway safety.

4.15 Can I appeal if my application is refused?

 Be careful not to proceed without approval, as you might find yourself obliged to restore the property to its original condition.

If you think the council's decision is unreasonable, you can appeal to the Secretary of State or (in Wales) to the National Assembly for Wales. Appeals must be made within six months of the date of the council's notice of decision. You can also appeal if the council does not issue a decision within eight weeks.

Appeals are intended as a last resort, and they can take several months to decide. It is often quicker to discuss with the council whether changes to your proposal would make it more acceptable. The planning authority will supply you with the necessary appeal forms.

 A free booklet *Planning Appeals – A Guide* is available from the Planning Inspectorate.

4.16 Before you start work

There are many kinds of alterations and additions to houses and other buildings that do not require planning permission. Whether or not you need to apply, you should think about the following before you start work.

4.16.1 What about neighbours?

Have neighbours any rights to complain?

Many of us live in close proximity to others and your neighbours should be the first individuals you talk to. What if your alteration infringes their access to light, or a view? Such disputes are notorious for causing bad feeling, but with a little consideration at an early stage you can avoid a good deal of unpleasantness later.

 If you do need to make a planning application the council will ask your neighbours for their views. If you or any of the people you are employing to do the work need to go onto a neighbour's property, you will, of course, need to obtain their consent before doing so.

It is always a good idea to talk to your neighbours about planned work to your property. They will probably be as concerned about how the work will affect them as you would be if you were in their position. If you carry out work that seriously overshadows a neighbour's window and that window has been there for 20 years or more, you may be affecting his or her 'right to light' and you could be open to legal action. It is best to consult a lawyer if you think you need advice about this. By simply modifying your proposals you could address some of your neighbour's worries. But even if you decide not to change what you want to do, it is usually better to have told your neighbours what you are proposing before you apply for planning permission and before any building work starts.

Plans for the local area can normally be viewed at the local town hall, but most planning applications will involve consultation with neighbours and statutory consultees such as the highway authority and the drainage authorities. The extent of consultation will, quite naturally, reflect on the nature and scale of the proposed development – together with its location. Applications to make an alteration to your property can also be refused because you live in an Area of Outstanding Natural Beauty, a National Park, or a Conservation Area or your property is listed. Any alterations to public utilities such as drains or sewers, or changes to public access such as footpaths will require consultation with the local council. The council will have to approve your plans. Even a sign on or above your property may need to be of a certain size or shape.

 New street names and house numbers and names need approval from the council.

Some properties may also be the home of a range of protected species such as bats or owls. These animals are protected by the Wildlife and Countryside Act 1981 and the Nature Conservancy Council must give approval to any work that may potentially disturb them. Likewise, many members of the public are extremely protective of trees that grow where they live. Tree Preservation Orders may control the extent to which you can fell or even prune a tree, even if it is on your property. Trees in Conservation Areas are particularly protected, and you will need to supply at least six weeks' notice before working on them.

4.16.2 What about design?

Everybody's taste varies and different styles will suit different types of property. Nevertheless, a well-designed building or extension is likely to be much more attractive to you and to your neighbours. It is also likely to add more value to your house when you sell it. It is therefore worth thinking carefully about how your property will look after the work is finished. Extensions often look better if they use the same materials and are in a similar style to the buildings that are already there – but good design is impossible to define and there may be many ways of producing a good result. In some areas, the council's planning department issues design guides or other advisory leaflets that may help you.

4.16.3 What about crime prevention?

You may feel that your home is secure against burglary and you may already have taken some precautions such as installing security locks to windows. However, alterations and additions to your house may make you more vulnerable to crime than you realize. For example, an extension with a flat roof, or a new porch, could give access to upstairs windows which previously did not require a lock. Similarly, a new window next to a drainpipe could give access. Ensure that all windows are secure. Also, your alarm may need to be extended to cover any extra rooms or a new garage. The Crime Prevention Officer at your local police station can provide helpful advice on ways of reducing the risk.

4.16.4 What about lighting?

If your plan includes putting in external lighting for security or other purposes, you need to make sure that others are not disturbed by the intensity and direction of the light. Excessive exposure to poorly designed lighting causes disturbance to many people, so, in particular, make sure that beams are **not** pointed directly at windows of other houses. Passive Infra-Red detectors (PIRs) and/or timing devices on security lights should be adjusted so that they minimize nuisance to neighbours and are set so that they are not triggered by traffic or pedestrians passing outside your property.

4.16.5 What about covenants?

Even if you do not need to apply for planning permission you may be required to get someone else's agreement before carrying out some kinds of work to your property because of a covenant or another restriction in the title to your property or condition in the lease. You can check this yourself or consult a lawyer. You will probably need to use the professional services of an architect or surveyor when planning a loft conversion. Their service should include considerations of planning control rules and any covenants.

4.16.6 What about listed buildings?

Buildings are listed because they are considered to be of special architectural or historic interest, and as a result require special protection. Listing protects the **whole** building, both inside and out, and possibly also adjacent buildings if they were erected before 1 July 1948. The prime purpose of having a building listed is to protect the building and its surroundings from changes that will materially alter the special historic or architectural importance of the building or its setting.

 Planning Policy Statement 5: Planning for the Historic Environment, sets out the government's planning policies on the conservation of the historic environment. It can be viewed at your planning office or downloaded from http://www. communities.gov.uk or ordered from The Stationery Office.

The list of buildings is prepared by the Department of Culture, Media and Sport and properties are scheduled into one of three grades, Grade I, Grade II* and Grade II, with Grade I being the highest grade. Over 90 per cent of all listed properties fall within Grade II. (In Scotland the grades are A, B and C.) All buildings erected prior to 1700 and substantially intact are listed, as are most buildings constructed between 1700 and 1840, although some selection does take place. The selection process is more discriminating for buildings erected since 1840 because so many more properties remain today. Buildings less than 30 years old are generally only listed if they are of particular architectural or historic value and are potentially under threat. Your district council holds a copy of the statutory list for public inspection, and this provides details on each of the listed properties.

Owner's responsibilities?

A photographic record of the property when it came into your possession may be useful, although you may also have inherited incomplete or unimplemented works from your predecessor, which you will become liable for.

If you are the owner of a listed building or come into possession of one, you are tasked with ensuring that the property is maintained in a reasonable state of repair. The council may take legal action against you if they have cause to believe that you are deliberately neglecting the property, or have carried out works without consent. Enforcement action may be instigated.

There is no statutory duty to effect improvements, but you must not cause the building to fall into any worse state than it was in when you became its owner. This may necessitate some works, even if they are just to keep the build- ing wind and watertight. However, you may need Listed Building Consent in order to carry these works out!

If you are selling a listed building you may wish to indemnify yourself against future claims: speak to your solicitor.

4.16.7 What about Conservation Areas?

Conservation Areas are 'areas of special architectural or historic interest the character and appearance of which it is desirable to preserve or enhance'.

Tighter regulations apply to developments in Conservation Areas and to devel- opments affecting listed buildings. Separate Conservation Area consent and/or Listed Building Consent may be needed in addition to planning consent and Building Regulation consent.

As the title indicates, these designations cover more than just a building or property curtilage and most local authorities have designated Conservation Areas within their boundary. Although councils are not required to keep any statutory lists, you can usually identify Conservation Areas from a local plan's 'proposals maps' and appendices. Some councils may keep separate records or even produce leaflets for individual areas.

The purpose of designating a Conservation Area is to provide the council with an additional measure of control over an area that they consider being of special historic or architectural value. This does not mean that development proposals cannot take place, or that works to your property will be automati- cally refused. It means, however, that the council will have regard to the effect of your proposals on the designation in addition to their normal assessment. The council may also apply this additional tier of assessment to proposals that are outside the designated Conservation Area boundary, but which may poten- tially affect the character and appearance of the area.

If you live or work in a Conservation Area, grants may be available towards repairing and restoring your home or business premises.

As a result, local planning authorities may ask for more information to accompany your normal planning application concerning proposals within (or adjoining) a Conservation Area. This may include:

*   a site plan to 1:1250 or 1:2500 scale showing the property in relation to the Conservation Area;
*   a description of the works and the effect (if any) you think they may have on the character and appearance of the Conservation Area;
*   a set of scale drawings showing the present and proposed situation, including building elevations, internal floor plans and other details as necessary.

For major works you may need to involve an architect with experience of works affecting Conservation Areas.

4.16.8 What is Conservation Area consent?

 Planning Policy Statement 5: Planning for the Historic Environment, sets out the government's planning policies on the preservation of the historic environment.

Development within Conservation Areas is dealt with under the normal planning application process, except where the proposal involves demolition. In this case you will need to apply for Conservation Area consent on the appropriate form obtainable from the planning department. Here again the council will assess the proposal against its effect on the special character and appearance of the designated area.

4.16.9 What about trees in Conservation Areas?

 Contact your council's landscape or tree officer for further information.

Nearly all trees in Conservation Areas are automatically protected, and it is necessary to obtain the council's approval for works to trees in Conservation Areas before they are carried out. Trees in Conservation Areas are generally treated in the same way as if they were protected by a Tree Preservation Order (TPO). There are certain exceptions (where a tree is dead or in a dangerous condition), but it is always advisable to seek the opinion of your council's tree officer to ensure your proposed works are acceptable. Even if you are certain that you do not need permission, notifying the council may save the embarrassment of an official visit if a neighbour contacts them to tell them what you are doing.

 It is an offence to carry out work on a tree with a TPO on it without prior permission.

If you wish to lop, top or fell a tree within a Conservation Area you must give six weeks' notice, in writing, to the local authority. This is required in order that they can check to see if the tree is already covered by a TPO, or consider whether it is necessary to issue a TPO to control future works on that tree.

4.16.10 What are Tree Preservation Orders?

Trees are possibly the biggest cause of upset in town and country planning, and many neighbours fall out over tree-related issues. They may be too tall, may block out natural light, have overhanging branches, shed leaves onto other property, or the roots may cause damage to property. When purchasing a property the official searches carried out by your solicitor should reveal the presence of a TPO on the property or whether your property is within a Conservation Area within which trees are automatically protected. However, not all trees are protected by the planning regulations system – but trees that have protection orders on them must not be touched unless specific approval is granted. Do not overlook the fact that a TPO could have been put on a tree on your land before you bought it, and is still enforceable.

Planning authorities have powers to protect trees by issuing a TPO, and this makes it an offence to cut down, top, lop, uproot, wilfully damage or destroy any protected tree(s) without first having obtained permission from the local authority. All types of tree can be protected in this way, whether as single trees or as part of a woodland, copse or other grouping of trees. Protection does not, however, extend to hedges, bushes or shrubs. TPOs are recorded in the local land charges register, which can be inspected at your council offices. The local authority regularly checks to see if trees on their list still exist and are in good condition. Civic societies and conservation groups also keep a close eye on trees. Before carrying out work affecting trees, you should check if the tree is subject to a TPO. If it is, you will need permission to carry out the work.

All trees in a Conservation Area are protected, even if they are not individually registered. If you intend to prune or alter a tree in any way you must give the local authority plenty of notice so they can make any necessary checks. A TPO will not prevent planning permission being granted for development. However, the council will take the presence of TPOs into account when reaching their decision. Even with a TPO in place it is possible to have a tree removed, if it is too decayed or dangerous, or if it stands in the way of a development; the local authority may consider its removal, but will normally want a similar tree put in or near its place.

If you have a tree on your property that is particularly desirable – either an uncommon species or a mature specimen – then you can request a TPO for it. However, this will mean that in years to come you, and others, will be unable to lop it, remove branches or fell it unless you apply for permission.

What are my responsibilities?

Trees covered by TPOs remain the responsibility of the landowner, both in terms of any maintenance that may be required from time to time and for any damage they may cause. The council must formally approve any works to a tree covered by a TPO. If you cut down, uproot or wilfully damage a protected tree or carry out works such as lopping or topping which could be likely to seriously damage or destroy the tree, there are fines on summary conviction of up to £20,000, or, on indictment, the fines are unlimited. Other offences concerning protected trees could incur fines of up to £2500.

What should I do if a protected tree needs lopping or topping?

 You may be required to plant a replacement tree if the protected tree is to be removed.

Although there are certain circumstances in which permission to carry out works to a protected tree are not required, it is generally safe to say that you should always write to your council to seek their permission before undertaking any works. You should provide details of the trees on which you intend to do work, the nature of that work – such as lopping or topping – and the reasons why you think this is necessary. The advice of a qualified tree surgeon may also be helpful (see *Yellow Pages*).

4.16.11 What about nature conservation issues?

Many traditional buildings, particularly farm buildings, provide valuable wildlife habitats for protected species such as barn owls and bats.

Planning permission will not normally be granted for conversion and reuse of buildings if protected species would be harmed. However, in many cases, careful attention to the timing and detail of building work can safeguard or recreate the habitat value of a particular building. Guidance notes prepared by Natural England (formerly English Nature) are available from councils or from their website (http://www.naturalengland.org.uk).

4.16.12 What about bats and their roosts?

 Information on bats and the law is included in the Natural England booklet *Focus on Bats* (ISBN 978-1-84754-019-5), which can be obtained free of charge from your local Natural England office (http://www.naturalengland.org.uk).

Bats make up nearly one-quarter of the mammal species throughout the world. Bats form social colonies; female bats gather together in maternity colonies for a few weeks during the summer to give birth to and rear their young. During the winter, clusters of hibernating bats gather together in sheltered places for long periods. There are two main types of roost that affect humans;

buildings (houses, churches, farms, ancient monuments and industrial buildings) are popular summer roosts, while underground caves (mines, cellars and tunnels) are popular hibernation roosts.

Some houses may hold roosts of bats or provide a refuge for other protected species. The Wildlife and Countryside Act 1981 gives special protection to **all** British bats because of their roosting requirements. Natural England or the Countryside Council for Wales (CCW) must be notified of any proposed action (e.g. remedial timber treatment, renovation, demolition and extensions) that is likely to disturb bats or their roosts. Natural England or CCW must then be allowed time to advise on how best to prevent inconvenience to both bats and householders.

The type of stone barns and traditional buildings found in the UK have lots of potential bat roosting sites, the most likely places being gaps in stone rubble walls, under slates or within beam joints. These sites can be used throughout the year by varying numbers of bats, but could be particularly important for winter hibernation. As a result, the following points should be followed when considering or undertaking any work on a stone barn or similar building, particularly where bats are known to be in the area.

 There is a £5000 fine per bat if they are killed, whether accidentally or not.

A survey for the presence of bats should be carried out by a member of the local bat group (contact via Natural England) before any work is done to a suitable barn during the summer bat breeding period.

 The pipistrelle, the smallest of the European bats, has been found lurking in many strange places, including vases, under floorboards and between the panes of double glass.

No pointing of walls should be carried out between mid-November and mid-April or during maternity roosting periods of the summer, to avoid potentially entombing any bats. When walls are to be pointed, areas of wall high up on all sides of the building should be left unpointed to preserve some potential roosting sites. If any bats are found whilst work is in progress, work should be stopped and Natural England contacted for advice on how to proceed.

If any timber treatment is carried out, only chemicals safe for use in bat roosts should be used. A list of suitable chemicals is available from Natural England on request. Any pretreated timber used should have been treated using the CCA method (copper–chrome–arsenic) which is safe for bats.

Work should not be commenced during the winter hibernation period (mid-November to mid-April). Any bats present during the winter are likely to be torpid (i.e. unable to wake up and fly away), and are therefore particularly vulnerable. In summer maternity roosts there are much higher numbers of bats (typically between 30 and 300), but they will be highly visible at feeding times.

4.16.13 What about barn owls?

Barn owls use barns and similar buildings as roosting sites in some areas. The owls are more obvious than bats and, therefore, perhaps easier to take into account. Barn owls are also fully protected by law and should not be disturbed during their breeding season. Special owl boxes can be incorporated into walls during building work, details of which can be obtained from Natural England.

4.17 What could happen if you don't bother to obtain planning permission?

If you build something which needs planning permission without obtaining permission first, you may be forced to put things right later, which could prove troublesome and extremely costly. You might even have to remove an unauthorized building.

4.17.1 Enforcement

If you think that works are being carried out without planning permission, or not in accord with approved plans and/or conditions of consent, then seek the advice of the local planning officer, who will then investigate and, if necessary take appropriate steps to deal with the problem. Conversely, if you are carrying out development works it is important that you stick to the approved plans and conditions. If changes become necessary, contact the development control staff before the changes are made.

4.18 How much does it cost?

A fee is required for the majority of planning applications, and the council cannot deal with your application until the correct fee is paid. The fee is not refundable if your application is withdrawn or refused.

 Work to provide access and/or facilities for disabled people to existing dwellings are exempt from these fees.

In most cases you will also be required to pay a fee when the work is commenced. These fees are dependent on the type of work that you intend to carry out. The fees outlined in Sections 14.18.1 to 14.18.9 are typical of the charges made by local authorities during 2008, when submitting an application and are an extract from *Town and Country Planning (Fees for Applications and Deemed Applications (Amendment) (England) Regulations 2008)*.

4.18.1 Householder applications

Outline applications (most types)	Site *not exceeding* 2.5 ha £335 per 0.1 ha (or part thereof) of site area, maximum £8285 (2.5 ha)
	Site *exceeding* 2.5 ha an additional £100 per each additional 0.1 ha, maximum £125,000

Full applications and reserved matters

Dwellings – **erection of new**	*Up to* 50 dwellings, £335 per dwelling-house, maximum £125,000
	Over 50 dwellings, £100 each additional unit, maximum £250,000
Dwellings – **alteration (including outline)**	£170 per dwelling-house, maximum £335
Approval of reserved matters where flat rate does **NOT** apply	A fee based on the amount of floorspace and/or number of dwelling houses involved
Flat rate (only when maximum fee has been paid)	£335

4.18.2 Industrial/retail and other buildings applications

Industrial/retail buildings	Where no additional floorspace is created £170
	Works not creating more than 40 m^2 of additional floorspace £170
Outline application (**see above**)	More than 40 m^2 but not more than 75 m^2 of additional floorspace £335
	Each additional 75 m^2 (or part thereof) £335, maximum £13,250 (53,750 m^2)
	Over 3750 m^2 £100 each additional 75 m^2 up to maximum fee £250,000
Plant and machinery (**erection, alteration, replacement**)	£335 per 0.1 ha (or part thereof) of the site area, maximum £16,565 (5 ha)
	Over 5 ha £100 each additional 0.1 ha up to maximum fee £25,000

4.18.3 Prior notice applications

Approvals for agricultural/forestry buildings/ operations and demolition of buildings and telecommunications works	£70

4.18.4 Agricultural applications

Agricultural buildings	Buildings not exceeding 465 m^2 **£70**
	Buildings exceeding 465 m^2 but more than 540 m^2 **£335**
	More than 540 m^2 **£335** for each additional 75 m^2 (or part thereof), maximum **£13,565** (4215 m^2)
	Over 4215 m^2 **£80** each additional 75 m^2 up to maximum fee **£250,000**
Erection of glasshouses/ polytunnels	Works not creating more than 465 m^2 **£70** (on land used for agriculture)
	Works creating more than 465 m^2 **£1870**

4.18.5 Concessionary fees and exemptions

Works to improve the disabled person's access to a public building, or to improve their access, safety, health or comfort at their dwelling-house	**No fee**
Applications by parish councils (all types)	**Half the normal fee**
Applications required by an Article 4 direction or removal of permitted development rights	**No fee**
Playing fields (for sports clubs etc.)	**£335**
Revised or fresh applications of the **same character** or **description** by the **same applicant** within **12 months** of **refusal**, or the expiry of the statutory 8 week period where the applicant has appealed to the Secretary of State on grounds of **non-determination. Withdrawn applications of the same character or description must be made within 12 months of making the earlier one**	**No fee**
Revised or fresh application of the **same character** or **description** within **12 months** of receiving permission	**No fee**
Duplicate applications made by the **same applicant for each application** submitted within 28 days of each other	**Full fee**
Alternative applications for one site submitted at the **same time**	**Highest of the fees applicable for each alternative and a sum equal to half the rest**
Development crossing local authority boundaries	**Only one fee paid to the authority having the larger site but calculated for the whole scheme and subject to a special ceiling**

4.18.6 Hazardous substances applications

Application for new consent	£200
New consent where maximum quantity specified exceeds twice the controlled quantity	£400
All other types of application	£250
Continuation of hazardous substances consent under Section 17(1) of the 1992 Regulations	£200

4.18.7 Legal applications

Application for a certificate of lawfulness for an existing use or operation	**Same fee payable** as if making a planning application operation
Application for a certificate of lawfulness for an existing activity in breach of planning condition(s)	£170
Application for a certificate of lawfulness for a proposed use or operation	**Half the fee payable** as if making a planning application

4.18.8 Advertisement applications

Adverts relating to the business on the premises	£95
Advance signs directing the public to a business (unless business can be seen from the sign's position)	£95
Other advertisements (e.g. hoardings)	£335

4.18.9 Other applications

Exploratory drilling for oil or natural gas	£335 per 0.1 ha (or part thereof) of site area, maximum £25,000 (7.5 ha)
	Over 7.5 ha £100 each additional 0.1 ha, maximum fee £250,000
Winning, working, storage of minerals etc., waste disposal	£170 per 0.1 ha (or part thereof) £25,315 (15 ha)
	Over 15 ha £170 each additional 0.1 ha, maximum fee £65,000
Car parks, service roads or other accesses (existing uses only)	£170
Other operations on land	£170 per 0.1 ha (or part thereof) of site area, maximum £1350 (15 ha)
Non-compliance with conditions	£170
Change of use to sub-division of dwellings	£335 per additional dwelling created, maximum £16,565 (50 units)
	Over 50 units £100 each additional unit, maximum fee £250,000
Other changes of use **except** waste or minerals	£335

Planning Portal – fee calculator

 For the online fee calculator, visit http://www.planningportal.gov.uk/planning/ usefultools.

The Planning Portal is the UK government's planning source. Within this site is a fee calculator. The fee calculator can help you by working out the cost of any particular planning application. The calculator asks a series of questions to help determine the total cost of an application, ranging from a simple house-holder development to large-scale schemes such as housing developments or industrial estates.

4.19 Sustainable homes

The Code for Sustainable Homes aims to create sustainable homes and reduce carbon emissions. The code was launched in 2006, and since then any devel-oper of any new home in England can choose to be assessed against the code.

4.19.1 How does the code work?

The code measures the sustainability of a home against design categories (see below) that rate the whole home as a complete package using stars (one star for entry level (i.e. above Building Regulations) and six stars for the highest level), depending on the extent to which it has achieved the required standards.

The code sets minimum standards for energy and water used at each level, and replaces the EcoHomes scheme in the UK. Notwithstanding the removal of the requirement for a Home Information Pack (HIP), the Code for Sustainable Homes is still operational and remains the UK government's national sustain-ability standard for new homes.

4.19.2 Is there any connection between 'sustainable homes' and 'sustainable locations'?

It should be noted that the code principally deals with sustainable homes as opposed to the *sustainability* of locations. This role will still be covered by the existing planning system, which is recognised as a means of ensuring that developments are located on sustainable sites and that developments are such that they assist in the reduction of the need to travel.

4.19.3 How many sustainable homes have been certified?

 More details concerning how many homes have been certified to the standards set out in the Code for Sustainable Homes can be found in the government statistics, *House Building: March Quarter 2010, England*, which is available at http://www. communities.gov.uk/publications/corporate/statistics/housebuildingq12010.

4.19.4 Do I still need to provide a Sustainability Certificate?

A Sustainability Certificate is no longer required. In May 2010 the UK government announced the suspension of the requirement for sellers to give a sustainability certificate to buyers of newly constructed homes. However, the Code for Sustainable Homes is still operational and remains the government's national sustainability standard for new homes.

4.20 Home Information Pack (HIP)

The UK government has announced the suspension of Home Information Packs (HIPs). Homes marketed for sale on or after 21 May 2010 will no longer require a HIP.

5

Requirements for planning permission and Building Regulations approval

Where the word 'original' is used in planning regulations this means the house as it was first built or as it was on 1 July 1948. Any extensions added since that date are counted towards the allowances.

You are allowed to make certain changes to your home without having to apply to the local council for permission as long as the work you are completing does not affect the external appearance of the building, These are called 'permitted development rights'. However, the majority of building work that you are likely to complete will require you to have planning permission and/or Buildings Regulations approval prior to work commencing.

This chapter shows whether planning permission and/or Building Regulations apply to particular projects (which are listed in alphabetical order) and Appendix 5A contains a useful summary of all the areas covered in this chapter. Be advised that there may be variations in the planning requirements, and to some extent the Building Regulations, from one area of the country to another, so you should consult your local authority for specific guidance for your area.

5.1 Advertising

Planning permission		Building Regulation approval	
No	If the advertisement is less than 0.3 m² and not illuminated	Possibly	Domestic adverts and signs are not usually subject to planning control, but consult your local planning officer for more details

Advertisement signs on buildings and on land often need planning consent. Some smaller signs (under 0.3 m²) and non-illuminated signs may not need consent, but it is always advisable to check with development control staff.

All measurements are taken externally.

You are allowed to display certain small signs at the front of residential premises, such as election posters, notices of meetings, jumble sales, car for sale etc., but business types of display and permanent signs may require planning permission. They may come under the category of 'advertising control' for which planning consent is required.

To display an advertisement bigger than $0.3\,\text{m}^2$ on the front of, or outside, your property you may need to apply for advertisement consent. This includes all types of sign, ranging from the house name or number, to information signs such as 'Beware of the dog'. Temporary notices up to $0.6\,\text{m}^2$ which relate to local events, such as fêtes and concerts, may be displayed for a short period. Estate agents' boards are covered by different rules but, in general, these should not be bigger than $0.5\,\text{m}^2$ on each side.

Illuminated signs and all advertising signs outside commercial premises need to be approved. Most local authorities can give advice, by way of booklets or leaflets, on what kinds of sign are allowed, not allowed or need approval. It is illegal to post notices on empty shops' windows, doors and buildings, and also on trees. This is commonly known as 'fly posting' and can carry heavy fines under the Town and Country Planning Act.

 You can get advice from the planning department of your local council; ask for a copy of the free booklet *Outdoor Advertisements and Signs: A Guide for Advertisers*. This can also be downloaded from the Planning Portal (http:// www.planningportal.gov.uk/permission/commonprojects/advertssigns).

5.2 Basements

Planning permission		Building Regulation approval	
No	Provided that: • it is not a separate unit; • the usage is not significantly changed; • a light well is not added; • it is not a listed building	Yes	Covering: • fire escapes; • ventilation; • ceiling height; • damp proofing; • wiring; • water supplies; • underpinning; • foundation work

The creation of living space in basements and its planning regime is evolving and under review at the time of writing. In all circumstances prior to carrying out any work you are advised to contact your local planning authority to discuss any local policy changes.

 Note: The Department for Communities and Local Government (DCLG) and the Basement Information Centre are currently developing updated guidance on this topic following the removal of 'Basements for Dwellings' guidance from the list of Approved Documents on 1st October 2010.

5.3 Biomass-fuelled appliances

Planning permission	Building Regulation approval
No Unless it is a listed building or within a Conservation Area or the work is external	Yes Particularly in relation to electrical and plumbing work, ventilation, noise and general safety

Any outside buildings for the storage of fuel or related equipment are subject to the regulations relating to outbuildings.

5.4 Building a conservatory

Planning permission	Building Regulation approval
No Provided that: • no more than half the area of land around the original house is taken up by other buildings; • it is not forward of the building line; • no part of the conservatory is higher than the original roof of the main building; • it does not extend beyond the rear wall by more than 3 m (4 m for detached houses); • a single-storey rear extension is no higher than 4 m; • the maximum depth of a rear extension of more than one storey is no more than 3 m (including ground floor); • the eaves height of the extension within 2 m of the boundary is not more than 3 m; • the eaves and ridge height of the extension is no higher than the existing house; • any side extension is single storey with a maximum height 4 m and width no more than half that of the original house; • the roof pitch of an extension higher than one storey matches that of the existing house; • there are no verandas, balconies or raised platforms; • on designated land there is no permitted development for rear extensions of more than one storey; no cladding of the exterior; no side extensions; • the building is not listed	No Provided that: • it is built at ground level and has a floor area less than 30 m^2; • the conservatory is separated from the house by external quality walls, doors or windows; • there is an independent heating system with separate temperature and ON/OFF controls; • glazing and any fixed electrical installations comply with the applicable Building Regulations requirements (see below)

A conservatory has to be separated from the rest of the house to be exempt (e.g. with patio doors).

Conservatories and sun lounges attached to a house are classed as permitted development, subject to the conditions above and provided that the glazing complies with the safety glazing requirements of the Building Regulations (Part N). Your local authority building control department or an approved inspector can supply further information on safety glazing. It is advisable to ensure that a conservatory is not constructed so that it restricts ladder access to windows serving a room in the roof or a loft conversion, particularly if that window is needed as an emergency means of escape in the case of fire. If you want a conservatory or sun lounge separated from the house, this needs planning consent under similar rules for outbuildings.

The regulations described here apply only to houses, and not to other buildings, maisonettes or flats. There may be a charge under the Community Infrastructure Levy if the development is over 100 m². More guidance that is applicable to the windows, doors and electrics in a conservatory is contained in the sections relating to these areas.

Another thing to keep in mind is your neighbours' reaction – always keep them informed of what is happening, and be prepared to alter the plans you had for locating the building if they object – it is better in the long run, believe me.

For the most up to date guidance use the conservatory guide on the Planning Portal.

5.5 Building an extension

Planning permission	Building Regulation approval
No Provided that: • no more than half the area of land around the original house is covered by additions or other buildings; • there is no extension forward of the building line; • it is not higher than the highest part of the roof; • it does not extend beyond the rear wall of the original house by more than 3 m (4 m if a detached house); • a single-storey rear extension is no higher than 4 m; • if it is more than one storey it is no deeper than 3 m beyond the rear wall of the original house, including the ground floor; • any eaves within 2 m of the boundary are no taller than 3 m; • the eaves and ridge are no higher than the existing house; • any side extensions are single storey and no wider than half the width of the original house and a maximum height of 4 m; • a two-storey extension is not closer to the rear boundary than 7 m;	Yes

Planning permission	Building Regulation approval
<ul><li>the roof pitch of extensions that are higher than one storey must match that of the existing house;</li><li>materials must be of similar appearance to those of the existing house;</li><li>there should be no verandas, balconies or raised platforms;</li><li>any side-facing windows on the upper floor are obscure-glazed and any opening are 1.7 m above the floor;</li><li>on designated land:<ul><li>no permitted development for rear extensions of more than one storey,</li><li>no cladding of the exterior,</li><li>no side extensions.</li></ul></li></ul>	

5.5.1 Do I need approval to build an extension to my house?

You need approval to build an extension if it would 'materially alter the appearance of the building'. Major alteration and extension nearly always need approval. However, some small extensions such as porches, garages and conservatories may be 'permitted development' and, therefore, do not need planning consent.

 Check with your local authority planning department if you are not sure.

If a building is extended, or undergoes a material alteration, the completed building must comply with the relevant requirements of the Approved Documents or, where this is not feasible, be 'no more unsatisfactory than before'. *In order to satisfy the requirements of the Regulations when building an extension, the following areas should also comply with the relevant Approved Documents:*

- doors and windows;
- drainage;
- electrics;
- energy efficiency;
- external walls;
- flooring;
- foundations;
- internal walls;
- kitchens and bathrooms;
- roofs;
- structural opening;
- ventilation;
- walls below ground level.

The area of windows, roof windows and doors in extensions should **not** be greater than 25 per cent of the floor area of the extension **plus** the area of any windows or doors that, as a result of the extension works, no longer exist or are no longer exposed.

You may also require planning permission if your house has previously been added to or extended, or if the original planning permission for your house imposed restrictions on future development (i.e. permitted development rights may have been removed by an 'Article 4 direction'. This is often the case with more recently constructed houses.)

 Any building that has been added to your property and which is more than 10 m³ in volume and that is within 5 m of your house is treated as an extension of the house and so reduces the allowance for further extensions without planning permission.

You **will** also require planning permission if you want to make additions or extensions to a flat or maisonette.

You will, therefore, need to apply for planning permission:

- If an extension to your house comes within 5 m of another building belonging to your house (i.e. a garage or shed). The volume of that building counts against the allowance given above.
- For all additional buildings which are more than 10 m³ in volume, if you live in a Conservation Area, a National Park, an Area of Outstanding Natural Beauty or the Norfolk Broads. Wherever they are in relation to the house, these buildings will be treated as extensions of the house and reduce the allowance for further extensions.
- For a terraced house, end-of-terrace house, or any house in a Conservation Area, a National Park, an Area of Outstanding Natural Beauty or the Norfolk Broads – where the volume of the original house would be increased by more than 10 per cent or 50 m³ (whichever is the greater).
- For any other type of house (i.e. detached or semi-detached) the volume of the original house would be increased by more than 15 per cent or 70 m³ (whichever is the greater). In any case the volume of the original house would be increased by more than 115 m³.
- For alterations to the roof, including dormer windows (but permission is not normally required for skylights).
- For extensions nearer to a highway than the nearest part of the original house (unless the house, as extended, would be at least 20 m away from the highway).
- To extend or add to your house so as to create a separate dwelling, such as self-contained living accommodation or a granny flat.
- If an extension to your house comes within 5 m of another building belonging to your house.
- To build an addition, which would be nearer to any highway than the nearest part of the original house, unless there would be at least 20 m between your house (as extended) and the highway. The term 'highway' includes all public roads, footpaths, bridleways and byways.
- If more than half the area of land around the original house would be covered by additions or other buildings – although you may not have built an extension to the house, a previous owner may have done so.

- If the extension or addition exceeds the certain limits on height and volume.
- If the extension is higher than the highest part of the roof of the original house, or any part of the extension is more than 4 m high and is within 2 m of the boundary of your property.

 You should measure the height of buildings from the ground level immediately next to it.

5.5.2 Extensions to non-domestic buildings

In accordance with Part M, an extension to a non-domestic building should now be treated in the same manner as a new building for compliance, which means that:

- there must be 'suitable independent access to the extension where reasonably practicable';
- if a building is to be extended, 'reasonable provision must be made within the extension for sanitary conveniences'.

 Note: This requirement does not apply if it is possible for people using the extension to gain access to and be able to use sanitary conveniences in the existing building.

If a building has a total useful floor area greater than $1000\,\text{m}^2$ and the proposed building work includes:

- an extension; or
- the initial provision of any fixed building services; or
- an increase to the installed capacity of any fixed building services;

then 'consequential improvements' should be made to improve the energy efficiency of the whole building. These will include:

- upgrading all thermal units that have a high U-value;
- replacing all existing windows (less display windows), roof windows, roof lights or doors (excluding high-usage entrance doors) within the area served by the fixed building service with an increased capacity;
- replacing any heating system that is more than 15 years old;
- replacing any cooling system that is more than 15 years old;
- replacing any air handling system that is more than 15 years old;
- upgrading any general lighting system that serves an area greater than $100\,\text{m}^2$ which has an average lamp efficacy of less than 40 lamp-lumens per circuit watt;
- installing energy metering;
- upgrading existing energy systems if they provide less than 10 per cent of the building's energy demand.

5.6 Building a fence, gate or wall

Planning permission		Building Regulation approval
Yes	If it is more than 1 m high and is a boundary enclosure adjoining a highway.	No
Yes	If it is more than 2 m (6 ft 6 in.) high elsewhere.	No

5.6.1 Do I need approval to build or alter a garden wall or boundary wall?

You will need to apply for Planning Permission if:

- your house is a listed building or in the curtilage of a listed building; or
- your right to put up or alter fences, walls and gates is removed by an article 4 direction or a planning condition; or
- the fence, wall or gate would be over 1 m high and next to a highway used for vehicles; or over 2 m high elsewhere; or
- the fence, wall or gate forms a boundary with a neighbouring listed building.

If the fence, wall or gate is classed as a 'party fence wall' then you must notify the adjoining owner of the work planned. In normal circumstances, the only restriction on walls and fences is the height allowed. This is in case its height might obscure a driver's view of other traffic, pedestrians or road users. If there is a valid reason for a wall or fence higher than the prescribed dimensions, then it is possible to get planning consent. There may be security issues that would support an application for a high fence. If it has no effect on other people's valid interests and does not impair any amenity qualities in an area, there is no reason why a request should be refused.

Some walls have historic value and they, as well as arches and gateways, can be listed. Modifications, extensions and removal of these must have planning consent.

You do not need to apply for planning permission to alter, improve or take down a fence, wall or gate unless you are in a Conservation Area.

You will need to apply for planning permission if you put an ornament on your gate post and together with the post it stands it higher than 1 m tall.

5.7 Building a hardstanding for a car, caravan or boat

Planning permission		Building Regulation approval	
No	Provided it is within your boundary, at, or near, ground level and does not require significant works of embanking or terracing	No	Unless you introduce steps where none existed before

5.7.1 Do I need to apply for planning permission to build a hardstanding for a car?

You do not need to apply for planning permission to build a hardstanding for a car provided that it is within your boundary and is not used for a commercial vehicle.

 You should check if there are any local covenants limiting changes in access to your premises or for a hardstanding and parking of vehicles on it.

Check local council rules, as there are different rules depending on what you use a hardstanding for. Provided there are no covenants limiting the installation of a hardstanding for parking of cars, caravans or boats, you do not need permission for a hardstanding on your own land or to gain access to it within the confines of your land. You would need permission for a hardstanding leading on to a public highway. However, there are still rules for commercial parking (e.g. taxis or commercial delivery vans), and a 'change of use' as a trade premises would probably need to be granted for this to be allowed.

Access from a new hardstanding to an unclassified roadway does not require planning permission. However, the busier the road, the less likely a new driveway or footway will be allowed to meet it.

 Your local authority highways department will be able to tell you if a road is classified or unclassified.

If the access crosses a pedestrian thoroughfare, pavement or roadside verge, then the planning department will gain approval from the highways department. If this is the case, highways approval is required in addition to planning consent. The basic principle is to maintain safety and eliminate hazards.

5.7.2 Do I need to apply for planning permission to build a hardstanding for a caravan and/or boat?

Some local authorities do not allow the parking of caravans or boats on driveways or hardstandings in front of houses. Check what the local rules are with your planning department, and if there is no restriction then you do not need to apply for permission.

There are no laws to prevent you, or your family, from making use of a parked caravan while it is on your land or drive, but you cannot actually live in it, as this would be classed as an additional dwelling. In addition, you cannot use a parked caravan for business use, as this would constitute a change of use of the property.

If you want to put a caravan on your land to lease out as holiday accommodation or for friends or family to stay in while they visit you, then this would require planning permission. Rules on siting of static caravans or mobile homes are quite stringent.

5.8 Building a new house

Planning permission	Building Regulation approval
Yes	Yes

5.8.1 Do I need planning permission to erect a new house?

All new houses or premises of any kind require planning permission.

Private individuals will normally only encounter this if they intend to buy a plot of land to build on, or buy land with existing buildings that they want to demolish to make way for a new property to be built.

 If you are hiring a professional, be sure to find out exactly who does what and that approval is obtained before going to too much expense, should a refusal arise.

In all cases like this, unless you are an architect or a builder, you **must** seek professional advice. If you are using a solicitor to act on your behalf in purchasing a plot on which to build, he will include the planning questions within all the other legal work, as well as investigating the presence of covenants, or existing planning consent together with other constraints or conditions.

There are plenty of substantial building projects that do not require any planning permission. However, it is a good idea to consult a range of people before you consider any work.

The architect, surveyor or contractor you hire will then need to take into account the planning requirements as part of their planning and design procedures. They will normally handle planning applications for any type of new development.

5.9 Building a porch

Planning permission	Building Regulation approval
No Unless: the floor area exceeds $3\,m^2$; any part is more than 3 m high; any part is less than 2 m from a boundary adjoining a highway	No As long as glazing and electrical installations comply with Building Regulations

5.9.1 Do I need planning permission for a porch?

 The regulations depend on previous works on the site (which can date back to 1948). You should always check with planning control staff if you are unsure.

A porch or conservatory built at ground level and less than $30\,m^2$ in floor area is exempt provided that the glazing complies with the safety glazing requirements of the Building Regulations (Part N). Your local authority building control department or an approved inspector can supply further information on safety glazing. It is advisable to ensure that a porch is not constructed so that it restricts ladder access to windows serving a room in the roof or a loft conversion, particularly if that window is needed as an emergency means of escape in the case of fire.

The permitted development allowances described here apply to houses, not flats, maisonettes or other buildings

5.10 Conversions

Planning permission		Building Regulation approval	
Yes	For flats – even where construction works may not be intended	Yes	Unless you are not proposing any building work to make the change
No	For loft conversions, as long as you do not alter or extend the roofspace	Yes	For further advice contact Building Control to discuss your proposal. You must also find out whether work you intend to carry out falls within The Party Wall etc. Act 1996
Yes	For conversion to shops and offices	Yes	

You will probably also need planning permission whether or not building work is proposed.

Note: Work on a loft conversion or a roof may affect bats. You need to consider protected species when planning work on this type. A survey may be needed, and if bats are using the building a licence may be needed.

5.10.1 Do I need approval to convert my buildings?

Do I need approval to convert my house into flats?

To subdivide your house into multiple units you need Planning Permission (and listed building/Conservation Area consent if appropriate) even where construction works may not be intended. Conversions of properties will require Building Regulations approval, liaison with fire Services and a licence under the Housing Act 2004.

Note: Permitted development between Class C4 to C3 and vice versa is allowed.

Do I need approval to convert my house to a shop?

You need approval to convert a house into a shop even if you are not proposing any building work to make the change. Permission is not required before work starts, but the majority of local authorities have designated shopping or commercial areas within their planning policies where they may consider such a change of use suitable. It is likely to be very difficult to gain permission outside these areas, such as in a wholly residential area.

Do I need approval to convert part or all of my shop to an office, cafe, public house or take away?

You will need approval for any material change of use. See Section 5.12 for details.

5.10.2 Converting an old building

Planning permission	Building Regulation approval
Yes	Yes

Do I need planning permission to convert an old building?

Throughout the UK there are many underused or redundant buildings, particularly farm buildings which may no longer be required, or suitable, for agricultural use. Such buildings of weathered stone and slate contribute substantially to the character and appearance of the landscape and the built environment. Their interest and charm stems from an appreciation of the functional requirements of the buildings, their layout and proportions, the type of building materials used and their display of local building methods and skills.

In most cases, traditional buildings are best safeguarded if their original use can be maintained. However, with changing patterns of land use and farming methods, changes of use or conversion may have to be considered. The conversion or reuse of traditional buildings may, in the right locations, assist in providing employment opportunities, housing for local people or holiday accommodation. Applicants and developers are encouraged to refer to the local plan for comprehensive guidance and to seek advice from a planning officer if further assistance is necessary.

All councils place the highest priority on good design and proposals. Those that fail to respect the character and appearance of traditional buildings will not be permitted. Sensitive conversion proposals should ensure that existing ridge and eave lines are preserved; new openings are avoided as far as possible; traditional matching materials are used and the impact of parking and garden areas is minimized. Buildings that are listed as being of 'special architectural or historic interest' require skilled treatment to conserve internal and external features.

In many instances, traditional buildings that are of simple, robust form with few openings may only be suitable for use as storage or workshops. Other uses, such as residential, may be inappropriate.

The Planning Portal has a useful guide on loft conversions.

5.11 Central heating

Planning permission		Building Regulation approval	
No	Unless it is a listed building	No	If electric and it is installed by an approved person and complies with Part P
		No	If gas, solid fuel or oil fired and it is installed by an approved person and complies with Part J

If the installation requires an outside flue it is usually permitted development unless the flue is more than 1 m above the highest part of the roof or the building is listed or in a Conservation Area.

Note: New standards apply if you replace your existing hot water central heating boiler. Because of the safety issues and the need for energy efficiency, any work to install a new boiler (or an Aga, Raeburn or cooker that also supplies central heating) needs Building Regulations approval. This is generally achieved by employing an installer who is registered under an approved scheme, who will follow the guidelines set out in Approved Document J. The new boiler must have a minimum efficiency of 86 per cent for gas and 85 per cent for oil, and any replacement gas boiler will probably have to be a condensing boiler unless there is sufficient reason why one cannot be installed.

5.12 Change of use

Planning permission		Building Regulation approval	
Possibly	Even if no building or engineering work is proposed	Yes	To certain changes of use, even though you may think that the work involved is not 'building work'
Working from home			
No	Provided that: • you still intend to use the building mainly as a private residence; • there will be no increase in traffic or people calling; • the activities you undertake are not considered unusual in a residential area; • the business does not disturb your neighbours at unreasonable hours or create other forms of nuisance such as noise or smells	No	Unless building work is carried out

The use of buildings or land for a different purpose may need consent even if no building or engineering works are proposed. Again, it is always advisable to check with development control staff.

When you are assessing use of your property, whatever business you carry out the key test is to identify if it is still mainly a home or has it become business premises. If you are in doubt you may apply to your council for a Certificate of Lawful Use for the proposed activity, to confirm it is not a change of use and still the lawful use.

5.12.1 What is meant by material change of use?

A material change of use is where there is a change in the purposes for which or the circumstances in which a building is used, so that after that change:

- the building is used as a dwelling, where previously it was not;
- the building contains a flat, where previously it did not;
- the building is used as a hotel or a boarding house, where previously it was not;
- the building is used as an institution, where previously it was not;
- the building is used as a public building, where previously it was not;
- a building no longer comes within the exemptions in Schedule 2 to the Building Regulations, where previously it did;
- the building, which contains at least one dwelling, contains a greater or lesser number of dwellings than it did previously;
- the building contains a room for residential purposes, where previously it did not;
- the building, which contains at least one room for residential purposes, contains a greater or lesser number of such rooms than it did previously; or
- the building is used as a shop, where previously it was not.

Whenever such changes occur, the building must be brought up to the standards required by all existing Approved Documents.

'Public building' means a building consisting of or containing:

- a theatre, public library, hall or other place of public resort;
- a school or other educational establishment;
- a place of public worship.

5.12.2 Use classes

The following table shows the types of use which may fall within each class of use.

Table 5.1 Use classes and types of use

Class	Description of types of use
A1 Shops	Shops, retail warehouses, hairdressers, undertakers, travel and ticket agencies, post offices (but not sorting offices), pet shops, sandwich bars, showrooms, domestic hire shops, dry cleaners, funeral directors and internet cafés
A2 Financial and professional services	Financial services such as banks and building societies, professional services (other than health and medical services), including estate and employment agencies and betting offices

Table 5.1 Use classes and types of use (Continued)

Class	Description of types of use
A3 Restaurants and cafés	For the sale of food and drink for consumption on the premises – restaurants, snack bars and cafés
A4 Drinking establishments	Public houses, wine bars or other drinking establishments (but not night clubs)
A5 Hot food takeaways	For the sale of hot food for consumption off the premises
B1 Business	Offices (other than those that fall within A2), research and development of products and processes, light industry appropriate in a residential area
B2 General industrial	Use for an industrial process other than one falling within class B1 (excluding incineration purposes, chemical treatment or landfill or hazardous waste)
B8 Storage or distribution	This class includes open-air storage
C1 Hotels	Hotels, boarding and guest houses where no significant element of care is provided (excludes hostels)
C2 Residential institutions	Residential care homes, hospitals, nursing homes, boarding schools, residential colleges and training centres
C2A Secure residential institution	Use for a provision of secure residential accommodation, including use as a prison, young offenders institution, detention centre, secure training centre, custody centre, short-term holding centre, secure hospital, secure local authority accommodation, or use as a military barracks
C3 Dwellinghouses	This class is formed of three parts:
C3(a)	Covers use by a single person or a family (a couple whether married or not, a person related to one another with members of the family of one of the couple to be treated as members of the family of the other), an employer and certain domestic employees (such as an au pair, nanny, nurse, governess, servant, chauffeur, gardener, secretary and personal assistant), a carer and the person receiving the care, and a foster parent and foster child
C3(b)	Up to six people living together as a single household and receiving care (e.g. supported housing schemes, such as those for people with learning disabilities or mental health problems)
C3(c)	Allows for groups of people (up to six) living together as a single household. This allows for those groupings that do not fall within the C4 definition of houses in multiple occupation, but which fell within the previous C3 use class, to be provided for (e.g. a small religious community may fall into this section, as could a homeowner who is living with a lodger)
C4 Houses in multiple occupation	Small, shared dwelling houses occupied by between three and six unrelated individuals, as their only or main residence, who share basic amenities such as a kitchen or bathroom
D1 Non-residential institutions	Clinics, health centres, crèches, day nurseries, day centres, schools, art galleries (other than for sale or hire), museums, libraries, halls, places of worship, church halls, law courts. Non-residential education and training centres

Table 5.1 Use classes and types of use (Continued)

Class	Description of types of use
D2 Assembly and leisure	Cinemas, music and concert halls, bingo and dance halls (but not night clubs), swimming baths, skating rinks, gymnasiums or areas for indoor or outdoor sports and recreations (except for motor sports, or where firearms are used)
Sui generis	Certain uses do not fall within any use class and are considered 'sui generis'. Such uses include: theatres, houses in multiple occupation, hostels providing no significant element of care, scrap yards, petrol filling stations and shops selling and/or displaying motor vehicles, retail warehouse clubs, nightclubs, launderettes, taxi businesses, amusement centres and casinos

Note: Changes to the Uses Classes Order in April 2010 split the old Class C3 into two separate classes (namely Class C3 'dwellinghouses' and Class C4 'houses in multiple occupancy'. A further amendment gave permitted development rights for change of use from C4 to C3, thereby allowing a change of use from a small-scale house in multiple occupation to a dwellinghouse, or vice versa, is now permitted development, and planning applications are not required. Councils have, however, been given the freedom to choose areas where landlords must submit a planning application to rent their properties to unrelated tenants.

5.12.3 Changes of use which do not require planning permission

Table 5.2 Changes of use not requiring planning permission

From	To
A2 (professional and financial services) when premises have a display window at ground level	A1 (shop)
A3 (restaurants and cafés)	A1 or A2
A4 (drinking establishments)	A1 or A2 or A3
A5 (hot food takeaways)	A1 or A2 or A3
B1 (business) (permission limited to change of use relating to not more than $235\,m^2$ of floor space)	B8 (storage and distribution)
B2 (general industrial)	B1 (business)
B2 (general industrial) (permission limited to change of use relating to not more than $235\,m^2$ of floor space)	B8 (storage and distribution)
B8 (storage and distribution) (permission limited to change of use relating to not more than $235\,m^2$ of floor space)	B1 (business)
C4 (houses in multiple occupation)	C3 (dwellinghouses)
Casinos (sui generis)	D2 (assembly and leisure)

The following table shows the changes of use, other than changes where both uses fall within the same class, where planning permission is not required

5.12.4 Material changes of use

Where there is a material change of use of a **whole** building to a hotel, boarding house, institution, public building or a shop (restaurant, bar or public house) the building must be upgraded, if necessary, so as to comply with Approved Document M1 (Access and Use).

If an existing building undergoes a change of use so that **part** of it can be used as a hotel, boarding house, institution, public building or a shop, the work being carried out must ensure that:

- people can gain access from the site boundary and any on-site car parking space;
- sanitary conveniences are provided in that part of the building or it is possible for people (no matter their disability) to use sanitary conveniences elsewhere in the building.

When a building is subject to a material change of use, then:

- any thermal element that is being retained should be upgraded;
- any existing window (including roof window or roof light) or door which separates a conditioned space from an unconditioned space (or the external environment) and which has a U-value that is worse than 3.3 W/m K should be replaced.

5.12.5 Material alterations

Material alterations (i.e. where work, or any part of it, would result in a building or controlled service or fitting not complying with a relevant requirement where previously it did, or make a previous compliance more unsatisfactory) should, in order to comply with the requirements for conservation of heat and energy, be made as follows.

Material alterations (domestic buildings)

If a building is subject to a material alteration by:

- substantially replacing a thermal element;
- renovating a thermal element;
- making an existing element part of the thermal envelope of the building (where previously it was not);
- providing a controlled fitting;
- providing (or extending) a controlled service;

then in addition to the requirements of Part L, all applicable requirements from the following Approved Documents must be taken into account:

- Part A (structure);
- Paragraph B1 (means of warning and escape);
- Part B2 (internal fire spread – linings);
- Paragraph B3 (internal fire spread – structure);
- Paragraph B4 (external fire spread);
- Paragraph B5 (access and facilities for the fire service);
- Part M (access to and use of buildings).

Material alterations (buildings other than dwellings)

When an existing element becomes part of the thermal element of a building (where previously it was not) and it has a U-value worse than $3.3 \, \text{W/m}^2 \, \text{K}$ it should be replaced (unless it is a display window or high-usage door).

5.12.6 Extensions, material alterations or a material change of use

Where any electrical installation work is classified as an extension, a material alteration or a material change of use, the work must consider and include:

- confirmation that the mains supply equipment is suitable and can carry the additional loads envisaged;
- the rating and the condition of existing equipment (belonging to both the consumer and the electricity distributor) are sufficient;
- the amount of additions and alterations that will be required to the existing fixed electrical installation in the building;
- the necessary additions and alterations to the circuits which feed them;
- the protective measures required to meet the requirements;
- the earthing and bonding systems are satisfactory and meet the requirements.

Note: Appendix C to Part P of the Building Regulations offers guidance on some of the older types of installations that might be encountered during alteration work, and Appendix D provides guidance on the application of the now harmonized European cable identification system.

5.12.7 What are the requirements relating to material change of use?

Where there is a material change of use of the whole of a building, any work carried out shall ensure that the building complies with the applicable requirements of the following paragraphs of Schedule 1:

(a) in all cases:

- means of warning and escape (B1)
- internal fire spread – linings (B2)

- internal fire spread – structure (B3)
- external fire spread – roofs (B4)(2)
- access and facilities for the fire service (B5)
- resistance to moisture (C1)(2)
- dwelling-houses and flats formed by material change of use (E4)
- ventilation (F1)
- sanitary conveniences and washing facilities (G4)
- bathrooms (G5)
- foul water drainage (H1)
- solid waste storage (H6)
- combustion appliances (J1, J2 and J3)
- conservation of fuel and power – dwellings (L1)
- conservation of fuel and power – buildings other than dwellings (L2)
- electrical safety (P1 and P2).

In the case of a building exceeding 15 m in height:

- external fire spread – walls (B4–(1)).

(b) in other cases:

Material change of use	Requirement	Approved Document
The building is used as a dwelling where previously it was not	Resistance to moisture	C2, E1, E2, E3
The public building consists of a new school	Acoustic conditions in schools	E4
The building contains a flat where previously it did not	Resistance to the passage of sound	E1, E2, E3
The building is used as a hotel or a boarding house, where previously it was not	Structure	A1, A2, A3, E1, E2, E3
The building is used as an institution, where previously it was not	Structure	A1, A2, A3
The building is used as a public building, where previously it was not	Structure	A1, A2, A3, E1, E2, E3
The building is not a building described in Classes I to VI in Schedule 2, where previously it was	Structure	A1, A2, A3
The building, which contains at least one room for residential purposes, contains a greater or lesser number of dwellings than it did previously	Structure	A1, A2, A3, E1, E2, E3
The building, which contains at least one dwelling, contains a greater or lesser number of dwellings than it did previously	Resistance to the passage of sound	E1, E2, E3

In some circumstances (particularly when a historic building is undergoing a material change of use and where the special characteristics of the building need to be recognized) it may **not** be practical to improve sound insulation to the standards set out in Part E1 or resistance to contaminants and water as set out in Part C. In these cases, the aim should be to improve the insulation and resistance where it is practically possible – always provided that the work does not prejudice the character of the historic building, or increase the risk of long-term deterioration to the building fabric and/or fittings.

 Note: BS 7913:1998 *The Principles of the Conservation of Historic Buildings* provides guidance on the principles that should be applied when proposing work on historic buildings.

5.12.8 Mixed use development

In mixed use developments the requirements of the Regulations may differ depending on whether the area concerned is part of a building used as a dwelling or part of a building which has a non-domestic use. In these cases the requirements for non-domestic use shall apply in any shared parts of the building.

5.12.9 Buildings suitable for conversion

Most local plans stipulate that conversion proposals 'should relate to buildings of traditional design and construction which enhance the natural beauty of the landscape' as opposed to 'non-traditional buildings, buildings of inappropriate design, or buildings constructed of materials which are of a temporary nature'.

Isolated buildings

Planning permission will not normally be granted for the conversion or re-use of isolated buildings. Exceptionally, permission may be given for such buildings to be used for small-scale storage or workshop uses or for camping purposes.
    An isolated building is normally:

*   a building, or part of a building, standing alone in the open countryside; or
*   a building, or part of a building, comprised within a group which otherwise occupies a remote location having regard to the disposition of other buildings within the locality, to the character of the surroundings, and to the nature and availability of access and essential services.

Assessing whether or not a particular building should be regarded as isolated may not always be straightforward, and in such instances early discussion with a planning officer at the National Park authority is advised.

Structural condition

 Planning permission will not normally be granted for reconstruction if substantial collapse occurs during work on the conversion of a building.

    Buildings proposed for conversion should be large enough to accommodate the proposed use without the necessity for major alterations, extension or reconstruction. In cases of doubt regarding the structural condition of any particular building, the authority will require the submission of a full structural survey to accompany a planning application. The authority can advise on this requirement and, if necessary, on persons who are suitably qualified to undertake such work and who practise locally.

 A list of local consulting engineers can be found in *Yellow Pages*.

Workshop conversions

Redundant farm buildings and buildings of historic interest are often well suited to workshop use, and such conversions normally require minimal alterations. Potential problems of traffic generation and unneighbourliness can usually be addressed by the imposition of appropriate conditions. The local authority will generally favourably consider proposals that make good use of traditional buildings by promoting local employment opportunities. In some instances grants may be available from other agencies to assist the conversion of buildings to workshop use.

Residential conversions

When reviewing proposals for converting a traditional building, the local authority will pay particular attention to the overall objectives of the housing policies of the local plan. If land that can be used for a new housing development is limited, residential conversions can make a valuable contribution to the local housing stock and support the social and economic well-being of rural communities. The local plan will require that residential conversions should, in most instances, contribute to the housing needs of the locality. Permission for such conversions are, in some districts, only granted subject to a condition restricting occupancy to local persons.

Renovation

Districts dedicate some areas as Environmentally Sensitive Areas (ESAs), and grants may be available towards the cost of renovating historical and important local buildings that have fallen into disrepair, or towards the cost of renovation works to retain agricultural buildings in farming use, so as to retain their importance as landscape features. Further advice on the workings of the scheme may be obtained from the ESA project officers or the local authority's building conservation officer.

 'Local persons' are normally defined as persons working, about to work, or having last worked in the locality or who have resided for a period of three years within the locality.

Applicants are strongly advised to employ qualified architects or designers in preparing conversion proposals. Informal discussions with a planning officer at an early stage in considering design solutions are also encouraged.

5.13 Decoration and repairs inside and outside a building

Planning permission		Building Regulation approval	
No	Unless it is a listed building or within a Conservation Area	No	Unless it is a listed building or within a Conservation Area. Consult your local authority
No	As long as the use of the house is not altered	Possibly	External walls are considered to be thermal elements, so approval is required if you want to re-render, replace cladding or insert insulation into a cavity wall. Consult your local authority
Yes	If the alterations are major, such as removing or part removing a load-bearing wall or altering the drainage system	Yes	If the alterations are major, such as removing or part removing a load-bearing wall or altering the drainage system

Generally speaking, you do not need to apply for planning permission:

- for repairs or maintenance;
- for minor improvements, such as painting your house or replacing windows;
- for internal alterations;
- for the insertion of windows, skylights or roof lights (but, if you want to create a new bay window, this will be treated as an extension of the house);
- for the installation of solar panels which do not project significantly beyond the roof slope (rules for listed buildings and houses in Conservation Areas are different, however);
- to re-roof your house (but additions to the roof are treated as extensions to the house).

 Occasionally, you may need to apply for planning permission for some of these works because your council has made an Article 4 direction withdrawing permitted development rights.

5.13.1 Do I need permission to carry out decoration or repairs?

Do I need approval to carry out repairs to my house, shop or office?

You do not need permission if the repairs are of a minor nature (e.g. replacing the felt to a flat roof, repointing brickwork, or replacing floorboards). However, if the repair work is major in nature (e.g. removing a substantial part of a wall and rebuilding it, or underpinning a building) you will require both planning permission and Building Regulations approval.

Do I need to apply for planning permission for internal decoration, repair and maintenance?

You do not need planning permission for internal decoration, repair and maintenance. Building regulations may apply to some works.

Do I need to apply for planning permission for external decoration, repair and maintenance?

Most external work does not need permission, provided it does not make the building any larger or affect the thermal element of the building.

Do I need approval to alter the position of a WC, bath etc. within my house, shop or flat?

If you are refitting a kitchen or bathroom with new units and fittings you do not need approval. However, fitting a bathroom or kitchen where there was not one before, or drainage and electrical work relating to a refit, will require Building Regulations approval.

Do I need approval to alter in any way the construction of fireplaces, hearths or flues within my house, shop or flat?

You do not need approval to fit or replace an external flue, chimney or soil and vent pipe. However, Building Regulations will apply if you are installing a flue.

Do I need to apply for planning permission if my property is a listed building?

If your property is a listed building, consent will probably be needed for **any** external work, especially if it will alter the visual appearance, or use alternative materials. You also may need planning permission to alter, repair or maintain a gate, fence, wall or other means of enclosure.

Do I need to apply for planning permission if my property is in a Conservation Area?

If the building undergoing repair or decoration is in a Conservation Area, or comes under any type of covenant restricting changes, you will probably need planning permission. You may also be restricted to replacing items such as roof tiles with the approved material, colour and texture, and have to use cast iron guttering rather than plastic etc.

Do I need approval to insert cavity wall insulation?

You need approval to insert new cavity wall insulation.

Do I need approval to apply cladding?

Building Regulations may apply if you want to re-render or replace timber cladding to external walls, depending on the extent of the work. Where 25 per cent or more of the wall is re-clad, re-rendered or re-plastered internally, or 25 per cent or more of the external wall is rebuilt, the thermal insulation will need to be improved and Building Regulations will apply.

5.14 Demolition

Planning permission		Building Regulation approval	
Yes	If it is a listed building or in a Conservation Area	Yes	Six weeks prior notice must be given to the local authority building control
Perhaps	The council may wish to agree with you how you propose to carry out the demolition	Yes	For a partial demolition to ensure that the remaining part of the house (or adjoining buildings/extensions) is structurally sound

You must have good reasons for knocking a building down, such as making way for rebuilding or improvement (which in most cases would be incorporated in the same planning application). Penalties are severe for demolishing something illegally. In any case you will require a 'prior approval application' or formal confirmation of whether the council wishes to agree with how you propose to carry out the demolition. You do not need to make a planning application to demolish a listed building or to demolish a building in a Conservation Area. However, you may need Listed Building or Conservation Area Consent.

Elsewhere, you will **not** need to apply for planning permission:

* to demolish a building such as a garage or shed of less than $50 \, m^3$; or
* if the demolition is urgently necessary for health and safety reasons; or
* if the demolition is required under other legislation; or
* where the demolition is on land that has been given planning permission for redevelopment; or
* to demolish a gate, fence, wall or other means of enclosure.

You are not allowed to begin any demolition work (even on a dangerous building) unless you have given the local authority notice of your intention and this has either been acknowledged by the local authority or the relevant notification period has expired. In this notice you will have to:

* specify the building to be demolished;
* state the reason(s) for wanting to demolish it;
* show how you intend to demolish it.

Copies of this notice will have to be sent to:

* the local authority;
* the occupier of any building adjacent to the building;
* British Gas;
* the area electricity board in whose area the building is situated.

Contact the local authority building control office during office hours, or the local authority emergency switchboard out of hours.

This regulation does not apply to the demolition of an internal part of an occupied building, or a greenhouse, conservatory, shed or prefabricated garage (that forms part of that building) or an agricultural building defined in Section 26 of the General Rate Act 1967.

5.14.1 What about dangerous buildings?

If a building, or part of a building or structure, is in such a dangerous condition (or is used to carry loads that would make it dangerous) then the local authority may apply to a magistrates' court to make an order requiring the owner:

* to carry out work to avert the danger;
* to demolish the building or structure, or any dangerous part of it, and remove any rubbish resulting from the demolition.

 The local authority can make an order restricting a building's use until such time as a magistrates' court is satisfied that all necessary works have been completed. These works are controllable by the local authority under Sections 77 and 78 of the Building Act 1984. In inner London the legislation is under the London Building (Amendment) Act 1939. This involves responding to all reported instances of dangerous walls, structures and buildings within each local authority's area on a 24-hour, 365-days-a-year basis.

If the building or structure poses a potential danger to the safety of people, the local authority will take the appropriate action to remove the danger. The local authority has powers to require the owners of buildings or structures to remedy the defects, or they can direct their own contractors to carry out works to make the building or structure safe. In addition, the local authority may provide advice on the structural condition of buildings to the fire brigade during fire-fighting.

If you are concerned that a building or other structure may be in a dangerous condition, then you should report it to the local council.

Emergency measures

In emergencies the local authority can make the owner take immediate action to remove the danger, or they can complete the necessary action themselves. In these cases, the local authority is entitled to recover from the owner such expenses reasonably incurred by them. For example:

* fencing off the building or structure;
* arranging for the building/structure to be watched.

5.14.2 Can I be made to demolish a dangerous building?

 Before complying with this notice, the owner must give the local authority 48 hours' notice of commencement.

If the local authority considers that a building is so dangerous that it should be demolished, they are entitled to issue a notice to the owner requiring the owner/occupier:

- to shore up any building adjacent to the building to which the notice relates;
- to weatherproof any surfaces of an adjacent building that are exposed by the demolition;
- to repair and make good any damage to an adjacent building caused by the demolition or by the negligent act or omission of any person engaged in it;
- to remove material or rubbish resulting from the demolition and clear the site;
- to disconnect, seal and remove any sewer or drain in or under the building;
- to make good the surface of the ground that has been disturbed in connection with this removal of drains etc.;
- in accordance with the Water Act 1945 (interference with valves and other apparatus) and the Gas Act 1972 (public safety), arrange with the relevant statutory undertakers (e.g. water board, British Gas or electricity supplier) for the disconnection of gas, electricity and water supplies to the building;
- to leave the site in a satisfactory condition following completion of all demolition work.

In certain circumstances, the owner of an adjacent building may be liable to assist in the cost of shoring up their part of the building and waterproofing the surfaces. It could be worthwhile checking this point with the local authority!

Under Section 80 of the Building Act 1984 anyone carrying out demolition work is required to notify the local authority. The local authority then has six weeks to respond with appropriate notices and consultation under Sections 81 and 82 of the Act (this does not apply to inner London).

Replacing a demolished building

If you decide to demolish a building, even one that has suffered fire or storm damage, it does not automatically follow that you will get Planning Permission to build a replacement.

5.15 Electrical work in the home or garden

Planning permission		Building Regulation approval	
No	Unless it is in a listed building or a Conservation Area	No	If it is installed by an approved person and complies with Part P and other relevant Building Regulations Approved Documents

5.15.1 Do I need approval to carry out electrical work?

Do I need approval to replace electric wiring?

You do not need planning permission, but in order to get Building Regulations approval:

- you must comply with Part P (and other relevant Building Regulations Approved Documents);
- you should meet the recommendations of the IET Wiring Regulations (i.e. BS 7671);
- your contract with the electricity supply company will have conditions about electrical safety which must not be broken. In particular, you should **not** interfere with the company's equipment, which includes the cables to your consumer unit up to and including the separate isolator switch, if provided.

Do I need approval to replace an existing electrical fitting?

Non-notifiable work (such as replacing an electrical fitting) can be completed by a DIY enthusiast (family member or friend) but **still** needs to be installed in accordance with the manufacturer's instructions and done in such a way that it does not present a safety hazard.

This work does **not** need to be notified to a local authority building control body (unless it is installed in an area of high risk such as a kitchen or a bathroom etc.) **but** all DIY electrical work (unless completed by a qualified professional) will still need to be checked, certified and tested by a competent electrician.

Do I need approval to install a new electrical circuit?

Any work that involves adding a new circuit to, in or around a dwelling will need to be either notified to the building control body (who will then inspect the work) or be carried out by a competent person who is registered under a government-approved Part P self-certification scheme.

Work involving any of the following will also have to be notified:

- consumer unit replacements;
- electric floor or ceiling heating systems;
- extra-low voltage lighting installations (other than pre-assembled, CE-marked lighting sets);
- garden lighting and/or power installations;
- installation of a socket outlet on an external wall;
- installation of outdoor lighting and/or power installations in the garden or that involves crossing the garden;
- installation of new central heating control wiring;
- Solar Photovoltaic (PV) power supply systems;
- small-scale generators (such as micro combined heat and power units).

Note: Where a person who is **not** registered to self-certify, intends to carry out the electrical installation, then a Building Regulation (i.e. a building notice or full plans) application will need to be submitted together with the appropriate fee, based on the estimated cost of the electrical installation. The building control body will then arrange to have the electrical installation inspected at first fix stage and tested upon completion.

5.16 Erecting aerials, satellite dishes, television and radio aerials, wind turbines, solar panels and flagpoles

Planning permission		Building Regulation approval	
Satellite dishes, aerials and antennae on buildings up to 15 m high			
No	Unless: • there are more than two antennas on the property overall; • a single antenna is more than 100 cm in any linear dimension; • if installing two antennas, one is more than 100 cm and the other is more than 60 cm in any linear dimension; • the cubic capacity of each antenna is more than 30 litres; • the antenna is fitted to a chimney stack and is greater than 60 cm in any linear dimension; • the antenna sticks out more than 60 cm above the roof line or chimney stack, whichever is the lower	No	But make sure that the fixing point is stable and the installation is safe
Possibly	If the building is in a designated area and the capacity of each antenna is greater than 35 litres or it is intended to install the antenna on a chimney, wall or roof slope visible from a road or Broads waterway	No	
Satellite dishes, aerials and antennae on buildings over 15 m high			
No	Unless: • there are more than four antennas on the property overall; • the size of any antenna is more than 130 cm in any linear dimension; • the cubic capacity of each antenna is more than 35 litres; • the antenna is fitted to a chimney stack and is greater than 60 cm in any linear dimension;	No	But make sure that the fixing point is stable and the installation is safe

Planning permission		Building Regulation approval	
	• the antenna sticks out more than 300 cm above the highest part of the roof line; • if the building is in a designated area and it is intended to install the antenna on a chimney, wall or roof slope visible from a road or Broads waterway		
Solar panels – roof and wall mounted			
No	The fitting of solar panels is classed as permitted development in many cases; however, they may not extend above the ridgeline or project more than 200 m	Yes	The ability of the roof to carry the weight of the panels needs to be checked and electrical work needs to be carried out by an approved contractor
Yes	Planning permission is required to fit solar panels to a listed building or in a Conservation Area		
Solar panels – stand alone			
No	Provided that there is only one installation, which is: • no higher than 4 m; • at least 5 m from boundaries; • no larger than 9 m² or 3 m wide and 3 m deep; • not installed within the boundary of a listed building; • not be visible from the highway if it is in a Conservation Area or a World Heritage Site		
Wind turbine			
Yes	If erecting a wind turbine, either stand alone or attached to a dwelling (consult your local planning officer)	No	Unless you intend to fit a wind turbine to your dwelling
Flagpole			
No	Flagpoles etc. erected in your garden are treated under the same rules as outbuildings, and cannot exceed 3 m in height	No	

 Note: These limits refer to buildings as a whole and not flats within them. If you wish to install any of the following on a flat you should contact the local authority.

You should get specific advice if you plan to install a large satellite dish or aerial, such as a short wave mast, as the rules differ between authorities.

Designated areas are:

- Areas of Outstanding Natural Beauty;
- Conservation Areas;
- National Parks;
- the Norfolk or Suffolk Broads;
- World Heritage Sites.

Unless it is a stand-alone antenna, flagpole or mast greater than 3 m in height you will not need to apply for permission.

Normally there is no need for planning permission for attaching an aerial or satellite dish to your house or its chimneys. However, if it rises significantly higher than the roof's highest point then it may contravene local regulations or covenants. Planning permission may be required to install a wind turbine, either free standing or attached to a dwelling. It is recommended that you seek the advice of your local planning authority in relation to wind turbines, as permissions vary depending on the region of the UK.

In certain circumstances, you will need to apply for planning permission to install a satellite dish on your house (see the Department for Communities and Local Government's free booklet *A Householder's Planning Guide for the Installation of Satellite Television Dishes*, which can be obtained from your local council). Conservation Areas have specific local rules on aerials and satellite dishes, so you need to approach your local planning department to find out the particular rules for your area. Certainly, if your house is a listed building, you may need Listed Building Consent to install a satellite dish on your house.

Remember, if you are a leaseholder, you may need to obtain permission from the landlord.

Satellite dish locator

The Planning Portal interactive house has a satellite dish locator. This provides a user-friendly way to check which parts of your house offer a suitable location for a satellite dish.

5.17 Felling or lopping trees

Planning permission	Building Regulation approval
No Unless the trees are protected by a Tree Preservation Order or you live in a Conservation Area	No

Many trees are protected by Tree Preservation Orders (TPOs), which mean that, in general, you need the council's consent to prune or fell them. Nearly all trees in Conservation Areas are automatically protected.

Ask the council for a copy of the free leaflet *Protected Trees: A Guide to Tree Preservation Procedures.*

5.18 Flats and maisonettes

There is a different planning regime for flats and maisonettes, and the permitted development rights which apply to many common projects for houses may not apply to flats. In particular:

Planning permission		Building Regulation approval	
Extensions to a ground floor flat			
Yes	You must apply for planning permission to extend a flat	Yes	You may also need to consult the Fire Service
To subdivide a house or single flat			
Yes		Yes	You may also need to consult the Fire Service and the property will need to be licensed under the Housing Act 2004
Loft conversion in a top floor flat			
No	Provided that it is internal works. Local requirements differ, so it is recommended that you check with the local planning authority	Yes	
Yes	If you intend to extend or alter the roof space	Yes	
Painting the exterior			
No	Unless you live in an area where an Article 4 direction applies	No	
Satellite dishes			
Perhaps	Planning permission may be required in certain circumstances. Check with the local council	No	
Windows			
Perhaps	To fit new double glazed windows	Yes	
No	If you are replacing like with like	Yes	

Note: Consult the local planning authority for advice before starting work on windows, because local policy and interpretation of the rules varies from council to council.

In addition to planning permission you may also require Listed Building or Conservation Area Consent, as work on a building that affects its special historic character without consent is a criminal offence. Local policy and interpretation of the rules covering windows in flats varies from council to council, and you are advised to contact your local planning authority for advice before starting work.

 Don't forget to get permission from your landlord, freeholder or management company if you are a leaseholder.

5.19 Garages and carports

Planning permission		Building Regulation approval	
Car port			
No		No	If it is an attached carport less than 30 m²
New garage – attached			
No	As long as the floor area is between 15 m² and 30 m²	Yes	If it is attached to an existing home
New garage – detached			
No	• The building is not forward of the principal elevation; • the building is single storey; • the building is no taller than 2.5 m if it is within 2 m of the boundary; • the building eaves are no taller than 2.5 m and the overall height no more than 4 m; • there is no veranda, balcony or raised platform; • no more than half the land around the original house would be covered by other buildings	No	If the floor area is no more than 15 m² or the floor area is between 15 m² and 30 m² and is at least 1 m from any boundary
Yes	If the building is listed or in a Conservation Area	No	If the floor area is no more than 15 m² or the floor area is between 15 m² and 30 m² and is at least 1 m from any boundary
Garage Conversion			
No	As long as the work is internal	Yes	In relation to the following: • infill garage door; • flooring; • walls below ground level; • ventilation; • doors and windows; • drainage; • electrics; • external walls; • internal walls; • roofs

Note: Sometimes permitted development rights have been removed with regard to garage conversion (particularly in new developments or Conservation Areas). You should contact your local planning department if this applies.

5.19.1 Do I need approval to build a garage extension to my house, shop or office?

A carport extension built at ground level, open on at least two sides and less than $30\,m^2$ in floor area, is exempt.

5.20 Hydroelectricity

Planning permission		Building Regulation approval	
Yes	Some form of environmental assessment will be essential for this type of project	Yes	Particularly for electrical installations

Note: This is a complex area and the Environment Agency must also be consulted about water extraction licences. More details can be found at http://www.energysavingtrust.org.uk/Generate-your-own-energy.

5.21 Heat pumps

Planning permission		Building Regulation approval	
Ground-source heat pumps			
No	Ground-source heat pups do not require planning permission unless the building is listed or in a Conservation area	Yes	It is advisable to contact an engineer who can provide the necessary advice
Air-source heat pumps			
Yes	Air-source heat pumps currently require planning permission	Yes	It is advisable to contact an engineer who can provide the necessary advice

Note: The government is consulting on allowing air-source heat pumps as permitted development.

5.22 Infilling

Planning permission		Building Regulation approval	
Possibly	Consult your local planning officer	Yes	If a new development

5.22.1 Can I use an unused, but adjoining, piece of land to build a house (e.g. build a new house on land that used to be a large garden)?

Often there may be no official grounds for denying consent, but residents and individuals can impose quite some delay. It is worth testing the likelihood of a successful application by talking to the neighbours and judging opinions. Planning consent is often quite difficult to obtain in these cases, as this sort of development normally causes a lot of opposition as it is in a settled residential area and people do not like change. New developments will undoubtedly also need to follow Building Regulations. This, and all site visits from inspectors, is normally arranged by your building contractor.

5.23 Installing a swimming pool

Planning permission		Building Regulation approval	
Possibly	Consult your local planning officer	Yes	For an indoor pool

Swimming pools and saunas are subject to special requirements specified in Part 6 of BS 7671:2001. Installing a covered swimming pool will be covered by the rules that apply to sheds and outbuildings.

5.24 Laying a patio, decking or a driveway

Planning permission		Building Regulation approval	
No	As long as: • no significant embanking or terracing works are required; • it is not a listed building	No	As long as alterations do not make access to the dwelling any less satisfactory than it was before
Decking			
No	As long as: • the decking is no more than 30 cm above the ground; • together with extensions, outbuildings etc. the decking covers no more than 50 per cent of the garden	No	As long as the structure does not require planning permission

Generally you do not need to apply for planning permission to install a patio or driveway. You may however, need approval from the local council if the pathway crosses a pavement. There are separate rules if you intend to pave over your front garden.

5.25 Micro combined heat and power

Planning permission		Building Regulation approval	
No	Unless it is a listed building or within a Conservation Area or the work is external	Yes	Particularly to electrical installations, plumbing work and in relation to Approved Documents L1A, L1B, L2A and L2B.

5.26 Oil storage tank

Planning permission		Building Regulation approval	
No	Provided that it is in the garden and has a capacity of not more than 3500 litres and **no** point is more than 3 m high and **no** part projects beyond the foremost wall of the house facing the highway	No	If it is installed by an approved person and complies with Building Regulations
Yes	Any container within the curtilage of a listed building will require planning permission		

Oil storage tanks, and the pipes connecting them to combustion appliances, should be constructed and protected so as to reduce the risk of the oil escaping and causing pollution.

5.27 Outbuildings

Planning permission		Building Regulation approval	
No	As long as: • the building is not forward of the principal elevation; • the building is single storey; • the building is no taller than 2.5 m if it is within 2 m of the boundary; • the building eaves are no taller than 2.5 m and the overall height no more than 4 m; • there is no veranda, balcony or raised platform; • no more than half the land around the original house would be covered by other buildings	No	Unless the floor area exceeds 15 m² and contains sleeping accommodation
Yes	If the building is listed or in a Conservation Area		

Planning permission		Building Regulation approval	
Polytunnels			
Yes	If it is a listed building or in a Conservation Area, National Park or Area of Outstanding Natural Beauty	Perhaps	Depending on work involved
No	As long as the polytunnel is no nearer to the road than the nearest part of the house (unless there is more than 20 m between the tunnel and the road) and is less than 3 m high	Perhaps	Depending on work involved

Many kinds of buildings and structures can be built in your garden or on the land around your house without the need to apply for planning permission. These can include sheds, garages, greenhouses, accommodation for pets and domestic animals (e.g. chicken houses), summer houses, swimming pools, ponds, sauna cabins, enclosures (including tennis courts) and many other kinds of structure. Outbuildings intended to go in the garden of a house do not normally require any planning permission, as long as they are associated with the residential amenities of the house and a few requirements are adhered to, such as position and size.

If your new building exceeds $10 \, m^3$ (and/or comes within 5 m of the house) it would be treated as an extension and would count against your overall volume entitlement.

Permission is required, however, for:

- any building or structure nearer to a highway than the nearest part of the original house, unless more than 20 m away from a highway;
- structures not required for domestic use;
- structures over 3 m high (or 4 m if it has a ridged roof);
- a propane gas (LPG) tank;
- a storage tank holding more than 3500 litres;
- a building or structure which would result in more than half of the grounds of your house being covered by buildings/structures.

You will also need to apply for planning permission if any of the following cases apply:

- you want to put up a building or structure which would be nearer to any highway than the nearest part of the original house, unless there would be at least 20 m between the new building and any highway (The term 'highway' includes public roads, footpaths, bridleways and byways.);
- more than half the area of land around the original house would be covered by additions or other buildings;
- the building or structure is not to be used for domestic purposes and is to be used instead, for example, for parking a commercial vehicle, running a business or for storing goods in connection with a business;

- you want to put up a building or structure which is more than 3 m high, or more than 4 m high if it has a ridged roof (measured from the highest ground next to it);
- if your house is a listed building and you want to put up a building or structure with a volume of more than 10 m^3.

Erecting any type of outbuilding can be a potential minefield, and it is best to consult with the local planning officer before commencing work.

5.27.1 External water storage tanks

An application is required under Building Regulations to check any new drainage which runs to a tank that recycles rainwater collected from the roof and ground. If you are considering installing an external water tank you should seek guidance from your local authority, especially if the tank is to be mounted on a roof.

5.27.2 Fuel storage tanks

Storage of oil, or any other liquids, especially petrol, diesel and chemicals, is strictly controlled and would not be allowed on residential premises. If you are considering installing an external oil storage tank for central heating use, then no planning permission is required, provided its capacity is no more than 3500 litres, it is no more than 3 m from the ground and it does not project beyond any part of a building facing a public thoroughfare. However Building Regulations will apply.

5.28 Paving your front garden

Planning permission		Building Regulation approval	
No	If the replacement driveway of any size uses permeable surfacing or if the rainwater is directed to a lawn or border to drain naturally	No	As long as alterations do not make access to the dwelling any less satisfactory than it was before
Yes	If the surface to be covered is more than 5 m^2 and you plan to lay a traditional impermeable driveway	No	As long as alterations do not make access to the dwelling any less satisfactory than it was before

Climate change is likely to increase the frequency of heavy rainfall and flooding in the UK. In 2007, the UK experienced serious flooding, which was in many cases caused by drains being unable to cope with the amount of rain water flowing into them. While paving over the garden of one home may not make a difference, the combined effect of many homes in a street or area paving over their front gardens increases the burden on the drains and can increase the risk of flooding.

Further details on permeable surfaces (such as loose gravel), hard permeable and porous surfaces, rain gardens and soakaways and wheel tracks can be found on the Planning Portal.

5.29 Planting a hedge

Planning permission		Building Regulation approval
No	Unless it obscures the view of traffic at a junction or access to a main road	No

You do not need planning permission for hedges or trees. However, if there is a condition attached to the planning permission for your property which restricts the planting of hedges or trees (e.g. on an 'open plan' estate or where a sight line might be blocked), you will need to obtain the council's consent to relax or remove the condition before planting a hedge or tree screen. If you are unsure about this, you can check with the planning department of your council.

Hedges should not be allowed to block out natural light, and the positioning of fast-growing hedges should be checked with your local authority. Recent incidents regarding hedging of the fast growing leylandii trees have led to changes in the planning rules, where hedges previously had no restrictive laws.

 Note: High hedges are dealt with under Part 8 of the Anti-social Behaviour Act 2003, which came into operation in England on 1 June 2005.

5.30 Plumbing

Planning permission		Building Regulation approval	
No	Unless it is a listed building	No	If it is installed by an approved person and complies with Part J
		Yes	If you use an unregistered installer or DIY you will need to get approval from the local authority building control

5.30.1 Do I need approval to install hot water storage within my house, shop or flat?

If you use an installer who is not a member of the approved competent person scheme you will need approval for work relating to the hot water storage.

5.31 Replacing windows and doors

Planning permission		Building Regulation approval	
No	Unless it is a listed building or in Conservation Area	Yes	If you use an unregistered installer or DIY you will need to get approval from the local authority building control
		No	If they are installed by an approved person (see Table 2.2) and comply with the requirements of Part F
Yes	To replace shop windows	Yes	To replace shop windows

5.31.1 Do I need approval to install replacement windows in my house?

You do not require approval to install replacement windows provided that:

- the window opening is not enlarged (if a larger opening is required, or if the existing frames are load-bearing, then a structural alteration will take place and approval is required);
- you do not remove those opening windows which are necessary as a means of escape in case of fire;
- the replacement of a window in an existing building is carried out by a person who is registered under the Fenestration Self-Assessment Scheme by FENSA Limited (http://www.fensa.org.uk/faqs.aspx).

5.31.2 Do I need approval to replace my shop front?

You need approval to replace a shop front and further information is available from your local building control or from the Glass and Glazing Federation (GGF) website (http://www.ggf.org.uk).

Note: As FENSA does not apply to commercial premises or new-build properties, replacement of windows in offices and other commercial premises (including the replacement of shop fronts) will, therefore, **all** require local authority building control approval.

5.32 Shops and change of use

There is a different planning regime for shops, and the permitted development rights which apply to many common projects for houses may not apply. In particular:

Planning permission	Building Regulation approval

Flats over shops

No	Provided that: • the space is in the same class of use as the shop or office to start with (either Class A1 or A2); • the space is not in a separate planning unit from the shop; • you will not change the outside appearance of the building; • if there is any display window at ground floor level, you do not incorporate any of the ground floor into the flat	Yes

Change from residential use to a shop

Yes	Planning permission is required as the intended use is in a different class within the planning system	Yes	The specific requirements include those concerned with escape and other fire precautions, hygiene, energy conservation, and access to and use of buildings

Convert a shop to a café or public house

Yes	Planning permission is required as the intended use is in a different class within the planning system	Perhaps	If the work being carried out involves building work

Convert a shop to an office or storage

Yes	Planning permission is required as the intended use is in a different class within the planning system	Perhaps	If the work being carried out involves building work

Changes to a shop's adverts or fascia

Yes	You must obtain consent, as it is required before an advertisement is displayed. There are limits on the size of adverts and duration for which they may be displayed	No

In all circumstances when you are changing the use of a building you should check with the local fire authority to see what 'on-going' fire precautions legislation will apply when the building is in use.

Many local authorities have policies designed to strike a balance of uses in an area, and some have specific polices to retain a level of retail uses. You do not require permission for the change of use before work starts but, if permission is refused, the work will have to be undone and the authority may take formal enforcement action over a change of use without planning permission.

5.33 Structural alterations inside

Planning permission		Building Regulation approval	
No	Unless it is a listed building or within a Conservation Area	Possibly	If it is a listed building or within a Conservation Area
Yes	If the alterations are major, such as removing or part removing a load-bearing wall or altering the drainage system	Yes	If you wish to build or remove an internal wall or make openings in an internal wall

5.33.1 Do I need approval to make internal alterations within my house?

You will need approval if your alterations are to the structure, such as the removal or part removal of a load-bearing wall, joist, beam or chimney breast, or would affect fire precautions of a structural nature either inside or outside your house. You also need approval if, in altering a house, work is necessary to the drainage system or to maintain the means of escape in case of fire.

5.33.2 Do I need approval to make internal alterations within my shop or office?

You need approval to make internal alterations to a shop or office.

Appendix 5A Basic requirements for planning permission and Building Regulation approval

The table below is a synopsis of the detailed section on basic requirements for planning permission and Building Regulation approval which is contained in Chapter 5. In all circumstances it is recommended that you talk to your local planning officer before contemplating any work. The cost of a local phone call could save you a lot of money (and stress) in the long term!

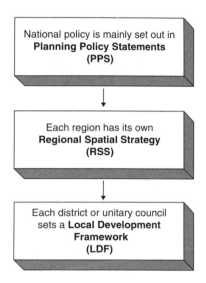

Figure 5.1 Planning responsibilities.

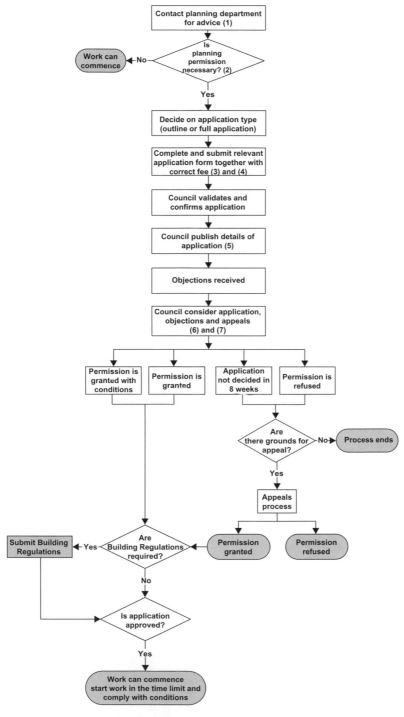

Figure 5.2 Flow chart of the planning permission process.

Type of work	Planning permission		Building Regulation approval	
Advertising	No	If the advertisement is less than 0.3 m² and not illuminated	Possibly	Domestic adverts and signs are not usually subject to planning control, but consult your local planning officer for more details
Basements	No	Provided that: • it is not a separate unit; • the usage is not significantly changed; • a light well is not added; • it is not a listed building	Yes	Covering: • fire escapes; • ventilation; • ceiling height; • damp proofing; • wiring; • water supplies; • underpinning; • foundation work
Biomass-fuelled appliances	No	Unless it is a listed building or within a Conservation Area or the work is external	Yes	Particularly in relation to electrical and plumbing work, ventilation, noise and general safety
Building a conservatory	No	Provided that: • no more than half the area of land around the original house is taken up by other buildings; • it is not forward of the building line; • no part of the conservatory is higher than the original roof of the main building; • it does not extend beyond the rear wall by more than 3 m (4 m for detached houses); • a single-storey rear extension is no higher than 4 m; • the maximum depth of a rear extension of more than one storey is no more than 3 m (including ground floor).	No	Provided that: • it is built at ground level and has a floor area less than 30 m²; • the conservatory is separated from the house by external quality walls, doors or windows; • there is an independent heating system with separate temperature and ON/OFF controls; • glazing and any fixed electrical installations comply with the applicable building regulations requirements

Type of work	Planning permission	Building Regulation approval
	• the eaves height of an extension within 2 m of the boundary is not more than 3 m; • the eaves and ridge height of the extension is no higher than the existing house; • any side extension is single storey with maximum height 4 m and a width no more than half that of the original house; • the roof pitch of an extension higher than one storey matches that of the existing house; • there are no verandas, balconies or raised platforms; • on designated land there is no permitted development for rear extensions of more than one storey; no cladding of the exterior; no side extensions; • the building is not listed	
Building an extension	No Provided that: • no more than half the area of land around the original house is covered by additions or other buildings; • there is no extension forward of the building line; • it is not higher than the highest part of the roof; • it does not extend beyond the rear wall of the original house by more than 3 m (4 m if a detached house); • a single-storey rear extension is no higher than 4 m; • if it is more than one storey it is no deeper than 3 m beyond the rear wall of the original house, including the ground floor;	Yes

Type of work	Planning permission	Building Regulation approval

	• any eaves within 2 m of the boundary are no taller than 3 m;
	• the eaves and ridge are no higher than the existing house;
	• any side extensions are single storey and no wider than half the width of the original house and a maximum height of 4 m;
	• a two-storey extension is not closer to the rear boundary than 7 m;
	• the roof pitch of extensions that are higher than one storey must match that of the existing house;
	• materials must be of similar appearance to those of the existing house;
	• there should be no verandas, balconies or raised platforms;
	• any side-facing windows on the upper floor are obscure-glazed and any opening are 1.7 m above the floor;
	• on designated land:
	  – no permitted development for rear extensions of more than one storey,
	  – no cladding of the exterior,
	  – no side extensions	

Building a fence, gate or wall	Yes	No
	If it is more than 1 m high and is a boundary enclosure adjoining a highway	
	Yes	No
	If it is more than 2 m (6 ft 6 in.) high elsewhere	

Type of work	Planning permission		Building Regulation approval	
Building a hardstanding for a car, caravan or boat	No	Provided it is within your boundary, at, or near, ground level and does not require significant works of embanking or terracing	No	Unless you introduce steps where none existed before
Building a new house	Yes		Yes	
Building a porch	No	Unless: • the floor area exceeds 3 m²; • any part is more than 3 m high; • any part is less than 2 m from a boundary adjoining a highway	No	As long as glazing and electrical installations comply with Building Regulations
Car port	No		No	If it is an attached carport less than 30 m²
Conversions	Yes	For flats – even where construction works may not be intended	Yes	Unless you are not proposing any building work to make the change
	No	For loft conversions, as long as you do not alter or extend the roofspace	Yes	For further advice contact Building Control to discuss your proposal. You must also find out whether work you intend to carry out falls within The Party Wall Act etc. 1996
	Yes	For conversion to shops and offices	Yes	
Central heating	No	Unless it is a listed building	No	If electric and it is installed by an approved person and complies with Part P
			No	If gas, solid fuel or oil fired and it is installed by an approved person and complies with Part J
Change of use	Possibly	Even if no building or engineering work is proposed	Yes	To certain changes of use, even though you may think that the work involved is not 'building work'

Type of work	Planning permission		Building Regulation approval	
Convert a shop to a café or public house	Yes	Planning permission is required as the intended use is in a different class within the planning system	Perhaps	If the work being carried out involves building work
Convert a shop to an office or storage	Yes	Planning permission is required as the intended use is in a different class within the planning system	Perhaps	If the work being carried out involves building work
Changes to a shop's adverts or fascia	Yes	You must obtain consent as it is required before an advertisement is displayed. There are limits on the size of adverts and the duration for which they may be displayed	No	
Change from residential use to a shop	Yes	Planning permission is required as the intended use is in a different class within the planning system	Yes	The specific requirements include those concerned with escape and other fire precautions, hygiene, energy conservation, and access to and use of buildings
Decking	No	As long as • the decking is no more than 30 cm above the ground; • together with extensions, outbuildings etc. the decking covers no more than 50 per cent of the garden	No	As longs as the structure does not require planning permission
Decoration and repairs inside and outside a building	No	Unless it is a listed building or within a Conservation Area	No	Unless it is a listed building or within a Conservation Area. Consult your local authority
	No	As long as the use of the house is not altered	Possibly	External walls are considered to be thermal elements, so approval is required if you want to re-render, replace cladding or insert insulation into a cavity wall. Consult your local authority

Type of work	Planning permission		Building Regulation approval	
		Yes		Yes
	If the alterations are major, such as removing or part removing a load-bearing wall or altering the drainage system		If the alterations are major such as removing or part removing a load-bearing wall or altering the drainage system	
Demolition	If it is a listed building or in a Conservation Area	Yes	Six weeks prior notice must be given to the local authority building control	Yes
	The council may wish to agree with you how you propose to carry out the demolition	Perhaps	For a partial demolition to ensure that the remaining part of the house (or adjoining buildings/ extensions) is structurally sound	Yes
Electrical work in the home or garden	Unless it is in a listed building or a Conservation Area	No	If it is installed by an approved person and complies with Part P and other relevant Building Regulations Approved Documents	No
Extensions to a ground floor flat	You must apply for planning permission to extend a flat	Yes	You may also need to consult the Fire Service	Yes
Flagpole	Flagpoles etc. erected in your garden are treated under the same rules as outbuildings, and cannot exceed 3 m in height	No		No
Flats over shops	Provided that: • the space is in the same class of use as the shop or office to start with (either Class A1 or A2); • the space is not in a separate planning unit from the shop; • you will not change the outside appearance of the building; • if there is any display window at ground floor level, you do not incorporate any of the ground floor into the flat	No		Yes

Type of work	Planning permission		Building Regulation approval	
Felling or lopping trees	No	Unless the trees are protected by a Tree Preservation Order or you live in a Conservation Area	No	
Garage – attached (new)	No	• As long as the floor area is between 15 m² and 30 m²	Yes	If it is attached to an existing home
Garage – detached (new)	No	• the building is not forward of the principal elevation; • the building is single storey; • the building is no taller than 2.5 m if it is within 2 m of the boundary; • the building eaves are no taller than 2.5 m and the overall height no more than 4 m; • there is no veranda, balcony or raised platform; • no more than half the land around the original house would be covered by other buildings	No	If the floor area is no more than 15 m² or the floor area is between 15 m² and 30 m² and is at least 1 m from any boundary
	Yes	If the building is listed or in a Conservation Area	No	If the floor area is no more than 15 m² or the floor area is between 15 m² and 30 m² and is at least 1 m from any boundary
Garage conversion	No	As long as the work is internal	Yes	In relation to the following: • infill garage door; • flooring; • walls below ground level; • ventilation; • doors and windows; • drainage; • electrics; • external walls; • internal walls; • roofs

Type of work	Planning permission		Building Regulation approval	
Hydroelectricity	Yes	Some form of environmental assessment will be essential for this type of project	Yes	Particularly for electrical installations
Heat pumps – ground source	No	Ground-source heat pups do not require planning permission unless it is a listed building or in a Conservation Area	Yes	It is advisable to contact an engineer who can provide the necessary advice
Heat pumps – air source	Yes	Air-source heat pumps currently require planning permission	Yes	It is advisable to contact an engineer who can provide the necessary advice
Infilling	Possibly	Consult your local planning officer	Yes	If a new development
Loft conversion in a top floor flat	No	Provided that it is internal works. Local requirements differ, so it is recommended that you check with the local planning authority	Yes	
	Yes	If you intend to extend or alter the roof space	Yes	
Micro combined heat and power	No	Unless the building is listed or within a Conservation Area or the work is external	Yes	Particularly to electrical installations, plumbing work and in relation to Approved Documents L1A, L1B, L2A and L2B
Oil storage tank	No	Provided that it is in the garden and has a capacity of not more than 3500 litres and no point is more than 3m high and no part projects beyond the foremost wall of the house facing the highway	No	If it is installed by an approved person and comply with Building Regulations
	Yes	Any container within the curtilage of a listed building will require planning permission		
Outbuildings	No	As long as: • the building is not forward of the principal elevation; • the building is single storey;	No	Unless the floor area exceeds 15 m^2 and contains sleeping accommodation

Type of work	Planning permission	Building Regulation approval
	• the building is no taller than 2.5 m if it is within 2 m of the boundary; • the building eaves are no taller than 2.5m; and the overall height no more than 4m; • there is no veranda, balcony or raised platform; • no more than half the land around the original house would be covered by other buildings	
	Yes — If the building is listed or in a Conservation Area	
Painting the exterior of a flat	No — Unless you live in an area where an Article 4 direction applies	No
Patio or driveway	No — As long as: • no significant embanking or terracing works are required; • it is not a listed building	No — As long as alterations do not make access to the dwelling any less satisfactory than it was before
Paving your front garden	No — If a replacement driveway of any size uses permeable surfacing or if the rainwater is directed to a lawn or border to drain naturally	No — As long as alterations do not make access to the dwelling any less satisfactory than it was before
	Yes — If the surface to be covered is more than 5 m² and you plan to lay a traditional impermeable driveway	No — As long as alterations do not make access to the dwelling any less satisfactory than it was before
Planting a hedge	No — Unless it obscures the view of traffic at a junction or access to a main road	No
Plumbing	No — Unless it is a listed building	No — If it is installed by an approved person and complies with Part J
		Yes — If you use an unregistered installer or DIY you will need to get approval from the local authority building control

Type of work	Planning permission		Building Regulation approval	
Replacing windows and doors	No	Unless it is a listed building or in a Conservation Area	Yes	If you use an unregistered installer or DIY you will need to get approval from the local authority building control
			No	If they are installed by an approved person (see Table 2.2) and comply with the requirements of Part F
	Yes	To replace shop windows	Yes	To replace shop windows
Satellite dishes, aerials and antennae on buildings up to 15m high	No	Unless: • there are more than two antennas on the property overall; • a single antenna is more than 100 cm in any linear dimension; • if installing two antennas, one is more than 100 cm and the other is more than 60 cm in any linear dimension; • the cubic capacity of each antenna is more than 30 litres; • the antenna is fitted to a chimney stack and is greater than 60 cm in any linear dimension; • the antenna sticks out more than 60 cm above the roof line or chimney stack, whichever is the lower	No	But make sure that the fixing point is stable and the installation is safe
	Possibly	If the building is in a designated area and the capacity of each antenna is greater than 35 litres or it is intended to install it on a chimney, wall or roof slope visible from a road or Broads waterway	No	

Type of work	Planning permission		Building Regulation approval
Satellite dishes, aerials and antennae on buildings over 15 m high	No	Unless: • there are more than four antennas on the property overall; • the size of any antenna is more than 130 cm in any linear dimension; • the cubic capacity of each antenna is more than 35 litres; • the antenna is fitted to a chimney stack and is greater than 60 cm in any linear dimension; • the antenna sticks out more than 300 cm above the highest part of the roof line; • if the building is in a designated area and it is intended to install the antenna on a chimney, wall or roof slope visible from a road or Broads waterway	No	But make sure that the fixing point is stable and the installation is safe
Satellite dishes on flats	Perhaps	Planning permission may be required in certain circumstances. Check with the local council	No	
Solar panels – roof and wall mounted	No	The fitting of solar panels is classed as permitted development in many cases; however, they may not extend above the ridgeline or project more than 200 m	Yes	The ability of the roof to carry the weight of the panels needs to be checked and electrical work needs to be carried out by an approved contractor
	Yes	Planning permission is required to fit solar panels to a listed building or in a Conservation Area	Yes	
Solar panels – stand alone	No	Provided that there is only one installation which is: • no higher than 4 m; • at least 5 m from boundaries; • no larger than 9 m^2 or 3 m wide and 3 m deep;	No	

Type of work	Planning permission	Building Regulation approval
	• not installed within boundary of a listed building; • not visible from the highway if it is in a Conservation Area or in a World Heritage Site	
Structural alterations inside	No — Unless it is a listed building or within a Conservation Area	Possibly — If it is a listed building or within a Conservation Area
	Yes — If the alterations are major, such as removing or part removing a load-bearing wall or altering the drainage system	Yes — If you wish to build or remove an internal wall or make openings in an internal wall
Subdividing a house or single flat	Perhaps	Yes — You may also need to consult the Fire Service and the property will need to be licensed under the Housing Act 2004
Swimming pool	Possibly — Consult your local planning officer	Yes — For an indoor pool
Windows in flats	Perhaps — To fit new double glazed windows	Yes
	No — If you are replacing like with like	Yes
Wind turbine	Yes — If erecting a wind turbine, either stand alone or attached to a dwelling (consult your local planning officer)	No — Unless you intend to fit a wind turbine to your dwelling
Working from home	No — Provided that the primary use of the property remains domestic	No — Unless building work is carried out

6

Meeting the requirements of the Building Regulations

How to use this chapter

This chapter has been designed to help you meet the requirements of the building regulations. We have adopted a foundations-up approach which is shown in Figures 6.1 and 6.2. In this chapter you will find separate sections providing all the requirements relating to specific projects such as cellars and roofs. Each section is then arranged to discuss the precise requirement and then how to meet the requirement. This last part of the section provides comprehensive details of **all** the Approved Documents relating to the area under discussion. Cross-sectional diagrams, tables and further amplification are also provided. The chapter is further supported by appendices, which cover specific areas in greater detail as entities in their own right rather than how they apply to a particular project (e.g. Appendix A – Access and Facilities for Disabled People). All the appendices are available on www.routledge.com/books/details/9780415809696/.

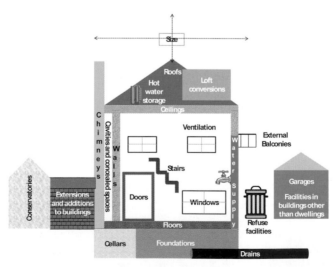

Figure 6.1 Diagrammatic representation of the contents of Chapter 6.

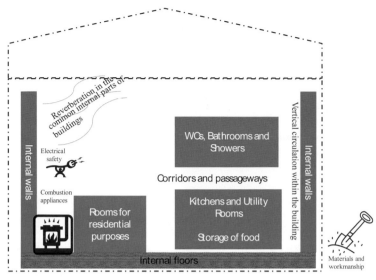

Figure 6.2 Diagrammatic representation of the contents of Chapter 6
– elements.

Owing to the huge amount of information contained in the Approved Document series, and in an effort to help you 'sort the wood from the trees', the following symbols have been used to help you:

In the margins you will find:

an important requirement or point

a good idea or suggestion

and within the text:

Notes are used to provide further amplification or information

Blue italic text is a quote from the relevant Act, Regulation or Approved Document

Background

Shaded boxes are used in this chapter to show either the full text of the Building Regulation's *legal requirements* or a paraphrased version of these requirements. The numbers in the right-hand column relate to the appropriate Approved Document and the paragraph number in the relevant guidance section of the Approved Document.

The latest edition of the Building Regulations is Building Regulations 2010 and the Building (Approved Inspectors etc.) Regulations 2010. These documents bring together all the amendments and Statutory Instruments which relate to the Building Regulations 2000 and the Building (Approved Inspectors etc.) Regulations 2000, respectively. The Building Regulations provide a practical guide to meeting the requirements of Schedule 1 and Regulation 7 of the Building Regulations 2010.

While there have been no substantive amendments to the requirements of either set of regulations, the definition of a *room for residential purposes* in Regulation 2 of the Building Regulations 2010 has been simplified from the original:

> *'Room for residential purposes' means a room, or a suite of rooms, which is not a dwelling-house or a flat and which is used by one or more persons to live and sleep and includes a room in a hostel, an hotel, a boarding house, a hall of residence or a residential home, but does not include a room in a hospital, or other similar establishment, used for patient accommodation.*

Please note there have been a number of changes to the numbering system relating to Building Regulations (e.g. Li (c) has become Regulation 40). These changes are discussed in detail in Chapter 2 and a full list of the changes is given in Appendix 2A at the end of Chapter 2.

Approved Documents

The documents listed in Table 6.1 have been approved and issued by the First Secretary of State for the purpose of providing practical guidance with respect to the requirements of the Building Regulations 2000 (as amended).

Compliance

There is no obligation to adopt any particular solution that is contained in an Approved Document. However, as the Approved Documents contain both the guidance on how to comply with the Regulation as well as extracts from the Regulations, you should be careful to ensure that you comply with the regulatory text. However, should a contravention of a requirement be alleged, if you have followed the guidance in the relevant Approved Documents that will be evidence tending to show that you have complied with the Regulations. If you have **not** followed the guidance, then that will be viewed as evidence tending to show that you have **not** complied with the requirements, and it will then be up to you, as the builder, architect and/or client, to demonstrate that you have satisfied the requirements of the Building Regulations.

Table 6.1 List of Approved Documents

Section	Title	Edition	Latest amendment
A	Structure	2004	
B	Fire safety	2006	2007
C	Site preparation and resistance to moisture	2004	
D	Toxic substances	1992	2002
E	Resistance to the passage of sound	2003	2004 and 2010
F	Ventilation	2010	
G	Sanitation, hot water safety and water efficiency	2010	
H	Drainage and waste disposal	2002	
J	Combustion appliances and fuel storage systems	2010	
K	Protection from falling, collision and impact	1998	2000
L	Conservation of fuel and power	2010	
M	Access to and use of buildings	2004	
N	Glazing – safety in relation to impact, opening and cleaning	1998	2000
P	Electrical safety	2006	
	Approved Document to support Regulation 7 – Materials and workmanship	1999	2000

Note: Electronic copies of Approved Documents may be downloaded from http://www.planningportal.gov.uk/buildingregulations/approveddocuments. Bound copies may be obtained from The Stationery Office (see http://www. thestationeryoffice.com).

Building Control Body (BCB) compliance

Fixed building services, including controls, should be commissioned by testing and adjusting as necessary to ensure that they use no more fuel and power than is reasonable in the circumstances.

If you are intending to carry out building work, it is best to **always** check with your Building Control Body (BCB) or one of their local authorities' Approved Inspectors first, so as to ensure that your/their proposals comply with Building Regulations!

Responsibility for compliance with Approved Documents

It is important to remember that, if you are the person (e.g. designer, builder, installer) carrying out building work to which any requirement of Building Regulations applies, then you are directly responsible for ensuring that the work complies with any such requirement.

The **building owner** also has a responsibility for ensuring compliance with Building Regulation requirements, and could be served with an enforcement notice in cases of non-compliance!

Materials and workmanship

As stated in the Building Regulations: *'Any building work which is subject to requirements imposed by Schedule 1 of the Building Regulations should, in accordance with Regulation 7, be carried out with proper materials and in a workmanlike manner'*.

In accordance with Regulation 8 of the Building Regulations, the requirements in Parts A–D, F–K and N and P (except for paragraphs G2, H2 and J7) of Schedule 1 to the Building Regulations do not require anything specific to be done except those things which are needed to secure reasonable standards of health and safety for persons in or about buildings (and any others who may be affected by buildings or matters connected with buildings).

* Paragraph G2 is excluded because it deals with water efficiency, and paragraphs H2 and J7 are excluded from Regulation 8 because they deal directly with the prevention of the contamination of water.
* Parts E and M (which deal, respectively, with resistance to the passage of sound, and access to and use of buildings) are excluded from Regulation 8 because they address the welfare and convenience of building users.
* Part L is excluded from Regulation 8 because it addresses the conservation of fuel and power.

What materials can I use?

Other than short-lived and unusable materials (see below), you may show that you have complied with Regulation 7 in a number of ways, such as by using:

* a product bearing CE marking in accordance with the Construction Products Directive (89/106/EEC), as amended by the CE Marking Directive (93/68/EEC), the Low Voltage Directive (2006/95/EC) and the EMC Directive (2004/108/EC);
* a product complying with an appropriate technical specification, such as:

    ○ a British Standard;
    ○ an alternative national technical specification of a Member State of the European Union or Turkey, or of another state signatory to the Agreement on the European Economic Area (EEA) that provides an equivalent level of safety and protection; or
    ○ a product covered by a national or European certificate issued by a European Technical Approval issuing body, provided the conditions of use are in accordance with the terms of the certificate.

 Note: Additional material to aid interpretation of the guidance is given in the Approved Documents (see http://www.planningportal.gov.uk/buildingregulations/approveddocuments).

Short-lived materials

Even if a plan for building work complies with the Building Regulations, if this work has been completed using short-lived materials (i.e. materials that are, in the absence of special care, liable to rapid deterioration) the local authority can:

*   reject the plans;
*   pass the plans subject to a limited use clause (on expiration of which they will have to be removed);
*   restrict the use of the building.

(Building Act 1984 Section 19)

Unsuitable materials

If, once building work has begun, it is discovered that it has been made using materials or components that have been identified by the Secretary of State (*or his nominated deputy*) as being unsuitable materials, the local authority has the power to:

*   reject the plans;
*   fix a period within which the offending work must be removed;
*   restrict the use of the building.

(Building Act 1984 Section 20)

If the person completing the building work fails to remove the unsuitable material or component(s), then that person is liable to be prosecuted and, on summary conviction, faces a heavy fine.

Technical specifications

Building Regulations may be made for specific purposes such as:

*   health and safety;
*   welfare and convenience of disabled people;
*   conservation of fuel and power;
*   prevention of waste or contamination of water.

These are aimed at furthering the protection of the environment, facilitating sustainable development, or the prevention and detection of crime.

Although the main requirements for health and safety are now covered by the Building Regulations, there are still some requirements contained in the Workplace (Health, Safety and Welfare) Regulations 1992 that may need to be considered, as they could contain requirements which affect building design. For further information see *Workplace (Health, Safety and Welfare) Regulations 1992. Approved Code of Practice L24*, published by HSE Books, 1992 (ISBN 0 7176 0413 6).

Standards and technical approvals, as well as providing guidance, also address other aspects of performance, such as serviceability and/or other characteristics related to health and safety not covered by the Building Regulations.

When an Approved Document makes reference to a named standard, the relevant version of the standard is the one listed at the end of that particular Approved Document. However, if this version of the standard has been revised or updated by the issuing standards body, the new version may be used as a source of guidance, provided it continues to address the relevant requirements of the Building Regulations.

The Secretary of State has agreed with the British Board of Agrément on aspects of performance that they need to assess in preparing their Agrément Certificates in order that the Board may demonstrate compliance of a product or system with the requirements of the Building Regulations. An Agrément Certificate issued by the Board under these arrangements will give assurance that the product or system to which the certificate relates (if properly used in accordance with the terms of the certificate) will meet the relevant requirements.

Independent certification schemes

Within the UK there are many product certification schemes that certify compliance with the requirements of a recognized standard or document that is suitable for the purpose and material being used. Certification bodies that approve such schemes will normally be accredited by The United Kingdom Accreditation Service (UCAS).

Standards and technical approvals

Standards and technical approvals provide guidance related to the Building Regulations and address other aspects of performance, such as serviceability, or aspects which, although they relate to health and safety, are not covered by the Regulations.

European pre-standards (ENV)

The British Standards Institution (BSI) will be issuing Pre-standard (ENV) Structural Eurocodes as they become available from the European Standards Organization, Comité Européen de Normalisation Electrotechnique (CEN).

For example, DD ENV 1992-1-1:1992 Eurocode 2: Part 1 and DD ENV 1993-1-1:1992 Eurocode 3: Part 1-1 *General Rules* and *Rules for Buildings in Concrete and Steel* were thoroughly examined over a period of several years and are now considered to provide appropriate guidance when used in conjunction with their national application documents for the design of concrete and steel buildings, respectively.

When other ENV Eurocodes have been subjected to a similar level of examination they may also offer an alternative approach to Building Regulation compliance and, when they are eventually converted into fully approved EN standards, they will be included as referenced standards in the Approved Document(s) concerned.

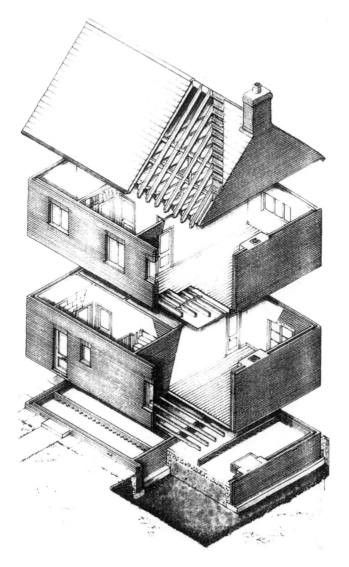

Figure 6.3 Brick-built house – typical components.

 Note: If a national standard is going to be replaced by a European harmonized standard, there will be a coexistence period during which either standard may be referred to. At the end of the coexistence period the national standard will be withdrawn.

House – construction

There are two main types of building in common use today: those made of brick and those made of timber. There are many different styles of brick-built houses and, equally, there are various methods of construction.

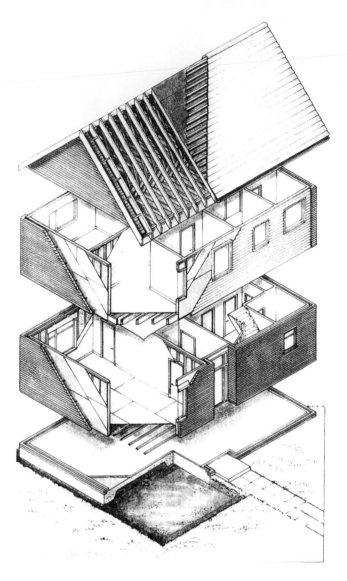

Figure 6.4 Timber-framed house – typical components.

Brickwork, as well as giving a building character, provides the main load-bearing element of a brick-built house. Timber-framed houses, on the other hand, are usually built on a concrete foundation with a 'strip' or 'raft' construction to spread the weight, and differ from their brick-built counterparts in that the main structural elements are timber frames. (See Figures 6.3 and 6.4.)

6.1 Foundations

To support the weight of the structure, most brick-built buildings are supported on a solid base called 'foundations'. Timber-framed houses are usually built on a concrete foundation, with a 'strip' or 'raft' construction to spread the weight.

6.1.1 Requirements

The building shall be constructed so that:

- *the combined dead, imposed and wind loads are sustained and transmitted by it to the ground, safely and without causing any building deflection/ deformation or ground movement that will affect the stability of any part of the building;*
- *ground movement caused by swelling, shrinkage or freezing of the subsoil, land-slip or subsidence will not affect the stability of any part of the building.*

(Approved Document A)

(1)   The ground to be covered by the building shall be reasonably free from any material that might damage the building or affect its stability, including vegetable matter, topsoil and pre-existing foundations.

(2)   Reasonable precautions shall be taken to avoid danger to health and safety caused by contaminants on or in the ground covered, or to be covered, by the building and any land associated with the building.

(3)   Adequate subsoil drainage shall be provided if it is needed to avoid:

- *(a)   the passage of ground moisture to the interior of the building;*
- *(b)   damage to the building, including damage through the transport of water-borne contaminants to the foundations of the building.*

(Approved Document C1)

 Note: For the purpose of this requirement, *contaminant* means any substance which is or may become harmful to persons or buildings, including substances that are corrosive, explosive, flammable, radioactive or toxic.

Potential problems

There may be known and/or recorded conditions of ground instability, such as geological faults, landslides, disused mines, and unstable strata of similar nature, which affect or may potentially affect a building site or its environs.

There may also be:

- unsuitable material, including vegetable matter, topsoil and pre-existing foundations;
- contaminants on or in the ground covered, or to be covered, by the building and any land associated with the building;
- groundwater.

These conditions should be taken into account before proceeding with the design of a building or its foundations.

What about hazards?

Hazards associated with the ground may include:

- chemical and biological contaminants;
- gas generation from biodegradation of organic matter;
- naturally occurring radioactive radon gas and gases produced by some soils and minerals;
- physical, chemical or biological hazards;
- underground storage tanks or foundations;
- unstable fill or unsuitable hardcore containing sulphate;
- the effects of vegetable matter, including tree roots.

In the most hazardous conditions, only the total removal of contaminants from the ground to be covered by the building can provide a complete remedy. In other cases, remedial measures can reduce the risks to acceptable levels. These measures should only be undertaken with the benefit of expert advice and where the removal would involve handling large quantities of contaminated materials – then you are advised to seek expert advice.

Even when these actions have been completed successfully, the ground to be covered by the building will **still** need to have at least 100 mm of concrete laid over it!

What about contaminated ground?

Potential building sites which are likely to contain contaminants can be identified at an early stage from planning records or from local knowledge (e.g. previous uses). In addition to solid and liquid contaminants, problems can also arise from natural contamination such as methane and the radioactive radon gas (and its decay product).

The following list contains examples of sites that are most likely to contain contaminants:

- asbestos works;
- ceramics, cement and asphalt manufacturing works;
- chemical works;
- dockyards and dockland;
- engineering works (including aircraft manufacturing, railway engineering works, shipyards, electrical and electronic equipment manufacturing works);
- gas works, coal carbonization plants and ancillary by-product works;
- industries making or using wood preservatives;
- landfill and other waste disposal sites;
- metal mines, smelters, foundries, steelworks and metal finishing works;
- munitions production and testing sites;
- oil storage and distribution sites;

- paper and printing works;
- power stations;
- railway land, especially larger sidings and depots;
- road vehicle fuelling, service and repair: garages and filling stations;
- scrap yards;
- sewage works, sewage farms and sludge disposal sites;
- tanneries;
- textile works and dye works.

If any signs of possible contaminants are present, the local authority's Environmental Health Officer should be told at once. If the officer confirms the presence of any of these contaminants (see Table 6.2) then he or she will require their removal or some prescribed action to be completed before any planning permission for building work can be sought.

What about gaseous contaminants?

Radon is a naturally occurring radioactive colourless and odourless gas which is formed in small quantities by radioactive decay wherever uranium and radium are found. It can move through the subsoil and then into buildings, and exposure to high levels over long periods increases the risk of developing lung cancer. Some parts of the country (in particular the West Country) have higher natural levels than elsewhere, and precautions against radon may be necessary.

 Note: Guidance on the construction of dwellings in areas susceptible to radon has been published by the Building Research Establishment in the report *Radon: Guidance on Protective Measures for New Dwellings.*

Landfill gas is generated by the action of anaerobic micro-organisms on biodegradable material in landfill sites, and generally consists of methane and carbon dioxide together with small quantities of VOCs (Volatile Organic Compounds), which give the gas its characteristic odour. It can migrate under pressure through the subsoil and through cracks and fissures into buildings.

Methane and carbon dioxide can also be produced by organically rich soils and sediments such as peat and river silts, and a wide range of VOCs can be present as a result of petrol, oil and solvent spillages.

Site preparation

Site investigation is now the recommended method for determining how much unsuitable material should be removed before commencing building work, and this will normally consist of a number of well-defined stages, for example:

| **Planning stage** | scope and requirements |
| **Desktop study** | historical, geological and environmental information about the site |

Site reconnaissance or walkover survey	identification of actual and potential physical hazards and the design of the main investigation
Main investigation and reporting	intrusive and non-intrusive sampling and testing to provide soil parameters

Risk assessment

The site investigation may identify certain risks, which will require a risk assessment. There are three types of assessment:

- preliminary (once the need for a risk assessment has been identified, and depending on the situation and the outcome);
- Generic Quantitative Risk Assessment (GQRA);
- Detailed Quantitative Risk Assessment (DQRA).

Each risk assessment should include:

Hazard identification	Developing the conceptual model by establishing contaminant sources, pathways and receptors. (This is the preliminary site assessment which consists of a desk study and a site walkover in order to gather sufficient information to obtain an initial understanding of the potential risks. An initial conceptual model for the site can then be based on this information.)
Hazard assessment	Identifying what pollutant linkages may be present and analysing the potential for unacceptable risks.
Risk estimation	Establishing the scale of the possible consequences by considering the degree of harm that may result and to which receptors.
Risk evaluation	Deciding whether the risks are acceptable or unacceptable – review all site data to decide whether estimated risks are unacceptable.

6.1.2 Meeting the requirement

General

Where the site is potentially affected by contaminants, a combined geotechnical and geo-environmental investigation should be considered.	C1.3

Hazard identification and assessment

A preliminary site assessment is required to provide information on the past and present uses of the site and surrounding area that may give rise to contamination (see Table 6.2).

C2.10

Table 6.2 Examples of possible contaminants

Signs of possible contaminants	Possible contaminant
Vegetation (absence, poor or unnatural growth)	Metals Metal compounds Organic compounds Gases (landfill or natural source)
Surface materials (unusual colours and contours may indicate wastes and residues)	Metals Metal compounds Oily and tarry wastes Asbestos Other mineral fibres Organic compounds, including phenols Combustible material, including coal and coke dust Refuse and waste
Fumes and odours (may indicate organic chemicals)	Volatile organic and/or sulphurous compounds from landfill or petrol/solvent spillage Corrosive liquids Faecal animal and vegetable matter (biologically active)
Damage to exposed foundations of existing buildings	Sulphates
Drums and containers (empty or full)	Various

The site assessment and risk evaluation should pay particular attention to the area of the site subject to building operations.

C2.11

The planning authority should be informed prior to any intrusive investigations or if any substance is found which was not identified in a preliminary statement about the nature of the site.

C2.12

Risks to buildings, building materials and services

The following hazards shall be considered:

- aggressive substances – including inorganic and   C2.23a
  organic acids, alkalis, organic solvents and inorganic
  chemicals such as sulphates and chlorides;
- combustible fill – including domestic waste, colliery   C2.23b
  spoil, coal, plastics, petrol-soaked ground, etc.
- expansive slags – e.g. blast furnace and steel-making   C2.23c
  slag;
- floodwater affected by contaminants – substances in   C2.23d
  the ground, waste matter or sewage.

Contaminated ground

The underlying geology of a potential site has to be   C2.3 and 2.4
considered, as natural contaminants may be present.
For example:

- naturally occurring heavy metals (e.g. cadmium and
  arsenic) originating in mining areas;
- gases (e.g. methane and carbon dioxide) originating
  in coal mining areas;
- organic-rich soils and sediments such as peat and
  river silts;
- radioactive radon gas – which can also be a problem
  in certain parts of the country.

Possible sulphate attack from some strata on   C2.5
concrete floor slabs and oversite concrete needs to be
considered.

Gaseous contaminants

Radon

All new buildings, extensions and conversions (whether   C2.39
residential or non-domestic), which are built in areas
where there may be high radon emissions, may need to
incorporate precautions against radon.

Landfill gas

Methane is an asphyxiant, will burn, and can explode in air.   C2
Carbon dioxide is non-flammable and toxic. Many of the
other components of landfill gas are flammable and some are
toxic. All will require careful analysis.

Risk assessment

A risk assessment should be completed for methane and other gases
particularly:

- on a landfill site or within 250 m of the boundary of a   C2.28a
  landfill site;
- on a site subject to the wide-scale deposition of biode-   C2.28b
  gradable substances (including made ground or fill);
- on a site that has been subject to a use that could give   C2.28c
  rise to petrol, oil or solvent spillages;
- in an area subject to naturally occurring methane,   C2.28d
  carbon dioxide and other hazardous gases (e.g.
  hydrogen sulphide).

During a site investigation for methane and other gases:

- measurements should be taken over a sufficiently   C2.30
  long period of time in order to characterize gas emissions fully;
- measurements should include periods when gas emis-   C2.30
  sions are likely to be higher, e.g. during periods of
  falling atmospheric pressure.

Gas risks (i.e. to human receptors) should be considered for:

- gas entering the dwelling through the substructure   C2.32
  (and building up to hazardous levels);
- subsequent householder exposure in garden areas   C2.32
  including outbuildings (e.g. garden sheds and greenhouses), extensions and garden features (e.g. ponds).

When land that is affected by contaminants is being
developed, 'receptors' (i.e. buildings, building materials
and building services, as well as people) are introduced
onto the site and it is necessary to break the pollutant
linkages. This can be achieved by:

- treating the contaminant (e.g. use of physical, chemical or biological processes to eliminate or reduce the contaminant's toxicity or harmful properties);   C2.7
- blocking or removing the pathway (e.g. isolating the contaminant beneath protective layers or installing barriers to prevent migration);
- protecting or removing the receptor (e.g. changing the form or layout of the development, using appropriately designed building materials, etc.).

A risk assessment based on the concept of a 'source–pathway–receptor' relationship, or pollutant linkage of a potential site (see Figure 6.5) should be carried out to ensure the safe development of land that is affected by contaminants.   C2.6

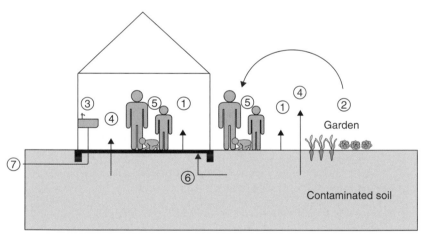

Possible pathways

Ingestion: of contaminants in soil/dust   ①
of contaminants in food   ②
of contaminants in water   ③

Inhalation: of contaminants in soil particles/dust/vapours   ④

Direct contact: with contaminants in soil/dust or water   ⑤

Attack on building structures   ⑥

Attack on services   ⑦

Figure 6.5 Conceptual model of a site showing a source–pathway–receptor relationship.

Risk estimation and evaluation

The detailed ground investigation: C2.13

• must provide sufficient information for the confirmation
  of a conceptual model for the site, the risk assessment
  and the design and specification of any remedial works;
• is likely to involve collection and analysis of soil, soil
  gas, surface and groundwater samples by the use of inva-
  sive and/or non-invasive techniques.

During the development of land affected by contaminants   C2.14
the health and safety of both the public and workers should
be considered.

Remedial measures

If the risks posed by the gas are unacceptable then these   C2.36
need to be managed through appropriate building remedial
measures.

Site-wide gas control measures may be required if the   C2.36
risks on any land associated with the building are deemed
unacceptable.

Consideration should be given to the design and layout of   C2.37
buildings to maximize the driving forces of natural ventilation.

For non-domestic buildings, expert advice concerning gas   C2.38
control measures should be sought as the floor area of such
buildings can be large and it is important to ensure that gas is
adequately dispersed from beneath the floor.

There is a need for continued maintenance and calibration of   C2.38
mechanical (as opposed to passive) gas control systems.

Sub-floor ventilation systems should be carefully designed to   C2.38
ensure adequate performance and should not be modified unless
subjected to a specialist review of the design.

Corrective measures

When building work is undertaken on sites affected by   C2.15
contaminants where control measures are already in place,
care must be taken not to compromise these measures.

Depending on the contaminant, three generic types of corrective measure can be considered: treatment, containment and removal.

 Note: The containment or treatment of waste may require a waste management licence from the Environment Agency.

Treatment

The choice of the most appropriate treatment process for a particular site is a highly site-specific decision for which specialist advice should be sought.	C2.16

Containment

In-ground vertical barriers may also be required to control lateral migration of contaminants.	C2.17
Cover systems involve the placement of one or more layers of materials over the site and may be used to: <ul><li>break the pollutant linkage between receptors and contaminants;</li><li>sustain vegetation;</li><li>improve geotechnical properties;</li><li>reduce exposure to an acceptable level.</li></ul>	C2.18
Imported fill and soil for cover systems should be assessed at source to ensure that it is not contaminated.	C2.20
The size and design of cover systems (particularly soil-based ones used for gardens) should take account of their long-term performance.	C2.20
Gradual intermixing due to natural effects and activities such as burrowing animals, gardening, etc., needs to be considered.	C2.20

Removal

Imported fill should be assessed at source to ensure that there are no materials that will pose unacceptable risks to potential receptors.	C2.21

Site preparation

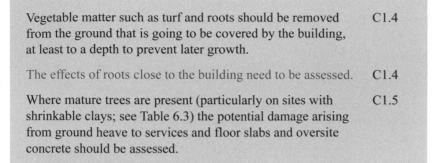

Vegetable matter such as turf and roots should be removed from the ground that is going to be covered by the building, at least to a depth to prevent later growth.   Cl.4

The effects of roots close to the building need to be assessed.   Cl.4

Where mature trees are present (particularly on sites with shrinkable clays; see Table 6.3) the potential damage arising from ground heave to services and floor slabs and oversite concrete should be assessed.   Cl.5

Table 6.3 Volume change potential for some common clays

Clay type	Volume change potential
Glacial till	Low
London	High to very high
Oxford and Kimmeridge	High
Lower lias	Medium
Gault	High to very high
Weald	High
Mercian mudstone	Low to medium

Building services such as below-ground drainage should be sufficiently robust or flexible to accommodate the presence of any tree roots.   Cl.6

Joints should be made so that roots will not penetrate them.   Cl.6

Where roots could pose a hazard to building services, consideration should be given to their removal.   Cl.6

On sites previously used for buildings, consideration should be given to the presence of other infrastructure (such as existing foundations, services and buried tanks, etc.) that could endanger persons in and about the building and any land associated with the building.   Cl.7

If the site contains fill or made ground, consideration should be given to its compressibility and its potential to collapse when wet.   Cl.8

Foundations

Table 6.4 provides guidance on determining the type of soil on which it is intended to lay a foundation.

Table 6.4 Types of subsoil

Type	Applicable field test
Rock (being stronger/denser than sandstone, limestone or firm chalk)	Requires at least a pneumatic or other mechanically operated pick for excavation
Compact gravel and/or sand	Requires a pick for excavation. Wooden peg 50 mm square in cross-section hard to drive beyond 150 mm
Stiff clay or sandy clay	Cannot be moulded with the fingers and requires a pick or pneumatic or other mechanically operated spade for its removal
Firm clay or sandy clay	Can be moulded by substantial pressure with the fingers and can be excavated with a spade
Loose sand, silty sand or clayey sand	Can be excavated with a spade. Wooden peg 50 mm square in cross-section can be easily driven
Soft silt, clay, sandy clay or silty clay	Fairly easily moulded in the fingers and readily excavated
Very soft silt, clay, sandy clay or silty clay	Natural sample in winter conditions exudes between the fingers when squeezed in fist

Subsoil drainage

Where the water table can rise to within 0.25 m of the lowest   C3.2
floor of the building, or where surface water could enter or
adversely affect the building, either the ground to be covered
by the building should be drained by gravity, or other effective
means of safeguarding the building should be taken.

If an active subsoil drain is cut during excavation and if it   C3.3
passes under the building it should be either:

- re-laid in pipes with sealed joints and have access points outside the building;
- re-routed around the building; or
- re-run to another outfall (see Figure 6.6).

Where contaminants are present in the ground, consideration   C3.7
should be given to subsoil drainage to prevent the
transportation of water-borne contaminants to the foundations
or into the building or its services.

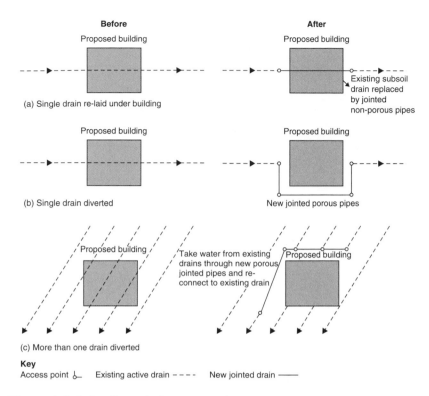

Before
After

Proposed building   Proposed building

(a) Single drain re-laid under building

Existing subsoil
drain replaced
by jointed
non-porous pipes

Proposed building   Proposed building

(b) Single drain diverted   New jointed porous pipes

Proposed building   Take water from existing   Proposed building
drains through new porous
jointed pipes and re-
connect to existing drain

(c) More than one drain diverted

Key
Access point ⌐   Existing active drain – – – –   New jointed drain ——

Figure 6.6 Subsoil cut during excavation.

Ground movement

Known or recorded conditions of ground instability, such as   A1/2 1.9
that arising from landslides, disused mines or unstable strata
should be taken into account in the design of the building and
its foundations.

Foundations – plain concrete

There should **not** be:

- non-engineered fill (see BRE Digest 427) or a wide   A1/2 2E1a
  variation in ground conditions within the loaded area;
- weaker or more compressible ground at such a depth   A1/2 2E1b
  below the foundation as could impair the stability of
  the structure.

The foundations should be situated centrally under the wall.	A1/2 2E2a
In non-aggressive soils, concrete should be composed of Portland cement to BS EN 197-1 and 197-2 and fine and coarse aggregate conforming to BS EN 12620 and the mix should either be:	A1/2 2E2b

- 50 kg of Portland cement to not more than 200 kg (0.1 m³) of fine aggregate and 400 kg (0.2 m³) of coarse aggregate; or
- Grade ST2 or Grade GEN I concrete to BS 8500-2.

For foundations in chemically aggressive soil conditions guidance in BS 8500-1:Part 1 and BRE Special Digest 1 should be followed.	(8.18)
The minimum thickness T of concrete foundation should be 150 mm or P, whichever is the greater, where P is derived using Table 6.4 and Figure 6.7.	A1/2 2E2c
	(8.18a)
Foundations stepped on elevation should overlap by twice the height of the step, by the thickness of the foundation, or 300 mm, whichever is greater (see Figure 6.8).	A1/2 2E2d

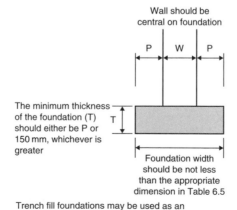

Wall should be central on foundation

P   W   P

The minimum thickness of the foundation (T) should either be P or 150 mm, whichever is greater

T

Foundation width should be not less than the appropriate dimension in Table 6.5

Trench fill foundations may be used as an alternative to strip foundations.

Figure 6.7 Foundation dimensions.

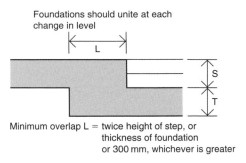

Foundations should unite at each change in level

Minimum overlap L = twice height of step, or
thickness of foundation
or 300 mm, whichever is greater

S should not be greater than T

Figure 6.8 Elevation of stepped foundation.

The overlap for trench fill foundations should be twice the height of the step or 1 m, whichever is greater.	A1/2 2E2d
Trench fill foundations may be used as an acceptable alternative to strip foundations.	A1/2 2E2c

Note: Steps in foundations should not be of greater height than the thickness of the foundation (see Figure 6.8).

Foundations for piers, buttresses and chimneys should project as shown in Figure 6.9.	A1/2 2E2f

The projection X should never be less than the value of P where there is no local thickening of the wall.

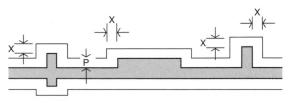

Projection X should not be less than P

Figure 6.9 Piers and chimneys.

Strip foundations

> The recommended minimum widths of strip foundations   A1/2 2E3
> shall be as indicated in Table 6.5.

Table 6.5 Minimum width of strip footings

Type of ground (including engineered fill)	Condition of ground	Field test applicable	Total load of load-bearing walling not more than (kN/linear metre)					
			20	30	40	50	60	70
			Minimum width of strip foundation (mm)					
I Rock	Not inferior to sandstone, limestone or firm chalk	Requires at least a pneumatic or other mechanically operated pick for excavation	In each case equal to the width of wall					
II Gravel or sand	Medium dense	Requires pick for excavation. Wooden peg 50 mm square in cross-section hard to drive beyond 150 mm	250	300	400	500	600	650
III Clay Sandy clay	Stiff Stiff	Can be indented slightly by thumb	250	300	400	500	600	650
IV Clay Sandy clay	Firm Firm	Thumb makes impression easily	300	350	450	600	750	850
V Sand Silty sand Clayey sand	Loose Loose Loose	Can be excavated with a spade. Wooden peg 50 mm square in cross-section can be easily driven	400	600	Note: Foundations on soil types V and VI do not fall within the provisions of this section if the total load exceeds 30 kN/m			
VI Silt Clay Sandy clay Clay or silt	Soft Soft Soft Soft	Finger pushed in up to 10 mm	450	650				
VII Silt Clay Sandy clay Clay or silt	Very soft Very soft Very soft Very soft	Finger easily pushed in up to 25 mm	Refer to specialist advice					

> Where strip foundations are founded on rock, the strip   A1/2 2E4
> foundations should have a minimum depth of 0.45 m
> to their underside to avoid the action of frost.

Note: This depth, however, will commonly need to be increased in areas subject to long periods of frost or in order to transfer the loading onto satisfactory ground.

> In clay soils subject to volume change on drying (i.e.   A1/2 2E4
> 'shrinkable clays' with a plasticity index greater than
> or equal to 10%), strip foundations should be taken to a
> depth where anticipated ground movements (caused by
> vegetation and trees on the ground) will not impair the
> stability of any part of the building.
>
> The depth to the underside of foundations on clay soils   A1/2 2E4
> should not be less than 0.75 m.

Disproportionate collapse

All buildings should be built so that their sensitivity to disproportionate collapse in the event of an accident is reduced.

> Buildings shall remain sufficiently robust to sustain a   A1/2 5.1
> limited extent of damage or failure, depending on the class
> of the building, without collapse (see below).

 Note:
(1) Buildings intended for more than one type of use should adopt the most onerous class.
(2) In determining the number of storeys in a building, basement storeys may be excluded provided that they meet the robustness requirements of Class 2B buildings.

Class 1

Building type and occupancy	Requirements
• Houses not exceeding four storeys. • Agricultural buildings. • Buildings into which people rarely go, provided no part of the building is closer to another building (or area where people go) than 1.5 times the building height.	Provided the building has been designed and constructed in accordance with Building Regulations and is in normal use, no additional measures are likely to be necessary.

Class 2A

Building type and occupancy	Requirements
• Five-storey single occupancy houses. • Hotels not exceeding four storeys. • Flats, apartments and other residential buildings not exceeding four storeys. • Offices not exceeding four storeys. • Industrial buildings not exceeding three storeys. • Retailing premises not exceeding three storeys of less than $2000\,m^2$ floor area in each storey. • Single-storey educational buildings. • All buildings not exceeding two storeys to which members of the public are admitted and which contain floor areas not exceeding $2000\,m^2$ at each storey.	Effective horizontal ties (or effective anchorage of suspended floors to walls) are required.

Class 2B

Building type and occupancy	Requirements
• Hotels, flats, apartments and other residential buildings greater than four storeys but not exceeding 15 storeys. • Educational buildings greater than one storey but not exceeding 15 storeys.	Effective horizontal ties need to be provided.
• Retailing premises greater than three storeys but not exceeding 15 storeys. • Hospitals not exceeding three storeys. • Offices greater than four storeys but not exceeding 15 storeys. • All buildings to which members of the public are admitted which contain floor areas exceeding $2000\,m^2$ but less than $5000\,m^2$ at each storey. • Car parking not exceeding six storeys.	Effective vertical ties need to be provided in all supporting columns and walls.

Or, alternatively, check that upon the notional removal of each supporting column and each beam supporting one or more columns, or any nominal length of

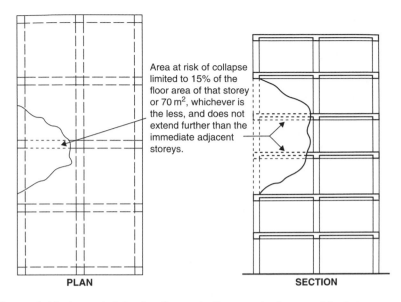

Figure 6.10 Area at risk of collapse in the event of an accident.

load-bearing wall (one at a time in each storey of the building) that the building remains stable and the area of floor at any storey at risk of collapse does not exceed 15% of the floor area of that storey or $70\,m^2$, whichever is smaller, and does not extend further than the immediate adjacent storeys (see Figure 6.10).

Where the notional removal of such columns and lengths of walls would result in damage in excess of the above limit, then such elements should be designed as a 'key element' (i.e. it should be capable of sustaining an accidental design loading of $34\,kN/m^2$) applied in the horizontal and vertical directions (in one direction at a time) to the member and any attached components (e.g. cladding, etc.).

Class 3

Building type and occupancy	Requirements
• All buildings defined above as Class 2A and 2B that exceed the limits on area and/or number of storeys.	A systematic risk assessment of the building should be undertaken taking into account all the normal hazards that may reasonably be foreseen, together with any abnormal hazards.

- Grandstands accommo-
  dating more than 5000
  spectators.
- Buildings containing
  hazardous substances
  and/or processes.

Critical situations for design should be
selected that reflect the conditions that
can reasonably be foreseen as possible
during the life of the building.

Protective measures should be
chosen and the detailed design of the
structure and its elements undertaken
in accordance with the following
recommendations:

- BS 5628: Part 1 – Structural use of
  unreinforced masonry
- BS 5950: Part 1 – Structural use of
  steelwork in building
- BS 8110: Parts 1 and 2 – Structural
  use of plain, reinforced and pre-
  stressed concrete.

For any building which does not fall into one of the classes listed above, or
where the consequences of collapse may warrant particular examination of the
risks involved, see one of the following reports, both of which are available on
the planning portal:

- *Guidance on Robustness and Provision against Accidental Actions*, dated
  July 1999, together with the accompanying BRE Report No. 200682 (http://
  www.planningportal.gov.uk/uploads/odpm/4000000002020.pdf);
- *Proposed Revised Guidance on Meeting Compliance with the Requirements of Building Regulation A3: Revision of Allot and Lomax Proposal*
  (http://www.planningportal.gov.uk/uploads/odpm/4000000002019.pdf).

Maximum floor area

No floor enclosed by structural walls on all sides shall   A1/2 (2C14)
exceed 70 m² (see Figure 6.11).

No floor with a structural wall on one side shall exceed   A1/2 (2C14)
36 m² (see Figure 6.11).

Maximum height of buildings

The maximum height of a building shall not exceed the   A1/2 (1C17)
heights given in Table 6.6 with regard to the relevant
wind speed.

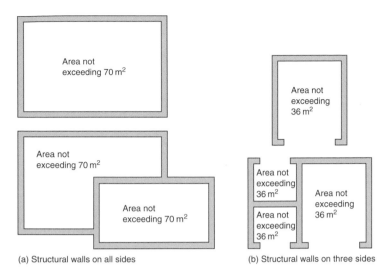

(a) Structural walls on all sides (b) Structural walls on three sides

Figure 6.11 Maximum floor area that is enclosed by structural walls.

Table 6.6 Maximum allowable height of building

Factor S	Country sites			Town sites		
	Distance to the coast			Distance to the coast		
	<10 km	15–50 km	>50 km	<10 km	15–50 km	>50 km
24	15	15	15	15	15	15
25	11.5	14.5	15	15	15	15
26	8	10.5	13	15	15	15
27	6	8.5	10	15	15	15
28	4.5	6.5	8	13.5	15	15
29	3.5	5	6	11	13	14.5
30	3	4	5	9	11	12.5
31		3.5	4	8	9.5	10.5
32		3	3.5	7	8.5	9.5
33			3	6	7.5	8.5
34				5	7	8
35				4	6	7
36				3	5.5	6
37					4.5	5.5
38					4	5
39					3	4
40						3

Heights of walls and storeys

> The measured height of a wall or a storey should be in   A1/2 2C18
> accordance with Figure 6.12.

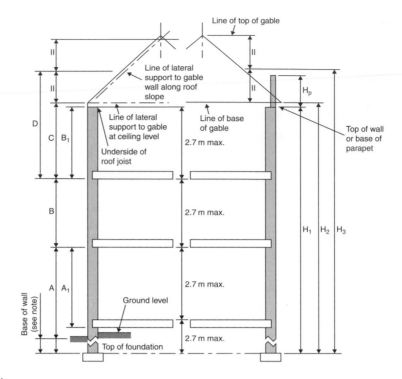

Key

(a) **Measuring Storey Heights**

A_1 is the ground storey height if the ground floor provides effective lateral support to the wall i.e. is adequately tied to the wall or is a suspended floor bearing on the wall.

A   is the ground storey height if the ground floor does not provide effective lateral support to the wall.

Note: If the wall is supported adequately and permanently on both sides by suitable compact material, the base of the wall for the purposes of the storey height may be taken as the lower level of this support. (Not greater than 3.7 m ground storey height.)

B   is the intermediate storey height.

B_1 is the top storey height for walls which do not include a gable.

C   is the top storey height where lateral support is given to the gable at both ceiling level and along the roof slope.

D   is the top storey height for the external walls which include a gable where lateral support is given to the gable only along the roof slope.

(b) **Measuring Wall Heights**

H_1 is the height of an external wall that does not include a gable.

H_2 is the height of an internal or separating wall which is built up to the underside of the roof.

H_3 is the height of an external wall which includes a gable.

H_p is the height of a parapet. If Hp is more than 1.2 m add H_p to H_1.

Figure 6.12 Method for measuring the heights of storeys and walls.

Imposed loads on roofs, floors and ceilings

> The imposed loads on roofs, floors and ceilings shall not exceed those shown in Table 6.7.   A1/2 (2C15)

Table 6.7 Imposed loads

Element	Distributed loads	Concentrated load
Roofs	$1.00\,kN/m^2$ for spans not exceeding 12 m $1.50\,kN/m^2$ for spans not exceeding 6 m	
Floors	$2.00\,kN/m^2$	
Ceilings	$0.25\,kN/m^2$	$0.9\,kN/m^2$

Structural safety

The safety of a structure depends on the successful combination of design and completed construction, particularly:

- the design – which should also: A1/2 0.2a

  - be based on identification of the hazards (to which the structure is likely to be subjected) and an assessment of the risks;
  - reflect conditions that can reasonably be foreseen during future use;

- loading – dead load, imposed load and wind load;   A1/2 0.2b
- the properties of materials used;   A1/2 0.2c
- the detailed design and assembly of the structure;   A1/2 0.2d
- safety factors;   A1/2 0.2e
- workmanship.   A1/2 0.2f

Basic requirements for stability

Adequate provision shall be made to ensure that the building is stable under the likely imposed and wind loading conditions.   A1/2 1A2

The overall size and proportioning of the building shall be limited according to the specific guidance for each form of construction.   A1/2 1A2a

The layout of walls (both internal and external) forming a robust three-dimensional box structure in plan shall be constructed according to the specific guidance for each form of construction.   A1/2 1A2b

The internal and external walls shall be adequately connected by either masonry bonding or by using mechanical connections.   A1/2 1A2c

> The intermediate floors and roof shall be constructed so   A1/2 1A2d
> that they:
>
> • provide local support to the walls;
> • act as horizontal diaphragms capable of transferring
>   the wind forces to buttressing elements of the building.

 Note: A traditional cut timber roof (i.e. using rafters, purlins and ceiling joists) generally has sufficient built-in resistance to instability and wind forces (e.g. from either hipped ends, tiling battens, rigid sarking, or the like). However, the need for diagonal rafter bracing equivalent to that recommended in BS 5268: Part 3:1998 or Annex H of BS 8103: Part 3:1996 for trussed rafter roofs, should be considered especially for single-hipped and non-hipped roofs of greater than 40° pitch to detached houses.

6.2 Buildings – size

6.2.1 Classification of purpose groups

Many of the provisions in Approved Documents are related to the use of the building. The classifications 'use' are termed purpose groups and represent different levels of hazard. They can apply to a whole building, or (where a building is compartmented) to a compartment in the building and the relevant purpose group should be taken from the main use of the building or compartment. Table 6.8 sets out the purpose group classification.

Table 6.8 Classification of purpose groups

Title	Group	Purpose for which the building or compartment of a building is intended to be used
Residential[(1)] (dwellings)	1(a)	Flat or maisonette.
	1(b)	Dwelling house which contains a habitable storey with a floor level which is more than 4.5 m above ground level.
	1(c)	Dwelling house which does not contain a habitable storey with a floor level which is more than 4.5 m above ground level.
Residential (institutional)	2(a)	Hospital, home, school or other similar establishment used as living accommodation for, or for the treatment, care or maintenance of persons suffering from disabilities due to illness or old age or other physical or mental incapacity, or under the age of five years, or place of lawful detention, where such persons sleep on the premises.
Other	2(b)	Hotel, boarding house, residential college, hall of residence, hostel, and any other residential purpose not described above.
Office	3	Offices or premises used for the purpose of administration, clerical work (including writing, book keeping, sorting papers, filing, typing, duplicating, machine calculating, drawing and the editorial preparation of matter for publication, police and

Table 6.8 (Continued)

Title	Group	Purpose for which the building or compartment of a building is intended to be used
		fire service work), handling money (including banking and building society work), and communications (including postal, telegraph and radio communications) or radio, television, film, audio or video recording, or performance (not open to the public) and their control.
Shop and business commercial	4	Shops or premises used for a retail trade (including the sale to members of the public of food or drink for immediate consumption and retail by auction, self-selection and over-the-counter wholesale trading, the business of lending books or periodicals for gain and the business of a barber or hairdresser) and premises to which the public is invited to deliver or collect goods in connection with their hire, repair or other treatment, or (except in the case of repair of motor vehicles) where they themselves may carry out such repairs or other treatments.
Assembly and recreation	5	Place of assembly, entertainment or recreation; including bingo halls, broadcasting, recording and film studios open to the public, casinos, dance halls; entertainment, conference, exhibition and leisure centres; funfairs and amusement arcades; museums and art galleries; non-residential clubs, theatres, cinemas and concert halls; educational establishments, dancing schools, gymnasia, swimming pool buildings, riding schools, skating rinks, sports pavilions, sports stadia; law courts; churches and other buildings of worship, crematoria; libraries open to the public, non-residential day centres, clinics, health centres and surgeries; passenger stations and termini for air, rail, road or sea travel; public toilets; zoos and menageries.
Industrial	6	Factories and other premises used for manufacturing, altering, repairing, cleaning, washing, breaking-up, adapting or processing any article; generating power or slaughtering livestock.
Storage and other non-industrial[2]	7(a)	Place for the storage or deposit of goods or materials (other than described under 7(b)) and any non-residential building not within any of the purpose groups 1 to 6.
	7(b)	Car parks designed to admit and accommodate only cars, motorcycles and passenger or light goods vehicles weighing no more than 2500 kg gross.

Notes:

(1) Includes any surgeries, consulting rooms, offices or other accommodation, not exceeding 50 m² in total, forming part of a dwelling and used by an occupant of the dwelling in a professional or business capacity.

(2) A detached garage not more than 40 m² in area is included in purpose group 1(c); as is a detached open carport of not more than 40 m², or a detached building which consists of a garage and open carport where neither the garage nor open carport exceeds 40 m² in area.

6.2.2 Requirements – size of residential buildings

The building shall be constructed so that the combined dead, imposed and wind loads are sustained and transmitted by it to the ground

- *safely;*
- *without causing such deflection or deformation of any part of the building (or such movement of the ground) as will impair the stability of any part of another building.*

(Approved Document A1)

The building shall be constructed so that ground movement caused by:

- *swelling, shrinkage or freezing of the subsoil; or*
- *landslip or subsidence (other than subsidence arising from shrinkage)*

will not impair the stability of any part of the building.

(Approved Document A2)

The maximum height of the building measured from the lowest finished ground level to the highest point of any wall or roof should be less than 15 m (see Figure 6.13).	A1/2 (2C4i)
The height of the building should not exceed twice the least width of the building (see Figure 6.13).	A1/2 (2C4ii)
The height of the wing H_2 should not be greater than twice the least width of the wing W_2 where the projection P exceeds twice the width W_2.	A1/2 (2C4iii)

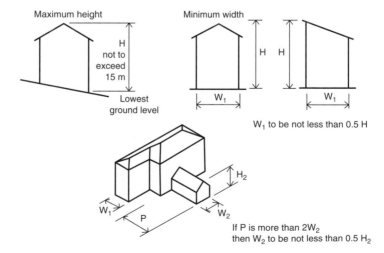

Figure 6.13 Residential buildings not more than three storeys.

Small single-storey non-residential buildings

> The height H should not exceed 3 m and the width (or   A1/2 (2C4b)
> greater length) should not exceed 9 m (see Figure 6.14).

Size of annexes

> The height H (as variously shown in Figure 6.15)   A1/2 (2C4b)
> should not exceed 3 m.

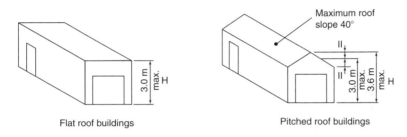

Figure 6.14 Size and proportion of non-residential buildings.

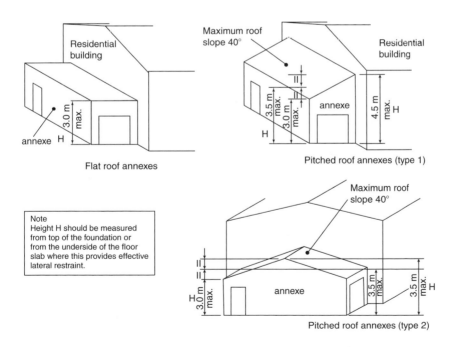

Figure 6.15 Size and proportion of non-residential annexes.

6.3 Ventilation

6.3.1 Requirements

There shall be adequate means of ventilation provided for people in the building.

(Approved Document F)

 A new Part F came into force in October 2010.

Ventilation is defined in the Building Regulations as *the supply and removal of air (by natural and/or mechanical means) to and from a space or spaces in a building.*

In addition to replacing 'stale' indoor air with 'fresh' outside air, the aim of ventilation is also to:

- provide outside air for breathing;
- control thermal comfort;
- limit the accumulation of moisture and pollutants from a building, which could, otherwise, become a health hazard to people living and/or working within that building;
- dilute and remove airborne pollutants (especially odours but also those that are released from materials and products used in the construction, decoration and furnishing of a building, and as a result of the activities of the building's occupants);
- control excess humidity;
- provide air for fuel burning appliances.

 Note: The requirements of the 2010 edition of Approved Document F have also been designed to deal with the products of tobacco smoking.

In general terms, all these aims can be met if the ventilation system:

- disperses residual pollutants and water vapour;
- extracts water vapour from wet areas where it is produced in significant quantities (e.g. kitchens, utility rooms and bathrooms);
- rapidly dilutes pollutants and water vapour produced in habitable rooms, occupiable rooms and sanitary accommodation;
- extracts pollutants from areas where they are produced in significant quantities (e.g. rooms containing processes or activities which generate harmful contaminants);
- is designed, installed and commissioned so that it:
  - is not detrimental to the health of the people living and/or working in the building;
  - helps maintenance and repair;
  - is reasonably secure;

- makes available, over long periods, a minimum supply of outdoor air for the occupants;
- minimizes draughts;
- provides protection against rain penetration.

Ventilation is also a means of controlling thermal comfort.

The aim of Approved Document F is to suggest to the designer the level of ventilation that should be sufficient for a particular situation as opposed to how it should be achieved. The designer is, therefore, free to use whatever ventilation system he considers most suitable for a particular building **provided** that it can be demonstrated that it meets the recommended performance criteria and levels concerning moisture, pollutants and air flow rates standards as shown in Table 6.9.

Table 6.9 Standards for performance-based ventilation

Type	Standard	Part
Intermittent extract fan	BS EN13141-4	Clause 4
Range hood	BS EN13141-3	Clause 4
Background ventilator (non-RH controlled)	BS EN 13141-1	Clauses 4.1 and 4.2
Background ventilator (RH controlled)	PrEN13141-9	Clauses 4.1 and 4.2
Passive stack ventilator	See Appendix D of Approved Document F	
Continuous mechanical extract ventilation (MEV) system	BS EN13141-6	Clause 4
Continuous mechanical supply and extract with heat recovery (MVHR)	PrEN13141-7	Clauses 6.1, 6.2 and 6.2.2
Single room heat recovery ventilator	PrEN13141-8	Clauses 6.1 and 6.2
RH, relative humidity.		

For further details and examples concerning 'Performance based ventilation', see Appendix A to Approved Document F.

Note: The 2010 edition of Approved Document F, Ventilation (replacing the 2006 edition and which came into force on 1 October 2010) includes the changes that have been made to the legal requirements of the Building Regulations 2000 and the Building (Approved Inspectors etc.) Regulations 2000. These stipulate that:

- All fixed mechanical ventilation systems, where they can be tested and adjusted, shall be commissioned and a commissioning notice given to the Building Control Body.

- For mechanical ventilation systems installed in new dwellings, air flow rates shall be measured on site and a notice given to the Building Control Body. This shall apply to intermittently-used extract fans and cooker hoods, as well as continuously running systems.
- The owner shall be given sufficient information about the ventilation system and its maintenance requirements so that the ventilation system can be operated to provide adequate air flow.

The technical guidance contained in Part F has also been changed as follows:

- Ventilation provisions have been increased for dwellings with a design air permeability tighter than or equal to $50\,Pa$ ($50\,N/m^2$).
- For passive stack ventilators, the stack diameter has been increased to 125 mm for all room types. Use of passive stack ventilation in inner wet rooms has been clarified.
- The guidance for ventilation when a kitchen or bathroom in an existing dwelling is refurbished has been clarified.

Exceptions

Part F provides guidance for those erecting new dwellings and buildings other than dwellings, or completing work on existing buildings. The exception from the ventilation requirements listed in Part F concern:

- buildings controlled under the Manufacture and Storage of Explosives Regulations 2005;
- buildings controlled under the Nuclear Installations Act 196;
- detached buildings of less than $30\,m^2$ floor area, designed shelters from nuclear, chemical or conventional weapons;
- buildings included in the schedule of monuments (maintained under section 1 of the Ancient Monuments and Archaeological Areas Act 1979);
- detached buildings into which people do not normally (or only occasionally) go;
- greenhouses (non retail);
- agricultural buildings – unless part of the building is used as a dwelling;
- temporary buildings not intended to remain in place for more than 28 days;
- ancillary buildings used for the disposal of buildings or building plots on site;
- site construction buildings;
- buildings on the site of mines and quarries which do not contain dwellings or are used as offices or showrooms;
- detached single-storey buildings, with less than $30\,m^2$ floor area that do not contain any sleeping accommodation;
- detached buildings of less than $15\,m^2$ floor area containing no sleeping accommodation;

- ground level building extension extensions (e.g. conservatories, porches, covered yards, covered ways, and/or carports) with a floor area less than $30\,m^2$.

Notification of work covered by the ventilation requirements

In most cases where it is proposed to carry out notifiable ventilation work on a building, the Building Control Body has to be notified in advance via a full plans application, a building or an initial notice given jointly with the approved inspector.

Competent person self-certification schemes

It is not necessary to notify a Building Control Body in advance of work which is to be carried out by a person registered with a competent person self-certification scheme for that type of work.

There are a number of competent person schemes for the installation of mechanical ventilation and air-conditioning systems in buildings (e.g. BESCA, CORGI, NAPIT, NICEIC); more details can be found at http://www.communities.gov.uk.

Emergency repairs

Where the work involves an emergency repair (e.g. to a failed fan), unless the person carrying out the work is registered under an appropriate competent person scheme, a notice has to be provided to the Building Control Body at the earliest opportunity. A completion certificate can then be issued in the normal way.

Minor works

Where the work is of a minor nature such as:

- replacement of parts, or the addition of an output or control device;
- provision of a self-contained mechanical ventilation or air-conditioning appliance which cannot be adjusted from their factory settings – as in the case of a cooker hood, a bathroom extract fan or a room air-conditioning unit.

The work need not be notified to the Building Control Body.

6.3.2 Meeting the requirements

External pollution

In urban areas, buildings are exposed to a large number of pollution sources from varying heights and upwind distances (i.e. long, intermediate and short range). Internal contamination from these pollution sources can have a detrimental effect on the buildings' occupants, and so it is very important to ensure that the ventilation system provided is sufficient and, above all, that the air intake cannot be contaminated.

Typical urban pollutants include:

- carbon monoxide, (CO);
- nitrogen dioxide, (NO2);
- sulphur dioxide, (SO2);
- ozone, (O3);
- particles (PM10);
- benzene;
- 1,3-butadiene;
- polycylic aromatic hydrocarbons (PAHs);
- ammonia;
- lead.

Typical emission sources include:

- building ventilation system exhaust discharges;
- combustion plant (e.g. heating appliances) running on conventional fuels;
- construction and demolition sites;
- discharges from industrial processes and other sources;
- other combustion type processes (e.g. waste incineration, thermal oxidation abatement schemes);
- road traffic, including traffic junctions and underground car parks;
- uncontrolled ('fugitive') discharges from industrial processes and other sources.

Indoor air pollutants

The maximum permissible level of indoor air pollutants is listed in Table 6.10.

Table 6.10 Maximum permissible level of indoor air pollutants

Pollutant	Permissible level
Nitrogen dioxide (NO$_2$)	Not exceeding: • 288 g/m^3 (150 ppb) – 1-hour average • 40 g/m^3 (20 ppb) – long-term average
Carbon monoxide (CO)	Not exceeding: • 100 mg/m^3 (90 ppm) – 15 minute averaging time • 60 mg/m^3 (50 ppm) – 30 minute averaging time (DOH, 2004) • 30 mg/m^3 (25 ppm) – 1 hour averaging time (DOH, 2004) • 10 mg/m^3 (10 ppm) – 8 hours averaging time (DOH, 2004)
Control of bio-effluents (body odours)	3.5 l/s per person
Total volatile organic compound (TVOC)*	Not exceeding: • 300 g/m^3 averaged over 8 hours

DOH, Department of Health.

Note: Mould growth can occur whether the dwelling is occupied or unoccupied, so the performance criteria for moisture (see Table 6.11) should be met at all times, regardless of occupancy. The other pollutants listed above (see Table 6.10) are harmful to the occupants **only** when the dwelling is occupied.

Table 6.11 Mould growth

Moving average speed	Surface water activity
1 month	65%
1 week	75%
1 day	85%

Extract ventilation concerns the removal of air directly from a space or spaces to outside. Extract ventilation may be by natural means such as passive stack ventilation (PSV) or by mechanical means (e.g. by an extract fan or central system).

 Extract fans lower the pressure in a building, which can cause the spillage of combustion products from open-flued appliances (i.e. the combustion gases may fill the room instead of going up the flue or chimney).

Requirements

All kitchens, utility rooms, bathrooms and sanitary accommodation shall be provided with extract ventilation to the outside, which is capable of operating either intermittently or continuously.   F5.5

The minimum extract airflow rates should be greater than that shown in Table 6.12.

Table 6.12 Extract ventilation rates

Room	Minimum intermittent extract rate	Continuous extract	
		Minimum high rate	Minimum low rate
Kitchen	30 l/s (adjacent to hob) or 60 l/s elsewhere	13 l/s	Total extract rate must be at least the whole building ventilation rate shown in Table 6.13
Utility room	30 l/s	8 l/s	
Bathroom	15 l/s	8 l/s	
Sanitary accommodation	6 l/s	6 l/s	

The whole building ventilation rate for habitable rooms in a dwelling should be greater than that shown in Table 6.13.	F5.6
The minimum ventilation rate (based on two occupants in the main bedroom and a single occupant in all other bedrooms) should be not less than 0.3 l/s per m².	F Table 5.1b

Table 6.13 Whole building ventilation rates

	Number of bedrooms				
Whole building ventilation rate (l/s)	1	2	3	4	5
	13	17	21	25	29

 Note: For greater occupancy, add 4 l/s per occupant.

 The provision of purge ventilation is required in each habitable room and should be capable of extracting a minimum of four air changes per hour (i.e. 4 ach) per room directly to the outside.

Ventilation effectiveness

Ventilation effectiveness is, as the term suggests, a measure of how well a ventilation system supplies air to the building's occupants. From an energy-saving perspective, the higher the level of ventilation effectiveness the more efficient the system will be in reducing pollutant levels in the occupant's breathing zone. As this can result in quite significant energy savings, it has to be considered when designing and installing ventilation systems.

As the designer cannot be absolutely certain of the future occupancy and/or use of the building in terms of seating plan, location of computers and printers etc., a ventilation effectiveness level of 1 (i.e. where the supply air is fully mixed with the room air before it is breathed by the occupants) should, similar to the designs and recommendations of Approved Document F, be assumed in their calculations.

 Note: For more details about ventilation effectiveness, see CIBSE Guide A.

Equivalent ventilator area for dwellings

 Note: 'Equivalent area' is defined as *the area of a sharp-edged orifice which air would pass at the same volume flow rate, under an identical applied pressure difference.*

Equivalent area is now considered a better measure of the aerodynamic performance of a ventilator than the previous free-area sizing of background ventilators. Primarily this is because 'free area' refers only to the physical size of the aperture of the ventilator and does not, therefore, accurately reflect the airflow performance of the ventilator.

Note: A European Standard (BS EN13141-1:2004) includes a method of measuring the equivalent area of background ventilator openings.

Designers should use the equivalent ventilator areas shown in Table 6.14 when designing systems using intermittent extract fans and background ventilators for multi-storey dwellings that are more than four storeys above ground level and which have more than one exposed façade.

Table 6.14 Equivalent ventilator area for dwellings

Total floor area (m²)	Equivalent ventilator area (mm²)				
	Number of bedrooms				
	1	2	3	4	5
≤50	25,000	35,000	45,000		
51–60	25,000	30,000	40,000		
61–70	30,000	30,000	35,000	45,000	55,000
71–80	35,000	35,000	35,000		
81–90	40,000	40,000	40,000		
91–100	45,000	45,000	45,000		
>100	Add 5000 mm² for every additional 10 m² floor area				

Notes:
- For single-storey dwellings up to four storeys above ground level, add 5000 mm².
- For an occupancy level greater than two persons in the main bedroom and one person in all other bedrooms, assume an extra bedroom for each additional person.
- For more than five bedrooms, add an additional 10,000 mm² per bedroom.

Ventilation air intakes

One method of achieving good indoor air quality is to reduce the amount of water vapour and/or air pollutants that are released into the indoor air, particularly those caused from construction and consumer products.

Air intakes that are located on a less polluted side of the building may be used for fresh air.

Note: Further information about control of emissions from construction products is available in BRE Digest 464.

Noise from ventilation systems

As the noise from ventilation systems can disturb the occupants of the building and in doing so affect their work effectiveness, the designer must consider methods of minimizing noise through careful design and use of quieter products such as sound-attenuating ventilation material.

Noise from a ventilation system may also disturb people who are outside the building, and so measures to minimize externally emitted noise should also be considered.

The installation and use of ventilation systems in buildings will also result in energy being used (e.g. to heat fresh air taken in from outside, to move air into, out of and/or around the building) and so consideration should always be given to using heat recovery devices, efficient types of fan motor and/or energy saving control devices in ventilation systems.

If a dwelling is near a busy road or by an airport etc., where the amount of external noise is likely to be intrusive, a sound-attenuator duct section may be fitted in the roof space just above the ceiling.

Provision of information

Sufficient information about the ventilation system and its maintenance requirements must be given to owners so that the ventilation system can be operated to provide adequate air flow.	F-4.47

Types of ventilation

Buildings are ventilated through a combination of *infiltration* and *purpose-provided ventilation*:

- *Infiltration* is the uncontrollable air exchange between the inside and outside of a building through a wide range of air leakage paths in the building structure.
- *Purpose-provided ventilation* is the controllable air exchange between the inside and outside of a building by means of a range of natural and/or mechanical devices.

The 2010 version of Approved Document F recommends that a reasonably high level of airtightness that is significantly tighter than the minimum target value recommended under Part L is archived, as all new buildings should be expected be better than the target value to some degree. Through good design and execution, domestic and non-domestic buildings can currently achieve an air permeability down to around 2–4 m^3/h per m^2 of envelope area at 50 pascal (Pa) pressure difference. Many buildings that are constructed nowadays are tighter than this, and it is envisaged that there will be an upward trend in the future owing to the UK's commitment to higher energy efficiency and lower carbon emissions.

The three main controllable ventilation methods are listed in Table 6.15.

Table 6.15 Ventilation methods

Method	Type	Why used	Remarks
Extract ventilation	Intermittent extract fans	In rooms where most water vapour and/or pollutants are released (e.g. cooking, bathing or photocopying)	This extract may be either intermittent or continuous and is aimed at minimizing the spread of vapour and pollutants to the rest of the building
Whole building ventilation	Trickle ventilators	To provide fresh air to the building, dilute and disperse residual water vapour/pollutants not dealt with by extract ventilation, and to remove water vapour and pollutants released by building materials, furnishings, activities and the presence of occupants	This type of ventilation provides continuous air ventilation exchange with a ventilation rate that can be reduced or ceased when the building is not occupied In some cases (e.g. when the building is reoccupied) it may be necessary to purge the air (see below)
Purge ventilation Note: Previously referred to as 'rapid' ventilation in the 1995 edition of Approved Document F	Windows	To assist in the removal of high concentrations of pollutants and water vapour released from occasional activities (e.g. painting and decorating) or accidental releases (e.g. smoke from burnt food or water spillage)	Purge ventilation is intermittent and may be used to improve thermal comfort and/or overheating in summer (see Approved Documents L1A (New dwellings) and L2A (New buildings other than dwellings))

Control of ventilation

Ventilation should be controllable so that it can maintain reasonable indoor air quality and avoid waste of energy. These controls can be either manual or automatic.

Manually controlled trickle ventilators (the most common type of background ventilators) can be located over the window frames, in window frames, just above the glass or directly through the wall (see Figure 6.21).

Trickle ventilators are intended to be normally left open in occupied rooms in dwellings.

In dwellings, humidity-controlled devices are available to regulate the humidity of the indoor air, and hence minimize the risk of condensation and mould growth. These are best installed as part of an extract ventilator in moisture-generating rooms (e.g. kitchen or *bathroom*). Humidity control is not appropriate for *sanitary accommodation*, where the dominant pollutant is normally odour.

Table 6.16 Ventilation systems for basements

Type of basement	Background ventilators and intermittent extract fans	Passive stack ventilation	Continuous mechanical extract	Continuous mechanical supply and extract with heat
Basement connected to the rest of the dwelling by an open stairway	Yes	Yes	Yes	Yes
Basement with a single exposed façade and dwelling above ground with more than one exposed façade		Yes	Yes	
Basements not connected to the rest of the dwelling by an open stairway	Yes	Yes	Yes	Yes
Dwelling above ground has no bedrooms	Yes	Yes	Yes	Yes
Dwelling comprises just a basement	Yes	Yes	Yes	Yes

Other types of *automatic control* (e.g. trickle ventilators, ventilation fans, dampers and air terminal devices) may also be suitable for regulating ventilation devices in dwellings.

Trickle ventilators with automatic controls should also have a manual override, so that the occupant can close the ventilator to avoid draughts. For pressure-controlled trickle ventilators that are fully open at typical conditions (e.g. 1 Pa pressure difference), only a manual close option is recommended.

In buildings other than dwellings, more sophisticated *automatic control* systems such as occupancy sensors (using local passive infrared detectors) or indoor carbon dioxide concentration sensors (using electronic carbon dioxide detectors) can be used as an indicator of occupancy level and, therefore, body odour.

6.3.3 Purge ventilation

Purge ventilation is a manually controlled type of ventilation that is used in rooms and spaces to rapidly dilute pollutants and/or water vapour. It can be achieved by natural means (e.g. an openable window or an external door) or by mechanical means (e.g. a fan).

Note: For further guidance on purge ventilations, see BS 5925:1991 *Code of Practice for Ventilation Principles and Designing for Natural Ventilation.*

6.3.4 Passive Stack Ventilation

Passive Stack Ventilation (PSV) is a ventilation device which uses ducts from terminals mounted in the ceiling of rooms to terminals on the roof, to extract air to the outside by a combination of the natural stack effect and the pressure effects of wind passing over the roof of the building.

The so-called 'stack effect' relies on the pressure differential between the inside and the outside of a building caused by differences in the density of the air due to an indoor/outdoor temperature difference.

Table 6.17 Passive Stack Ventilation

Room	Internal duct diameter (mm)	Internal cross-sectional area (mm)
Kitchen	125	12,000
Utility room	100	8,000
Bathroom	100	8,000
Sanitary accommodation	80	5,000

For sanitary accommodation **only**, purge ventilation may be used **provided** that security is not an issue.

 Note: Open-flued appliances may provide sufficient extract ventilation when in operation and can be arranged to provide sufficient ventilation when not firing.

Design

The design and installation of PSV systems is crucial to their operation, and Figure 6.16 shows the preferred option for kitchen and bathroom ducts with ridge terminals.

Another option (see Figure 6.17) is to have the kitchen and bathroom ducts penetrating the roof and extend its terminals to ridge height.

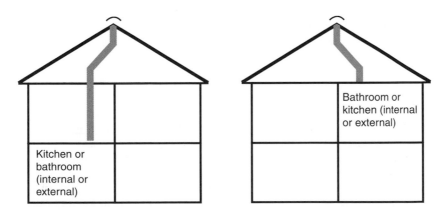

Figure 6.16 Preferred PSV system layouts.

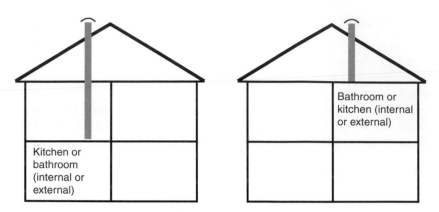

Figure 6.17 Alternative PSV system layouts.

Requirements

In designing PSV systems, the following requirements shall be met (see Figure 6.18):

Figure 6.18 PSV offset requirements.

Dwellings which only have a single exposed façade should be designed so that the habitable rooms are on the exposed façade and are capable of achieving cross-ventilation.	F1-Table 5.2b
PSV extract terminals should be located in the ceiling or on a wall less than 400 mm below the ceiling.	F1-Table 5.2b

 Instead of PSV, an open-flued appliance may provide sufficient extract ventilation for the room in which it is located when in operation and can be arranged to provide sufficient ventilation when not firing.

Background ventilators should be located in all rooms
with external walls except the rooms where a PSV is
located

F1-Table 5.2b

Controls

- Should be set up to operate without occupant
  intervention.
- May have automatic controls (e.g. sensors for
  humidity, occupancy/usage, pollutant release).
- Humidity controls should not be used for
  sanitary accommodation as odour is the main
  pollutant.
- In kitchens, any automatic control must
  provide sufficient flow during cooking with
  fossil fuel (e.g. gas) to avoid build-up of
  combustion products.
- Background ventilators may be either manually
  adjustable or automatically controlled.

F1-Table 5.2b

Common outlet terminals and/or branched ducts shall
not be used for wet rooms (e.g. the kitchen, bathroom,
utility room and/or WCs).

F

Ducts should have no more than one offset (i.e. bend)
and ideally these should be 'swept' at an angle of no
more than 40° to the vertical.

F

If a duct penetrates the roof more than 0.5 m from the
roof ridge, then it must extend above the roof slope to
at least the height of the roof ridge.

F

If tile ventilators are used on the roof slope they must
be positioned no more than 0.5 m from the roof ridge.

F

Separate ducts shall be taken from the ceilings of wet
rooms to separate terminals on the roof.

F

Ceiling extract grilles should have a free area, not less
than the duct cross-sectional area (when in the fully
open position if adjustable).

F

Ducts should be insulated in the roof space and other
unheated areas with at least 25 mm of a material
having a thermal conductivity of 0.04 W/m K.

F

If a duct extends above the roof level, then that section of the duct should be insulated or be fitted with a condensation trap just below roof level.	F
If a conversion fitting is required to connect the duct to the terminal then the duct cross-section area must be maintained (or exceeded) throughout the conversion fitting.	F
PSVs for dwellings that are situated near a significantly taller building (i.e. more than 50% taller), should be at least five times the difference in height **away** from the taller building (i.e. if the difference in height is 10 m, then the PSV should not be installed in a dwelling within 50 m of the taller building).	F
Outlet terminals should have a free area that is not less than the duct cross-section area.	F
Terminals should not allow ingress of large insects or birds and should be designed so that rain is not likely to enter the duct and run down into the dwelling.	F
Terminals should be designed so that any condensation forming inside it cannot run down into the dwelling but will run off onto the roof.	F

 Note: An amendment to the existing European Standard 13141-5:2004 for the ventilation of buildings. is currently under discussion which (it is anticipated) will suggest that terminals, with any necessary conversion fittings, should have an overall static pressure loss (upstream duct static minus test room static) equivalent to no more than four times the mean duct velocity pressure when measured at a static pressure difference of 10 Pa.

Installation

Location of passive stack ventilators

PSV extract terminals should be located in the ceiling or on a wall less than 400 mm below the ceiling.	F
Background ventilators should not be within the same room as a PSV terminal.	F
If a PSV is located in a protected stairway of a dwelling, it shall not allow smoke or fire to spread into the stairway.	F

The duct length should be just sufficient to fit between the ceiling   F
grille and the outlet terminal.

Flexible ducting should be fully extended but not taut.   F

Allow approximately 300 mm extra to make smooth bends in an offset
system.

Ducting should: F

- be properly supported along its entire length (remembering
  that flexible ducting generally requires more support than rigid
  ducting);
- be run straight without any distortion or sagging;
- not have any kinks at bends or connections with ceiling grilles
  and outlet terminals;
- be securely fixed to the roof outlet terminal so that it cannot
  sag or become detached.

In roof spaces: F

- ducts should, ideally, be secured to a wooden strut that is
  securely fixed at both ends;
- flexible ducts should be allowed to curve gently at each end of
  the strut.

For stability, rigid ducts should be used for any outside part of the   F
PSV system that is above the roof slope. To provide stability, they
should also project down into the roof space far enough to allow
firm support.

Fire precautions

In dwellings with three or more storeys and blocks of flats, PSV ducts should
not impede fire escape routes.

Common stairs

If a stair serves a place that is a special fire hazard, the   B1 2.47
lobby or corridor should have not less than 0.4 m^2
permanent ventilation or be protected by a mechanical
smoke control system.

Ducting

Precautions must be taken where ducting passes through a   FI-4.37
fire-resisting wall, floor or fire compartment

External escape stairs

Protected lobbies (with less than $0.4\,m^2$ permanent ventilation) should be provided between an escape stairway and a place of special fire hazard.	B1 4.35 (V2)

Flues

If a flue (or a duct containing a flue and/or ventilation duct) passes through a compartment wall or compartment floor (or is built into a compartment wall) then each wall of the flue or duct should have a fire resistance of at least half that of the wall or floor.	B3 10.16 (V2)

Mechanical ventilation

Smoke control of common escape routes by mechanical ventilation is permitted provided that it meets the requirements of BS EN12101-6:2005.	B1 2.27
Mechanical ventilation systems should be designed to ensure that:	B1 5.46 (V2)
• ductwork does not assist in transferring fire and smoke through the building;	B1 2.18 (V2) B3 10.2 (V2)
• exhaust points are sited away from final exits, combustible building cladding or roofing materials and openings into the building;	
• recirculated air serving a stairway and/or entrance hall (as well as other areas) should be designed to shut down on the detection of smoke within the system.	

Protected escape routes

Protected lobbies (with not more than $0.4\,m^2$ permanent ventilation) should be provided between an escape stairway and a place of special fire hazard.	B1 4.35 (V2)
Ventilation ducts supplying or extracting air directly to or from a protected escape route, should **not** also serve other areas.	B1 5.47 (V2)

Separate ventilation systems should be provided for each protected stairway.	B1 5.47 (V2)

Ventilation ducts

Ventilation ducts supplying or extracting air directly to or from a protected stairway, should not also serve other areas.	B1 2.17 (V1)
Ventilation ducts supplying or extracting air directly to or from a protected stairway or entrance hall, should not also serve other areas.	B1 2.18 (V2) B1 5.47
Separate ventilation systems should be provided for each protected stairway.	B1 5.47 (V2)

Operation in hot weather

Although PSV units should be capable of extracting sufficient air from wet rooms during the winter, in the summer months (i.e. when the temperature difference between the internal and external air is considerably reduced) they may not. To guard against this happening, purge ventilation should also be provided in wet rooms.

6.3.5 Installation of fans in dwellings

The three fan types most commonly used in domestic applications are:

- axial fans
- centrifugal fans
- in-line fans.

Axial fans

The axial fan is the most common form of fan, and can be mounted on the wall, window (i.e. through a suitable glazing hole) or in the ceiling (e.g. in a bathroom).

For wall and window mounting applications up to 350 mm thick, use a short length of rigid round duct or a flexible duct pulled taut.

For bathrooms, 100 mm diameter fans can be used as an axial fan in the ceiling with a short (1.5 m maximum) length of flexible duct with (a maximum) two 90° bends.

 Note: The duct must be pulled taut and the discharge terminal should have at least 85 per cent free area of the duct diameter.

Centrifugal fans

Centrifugal fans (because they develop greater pressure) permit longer lengths of ducting to be used, and so can be used for most wall and/or window applications in high-rise (i.e. above three storeys) buildings or in exposed locations to overcome wind pressure.

Most centrifugal fans are designed with 100 mm diameter outlets, which enables them to be connected to a wide variety of duct types.

 Wall/ceiling-mounted centrifugal fans that are designed to achieve 60 l/s for kitchens should not be ducted further than 3 m and should have no more than one 90° bend. Fans designed to achieve 15 l/s for bathrooms should not be ducted further than 6 m and should have no more than two 90° bends.

In-line fans

There are two types of in-line fan available:

- in-line axial fans which have to be installed with the shortest possible duct length to the discharge terminal;
- in-line mixed flow fans which have the characteristics of both axial and centrifugal fans and can, therefore, be used with longer lengths of ducting.

Both types can be used for bathrooms (100 mm diameter), utility rooms (125 mm diameter) and kitchens (150 mm diameter).

Intermittent extract fans

Minimum extract airflow rates for intermittent extract   F Appendix A
fans should be greater than that shown in Table 6.18.

Table 6.18 Extract ventilation rates

Room	Minimum intermittent extract rate	Continuous extract	
		Minimum high rate	Minimum low rate
Kitchen	30 l/s (adjacent to hob) or 60 l/s elsewhere	13 l/s	Total extract rate must be at least the whole building ventilation rate shown in Table 6.12
Utility room	30 l/s	8 l/s	
Bathroom	15 l/s	8 l/s	
Sanitary accommodation	6 l/s	6 l/s	

Note: For sanitary accommodation, a purge ventilation system may be used, and in wet rooms a heat recovery ventilator may be used instead of a conventional fan, provided that it has the same extract rate.

Fan terminals

When installing fans:

> Ensure that the free area of the grill opening of a room terminal   F
> extract grille and/or discharge terminal has a minimum of 85% of
> the free area of the ducting being used.

Note: In these cases (only), the equivalent area may be assumed to be equal to the free area.

6.3.6 Ventilation systems – dwellings without basements

The following systems may be used in dwellings without basements:

- background ventilators and intermittent extract fans;
- Continuous Mechanical Extract (MEV);
- continuous Mechanical Supply and Extract with Heat Recovery (MVHR);
- Passive Stack Ventilation (PSV).

Background ventilators and intermittent extract fans

The need for background ventilators will depend on the air permeability or air tightness of a building.

Note: Air permeability is defined as *the average volume of air (in cubic metres per hour) that passes through unit area of the building envelope (in square metres) when subject to an internal to external pressure difference of 50 Pa.*

Location of background ventilators in rooms

> Background ventilators should be located to avoid   F1-Table 5.2a
> draughts (e.g. typically 1.7 m above floor level).
>
> Background ventilators should be located in all   F1-Table 5.2a
> rooms with external walls, with at least 5000 mm^2
> equivalent area in each habitable room and
> 2500 mm^2 equivalent area in each wet room.
>
> To ensure good transfer of air throughout the dwelling   F1-Table 5.2a
> there should be an undercut of minimum area
> 7600 mm^2 in all internal doors above the floor finish.

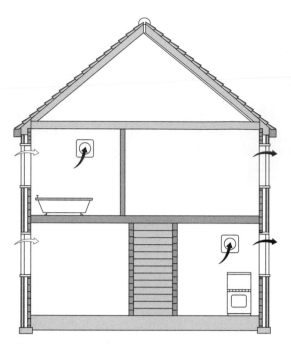

Figure 6.19 Background ventilators and intermittent extract fans.

Controls

Intermittent extract ventilators: F1-Table 5.2a

- may be operated manually and/or automatically by a sensor (e.g. humidity, occupancy/usage, pollutant release;
- humidity controls should not be used for *sanitary accommodation* as odour is the main pollutant;
- in kitchens, any *automatic control* must provide sufficient flow during cooking with fossil fuel (e.g. gas) to avoid build-up of combustion products;
- any automatic control should have a manual override to allow the occupant to turn the extract on;
- in a room with no openable window (i.e. an internal room) an intermittent extract fan should have a 15 minute overrun.

 In rooms with no natural light, the fans could be controlled by the operation of the main room light switch.

Background ventilators may be either manually adjustable or automatically controlled.	F1-Table 5.2a
Manual controls should be within reasonable reach of the occupants.	F1-Table 5.2a
Fans should be quiet so as not to discourage their use by occupants.	F1-Table 5.2a
Background ventilators for dwellings with a single exposed façade should be located at both high (typically 1.7 m above floor level) and low (i.e. at least 1.0 m below the high ventilators) positions in the façade (see Figure 6.20).	F1-Table 5.2a

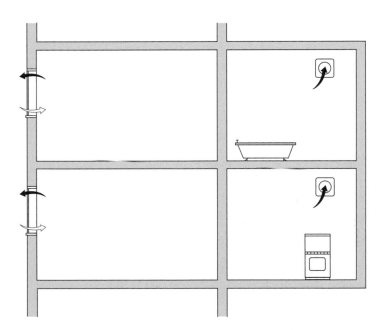

Figure 6.20 Single-sided ventilation.

Dwellings with only a single exposed façade should be designed so that the habitable rooms are on the exposed façade in order to achieve cross-ventilation.	F Table 5.2b
Background ventilators should be at least 0.5 m from an extract fan.	F

Background ventilators should be located so as to avoid draughts (e.g. typically 1.7 m above floor level).   F

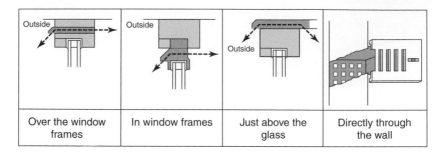

Over the window frames	In window frames	Just above the glass	Directly through the wall

Figure 6.21 Background ventilation systems.

Controllable background ventilators with a minimum equivalent area of 2500 mm² shall be fitted in each room (except wet rooms from which air is extracted).   F Table 5.2c

Background ventilators may be manually adjustable or automatically controlled.   F Table 5.2c

Windows with night latches should **not** be used as they are more liable to draughts as well as being a potential security risk.   F

Trickle ventilators

Manually controlled trickle ventilators are widely used for background ventilation and these can be located as shown in Figure 6.21.

To avoid cold draughts, trickle ventilators are normally positioned 1.7 m above floor level, and usually include a simple control (such as a flap) to allow users to shut off the ventilation according to personal choice or external weather conditions. Nowadays, pressure-controlled trickle ventilators that reduce the air flow according to the pressure difference across the ventilator are available to reduce draught risks during windy weather.

Trickle ventilators are normally left open in occupied rooms in dwellings.

Trickle ventilators that include an automatic control should be capable of being manually overridden so that they can be opened by the occupant when required.   F Table 5.2c

Pressure-controlled trickle ventilators that, under normal conditions, are left open (e.g. 1 Pa pressure difference) should only be capable of being manually closed.	F 4.20
Trickle ventilators etc. should be clearly marked with their equivalent area (measured according to BS EN13141-1:2004) either by means of a stamp or an indelibly printed self-adhesive label.	F 4.26
All fans should operate quietly at their minimum (i.e. normal) rate so as not to disturb the occupants of the building.	F Table 5.2c

Continuous mechanical extract

This system may consist of either a central extract system or individual room fans, or a combination of both.

To calculate the required extract rate, first determine the **whole building ventilation rate** from Table 6.19.

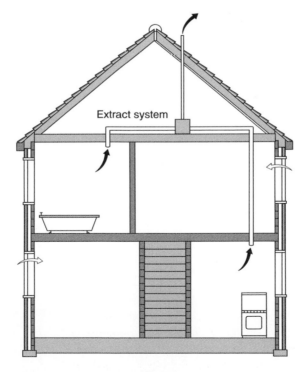

Figure 6.22 Continuous mechanical extract.

Table 6.19 Whole building ventilation rates

	Number of bedrooms				
	1	2	3	4	5
Whole building ventilation rate (l/s)	13	17	21	25	29

Then work out the **whole dwelling air extract rate** (at maximum operation) by summing the individual room rates from Table 6.20.

Table 6.20 Whole dwelling air extract rate

Room	Minimum intermittent extract rate (l/s)	Minimum high continuous extract rate (l/s)
Kitchen	30 (adjacent to hob) 60 elsewhere	13
Utility room	30	8
Bathroom	15	8
Sanitary accommodation	6	6

For sanitary accommodation **only**, purge ventilation may be used provided that security is not an issue.

Requirements

Extract should be from each wet room.	F1-Table 5.2c
Cooker hoods should be 650 mm to 750 mm above the hob surface.	F1-Table 5.2c
Mechanical extract terminals and fans should be installed as high as is practicable and preferably less than 400 mm below the ceiling.	F1-Table 5.2c

 Where ducts etc. are provided in a dwelling with a protected stairway, precautions may be necessary to avoid the possibility of the system allowing smoke or fire to spread into the stairway.

Air transfer

To ensure good transfer of air throughout the dwelling, there should be an undercut of minimum area 7600 mm^2 in all internal doors above the floor finish.	F1-Table 5.2c

Controls

Should be set up to operate without occupant intervention, but may have manual or automatic controls to select the boost rate. — F1-Table 5.2c

Any manual boost controls should be provided locally to the spaces being served (e.g. bathrooms and kitchen), as provision of a single centrally located switch may result in fans being left in an inappropriate mode of operation. — F1-Table 5.2c

Automatic controls could include sensors for humidity, occupancy/usage and pollutant release.

Humidity controls should not be used for sanitary accommodation as odour is the main pollutant.

In kitchens, any automatic control must provide sufficient flow during cooking with fossil fuel (e.g. gas) to avoid build-up of combustion products. — F1-Table 5.2c

Where manual controls are provided, they should be within reasonable reach of the occupants. — F1-Table 5.2c

Operation

The maximum ('boost') rate should be the greater of the whole building ventilation rate or the whole dwelling air extract rate. — F Table 5.2c

The maximum individual room extract rates should be at least those given in Table 6.19. — F Table 5.2c

The minimum air supply rate should be at least the whole building ventilation rate. — F Table 5.2c

Note: Extract terminals located on the prevailing windward façade should be protected against the effects of wind by using ducting to another façade, using a constant-volume flow rate unit or a central extract system.

All fans should operate quietly at their minimum (i.e. normal) rate so as not to disturb the occupants of the building. — F Table 5.2c

Ventilation devices designed to work continuously: F Tables 5.2a
to 5.2d

- shall be set up to operate without occupant
  intervention;
- may have a manual control to select maximum
  'boost';
- may have automatic controls such as humidity
  control (but not if used for sanitary accommodation), occupancy/usage sensor, moisture/pollutant
  release detector etc.

Automatic controls for ventilators that are designed   F Tables 5.2a
to work continuously in kitchens must be capable   to 5.2d
of providing sufficient flow during cooking with
fossil fuels (e.g. gas) so as to avoid the build-up of
combustion products.

Continuous mechanical supply and extract with heat recovery

To calculate the air flow rate of a building using continuous Mechanical
Supply and Extract with Heat Recovery (MVHR), first determine the **whole
building ventilation rate** from Table 6.19, then, depending on whether it is a

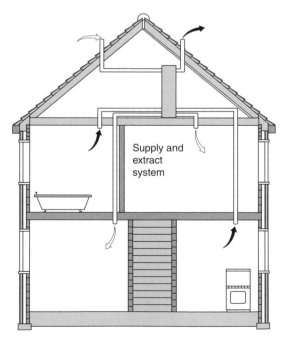

Supply and
extract
system

Figure 6.23 Continuous mechanical supply – with heat recovery.

multi-storey or single-storey building, subtract the gross internal volume of dwelling heated space (m^2) as follows:

Multi-storey dwelling	whole building ventilation rate – 0.04 × gross internal volume
Single-storey dwelling	whole building ventilation rate – 0.06 × gross internal volume

Next, work out the whole dwelling air extract rate at maximum operation by summing the individual room rates from Table 6.20.

Requirements

Extract should be from each wet room.	F1-Table 5.2d
Air should normally be supplied to each habitable room.	F1-Table 5.2d
The total supply air flow should usually be distributed in proportion to the habitable room volumes.	F1-Table 5.2d
Recirculation by the system of moist air from the wet rooms to the habitable rooms should be avoided.	F1-Table 5.2d
Cooker hoods should be 650 mm to 750 mm above the hob surface (or follow the manufacturer's instructions).	F1-Table 5.2d
Mechanical extract terminals and fans should be installed as high as is practical and preferably less than 400 mm below the ceiling.	F1-Table 5.2d
Mechanical supply terminals should be located and directed to avoid draughts.	F1-Table 5.2d
Where ducts etc. are provided in a dwelling with a protected stairway, precautions may be necessary to avoid the possibility of the system allowing smoke or fire to spread into the stairway.	F1-Table 5.2d

Controls

All controls should be set up to operate without occupant intervention, but may have manual or automatic controls to select the boost rate.	F1-Table 5.2d

Any manual boost controls should be provided locally to the spaces being served (e.g. bathrooms and kitchen) to guard against a single centrally located switch being left on by mistake.	F1-Table 5.2d
Automatic controls could include sensors for humidity, occupancy/usage and pollutant release.	F1-Table 5.2d
Humidity controls should not be used for sanitary accommodation as odour is the main pollutant.	F1-Table 5.2d
In kitchens, any automatic control must provide sufficient flow during cooking with fossil fuel (e.g. gas) to avoid build-up of combustion products.	F1-Table 5.2d
Where manual controls are provided, they should be within reasonable reach of the occupants.	F1-Table 5.2d

Operation

The maximum ('boost') rate should be the greater of the whole building ventilation rate or the whole dwelling air extract rate.	F Table 5.2d
The maximum individual room extract rates should be at least those given in Table 6.19.	F Table 5.2d
The minimum air supply rate should be at least the whole building ventilation rate.	F Table 5.2d

Single room heat recovery ventilator

If a Single Room Heat Recovery Ventilator (SRHRV) is used to ventilate a habitable room, to calculate the air flow rate, first determine the **whole building ventilation rate** from Table 6.19, and then work out the room supply rate using the following formula:

$$\frac{\text{Whole building ventilation rate} \times \text{room volume}}{\text{Total volume of all habitable rooms}}$$

When working out the continuous mechanical extract for a whole building, which also includes a room ventilated by an SRHRV, the following formula should be used:

$$\frac{\text{Whole building ventilation rate} \times \text{room volume} - \text{SRHRV rate}}{\text{Total volume of all habitable rooms}}$$

Mechanical intermittent extract

As odour is the main pollutant, humidity controls should **not** be used for intermittent extract in sanitary accommodation.	F Tables 5.2a to 5.2d
Ventilators equipped with intermittent extract shall be capable of being operated manually and/or automatically by a sensor (e.g. humidity sensor, occupancy/usage sensor, moisture/pollutant release detector etc.).	F Tables 5.2a to 5.2d
All ventilator automatic controls **must** be provided with a manual override to allow the occupant to turn the extract on.	F Tables 5.2a to 5.2d
Automatic controls for ventilators used in kitchens **must** be capable of providing sufficient flow during cooking with fossil fuels (e.g. gas) so as to avoid the build-up of combustion products.	F Tables 5.2a to 5.2d
If a fan is installed in an internal room without an openable window, then the fan should have a 15 minute overrun.	F Tables 5.2a to 5.2d
In rooms with no natural light, fans could be controlled by the operation of the main room light switch.	F Tables 5.2a to 5.2d

Note: In dwellings, humidistat controls should be available to regulate the humidity of the indoor air and to minimize the risk of condensation and mould growth. Humidistats are normally installed as part of an extract ventilator, especially in moisture-generating rooms such as a kitchen or a bathroom. They should **not** be used for sanitary accommodation, where the dominant pollutant is usually odour.

6.3.7 Ventilation systems – basements

If a basement is connected to the rest of the dwelling by a large permanent opening, such as an open stairway, then the whole dwelling, including the basement, should be treated as a multi-storey dwelling and ventilated in a similar manner to dwellings without basements.	F1-5.11

If the basement has a single exposed façade, while the rest of the dwelling above ground has more than one exposed façade, then passive stack ventilation or continuous mechanical extract should be used.   F1-5.11

For basements that are not connected to the rest of the dwelling by a large permanent opening, then: F1-5.12

- the part of the dwelling above ground should be considered separately;
- the basement should be treated as if it were a single-storey dwelling above ground.

If the part of the dwelling above ground has no bedrooms, then for the purpose of ventilation requirements: F1-5.12

- assume that the dwelling has one bedroom; and
- treat the basement as a single-storey dwelling (with one bedroom) as if it were above ground.

If a dwelling only compromises a basement, then it should be treated as if it were a single-storey dwelling (with one bedroom) above ground.   F1-5.13

6.3.8 Ventilation of habitable rooms through another room or a conservatory

Habitable rooms without an openable window shall be either ventilated through another habitable room or through a conservatory.   F1-5.11

6.3.9 Ventilation through another room

Habitable rooms without an openable window may be ventilated through another habitable room (see Figure 6.24) provided that the other room has: F1-5.15

- purge ventilation; and
- an $8000\,mm^2$ background ventilator; and
- there is a permanent opening between the two rooms.

Table 6.21 Ventilation systems for basements

Type of basement	Background ventilators and intermittent extract fans	Passive stack ventilation	Continuous mechanical extract	Continuous mechanical supply and extract with heat
Basement connected to the rest of the dwelling by an open stairway	Yes	Yes	Yes	Yes
Basement with a single exposed façade and dwelling above ground with more than one exposed façade		Yes	Yes	
Basements not connected to the rest of the dwelling by an open stairway	Yes	Yes	Yes	Yes
Dwelling above ground has no bedrooms	Yes	Yes	Yes	Yes
Dwelling comprises just a basement	Yes	Yes	Yes	Yes

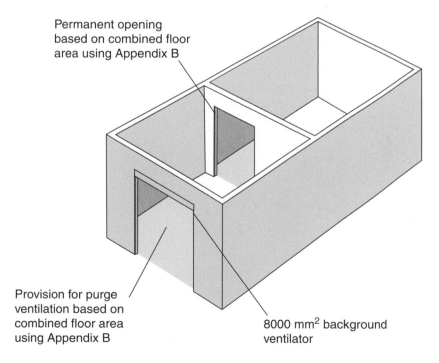

Permanent opening based on combined floor area using Appendix B

Provision for purge ventilation based on combined floor area using Appendix B

8000 mm² background ventilator

Figure 6.24 Two habitable rooms treated as a single room for ventilation purposes.

6.3.10 Ventilation through a conservatory

Habitable rooms without an openable window may be
ventilated through a conservatory (see Figure 6.25) provided
that that conservatory has:

F1-5.16

- purge ventilation; and
- an 8000 mm² background ventilator; and
- there is a closable opening between the room and the
  conservatory that is equipped with:

  - purge ventilation, and
  - an 8000 mm² background ventilator.

Note: If a building contains both living accommodation and space to be used for commercial purposes (e.g. workshop or office), the whole unit should be treated as a dwelling for the purposes of this Approved Document, as long as the commercial part could revert to domestic use.

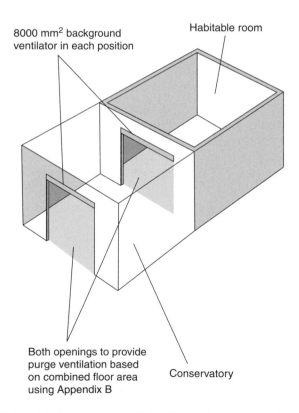

8000 mm² background ventilator in each position

Habitable room

Both openings to provide purge ventilation based on combined floor area using Appendix B

Conservatory

Figure 6.25 A habitable room ventilated through a conservatory.

6.3.11 Ventilation systems – buildings other than dwellings

Fresh air supplies should be protected from contaminants that would be injurious to health.　　F1-6.3

Offices

All office sanitary accommodation, washrooms and food and beverage preparation areas shall be provided with intermittent air extract ventilation capable of meeting the requirements of Table 6.22.　　F1-6.10

Table 6.22 Extract ventilation rate

Room	Air extract rate
Rooms containing printers and photocopiers in substantial use (greater than 30 minutes per hour)	20 l/s per machine during use
Office sanitary accommodation and washrooms	15 l/s per shower/bath 6 l/s per WC/urinal
Food and beverage preparation areas (not commercial kitchens)	15 l/s with microwave and beverages only 30 l/s adjacent to the hob with cooker(s) 60 l/s elsewhere with cooker(s)
Specialist buildings and spaces (e.g. commercial kitchens, fitness rooms)	See Table 6.3 of Approved Document F: 2010, pages 36–37

Extract fans that are located in an internal room which does not have an openable window, should have a 15 minute overrun.　　F1-Table 6.2c

Extract ventilators should be located as high as practicable and preferably not less than 400 mm below the ceiling level.　　F1-Table 6.2b

PSVs should be located in the ceiling of the room.　　F1-Table 6.2b

Purge ventilation shall be provided in each office.　　F1-6.12

Purged air should be taken directly to outside and should not be recirculated to any other part of the building.　　F1-6.12

PSVs can be used as an alternative to a mechanical extract fan for office sanitary and washrooms and food preparation areas.　　F1-Table 6.2a

PSV controls can be either manual or automatic.	F1-Table 6.2c
The controls for extract fans can be either manual or automatic.	F1-Table 6.2c
Printers and photocopiers that are being used in large numbers and which are in almost constant use (i.e. greater than 30 minutes per hour) shall:	F1-6.10

- be located in a separate room;
- have extract facilities capable of providing an extract rate greater than 20 l/s per machine, during use (see Table 6.22).

The whole building ventilation rate for the supply of air to the offices should be greater than 10 l/s per person (see Table 6.23).	F1-6.11

Table 6.23 Whole-building ventilation rate for air supply to offices

	Air supply rate
Total outdoor air supply rate for offices (no-smoking and no significant pollutant sources)	10 l/s per person

The following air flow rates can mainly be provided by natural ventilation.

Modular and portable buildings

Other types of building

The ventilation requirements for other buildings (such as assembly halls, broadcasting studios, computer rooms, factories, hospitals, hotels, museums, schools, sports centres and warehouses etc.) are listed in Table 6.3 of Approved Document F, which also provides a link to the relevant controlling Acts of Parliament, Statutory Instruments, British Standards, and Chartered Institution of Building Services Engineers (CIBSE) and Health and Safety Executive (HSE) standards, practices and recommendations.

Sensors

Ventilation in buildings **other than** dwellings is dependent on occupancy levels, and currently there are some very sophisticated automatic control systems available, such as local passive infrared detectors and electronic carbon dioxide detectors.

6.3.12 Ventilation systems – car parks

Ventilation is the important factor and, as heat and smoke cannot be dissipated so readily from a car park that is not open-sided, fewer concessions are made.

Underground car parks, enclosed car parks and multi-storey car parks should be designed to limit the concentration of carbon monoxide to not more than 30 ppm (parts per million) averaged over an 8-hour period and peak concentrations.	F1-6.18a
Ramps and exits shall not go above 90 ppm for periods not exceeding 15 minutes.	F1-6.18b
Naturally ventilated car parks shall have openings at each car parking level: • at least 1/20 of the floor area at that level; • with a minimum of 25% on each of two opposing walls.	F1-6.19a
Mechanically ventilated car parks can have either natural ventilation openings that are not less than 1/40 of the floor area or a mechanical ventilation system capable of at least three air changes per hour (3 ach).	F1-6.2bi
Mechanically ventilated basement car parks shall be capable of at least six air changes per hour (6 ach).	F1-6.2bi
Mechanically ventilated exits and ramps (i.e. where cars queue inside the building with engines running) shall be capable of at least ten air changes per hour (10 ach).	F1-6.2bii

6.3.13 General requirements

To ensure good transfer of air throughout the dwelling there shall be an undercut of 7600 mm^2 (minimum) in all internal doors above the floor finish (equivalent to an undercut of 10 mm for a standard 760 mm width door).	F Tables 5.2a to 5.2d
Adequate replacement air must also be available (e.g. a 10 mm gap under the door or equivalent).	F Tables 5.2a to 5.2d
All ducting that passes through a fire-stopping wall or fire compartment shall meet the requirements of Approved Document B of the Building Regulations.	F Tables 5.2a to 5.2d

Duct runs should be straight, with as few bends and kinks as possible to minimize system resistance.	F Tables 5.2a to 5.2d
Horizontal ducting (including ducting in walls) should be arranged to slope slightly downwards away from the fan to prevent backflow of any moisture.	F Tables 5.2a to 5.2d
Fans and/or ducting placed in, or passing through, an unheated void or loft space should be insulated to reduce the possibility of condensation forming.	F Tables 5.2a to 5.2d
The inner radius of any bend should be greater than or equal to the diameter of the ducting being used (see Figure 6.26).	F Tables 5.2a to 5.2d

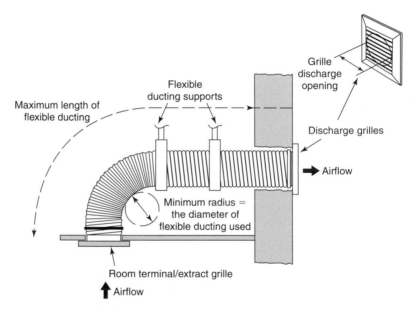

Figure 6.26 Correct installation of ducting.

Vertical duct rises may need to be fitted with a condensation trap in order to prevent the backflow of any moisture.	F Tables 5.2a to 5.2d
The circular profile of a flexible duct should be maintained throughout the full length of the duct run (see Figure 6.26).	F Tables 5.2a to 5.2d

If a back-draught device is used it may be incorporated into the fan itself. — F Tables 5.2a to 5.2d

Flexible ducting should be installed without any peaks or troughs (see Figure 6.27). — F Tables 5.2a to 5.2d

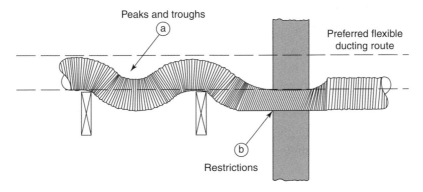

Peaks and troughs
(a)

Preferred flexible ducting route

(b)
Restrictions

Figure 6.27 Incorrect installation of ducting.

Access

There should be reasonable access to ventilation systems to enable changing filters, replacing defective components, cleaning duct work and other maintenance activities. — F Tables 5.2a to 5.2d

Accessibility of controls

Ventilators that are provided with manual controls (e.g. pull cords, operating rods etc.) should be: — F Tables 5.2a to 5.2d

- within reasonable reach of occupants
- located in accordance with the guidance for Requirements of Approved Document N3 (*Safe Opening and Closing of Windows*) as detailed below.

Where controls can be reached without leaning over an obstruction, they should not be more than 1.9 m above the floor. Where there is an obstruction the control should be lower (e.g. not more than 1.7 m where there is a 600 mm deep obstruction). — N3 3.2

Where controls cannot be positioned within safe reach from a permanent stable surface, a safe means of remote operation, such as a manual or electrical system, should be provided.	N3 3.2
Where there is a danger of the operator or other person falling through a window above ground floor level, suitable opening limiters should be fitted or guarding should be provided.	N3 3.3

Note: Although Requirement N3 only applies to work places, for ventilation purposes this requirement also applies to dwellings.

Room Thermostats

Room thermostats located in an individual flat with an internal protected stairway or entrance hall should be mounted in the living room at a height between 1370 mm and 1830 mm and its maximum setting should not exceed 27°C.	B1 2.18 (V2)

Access for maintenance

Buildings other than dwellings should include: • reasonable access for the purpose of replacing filters, fans and coils; and • availability of access points for cleaning duct work.	F 6.6
Central plant rooms should include adequate space for the maintenance of the plant (see Figure 6.28).	F 6.7

Combustion appliances

A room containing an open-flued appliance may need permanently open air vents.	J 1.4
If open-flued combustion appliances and extract fans are going to be installed, then the combustion appliance should be capable of operating safely – whether or not the fans are running.	F 1.3

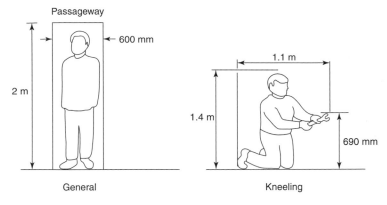

Figure 6.28 Access space in central plant rooms.

Exhaust outlets

Exhaust outlets should be located so that re-entry, or ingestion, into the building and/or other nearby buildings, is minimized. This can be achieved by ensuring that:

exhausts:	F Tables 5.2a to 5.2d
• are located downstream of air intakes which are located in a prevailing wind direction;	
• do not discharge into courtyards, enclosures or architectural screens;	
• stacks discharge vertically upwards with sufficient height to clear surrounding buildings and avoid a downwash occurring.	F Tables 5.2a to 5.2d

 Note: Where possible, pollutants from stacks should be grouped together and discharged vertically upwards.

External doors

The height times the width of an external door (including patio doors) should be at least 1/20 of the floor area of the room.	F Tables 5.2a to 5.2d

If a room contains more than one external door (or a combination of at least one external door and at least one openable window) then the areas of **all** the opening parts may be added together to achieve the required floor area.

Note: See Appendix C of Approved Document F for example calculations for ventilator sizing for dwellings using:

* background ventilators and intermittent extract fans;
* continuous mechanical extract;
* continuous mechanical supply and extract with heat recovery;
* passive stack ventilation.

Location of ventilation devices in rooms

Openings in compartment walls and/or compartment floors should be limited to those for the passage of pipes, ventilation ducts, service cables, chimneys, appliance ventilation ducts or ducts encasing one or more flue pipes.	B3 8.34 (V2)
Cooker hoods should be 650 mm to 750 mm above the hob surface.	F Tables 5.2a to 5.2d
Ducts etc. that are located in a protected stairway of a dwelling shall not allow smoke or fire to spread into the stairway.	F Tables 5.2a to 5.2d
Mechanical extract terminals and fans should be located as high as possible.	F Tables 5.2a to 5.2d
Mechanical supply terminals should be located and directed to avoid draughts.	F Tables 5.2a to 5.2d
Recirculation (i.e. by the system) of moist air from wet rooms to habitable rooms should be avoided.	F Tables 5.2a to 5.2d

Windows

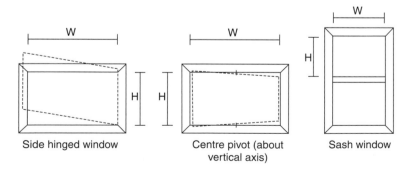

Side hinged window   Centre pivot (about vertical axis)   Sash window

Figure 6.29 Window dimensions.

Note: The window opening area is the dimensions of the open area (i.e. height (H) × width (W)).

> The height times width of the opening part of hinged or pivot windows that are designed to open **more** than 30° and/or sliding sash windows should be at least 1/20 of the floor area of the room.
> F App B
>
> The height times the width of the opening part of hinged or pivot windows designed to open **less** than 30° should be at least 1/10 of the floor area of the room.
> F App B
>
> If a room contains more than one openable window, then the areas of **all** the opening parts may be added together to achieve the required floor area.

If the room contains more than one openable window, the areas of all the opening parts may be added to achieve the required proportion of the floor area.

Protected shaft

> A protected shaft conveying piped flammable gas should have ventilation openings to the outside air at high and low level in the shaft.
> B2 8.41 (V2)

Note: Generally speaking, an external wall of a protected shaft does not need to have fire resistance (but see BS 5588-5:2004 for fire resistance of external walls of fire-fighting shafts).

6.3.14 Work on existing buildings

Under Regulation 3(1) and 3(1A) of the Building Regulations 2000 (as amended), **windows are a controlled fitting**. These clauses, therefore, make it mandatory that when windows in an existing building are replaced the replacement work:

- **shall** comply with the requirements of Approved Document L and Approved Document N;
- **shall** not have a worse level of compliance with other applicable Approved Documents of Schedule 1 (in particular Approved Documents B, F and J).

Replacement windows

> All replacement windows should include trickle ventilators or have an equivalent background ventilation opening in the same room.
> F 7.4

Ventilation openings should not be smaller than the original   F 7.6
opening and it should be controllable.

Where there was no previous ventilation opening, or where   F 7.6
the size of the original ventilation opening is not known, the
replacement window(s) shall be greater than the minimum
requirements shown in Tables 6.24 and 6.25.

Table 6.24 Equivalent areas for replacement windows – dwellings

Type of room	Equivalent area
Habitable rooms	5000 mm²
Kitchen	2500 mm²
Utility room	2500 mm²
Bathroom (without a WC)	2500 mm²

Table 6.25 Equivalent areas for replacement windows – buildings other than dwellings

Type of room	Equivalent area
Occupiable rooms with floor areas <10 m²	2500 mm²
Occupiable rooms with floor areas >10 m²	250 mm² per m² of floor area
Kitchens (domestic type)	2500 mm²
Bathrooms and shower rooms	2500 mm² per bath or shower
Sanitary accommodation (and/or washing facilities)	2500 mm² per WC

The addition of a habitable room

The general ventilation rates for an additional habitable room (not including a conservatory) to an existing building may be achieved by using background ventilators, heat recovery ventilators and/or purge ventilation.

A single room heat recovery ventilator may be used to ventilate an additional habitable room (F 7.8b).

Additional requirements for background ventilators

If the additional room is connected to an existing habitable   F 7.8a(i)
room which now has no windows opening to outside, then
the ventilation opening (or openings) shall be greater than
8000 mm² equivalent area.

If the additional room is connected to an existing habitable room which still has windows opening to outside, but with a total background ventilator equivalent area less than 5000 mm² equivalent area, then the ventilation opening (or openings) shall be greater than 8000 mm² equivalent area.

F 3.7a(ii)

If the additional room is connected to an existing habitable room which still has windows opening to outside, but with a total background ventilator equivalent area of at least 5000 mm² equivalent area, then there should be:

F 3.7a(iii)

- background ventilators of at least 8000 mm² equivalent area between the two rooms; and
- background ventilators of at least 8000 mm² equivalent area between the additional room and outside.

The addition of a wet room to an existing building

Internal doors between the wet room and the existing building should have an undercut of at least minimum area 7600 mm² (equivalent to an undercut of 10 mm above the floor finish for a standard 760 mm width door).

F 7.13

Whole building and extract ventilation can be provided by:

F 7.12

- intermittent extract and a background ventilator of at least 2500 mm² equivalent area; or
- single room heat recovery ventilator; or
- passive stack ventilator; or
- continuous extract fan.

The addition of a conservatory to an existing building

The general ventilation rate for conservatories with floor area greater than 30 m² conservatory (and adjoining rooms) can be achieved by the use of background ventilators.

F 7.16-7.18

Refurbishing a kitchen or bathroom in an existing dwelling

> If any of the work being carried out in the kitchen or bathroom   F 7.23
> of an existing building is 'building work' (as defined in
> Regulation 3 of the Building Regulations and there is an
> existing extract fan or passive stack ventilator (or cooker hood
> extracting to outside in the kitchen) this will suffice (always
> assuming that it is in good working order!).

Historic buildings

Ventilation systems should **not** introduce new or increased technical risk, or in any other way prejudice the use or character of the building – particularly historic buildings that are:

- listed buildings;
- buildings in Conservation Areas;
- buildings which are of architectural and historical interest;
- buildings which are of architectural and historical interest within National Parks, Areas of Outstanding Natural Beauty, historic parks and gardens, registered battlefields, the curtilages of scheduled ancient monuments, and World Heritage Sites; and
- buildings of traditional construction with permeable fabric that both absorbs and readily allows the evaporation of moisture,

as these are exempt from compliance with the requirements of the Building Regulations.

When undertaking work on, or in connection with, a building that falls within one of the classes listed above, the aim should be to provide adequate ventilation as far as is reasonable and practically possible – without damaging the character of the building or increasing the risk of long-term deterioration of the building fabric or fittings

Many books have been written about the problems related to restoring historic buildings and, before considering any work of this nature, you would be advised to seek the advice of the local planning authority's conservation officer, particularly if you are contemplating:

- the restoration of a historic building that had been subject to previous inappropriate alteration (e.g. replacement windows, doors and roof-lights);
- rebuilding a former historic building following a fire or major demolition;
- making the building's fabric able to 'breathe', in order to control moisture and potential long-term decay.

In all cases:

The overall aim should be to improve ventilation of a historic   F 3.11
building without:

- having a detrimental influence on the character of the building;
- increasing the risk of long-term deterioration of the building's fabric or fittings.

6.4 Drainage

6.4.1 The requirement (Building Act 1984 Sections 21 and 22)

Drains – fire protection

Drains should also provide a degree of fire protection as shown by the following requirement:

- *all openings in fire-separating elements shall be suitably protected in order to maintain the integrity of the continuity of the fire separation,*
- *any hidden voids in the construction shall be sealed and subdivided to inhibit the unseen spread of fire and products of combustion, in order to reduce the risk of structural failure, and the spread of fire.*

(Approved Document B3)

Foul water drainage

The foul water drainage system shall:

- *convey the flow-off foul water to a foul water outfall (i.e. sewer, cesspool, septic tank or settlement (i.e. holding) tank),*
- *minimise the risk of blockage or leakage,*
- *prevent foul air from the drainage system from entering the building under working conditions,*
- *be ventilated,*
- *be accessible for clearing blockages,*
- *not increase the vulnerability of the building to flooding.*

(Approved Document H1)

Wastewater treatment systems and cesspools

Wastewater treatment systems shall:

- *have sufficient capacity to enable breakdown and settlement of solid matter in the wastewater from the buildings;*
- *be sited and constructed so as to prevent overloading of the receiving water.*

*Cesspools shall have sufficient capacity to store the foul water from the build-
ing until they are emptied.*

*Wastewater treatment systems and cesspools shall be sited and constructed
so as not to:*

* *be prejudicial to health or a nuisance;*
* *adversely affect water sources or resources;*
* *pollute controlled waters;*
* *be in an area where there is a risk of flooding.*

*Septic tanks and wastewater treatment systems and cesspools are constructed
and sited so as to:*

* *have adequate ventilation;*
* *prevent leakage of the contents and ingress of subsoil water;*
* *have regard to water table levels at any time of the year and rising
groundwater levels.*

Drainage fields are sited and constructed so as to:

* *avoid overloading of the soakage capacity, and*
* *provide adequately for the availability of an aerated layer in the soil at all
times.*

(Approved Document H2)

Rainwater drainage

Rainwater drainage systems shall:

* *minimize the risk of blockage or leakage;*
* *be accessible for clearing blockages;*
* *ensure that rainwater soaking into the ground is distributed sufficiently
so that it does not damage foundations of the proposed building or any
adjacent structure;*
* *ensure that rainwater from roofs and paved areas is carried away from
the surface either by a drainage system or by other means;*
* *ensure that the rainwater drainage system carries the flow of rainwater
from the roof to an outfall (e.g. a soakaway, a watercourse, a surface
water or a combined sewer).*

(Approved Document H3)

Building over existing sewers

Building or extension or work involving underpinning shall:

* *be constructed or carried out in a manner which will not overload or
otherwise cause damage to the drain, sewer or disposal main either
during or after the construction;*
* *not obstruct reasonable access to any manhole or inspection chamber on
the drain, sewer or disposal main;*

- in the event of the drain, sewer or disposal main requiring replacement, not unduly obstruct work to replace the drain, sewer or disposal main, on its present alignment;
- reduce the risk of damage to the building as a result of failure of the drain, sewer or disposal main.

(Approved Document H4)

Separate systems for drainage

Separate systems of drains and sewers shall be provided for foul water and rainwater where:

(a) the rainwater is not contaminated; and
(b) the drainage is to be connected either directly or indirectly to the public sewer system and either –

    (i) the public sewer system in the area comprises separate systems for foul water and surface water; or

    (ii) a system of sewers which provides for the separate conveyance of surface water is under construction either by the sewerage undertaker or by some other person (where the sewer is the subject of an agreement to make a declaration of vesting pursuant to section 104 of the Water Industry Act 1991).

(Approved Document H5)

Solid waste storage shall be:

- designed and sited so as not to be prejudicial to health,
- of sufficient capacity having regard to the quantity of solid waste to be removed and the frequency of removal,
- sited so as to be accessible for use by people in the building and of ready access from a street for emptying and removal.

(Approved Document H6)
(Building Act 1984 Section 84)

You are required by the Building Act 1984 to ensure that all courts, yards and passageways giving access to a house, industrial or commercial building (not maintained at public expense) are capable of allowing satisfactory drainage of its surface or subsoil to a proper outfall.

The local authority can require the owner of any of the buildings to complete such works as may be necessary to remedy the defect.

All plans for building work need to show that drainage of refuse water (e.g. from sinks) and rainwater (from roofs) have been adequately catered for. Failure to do so will mean that these plans will be rejected by the local authority. All plans for buildings must include at least one water or earth closet **unless** the local authority is satisfied that one is not required (e.g. in a large garage separated from the house).

If you propose using an earth closet, the local authority cannot reject the plans unless they consider that there is insufficient water supply to that earth closet.

What are the rules about drainage? (Building Act 1984 Section 59)

The Building Act requires that all drains are connected either with a sewer (unless the sewer is more than 120 ft away or the person carrying out the building work is not entitled to have access to the intervening land) or is able to discharge into a cesspool, settlement tank or other tank designed for the reception and/or disposal of foul matter from buildings.

The local authorities view this requirement very seriously and will need to be satisfied that:

• satisfactory provision has been made for drainage;
• all cesspools, private sewers, septic tanks, drains, soil pipes, rain water pipes, spouts, sinks or other appliances are adequate for the building in question;
• all private sewers that connect directly or indirectly to the public sewer are not capable of admitting subsoil water;
• the condition of a cesspool is not detrimental to health, or does not present a nuisance;
• cesspools, private sewers and drains previously used, but now no longer in service, do not prejudice health or become a nuisance.

This requirement can become quite a problem if it is not recognized in the early planning stages and so it is always best to seek the advice of the local authority. In certain circumstances, the local authority might even help to pay for the cost of connecting you up to the nearest sewer!

The local authority has the authority to make the owner renew, repair or cleanse existing cesspools, sewers and drains etc.

Can two buildings share the same drainage?

Usually the local authority will require every building to be drained separately into an existing sewer but in some circumstances they may decide that it would be more cost effective if the buildings were drained in combination. On occasions, they might even recommend that a private sewer is constructed.

What about ventilation of soil pipes? (Building Act 1984 Section 60)

A major requirement of the Building Regulations is that all soil pipes from water closets shall be properly ventilated and that no use shall be made of:

• an existing or proposed pipe designed to carry rain water from a roof to convey soil and drainage from a sanitary convenience;
• an existing pipe designed to carry surface water from a premises to act as a ventilating shaft to a drain or a sewer conveying foul water.

What happens if I need to disconnect an existing drain?
(Building Act 1984 Section 62)

If, in the course of your building work, you need to:

- reconstruct, renew or repair an existing drain that is joined up with a sewer or another drain;
- alter the position of an existing drain that is joined up with a sewer or another drain;
- seal off an existing drain that is joined up with a sewer or another drain,

then, provided that you give 48 hours' notice to the local authority, the person undertaking the reconstruction may break open any street for this purpose.

You do not need to comply with this requirement if you are demolishing an existing building.

Can I repair an existing water closet or drain? (Building
Act 1984 Section 63)

Repairs can be carried out to water closets, drains and soil pipes, but if that repair or construction work is prejudicial to health and/or a public nuisance, then the person who completed the installation or repair is liable, on conviction, to a heavy fine.

In the Greater London area, a 'water closet' can **also** be taken to mean a 'urinal'.

Can I repair an existing drain? (Building Act 1984 Section 61)

Only in extreme emergencies are you allowed to repair, reconstruct or alter the course of an underground drain that joins up with a sewer, cesspool or other drainage method (e.g. septic tank).

If you have to carry out repairs etc. in an emergency, then make sure that you do **not** cover over the drain or sewer without notifying the local authority of your intentions!

6.4.2 Meeting the requirement

Enclosures for drainage and/or water supply pipes

The enclosure should:	B3 7.6–7.9 (V1)
• be bounded by a compartment wall or floor, an outside wall, an intermediate floor, or a casing; • have internal surfaces (except framing members) of Class 0;	B3 10.7 (V2)

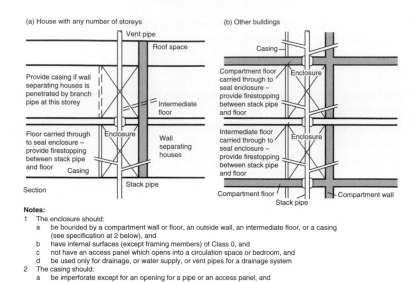

Figure 6.30 Enclosure for drainage or water supply pipes.

- not have an access panel which opens into a circulation space or bedroom;
- be used only for drainage, or water supply, or vent pipes for a drainage system.

The casing should: B3 7.6–7.9 (V1)

- be imperforate except for an opening for a pipe   B3 10.7 (V2)
  or an access pane
- not be of sheet metal;
- have (including any access panel) not less than
  30 minutes' fire resistance.

The opening for a pipe, either in the structure or   B3 7.6–7.9 (V1)
the casing, should be as small as possible and fire-   B3 10.7 (V2)
stopped around the pipe.

Protection of openings for pipes

Pipes which pass through a compartment wall or compartment floor (unless the pipe is in a protected shaft), or through a cavity barrier, should conform to one of the following alternatives:

Proprietary seals (any pipe diameter) that maintain the fire resistance of the wall, floor or cavity barrier.	B3 7.6–7.7 (V1) B3 10.5–10.6 (V1)
Pipes with a restricted diameter should be used where fire-stopping is used around the pipe, keeping the opening as small as possible	B3 7.6–7.8 (V1) B3 10.5–10.7 (V2)
Sleeving – a pipe of lead, aluminium, aluminium alloy, fibre-cement or UPVC, with a maximum nominal internal diameter of 160 mm, may be used with a sleeving of non-combustible pipe as shown in Figure 6.31.	B3 7.6–7.9 (V1) B3 10.5–10.8(V2)

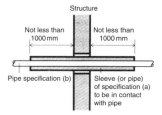

Figure 6.31 Pipes penetrating a structure.

Foul water drainage

The capacity of the system should be large enough to carry the expected flow at any point (BS 5572, BS 8301).	H1 (0.1)
All pipes, fittings and joints should be capable of withstanding an air test of positive pressure of at least 38 mm water gauge for at least 3 minutes.	H1 (1.38)
Every trap should maintain a water seal of at least 25 mm.	H1 (1.38)

Traps

| All points of discharge into the system should be fitted with a trap (e.g. a water seal) to prevent foul air from the system entering the building. | H1 (1.3–1.4) |
| All traps should be fitted directly over an appliance and should be removable or be fitted with a cleaning eye. | H1 (1.6) |

Branch discharge pipes

Branch pipes should either discharge into another branch pipe or a discharge stack (unless the appliances discharge into a gully on the ground floor or at basement level).	H1 (1.5)
If the appliances are on the ground floor, the pipe(s) may discharge to a stub stack, discharge stack, directly to a drain, or (if the pipe carries only waste water) to a gully.	H1 (1.5–1.17) H1 (1.11) H1 (1.30)
A branch pipe from a ground floor closet should only discharge directly to a drain if the depth from the floor to the drain is 1.3 m or less (see Figure 6.32).	H1 (1.9)

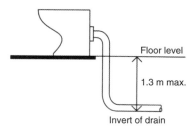

Floor level

1.3 m max.

Invert of drain

Figure 6.32 Direct connection of ground floor WC to drain.

A branch pipe serving any ground floor appliance may discharge direct to a drain or into its own stack.	H1 (A5)
A branch pipe should not discharge into a stack in a way which could cause cross-flow into any other branch pipe (see Figure 6.33).	H1 (1.10)

Branch discharge pipes

A branch discharge pipe should not discharge into a stack lower than 450 mm above the invert of the tail of the bend at the foot of the stack in single dwellings up to three storeys (see Figure 6.34).	H1 (1.8) H1 (A3, A4) H1 (1.21)

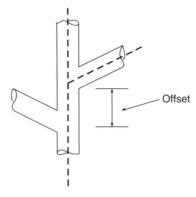

Figure 6.33 Branch connections.

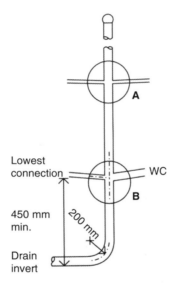

Figure 6.34 Branch discharge stack.

Branch discharge pipes

Branch pipes may discharge into a stub stack.	H1 (1.12)
	H1 (1.30)
A branch pipe discharging to a gully should terminate between the grating or sealing plate and the top of the water seal.	H1 (1.13)
Bends in branch pipes should be avoided if possible (see Figure 6.35).	H1 (1.16)

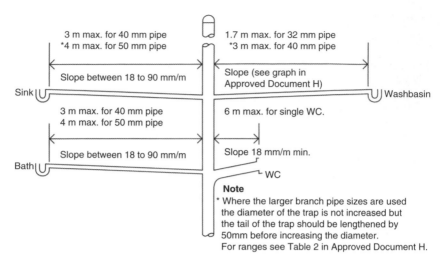

Figure 6.35 Branched connections.

Junctions on branch pipes should be made with a sweep of 25 mm radius or at 45°.	H1 (1.17)
Rodding points should be provided to give access to any lengths of discharge pipes which cannot be reached by removing traps or appliances with integral traps.	H1 (1.25) H1 (1.6)
A branch pipe discharging to a gully should terminate between the grating or sealing plate and the top of the water seal.	H1 (1.13)
Condensate drainage from boilers may be connected to sanitary pipework provided the connection is made to an internal stack with a 75 mm condensate trap.	H1 (1.14)
If the connection is made to a branch pipe, the connection should be made downstream of any sink waste connection.	H1 (1.14)
All sanitary pipework receiving condensate should be made from materials resistant to a pH value of 6.5 or lower and be installed in accordance with BS 6798.	H1 (1.14)
Pipes serving a single appliance should have at least the same diameter as the appliance trap (see Table 6.26).	

A separate ventilating stack is only likely to be preferred where the numbers of sanitary appliances and their distance to a discharge stack are large.

Table 6.26 Minimum trap sizes and seal depths

Appliance	Diameter of trap (mm)	Depth of seal (mm of water or equivalent)
Washbasin	32	75
Bidet		
Bath	40	50
Shower		
Food waste disposal unit	40	75
Urinal bowl		
Sink		
Washing machine		
Dishwashing machine		
WC pan (outlet 80 mm)	75	50
WC pan (outlet 80 mm)	100	50

Branch ventilation stacks

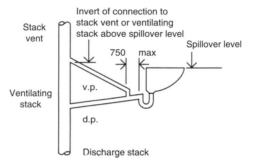

Figure 6.36 Branch ventilation pipes.

Should be connected to the discharge pipe within 750 mm of the trap and should connect to the ventilating stack or the stack vent, above the highest 'spillover' level of the appliances served.	H1 (1.22)
The ventilating pipe should have a continuous incline from the discharge pipe to the point of connection to the ventilating stack or stack vent.	H1 (1.22)
Branch ventilating pipes which run direct to outside air should finish at least 900 mm above any opening into the building nearer than 3 m.	H1 (1.23)
A dry stack may provide ventilation for branch ventilation pipes as an alternative to carrying them to outside air or to a ventilated discharge stack (ventilated system).	H1 (A7 and 1.21)

Ventilation stacks serving buildings with not more than 10 storeys and containing only dwellings should be at least 32 mm diameter (for all other buildings see paragraph H1 (1.29)).	H1 (A8) H1 (1.21 and 1.29)
A separate ventilating stack is only likely to be preferred where the numbers of ventilating pipes and their distance to a discharge stack are large.	H1 (1.19) H1 (Table 2)

Discharge stacks

All stacks should discharge to a drain.	H1 (1.26)
The bend at the foot of the stack should have as large a radius (i.e. at least 200 mm) as possible.	H1 (1.26)
Discharge stacks should be ventilated.	H1 (1.29)
Offsets in the 'wet' portion of a discharge stack should be avoided.	H1 (1.27)
Stacks serving urinals should be not less than 50 mm.	H1 (1.28)
Stacks serving closets with outlets less than 80 mm should be not less than 75 mm.	H1 (1.28)
Stacks serving closets with outlets greater than 80 mm should be not less than 100 mm.	H1 (1.28)
The internal diameter of the stack should be not less than that of the largest trap or branch discharge pipe.	H1 (1.28)
Ventilating pipes open to outside air should finish at least 900 mm above any opening into the building within 3 m and should be fitted with a perforated cover or cage (see Figure 6.37) which should be metal if rodent control is a problem.	H1 (1.31)
Ventilating pipes open to outside air should finish at least 900 mm above any opening into the building within 3 m.	H1 (1.31)
Air admittance valves should be located in areas that have adequate ventilation.	H1 (1.33)
Air admittance valves should not be used outside buildings or in dust-laden atmospheres.	H1 (1.33)
Rodding points should be provided in discharge stacks.	H1 (1.34)

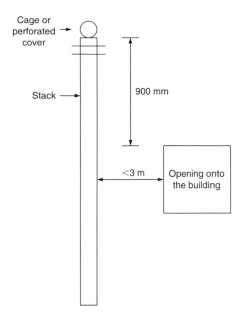

Figure 6.37 Termination of ventilation stacks.

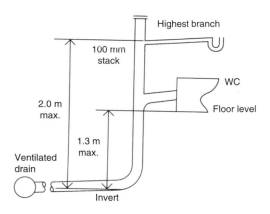

Figure 6.38 Stub stack.

Pipes should be firmly supported without restricting thermal movement.	H1 (1.35)
Sanitary pipework connected to WCs should not allow light to be visible through the pipe wall, as this is believed to encourage damage by rodents.	H1 (1.36)

Drainage serving kitchens in commercial hot food premises should be fitted with a grease separator complying with prEN 1825-1.

Foul drainage

Some public sewers may carry foul water and rainwater in the same pipe. If the drainage system is also to carry rainwater to such a sewer these combined systems should not be capable of discharging into a cesspool or septic tank.	H1 (2.1)
Foul drainage should be connected to either:	
• a public foul or combined sewer (wherever this is reasonably practicable);	H1 (2.3)
• an existing private sewer that connects with a public sewer; or	H1 (2.6)
• a wastewater treatment system or cesspool.	H1 (2.7)
Combined and rainwater sewers shall be designed to surcharge (i.e. the water level in the manhole rises above the top of the pipe) in heavy rainfall.	H1 (2.8)
Basements containing sanitary appliances, where the risk of flooding due to surcharge of the sewer is possible, should either use an anti-flooding valve (if the risk is low) or be pumped.	H1 (2.9) H1 (2.36–2.39) H1 (2.10)
For other low-lying sites (i.e. not basements) where the risk is considered low, a gully (at least 75 mm below the floor level) can be dug outside the building.	
Anti-flooding valves should preferably be a double valve type that complies with prEN 13564.	H1 (2.11)
The layout of the drainage system should be kept simple.	H1 (2.13)

 Pipes should (wherever possible) be laid in straight lines. Changes of direction and gradient should be minimized.

Access points should be provided only if blockages could not be cleared without them.	H1 (2.13)
Connections should be made using prefabricated components.	H1 (2.15)

Connection of drains to other drains or private or public sewers and of private sewers to public sewers should be made obliquely, or in the direction of flow.	H1 (2.14)
The system should be ventilated by a flow of air.	H1 (2.18) H1 (1.27–1.29)
Ventilating pipes should not finish near openings in buildings.	H1 (2.18) H1 (1.31)
Pipes should be laid to even gradients and any change of gradient should be combined with an access point.	H1 (2.19) H1 (2.49)
Pipes should also be laid in straight lines where practicable.	H1 (2.20) H1 (2.49)

Rodent control

If the site has been previously developed, the local authority should be consulted to determine whether any special measures are necessary for control of rodents. Special measures which may be taken include the following:

Sealed drainage – access covers to the pipework should be in the inspection chamber instead of an open channel.	H1 (2.22a)
Intercepting traps – should be of the locking type that can be easily removed from the chamber surface and securely replaced.	H1 (2.22b)
Rodent barriers – including enlarged sections on discharge stacks to prevent rats climbing, flexible downward facing fins in the discharge stack, or one-way valves in underground drainage.	H1 (2.22c)
Metal cages on ventilator stack terminals – to discourage rats from leaving the drainage system.	H1 (2.22d) H1 (1.31)
Covers and gratings to gullies – used to discourage rats from leaving the system.	H1 (2.22e)
During construction, drains and sewers that are left open should be covered when work is not in progress to prevent entry by rats.	H1 (2.56)

Disused drains or sewers less than 1.5 m deep that are in   H1 (B18)
open ground should as far as is practicable be removed.
Other pipes should be sealed at both ends (and at any
point of connection) and grout filled to ensure that rats
cannot gain access.

Protection from settlement

- A drain may run under a building if at least 100 mm of   H1 (2.23)
  granular or other flexible filling is provided round the
  pipe.

- Where pipes are built into a structure (e.g. inspection   H1 (2.24)
  chamber, manhole, footing, ground beam or wall) suitable measures (such as using rocker joints or a lintel)
  should be taken to prevent damage or misalignment
  (see Figures 6.39 and 6.40).

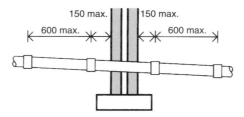

Figure 6.39 Pipe embedded in the wall. Short length of pipe embedded in a wall with joints within 150 mm of either wallface. Additional rocker pipes (max. length 600 mm) with flexible joints are then added.

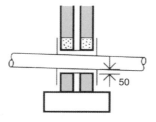

Figure 6.40 Pipe shielded by a lintel. Both sides are masked with rigid sheet material (to prevent entry of fill or vermin) and the void is filled with a compressible sealant to prevent entry of gas.

The depth of cover will usually depend on the levels of the connections to the system, the gradients at which the pipes should be laid and the ground levels.

H1 (2.27)

H1 (2.41–2.45)

All drain trenches should not be excavated lower than the foundations of any building nearby (see Figure 6.41).

H1 (2.25)

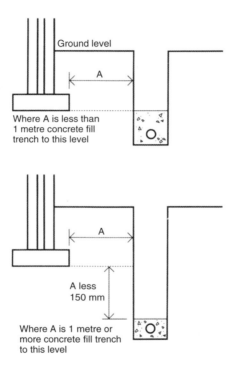

Ground level

A

Where A is less than 1 metre concrete fill trench to this level

A

A less 150 mm

Where A is 1 metre or more concrete fill trench to this level

Figure 6.41 Pipe runs near buildings.

Pipe gradients and sizes

Drains should have enough capacity to carry the anticipated maximum flow (see Table 6.27).

H1 (2.29)

Sewers (i.e. a drain serving more than one property) should have a minimum diameter of 100 mm when serving 10 dwellings or diameter of 150 mm if more than 10.

H1 (2.30)

Table 6.27 Flow rates from dwellings

Number of dwellings	Flow rate (l/s)
1	2.5
5	3.5
10	4.1
15	4.6
20	5.1
25	5.4
30	5.8

Drains carrying foul water should have an internal diameter of at least 75 mm.	H1 (2.33)
Drains carrying effluent from a WC or trade effluent should have an internal diameter of at least 100 mm.	H1 (2.33)

Pumping installations

Where gravity drainage is impracticable, or protection against flooding due to surcharge in downstream sewers is required, a pumping installation will be needed.	H1 (2.36)
Where foul water drainage from a building is to be pumped, the effluent receiving chamber should be sized to contain 24-hour inflow to allow for disruption in service.	H1 (2.39)
The minimum daily discharge of foul drainage should be taken as 150 litres per head per day for domestic use.	H1 (2.39)

Materials for pipes and jointing

To minimize the effects of any differential settlement, pipes should have flexible joints.	H1 (2.40)
All joints should remain watertight under working and test conditions.	H1 (2.40)
Nothing in the pipes, joints or fittings should project into the pipe line or cause an obstruction.	H1 (2.40)
Different metals should be separated by non-metallic materials to prevent electrolytic corrosion.	H1 (2.40)

Bedding and backfill

The choice of bedding and backfill depends on the depth   H1 (2.41)
at which the pipes are to be laid and the size and strength
of the pipes.

Table 6.28 Materials for below-ground gravity drainage

Material	British Standard
Rigid pipes	
Vitrified clay	BS 65, BSEN 295
Concrete	BS 5911
Grey iron	BS 437
Ductile iron	BSEN 598
Flexible pipes	
UPVC (unplasticized poly vinyl chloride)	BSEN 1401
PP (polypropylene)	BSEN 1852
Structured walled plastic pipes	BSEN 13476

 Special precautions should be taken to take account of the effects of settlement where pipes run under or near buildings.

The depth of the pipe cover will usually depend on the levels of the connections to the system and the gradients at which the pipes should be laid and the ground levels.

 Pipes need to be protected from damage, particularly pipes which could be damaged by the weight of backfilling.

Rigid pipes should be laid in a trench as shown in   H1 (2.42)
Figure 6.42.

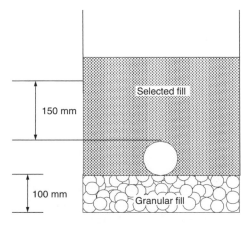

150 mm

100 mm

Selected fill

Granular fill

Figure 6.42 Bedding for rigid pipes.

Flexible pipes shall be supported to limit deformation under load.   H1 (2.44)

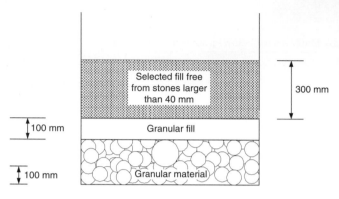

Figure 6.43 Bedding for flexible pipes.

Flexible pipes with very little cover shall be protected from damage by a reinforced cover slab with a flexible filler and at least 75 mm of granular material between the top of the pipe and the underside of the flexible filler below the slabs (see Figure 6.44).   H1 (2.42–2.44)

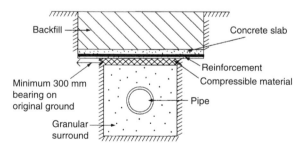

Figure 6.44 Protection of pipes laid in shallow depths.

Trenches may be backfilled with concrete to protect nearby foundations. In these cases a movement joint (as shown in Figure 6.45) formed with a compressible board should be provided at each socket or sleeve joint.   H1 (2.45)

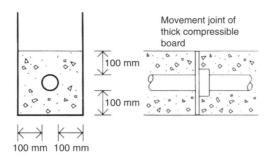

Figure 6.45 Joints for concrete encased pipes.

Access points

Access should be provided to long runs.	H1 (2.50)
Sufficient and suitable access points should be provided for clearing blockages from drain runs that cannot be reached by any other means.	H1 (2.46)
Access points should be provided:	H1 (2.49)
• on or near the head of each drain run; • at a bend; • at a change of gradient or pipe size; • at a junction.	
Access points should be either:	H1 (2.48)
• rodding eyes – capped extensions of the pipes; • access fittings – small chambers on (or an extension of) the pipes but not with an open channel; • inspection chambers – chambers with working space at ground level; • manholes – deep chambers with working space at drain level.	
Access points should be constructed so as to resist the ingress of ground water or rainwater.	H1 (2.52)
Inspection chambers and manholes should have removable non-ventilating covers of durable material (such as cast iron, cast or pressed steel, precast concrete or UPVC).	H1 (2.54)
Access points to sewers (serving more than one property) should be located in places where they are accessible and apparent for use in an emergency (e.g. highways, public open space, unfenced front gardens, and shared or unfenced driveways).	H1 (2.51)

Inspection chambers and manholes in buildings should have mechanically fixed airtight covers unless the drain itself has watertight access covers.	H1 (2.54)
Manholes deeper than 1 m should have metal step irons or fixed ladders.	H1 (2.54)

General

Drains and sewers should be protected from damage by construction traffic and heavy machinery.	H1 (2.57)
Heavy materials should not be stored over drains or sewers.	H1 (2.57)
After laying (including any necessary concrete or other haunching or surrounding and backfiring) gravity drains and private sewers should be tested for watertightness.	H1 (2.59)
All pipework carrying greywater for reuse should be clearly marked with the word 'GREYWATER'.	H1 (A11)
Material alterations to existing drains and sewers are subject to (and covered by) the Building Regulations.	H1 (b7)
Repairs, reconstruction and alterations to existing drains and sewers should be carried out to the same standards as new drains and sewers.	H1 (B15)

Wastewater treatment systems and cesspools

A notice giving information as to the nature and frequency of maintenance required for the cesspool or wastewater treatment system to continue to function satisfactorily should be displayed within each of the buildings.

The use of non-mains foul drainage, such as wastewater treatment systems, septic tanks or cesspools, should only be considered where connection to mains drainage is **not** practicable.

Any discharge from a wastewater treatment system is likely to require consent from the Environment Agency.

Septic tanks

Septic tanks with some form of secondary treatment (e.g. from a drainage field/ mound, or constructed wetland such as a reed bed) will normally be the most economic means of treating wastewater from small developments (e.g. one to three dwellings). They provide suitable conditions for the settlement, storage and partial decomposition of solids, which need to be removed at regular intervals.

Septic tanks should be sited at least 7 m from any habitable parts of buildings, and preferably down a slope.	H2 (1.16)
Septic tanks should only be used in conjunction with a form of secondary treatment (e.g. a drainage field, drainage mound or constructed wetland).	H2 (1.15)
Septic tanks should be sited within 30 m of a vehicle access to enable the tank to be emptied and cleaned without hazard to the building occupants and without the contents being taken through a dwelling or place of work.	H2 (1.17 and 1.64)
Septic tanks and settlement tanks should have a capacity below the level of the inlet of at least 2700 litres (2.7 m³) for up to four users. This size should be increased by 180 litres for each additional user.	H2 (1.18)
Septic tanks may be constructed in brickwork or concrete (roofed with heavy concrete slabs) or factory-manufactured septic tanks (made out of glass reinforced plastics, polyethylene or steel) can be used.	H2 (1.19–20 and 1.65–66)
The brickwork should consist of engineering bricks at least 220 mm thick. The mortar should be a mix of 1:3 cement/sand ratio and in situ concrete should be at least 150 mm thick of C/25/P mix (see BS 5328).	H2 (1.20 and 1.66)
Septic tanks should be ventilated.	H2 (1.21)
Septic tanks should incorporate at least two chambers or compartments operating in series.	H2 (1.22)
Septic tanks should be provided with access for emptying and cleaning.	H2 (1.24)
A notice should be fixed within the building describing the necessary maintenance.	H2 (1.25)
Septic tanks should be inspected monthly to check they are working correctly.	H2 (A.11)
Septic tanks should be emptied at least once a year.	H2 (A.13)

Cesspools

A cesspool is a watertight tank, installed underground, for the storage of sewage. No treatment is involved.

Cesspools should be sited at least 7 m from any habitable parts of buildings and preferably downslope.	H2 (1.58)
Cesspools should be provided with access for emptying and cleaning.	H2 (1.60)
Cesspools should be inspected fortnightly for overflow.	H2 (A.20)
Cesspools should be emptied on a monthly basis by a licensed contractor.	H2 (1.60) H2 (A.21)
A filling rate of 150 litres per person per day is assumed and if the cesspool does not fill within the estimated period, the tank should be inspected for leakage.	H2 (A.22)
Cesspools should be ventilated.	H2 (1.63)
The inlet of a cesspool should be provided with access for inspection.	H2 (1.67)
Cesspools and settlement tanks (if they are to be desludged using a tanker) should be sited within 30 m of a vehicle access.	H2 (1.64)
Cesspools and settlement tanks should prevent leakage of the contents and ingress of subsoil water.	H2 (1.63)
Cesspools should have a capacity below the level of the inlet of at least 18 000 litres (18 m³) for 2 users increased by 6800 litres (6.8 m³) for each additional user.	H2 (1.61)
Cesspools, septic tanks and settlement tanks may be constructed in brickwork, concrete, or glass reinforced concrete.	H2 (1.65–1.66)
Factory-made cesspools and septic tanks are available in glass reinforced plastic, polyethylene or steel.	
The brickwork should consist of engineering bricks at least 220 mm thick. The mortar should be a mix of 1:3 cement/sand ratio and in-situ concrete should be at least 150 mm thick of C/25/P mix (see BS 5328).	H2 (1.66)

Cesspools should be covered (with heavy concrete slabs) and ventilated.

Cesspools should have no openings except for the inlet, access for emptying and ventilation.   H2 (1.62)

Cesspools should be inspected fortnightly for overflow and emptied as required.   H2 (A.20)

Packaged treatment works

This term is applied to a range of systems designed to treat a given hydraulic and organic load using prefabricated components which can be installed with minimal site work. They are capable of treating effluent more efficiently than septic tank systems and this normally allows the product to be discharged directly to a watercourse.

The discharge from the wastewater treatment plant should be sited at least 10 m away from watercourses and any other buildings.   H2 (1.54)

Regular maintenance and inspection should be carried out in accordance with the manufacturer's instructions.   H2 (A.17)

Drainage fields and mounds

Drainage fields (or mounds) serving a wastewater treatment plant or septic tank should be located: H2 (1.27)

- at least 10 m from any watercourse or permeable drain;
- at least 50 m from the point of abstraction of any groundwater supply;
- at least 15 m from any building;
- sufficiently far from any other drainage fields, drainage mounds or soakaways so that the overall soakage capacity of the ground is not exceeded.

No water supply pipes or underground services other than those required by the disposal system itself should be located within the disposal area.   H2 (1.29)

No access roads, driveways or paved areas should be located within the disposal area.   H2 (1.30)

The ground water table should not rise to within 1 m of the invert level of the proposed effluent distribution pipes.   H2 (1.33)

An inspection chamber should be installed between the septic tank and the drainage field.	H2 (1.43)
Constructed wetlands should not be located in the shade of trees or buildings.	H2 (1.47)
The drainage field/mound should be checked on a monthly basis to ensure that it is not waterlogged and that the effluent is not backing up towards the septic tank.	H2 (A.15)

Under Section 50 (*overflowing and leaking cesspools*) of the Public Health Act 1936 action could be taken against a builder who had caused the problem, and **not** just against the owner.

Under Section 59 (*drainage of building*) of the Building Act 1984, local authorities can require either the owner or the occupier to remove (or otherwise make innocuous) any disused cesspool, septic tank or settlement tank.

Greywater and rainwater tanks

Greywater and rainwater tanks should: • prevent leakage of the contents and ingress of subsoil water; • be ventilated; • have an anti-backflow device; • be provided with access for emptying and cleaning.	H2 (1.70)

Rainwater drainage

The capacity of the drainage system should be large enough to carry the expected flow at any point in the system.	H3 (0.3)
Rainwater or surface water should not be discharged to a cesspool or septic tank.	H3 (0.6)

Gutters and rainwater pipes

Although this part of the Building Regulations only actually applies to draining the rainfall from areas of $6\,m^2$ or more (unless they receive a flow from a rainwater pipe or from paved and/or other hard surfaces), each case should be considered separately and a decision made. This particularly applies to small roofs and balconies. Table 6.29 shows the largest effective area that should be drained into the gutter sizes most often used.

For eaves gutters the design rainfall intensity should be 0.021 l/s/m^2. In some cases, eaves drop systems may be used.

Table 6.29 Gutter and outlet sizes

Max effective roof area (m^2)	Gutter size (mm diameter)	Outlet size (mm diameter)	Flow capacity (l/s)
6.0	—	—	—
18.0	75	50	0.38
37.0	100	63	0.78
53.0	115	63	1.11
65.0	125	75	1.37
103.0	150	89	2.16

 Gutters should be laid with any fall towards the nearest outlet.

Gutters should be laid so that any overflow in excess of the design capacity (e.g. above normal rainfall) will be discharged clear of the building.   H3 (1.7)

Rainwater pipes should discharge into a drain or gully (but may discharge to another gutter or onto another surface if it is drained).   H3 (1.8)

Any rainwater pipe which discharges into a combined system should do so through a trap.   H3 (1.8)

The size of a rainwater pipe should be at least the size of the outlet from the gutter.   H3 (1.10)

A down pipe which serves more than one gutter should have an area at least as large as the combined areas of the outlets.   H3 (1.10)

On flat roofs, valley gutters and parapet gutters additional outlets may be necessary.   H3 (1.7)

Where a rainwater pipe discharges onto a lower roof or paved area, a pipe shoe should be fitted to divert water away from the building.   H3 (1.9)

 Gutters and rainwater pipes should be firmly supported without restricting thermal movement.

The materials used should be of adequate strength and durability, and: H3 (1.16)

• all gutter joints should remain watertight under working conditions;

- pipework in siphonic roof drainage systems should be able to resist to negative pressures in accordance with the design;
- gutters and rainwater pipes should be firmly supported;
- different metals should be separated by non-metallic material to prevent electrolytic corrosion.

Drainage of paved areas

Surface gradients should direct water draining from a paved area away from buildings.	H3 (2.2)
Gradients on impervious surfaces should be designed to permit the water to drain quickly from the surface.	H3 (2.3)
A gradient of at least 1 in 60 is recommended.	
Paths, driveways and other narrow areas of paving should be free draining to a pervious area such as grassland, provided that: - the water is not discharged adjacent to buildings where it could damage foundations; and - the soakage capacity of the ground is not overloaded.	H3 (2.6)
Where water is to be drained onto the adjacent ground the edge of the paving should be finished above or flush with the surrounding ground to allow the water to run off.	H3 (2.7)
Where the surrounding ground is not sufficiently permeable to accept the flow, filter drains may be provided.	H3 (2.8 and 3.33)
Pervious paving should not be used where excessive amounts of sediment are likely to enter the pavement and block the pores.	H3 (2.11)
Pervious paving should not be used in oil storage areas, or where runoff may be contaminated with pollutants.	H3 (2.12)
Gullies should be provided at low points where water would otherwise pond.	H3 (2.15)
Gully gratings should be set approximately 5 mm below the level of the surrounding paved area in order to allow for settlement.	H3 (2.16)

Provision should be made to prevent silt and grit   H3 (2.17)
entering the system, either by provision of gully pots
of suitable size, or catchpits.

Surface water drainage

Discharge to a watercourse may require consent from the Environment Agency, who may limit the rate of discharge. Where other forms of outlet are not practicable, discharge should be made to a sewer (H3 (3.2–3.3)). For design purposes a rainfall interval of $0.014\,l/s/m^2$ can be assumed as normal.

Some drainage authorities have sewers that carry both foul water and rainwater (i.e. combined systems) in the same pipe. Where they do, they can allow rainwater to discharge into the system if the sewer has enough capacity to take the added flow. Some private sewers (drains serving more than one property) also carry both foul water and rainwater. If a sewer (or private sewer) operated as a combined system does not have enough capacity, the rainwater should be run in a separate system with its own outfall.

Surface water drainage should discharge to a soakaway   H3 (3.2)
or other infiltration system where practicable.

Surface water drainage connected to combined sewers   H3 (3.7)
should have traps on all inlets.

Drains should be at least 75 mm diameter.   H3 (3.14)

Where any materials that could cause pollution are   H3 (3.21)
stored or used, separate drainage systems should be
provided.

On car parks, petrol filling stations or other areas   H3 (3.22)
where there is likely to be leakage or spillage of   H3 (A)
oil, drainage systems should be provided with oil
interceptors.

Separators should be leak tight and comply with   H3 (A.9–10)
the requirements of the Environment Agency and
prEN858.

Infiltration devices (including soakaways, swales,   H3 (3.23–26)
infiltration basins, and filter drains) should not be built:

- within 5 m of a building or road or in areas of
  unstable land;
- in ground where the water table reaches the bottom
  of the device at any time of the year;

• sufficiently far from any drainage fields, drainage mounds or other soakaways;	
• where the presence of any contamination in the runoff could result in pollution of groundwater source or resource.	
Soakaways should be designed to a return period of once in ten years.	H3 (3.27)
Soakaways for areas less than 100 m² shall consist of square or circular pits, filled with rubble or lined with dry jointed masonry or perforated ring units.	H3 (3.26)
Soakaways serving larger areas shall be lined pits or trench type soakaways.	
The storage volume should be calculated so that, over the duration of a storm, it is sufficient to contain the difference between the inflow volume and the outflow volume.	H3 (3.29)
Soakaways serving larger areas should be designed in accordance with BS EN 752-4.	H3 (3.30)

Under Section 85 (*offences concerning the polluting of controlled waters*) of the Water Resources Act 1991 it is an offence to discharge any noxious or polluting material into a watercourse, coastal water or underground water. Most surface water sewers discharge to watercourses.

Under Section 111 (*restrictions on use of public sewers*) of the Water Industry Act 1991 it is an offence to discharge petrol into any drain or sewer connected to a public sewer.

Building over existing sewers

Where it is proposed to construct a building over or near a drain or sewer shown on any map of sewers, the developer should consult the owner of the drain or sewer.	H4 (0.3)
A building constructed over or within 3 m of any:	H4 (1.2)
• rising main;	
• drain or sewer constructed from brick or masonry;	
• drain or sewer in poor condition;	
shall not be constructed in such a position unless special measures are taken.	

Buildings or extensions should not be constructed over a manhole or inspection chamber or other access fitting on any sewer (serving more than one property).	H4 (1.3)
A satisfactory diversionary route should be available so that the drain or sewer could be reconstructed without affecting the building.	H4 (1.4)
The length of drain or sewer under a building should not exceed 6 m except with the permission of the owners of the drain or sewer.	H4 (1.5)
Buildings or extensions should not be constructed over or within 3 m of any drain or sewer more than 3 m deep, or greater than 225 mm in diameter except with the permission of the owners of the drain or sewer.	H4 (1.60)
Where a drain or sewer runs under a building at least 100 mm of granular or other suitable flexible filling should be provided round the pipe.	H4 (1.9)
Where a drain or sewer running below a building is less than 2 m deep, the foundation should be extended locally so that the drain or sewer passes through the wall.	H4 (1.10)
Where the drain or sewer is more than 2 m deep to invert and passes beneath the foundations, the foundations should be designed with a lintel spanning over the line of the drain or sewer. The span of the lintel should extend at least 1.5 m either side of the pipe and should be designed so that no load is transmitted onto the drain or sewer.	H4 (1.12)
A drain trench should not be excavated lower than the foundations of any building nearby.	H4 (1.13)

Separate systems for drainage

Separate systems of drains and sewers shall be provided for foul water and rainwater where:

(a)   the rainwater is not contaminated; and
(b)   the drainage is to be connected either directly or indirectly to the public sewer system, which has separate systems for foul water and surface water.

Solid waste storage

Although the requirements of the Building Regulations do not cover the recycling of household and other waste, H6 sets out general requirements for solid waste storage.

For domestic developments space should be provided for storage of containers for separated waste (i.e. waste that can be recycled is stored separately from waste that cannot) and having a combined capacity of 0.25 m² per dwelling.	H6 (1.1)
In low-rise domestic developments (houses, bungalows and flats up to the 4th floor) any dwelling should have, or have access to, a location with at least two movable, individual or communal waste containers.	H6 (1.2)
In multi-storey domestic developments, dwellings above the fourth storey should share a container fed by a chute unless siting or operation of a chute is impracticable.	H6 (1.6)
In such a case a satisfactory management arrangement for conveying refuse to the storage area should be assured.	
In multi-storey domestic developments, dwellings up to the fourth floor may each have their own waste container or may share a waste container.	H6 (1.5)
For waste containers up to 250 litres, steps should be avoided between the container store and collection point wherever possible.	H6 (1.10)
Containers and chutes should be sited so that householders are not required to carry refuse further than 30 m.	H6 (1.8)
Containers should be within 25 m of the vehicle access.	H6 (1.8)
Containers should be sited so that they can be collected without being taken through a building, unless it is a garage, carport or other open covered space.	H6 (1.10)

 This provision applies only to new buildings.

The collection point should be reasonably accessible to the size of waste collection vehicles typically used by the waste collection authority.	H6 (1.11)

External storage areas for waste containers should be away from windows and ventilators and preferably be in the shade or under a shelter.   H6 (1.12)

Storage areas should not interfere with pedestrian or vehicle access to buildings.   H6 (1.12)

Where enclosures, compounds or storage rooms are provided they should allow room for filling and emptying and provide a clear space of 150 mm between and around the containers.   H6 (1.13)

Enclosures, compounds or storage rooms for communal containers should be a minimum of 2 m high.   H6 (1.13)

Enclosures for individual containers should be sufficiently high to allow the lid to be opened for filling.   H6 (1.13)

The enclosure should be permanently ventilated at the top and bottom and should have a paved impervious floor.   H6 (1.13)

Communal storage areas should have provision for washing down and draining the floor into a system suitable for receiving a polluted effluent.   H6 (1.14)

Gullies should incorporate a trap that maintains a seal even during prolonged periods of disuse.   H6 (1.14)

Any room (or compound) for the open storage of waste should be secure to prevent access by vermin.   H6 (1.15)

Where storage rooms are provided, separate rooms should be provided for the storage of waste that cannot be recycled, and waste that can be recycled.   H6 (1.16)

Where the location for storage is in a publicly accessible area or in an open area around a building (e.g. a front garden) an enclosure or shelter should be considered.   H6 (1.17)

In high-rise domestic developments, where chutes are provided they should be at least 450 mm in diameter and should have a smooth non-absorbent surface and close-fitting access doors at each storey that has a dwelling and be ventilated at the top and bottom.   H6 (1.18)

6.5 Water supplies

6.5.1 The requirement (Building Act 1984 Sections 25 and 69)

The Building Act stipulates that plans for proposed buildings will ensure that all occupants of the house will be provided with a supply of 'wholesome water, sufficient for their domestic purposes'. This can be achieved by either:

- connecting the house to water supplies from the local water authority (normally referred to as the *statutory water undertaker*);
- taking water into the house by means of a pipe (e.g. from a local recognized supply);
- providing a supply of water within a reasonable distance from the house (e.g. such as from a well).

If an occupied house is not within a reasonable distance of a supply of 'wholesome water' or if the local authority is not satisfied that the water supply is capable of supplying 'wholesome water', then it can give notice that the owner of the building must provide water within a specified time. It also has the authority to prohibit the building from being occupied.

Cold water supply

(1) There must be a suitable installation for the provision of:

    (a) wholesome water to any place where drinking water is drawn off;

    (b) wholesome water or softened wholesome water to any washbasin or bidet provided in or adjacent to a room containing a sanitary convenience;

    (c) wholesome water or softened wholesome water to any washbasin, bidet, fixed bath or shower in a bathroom; and

    (d) wholesome water to any sink provided in any area where food is prepared.

(2) There must be a suitable installation for the provision of water of suitable quality to any sanitary convenience fitted with a flushing device.

(Approved Document G1)

What is wholesome water?

Water supplied for such domestic purposes as cooking, drinking, food preparation or washing and/or to premises in which food is produced is regarded as **wholesome water** – provided that the water does not contain:

- any micro-organism or parasite; or
- any substance whose concentration or value would constitute a potential danger to human health.

Water supplied to the building by a statutory water undertaker or a licensed water supplier may be assumed to be wholesome water.

What is wholesome softened water?

If all of requirements for wholesome water are met other than its sodium content, then it will be considered to be wholesome softened water, which should **not** be used as drinking water or used in an area where food is prepared.

What about alternative sources of water?

Water from alternative sources, such as:

*   water abstracted from wells, springs, bore-holes or water courses;
*   harvested rainwater;
*   reclaimed greywater;
*   reclaimed industrial process water;

may be used in dwellings for sanitary conveniences, washing machines and irrigation, **provided** the appropriate risk assessment has been carried out and appropriate measures to minimize the impact on water quality from:

*   failure of any components;
*   failure due to lack of maintenance;
*   power failure; and
*   any other measures identified in a risk assessment

have been taken.

Water efficiency

Reasonable provision must be made by the installation of fittings and fixed appliances that use water efficiently for the prevention of undue consumption of water.

(Approved Document G2)

 Requirement G2 applies only when a dwelling is:

(a) erected; or
(b) formed by a material change of use of a building.

Water efficiency of new dwellings

17K. (1) The potential consumption of wholesome water by persons occupying a dwelling to which this regulation applies must not exceed 125 liters per person per day, calculated in accordance with the methodology set out in the document 'The Water Efficiency Calculator for New Dwellings'.

(2) This regulation applies to a dwelling which is –
    (a) erected; or
    (b) formed by a material change of use of a building within the meaning of regulation 5(a) or (b).

Wholesome water consumption calculation

20E. (1) Where regulation 17K applies, the person carrying out the work must give the Local Authority a notice which specifies the potential consumption of wholesome water per person per day calculated in accordance with the methodology referred to in that regulation in relation to the completed dwelling.

(2)   The notice shall be given to the Local Authority not later than five days after the work has been completed.

Building (Approved Inspectors) Regulations 2000

12E. (1) Where regulation 17K of the Principal Regulations applies to work which is the subject of an initial notice, the person carrying out the work must give the approved inspector a notice which specifies the potential consumption of wholesome water per person per day calculated in accordance with the methodology referred to in that regulation in relation to the completed dwelling.

(2)   The notice shall be given to the approved inspector not later than –
  (a)   five days after the work has been completed; or
  (b)   the date on which, in accordance with Regulation 18, the initial notice ceases to be in force, whichever is the earlier.

What happens if there is more than one property?

Where the local authority is satisfied that two or more houses can most conveniently be joined by means of a joint supply, they may give notice accordingly.

Can I ask a local water company to provide me with a supply of water?

Water is provided to a house by a 'statutory water undertaker' and separate water rates are paid for use of this water. There are currently 21 regional monopoly water companies in England and Wales. Ten provide both water and sewerage services, while the remainder provide water services only. They are regulated by Ofwat and operate under the provisions of the Water Industry Acts 1991 and 1999, the Water Resources Act 1991, the Water Act 2003 and the Flood and Water Management act 2010. The supply of water intended for human consumption is governed in England by the Private Water Supplies Regulations 2009 (SI 2009/3101) and in Wales by the Private Water Supplies (Wales) Regulations (SI 2010/66).

 If there is no water main near your property you can ask a local water company to provide a connection to a company water main for domestic purposes. In this case the company is entitled to recover reasonable costs from you.

There are different rules relating to the connection of a property to the water main. Both domestic and non-domestic properties can be provided with water for either domestic purposes (washing, cooking, sanitary facilities) and non-domestic purposes (garden taps etc.). If you want a new connection for an existing property, or if there is no water main to the property, you can request a connection to the water main

You will have to pay an infrastructure charge to connect a new property to the water main. The maximum charge set for 2010–2011 as set by OFWAT was £312.19.

Where two or more houses are supplied with water by a common pipe belonging to the owners or occupiers of those houses, the local undertaker may, when necessary, repair or renew the pipe and recover any expenses reasonably incurred by them from the owners or occupiers of the houses.

Notifiable work

If the type of work is mainly of a minor nature (e.g. replacement of a part) and there is no significant risk to health, safety, water efficiency or energy efficiency, then it is deemed to be non-notifiable.
   On the other hand:

- adding an output or control device to an existing cold water supply or replacing any part of the existing water system is notifiable building work;
- replacing a sanitary convenience with one that uses no more water than the one it replaces and does not include any work to:

  - underground drainage;
  - the hot or cold water system; or
  - above-ground drainage;

but **could** prejudice the health and safety of any person on completion of work, is also notifiable building work.

6.5.2 Meeting the requirement

Cold water supplies

Cold water supply shall:

> - be reliable;   G1
> - be wholesome;
> - have a pressure and flow rate sufficient for the operation of all appliances and locations planned in the building;
> - convey wholesome water or softened wholesome water without waste, misuse, undue consumption or contamination of water.

Sanitary conveniences

For sanitary conveniences the cold water supply shall be:

> • reliable;   G1
> • either wholesome, softened wholesome or of suitable quality;
> • have a pressure and flow rate sufficient for the operation of the
>   sanitary appliances.

Wholesome and wholesome softened water

Water supplied to the building by a statutory water undertaker or a licensed water supplier may be assumed to be wholesome water.

Buildings supplied with water from a source **other than** a water undertaker or licensed water supplier may be considered to be wholesome, **provided** that it meets the criteria set out in the Private Water Supplies Regulations 2009 (SI 2009/3101) in England or the Private Water Supplies (Wales) Regulations (SI 2010/66) in Wales.

Wholesome water which has been treated by a water softener or a water softening process to adjust the content of hardness minerals may be classified as wholesome water provided that it still complies with the requirements for wholesome water. Non-compliant treated water shall be classified as **wholesome softened water**.

> The estimated consumption of wholesome water of a new   G2.3
> dwelling should be not more than 125 litres/head/day (l/hid).

Alternative sources of water

Wholesome water need not be used for toilet flushing and irrigation etc. In these circumstances, water from alternative sources may be used. These include:

• water abstracted from wells, springs, bore-holes or water courses;
• harvested rainwater;
• reclaimed greywater; and
• reclaimed industrial process water.

In all cases:

> the water obtained from an alternative source shall incorporate   G1.7
> measures to minimize the impact on water quality from:
>
> • failure of any components;
> • failure due to lack of maintenance;
> • power failure; and
> • any other measures identified in a risk assessment.

| Any system/unit used to supply dwellings with water from alternative sources shall be subject to a risk assessment by the system designer and manufacturer. | G1 .4 |

Cold water storage systems

| The cold water storage cistern into which the vent pipe discharges should be supported on a flat, level, rigid platform. The platform should extend a minimum of 150 mm in all directions beyond the edge of the maximum dimensions of the cistern. | G3.15 |
| The cistern should be accessible for maintenance, cleaning and replacement. | G3.16 |

Protection of openings for pipes

Pipes that pass through a compartment wall or compartment floor (unless the pipe is in a protected shaft), or through a cavity barrier, should incorporate one of the following alternatives:

Enclosures for drainage and/or water supply pipes	
Proprietary seals (any pipe diameter) that maintain the fire resistance of the wall, floor or cavity barrier.	B3 7.6–7.7 (V1) B3 10.6 (V2)
Pipes with a restricted diameter where fire-stopping is used around the pipe, keeping the opening as small as possible.	B3 7.6 and 7.8 (V1) B3 10.5 and 10.7 (V2)
Sleeving – a pipe of lead, aluminium, aluminium alloy, fibre-cement or UPVC, with a maximum nominal internal diameter of 160 mm, may be used with a sleeving of non-combustible pipe as shown in Figure 6.17.	B3 7.6 and 7.9 (V1) B3 10.5 and 10.8 (V2)

6.6 Cellars and basements

6.6.1 The requirement (Building Act 1984 Section 74)

Unless you have the consent of the local authority, you are not allowed to construct a cellar or room *in (or as part of) a house, an existing cellar, a shop, inn, hotel or office if the floor level of the cellar or room is lower than the ordinary level of the subsoil water on, under or adjacent to the site of the house, shop, inn, hotel or office.*

This does not apply to:

- the construction of a cellar or room carried out in accordance with plans deposited on an application under the Licensing Act 1964;
- the construction of a cellar or room in connection with a shop, inn, hotel or office that forms part of a railway station.

If the owner of the house, shop, inn, hotel or office allows a cellar or room forming part of it to be used in a manner that he knows to be in contravention of the Building Regulations, he is liable, on summary conviction, to a fine.

Fire precautions

The building shall be provided with:

- sufficient internal fire mains and other facilities to assist firefighters in their tasks;
- adequate means for venting heat and smoke from a fire in a basement.

Ventilation

There shall be adequate means of ventilation provided for people in the building.

(Approved Document F)

A new Part F came into force on 1 October 2010.

6.6.2 Meeting the requirements

Owing to the risk that a single stairway may be blocked by smoke from a fire in the basement or ground storey:

• basement storeys that contain a habitable room shall be provided with either:	B1 2.13 (V1)
– an external door or window suitable for egress from the basement; or	B1 2.6 (V2)
– a protected stairway leading from the basement to a final exit;	
• final exits shall be sited so that they are clear of any risk from fire or smoke in a basement;	B1 5.34 (V2)
• in non-residential, purpose group buildings (such as office, shop and commercial, assembly and recreation, industrial, storage etc.), the following floors shall be constructed as compartment walls and compartment floors:	B2 8.18c (V2) B2 8.18d (V2)
– the floor of every basement storey (except the lowest floor) greater than 10 m below ground level (see Figure 6.46);	
– the floor of the ground storey (see Figure 6.47).	

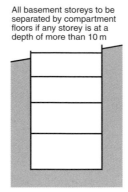

All basement storeys to be
separated by compartment
floors if any storey is at a
depth of more than 10 m

Figure 6.46 Compartment floors – deep basements.

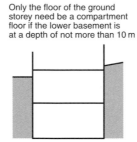

Only the floor of the ground
storey need be a compartment
floor if the lower basement is
at a depth of not more than 10 m

Figure 6.47 Compartment floors – shallow basements.

 Note: If the building has one or more basements, and with the exception of small premises (see paragraph 3.1 of V2).

Emergency egress windows and external doors

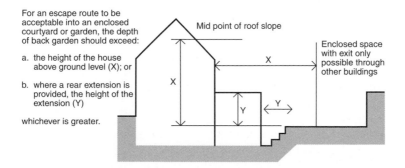

For an escape route to be
acceptable into an enclosed
courtyard or garden, the depth
of back garden should exceed:

a. the height of the house
   above ground level (X); or

b. where a rear extension is
   provided, the height of the
   extension (Y)

whichever is greater.

Mid point of roof slope

Enclosed space
with exit only
possible through
other buildings

Figure 6.48 Ground or basement storey exit into an enclosed space.

> The window or door should enable the person escaping to reach a place free from danger of fire (e.g. a courtyard or back garden which is at least as deep as the dwelling-house is high – see Figure 6.48), and: B1 2.8 (V1)
>
>   B1 2.9 (V2)
>
> - the window should be at least 450 mm high and 450 mm wide and have an unobstructed openable area of at least 0.33 m²; and
> - the bottom of the openable area should be not more than 1100 mm above the floor.

Notes:

(1) Approved Document K (*Protection from Falling, Collision and Impact*) specifies a minimum guarding height of 800 mm, except in the case of a window in a roof where the bottom of the opening may be 600 mm above the floor.

(2) Locks (with or without removable keys) and stays may be fitted to egress windows, provided that the stay is fitted with a child resistant release catch.

(3) Windows should be designed so that they remain in the open position without needing to be held open by the person making their escape.

Basement stairways

Because basement stairways are more likely to be filled with smoke and heat than are stairs in ground and upper storeys:

> - the flights and landings of an escape stair shall be constructed using materials of limited combustibility, particularly if it is within a basement storey;   B1 5.19 (V2)
> - the basement should be served by a separate stair.   B1 2.44 (V2)
>   B1 4.42 (V2)

Note: If an escape stair forms part of the **only** escape route from an upper storey of a large building, it should **not** be continued down to serve any basement storey. Other stairs may connect with the basement storey(s) if there is a protected lobby or a protected corridor between the stair(s) and accommodation at each basement level.

Lifts

In basements:

> - the lift should be approached only by a protected lobby or protected corridor (unless it is within the enclosure of a protected stairway);   B1 5.43 (V2)

- lift entrances should be separated from the floor area on every storey by a protected lobby;   B1 5.42 (V2)

- lift shafts should not be continued down to serve any basement storey if it is: B1 5.44 (V2)

  - in a building served by only one escape stair;
  - within the enclosure to an escape stair which is terminated at ground level.

Access and facilities for the fire service

Buildings with a basement more than 10 m below the fire and rescue service vehicle access level should be provided with at least two fire-fighting shafts containing fire-fighting lifts (see Figure 6.49).   B5 17.2 (V2) B5 17.8 (V2)

Buildings with two or more basement storeys, each exceeding 900 m^2 in area, should be provided with fire-fighting shaft(s), which need not include firefighting lifts.   B5 17.4 (V2)

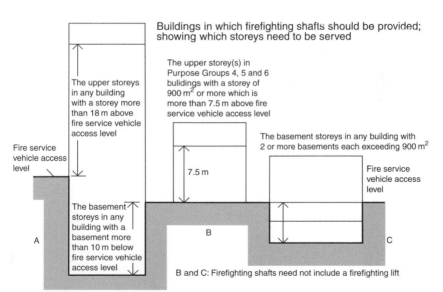

Buildings in which firefighting shafts should be provided; showing which storeys need to be served

The upper storey(s) in Purpose Groups 4, 5 and 6 bulidings with a storey of 900 m^2 or more which is more than 7.5 m above fire service vehicle access level

The upper storeys in any building with a storey more than 18 m above fire service vehicle access level

Fire service vehicle access level

The basement storeys in any building with a basement more than 10 m below fire service vehicle access level

7.5 m

The basement storeys in any building with 2 or more basements each exceeding 900 m^2

Fire service vehicle access level

A   B   C

B and C: Firefighting shafts need not include a firefighting lift

A. Firefighting shafts should include firefighting lift(s)

Note: Height excludes any top storey(s) consisting exclusively of plant rooms.

Figure 6.49 Provision of fire-fighting shafts.

Venting of heat and smoke from basements

The building should be provided with adequate means for venting heat and smoke from a fire in a basement.	B5 (V2)
Where practicable, each basement space should have one or more smoke outlets (see Figure 6.50).	B5 18.3 (V2)
Outlet ducts or shafts, including any bulkheads over them (see Figure 6.50), should be enclosed in non-combustible construction having a greater fire resistance than the element that they pass through.	B5 18.15 (V2)

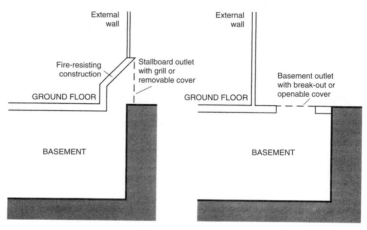

Figure 6.50 Fire-resisting construction for smoke outlet shafts.

Smoke outlets connected to the open air should be provided from every basement storey, except for a basement in a single family dwelling.	B5 18.4 (V2)
Smoke outlets (also referred to as smoke vents) should be:	
• available so as to provide a route for heat and smoke to escape to the open air from the basement area;	B5 18.2 (V2)
• sited at high level, either in the ceiling or in the wall of the space they serve;	B5 18.7 (V2) B5 18.3 (V2)
• evenly distributed around the perimeter to discharge in the open air outside the building.	B5 18.3 (V2)
A system of mechanical extraction may be provided as an alternative to natural venting to remove smoke and heat from basements, **provided** that the basement storey(s) is fitted with a sprinkler system.	B5 18.13 (V2)

Where there are natural smoke outlet shafts from dif-   B5 18.16 (V2)
ferent compartments to the same or different basement
storeys, they should be separated from each other by a
non-combustible construction.

Ventilation

If a basement is connected to the rest of the dwelling by a large   F 5.11
permanent opening such as an open stairway, then the whole
dwelling including the basement should be treated as a multi-
storey dwelling and ventilated in a similar manner to dwellings
without basements.

If the basement has a single exposed façade, while the rest of   F 5.11
the dwelling above ground has more than one exposed façade,
then Passive Stack Ventilation (PSV) or continuous
mechanical extract should be used.

If the basement is not connected to the rest of the dwelling by a   F 5.12
large permanent opening, then:

- the part of the dwelling above ground should be considered
  separately; and
- the basement should be treated as a single-storey dwelling,
  as if it were above ground.

If the part of the dwelling above ground has no bedrooms, then
for the purpose of ventilation requirements:

Table 6.30 Ventilation systems for basements

Type of basement	Background ventilators and intermittent extract fans	Passive stack ventilation	Continuous mechanical extract	Continuous mechanical supply and extract with heat
Basement connected to the rest of the dwelling by an open stairway	Yes	Yes	Yes	Yes
Basement with a single exposed façade and dwelling above ground with more than one exposed façade		Yes	Yes	
Basements not connected to the rest of the dwelling by an open stairway	Yes	Yes	Yes	Yes
Dwelling above ground has no bedrooms	Yes	Yes	Yes	Yes
Dwelling comprises just a basement	Yes	Yes	Yes	Yes

- assume that the dwelling has one bedroom; and
- treat the basement as a single-storey dwelling (with one bedroom) as if it were above ground.

If a dwelling only compromises a basement, then it should be treated as if it were a single-storey dwelling (with one bedroom) above ground.   F 5.13

Mechanically ventilated basement car parks shall be capable of at least six air changes per hour (6 ach).   F 6.20bi

6.7 Floors

The ground floor of a building is either solid concrete or a suspended timber type. With a concrete floor, a damp-proof membrane is laid between walls. With timber floors, sleeper walls of honeycomb brickwork are built on over-site concrete between the base brickwork; a timber sleeper plate rests on each wall and timber joists are supported on them. Their ends may be similarly supported, let into the brickwork or suspended on metal hangers. Floorboards are laid at right angles to joists. First-floor joists are supported by the masonry or hangers.

Similar to a brick-built house, the floors in a timber-framed house are either solid concrete or suspended timber. In some cases, a concrete floor may be screeded or surfaced with timber or chipboard flooring. Suspended timber floor joists are supported on wall plates and surfaced with chipboard.

6.7.1 Requirements

The building shall be constructed so that the combined dead, imposed and wind loads are sustained and transmitted by it to the ground:

- *safely;*
- *without causing such deflection or deformation of any part of the building (or such movement of the ground) as will impair the stability of any part of another building.*

(Approved Document A1)

The building shall be constructed so that ground movement caused by:

- *swelling, shrinkage or freezing of the subsoil; or*
- *landslip or subsidence (other than subsidence arising from shrinkage) will not impair the stability of any part of the building.*

(Approved Document A2)

Fire precautions

As a fire precaution, the spread of flame over the internal linings of a building and the amount of heat released from internal linings shall be restricted.

- *all loadbearing elements of structure of the building shall be capable of withstanding the effects of fire for an appropriate period without loss of stability;*
- *ideally the building should be subdivided by elements of fire-resisting construction into compartments;*
- *all openings in fire-separating elements shall be suitably protected in order to maintain the integrity of the continuity of the fire separation;*
- *any hidden voids in the construction shall be sealed and subdivided to inhibit the unseen spread of fire and products of combustion, in order to reduce the risk of structural failure, and the spread of fire.*

(Approved Document B3)

The floors of the building shall adequately protect the building and people who use the building from harmful effects caused by:

- *ground moisture;*
- *precipitation and wind-driven spray;*
- *interstitial and surface condensation; and*
- *spillage of water from or associated with sanitary fittings or fixed appliances.*

(Approved Document C2)

Airborne and impact sound

Dwellings shall be designed so that the noise from domestic activity in an adjoining dwelling (or other parts of the building) is kept to a level that:

- *does not affect the health of the occupants of the dwelling;*
- *will allow them to sleep, rest and engage in their normal activities in satisfactory conditions.*

(Approved Document E1)

Dwellings shall be designed so that any domestic noise that is generated internally does not interfere with the occupants' ability to sleep, rest and engage in their normal activities in satisfactory conditions.

(Approved Document E2)

Domestic buildings shall be designed and constructed so as to restrict the transmission of echoes.

(Approved Document E3)

Schools shall be designed and constructed so as to reduce the level of ambient noise (particularly echoing in corridors).

(Approved Document E4)

Note: The normal way of satisfying Requirement E4 will be to meet the values for sound insulation, reverberation time and internal ambient noise which are given in Section 1 of Building Bulletin 93, *The Acoustic Design of Schools*, produced by Department for Education and Skills (DfES) and published by The Stationery Office (ISBN: 0 11 271105 7).

Ventilation

There shall be adequate means of ventilation provided for people in the building.

(Approved Document F)

A new Part F came into force on 1 October 2010.

Conservation of fuel and power

Reasonable provision shall be made for the conservation of fuel and power in buildings by:

(a) limiting heat gains and losses –

(i) through thermal elements and other parts of the building fabric; and
(ii) from pipes, ducts and vessels used for space heating, space cooling and hot water services;

(b) providing fixed building services which –

(i) are energy efficient;
(ii) have effective controls; and
(iii) are commissioned by testing and adjusting as necessary to ensure they use no more fuel and power than is reasonable in the circumstances; and

(c) providing to the owner sufficient information about the building, the fixed building services and their maintenance requirements so that the building can be operated in such a manner as to use no more fuel and power than is reasonable in the circumstances.

(Approved Document L)

A new Part L came into force on 1 October 2010.

6.7.2 The use of Robust Standards

Background

The 2003 edition of Part E of the Building Regulations (i.e. *Resistance to the Passage of Sound*) involves Pre-completion Sound Testing (PCT) for certain types of home. In an attempt to eliminate the risk of any remedial work being required to completed floor and/or wall constructions (together with the potential for delays in completing the property) the House Builders Federation

(HBF) suggested that a series of construction solutions (called Robust Details) should be developed as an alternative to PCT.

This approach was agreed, and in May 2003 the (then) Office of the Deputy Prime Minister (ODPM) published the first batch of Robust Detail proposals for public consultation. At the same time, the introduction of PCT in new homes was postponed until July 2004 – on the assumption that the Robust Details scheme would eventually receive ministerial approval.

In January 2004 the minister responsible for Building Regulations announced that he would allow Robust Details to be used as an alternative to PCT, and that it would take effect from 1 July 2004 (i.e. to coincide with the introduction of PCT). Under a Memorandum of Understanding (MOU) with the ODPM, a limited company (Robust Details Ltd) was set up to approve, manage, monitor and promote the use of Robust Details as a method of satisfying Building Regulations.

What is a Robust Detail?

Robust Details provide builders with a choice of possible construction solutions that have been proven to outperform the standards of Part E, thus eliminating the need for routine PCT!

A Robust Detail is only used in connection with Part E and is defined as '*a separating wall or floor (of concrete, masonry, timber, steel or steel-concrete composite) construction, which has been assessed and approved by Robust Details Limited*'.

How are Robust Details approved?

In order to be approved, each Robust Detail must:

* be capable of consistently exceeding the performance standards given in Approved Document E to the Building Regulations;
* be practical to construct on site;
* be reasonably tolerant to workmanship.

How can Robust Details be used?

Builders are only permitted to use Robust Details instead of PCT **if** the plots concerned have been registered in advance with Robust Details Ltd.

Once a plot has been registered, Robust Details Ltd will provide the relevant registration documentation (which will be accepted by all building control bodies as evidence that the builder is entitled to use Robust Details instead of PCT). The builder will then need to select the Robust Detail specific to the walls and/or floors they wish to build from the Robust Details Handbook (available from Robust Details Ltd) and produce a sitework checklist to show how they are going to ensure that building work is carried out exactly in accordance with the Robust Detail specifications.

Will there be more Robust Details?

Trade associations, manufacturers or other interested parties may submit applications for new Robust Details, and these will be evaluated and, if found acceptable, approved and published.

Where can I obtain more information?

More information is available from Robust Details Ltd (customerservice@ robustdetails.com) and full contact information is in our useful contact names and addresses section.

6.7.3 Meeting the requirements

General

Floors next to the ground should: C4.2

- resist the passage of ground moisture to the upper surface of the floor;
- not be damaged by moisture from the ground;
- not be damaged by groundwater;
- resist the passage of ground gases.

Floors next to the ground and floors exposed from below should be designed and constructed so that their structural and thermal performances are not adversely affected by interstitial condensation.   C4.4

All floors should be designed so they do not promote surface condensation or mould growth.   C4.5

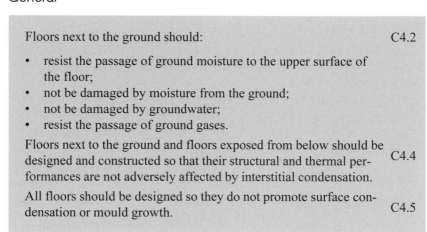

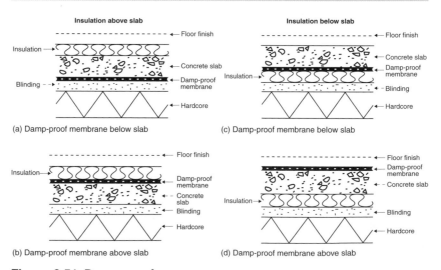

Figure 6.51 Damp-proof courses.

Ground supported floors exposed to moisture from the ground

Unless subjected to water pressure, the ground of a ground supported floor should be covered with dense concrete laid on a hardcore bed and a damp-proof membrane as shown in Table 6.31.

Table 6.31 Ground supported floor – construction

Hardcore	Well compacted, no greater than 600 mm deep, of clean, broken brick or similar inert material, free from materials including water-soluble sulphates in quantities which could damage the concrete	C4.7a
Concrete	At least 100 mm thick to mix ST2 in BS 8500 or (if there is embedded reinforcement) to mix ST4 in BS 8500	C4.7b
Damp-proof membrane	Above or below the concrete which is continuous with the damp-proof courses in walls and piers etc.	C4.7c
	If below the concrete, the membrane could be formed with a sheet of polyethylene, at least 300 mm thick with sealed joints and laid on a bed of material that will not damage the sheet	C4.8
	If laid above the concrete, the membrane may be either polyethylene (but without the bedding material) or three coats of cold applied bitumen solution or similar moisture and water vapour resisting material	C4.9
	In each case it should be protected either by a screed or a floor finish, unless the membrane is pitch mastic or similar material which will also serve as a floor finish	C4.9
Insulation	If placed beneath floor slabs, it should have sufficient strength to resist the weight of the slab, the anticipated floor loading as well as any possible overloading during construction	C4.10
	If placed below the damp-proof membrane, it should have low water absorption and (if considered necessary) should be resistant to contaminants in the ground	C4.10
Timber floor finish	If laid directly on concrete, it may be bedded in a material which can also serve as a damp-proof membrane	C4.11
	Timber fillets that are laid in the concrete as a fixing for a floor finish should be treated with an effective preservative unless they are above the damp-proof membrane	C4.11

 Note: Suitable insulation may also be incorporated.

 Some schools of thought believe that there is also a need for an additional damp-proof membrane on top of the insulation to combat interstitial condensation.

This, however, then begs the question 'How can this moisture escape?'! Moisture would, presumably, just sit where it is generated and, if interstitial moisture is not controlled by a vapour membrane, then it will surely eventually migrate into the concrete or the insulation.

These points have been put to the Department for Communities and Local Government (DCLG), but unfortunately it has been unable to offer any definite answer – saying that '*the intention of Approved Documents is to provide guidance to the more common building situations and as there may be alternative ways of achieving compliance with the requirements, there is no obligation*

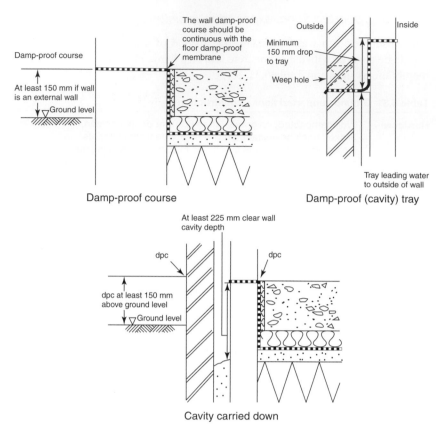

Figure 6.52 Ground supported floor – construction.

to adopt any particular solution contained in an Approved Document – if the builder prefers to meet the relevant requirement in some other way.' One of my readers has said that he prefers to employ the insulation below the slab and place a damp-proof membrane between the insulation and the blinding wherever possible (if only for ease of construction), which sounds like a very logical solution.

Suspended timber ground floors exposed to moisture from the ground

Any suspended timber floor next to the ground should:

- ensure that the ground is covered so as to resist moisture and prevent plant growth; C4.13a

- have a ventilated air space between the ground covering and the timber; C4.13b

- have damp-proof courses between the timber and any mate- C4.13c
  rial which can carry moisture from the ground.

Unless covered with a highly vapour-resistant floor finish, a suspended timber floor next to the ground may be built as shown in Figure 6.53 and as follows:

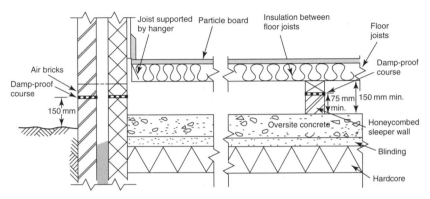

Figure 6.53 Suspended timber floor – construction.

Hardcore	A bed of clean, broken brick or any other inert material free from materials including water-soluble sulphates in quantities which could damage the concrete.	C4.14
Concrete	A ground covering of unreinforced concrete at least 100 mm thick to mix ST 1 in BS 8500 or laid on at least 300 m polyethylene sheet with sealed joints (and itself laid on a bed of material which will not damage the sheet).	C4.14a(i) C4.14a(ii)

> **Note:** To prevent water collecting on the ground covering, either the top should be entirely above the highest level of the adjoining ground or, on sloping sites, consideration should be given to installing drainage on the outside of the upslope side of the building.

Ventilation	There should be a ventilated air space at least 75 mm from the ground (and covering the underside of any wall plates) and at least 150 mm from the underside of the suspended timber floor.	C4.14b
	Two opposing external walls should have ventilation openings placed so that the ventilating air will have a free path between opposite sides and to all parts.	C4.14b

	Ventilating openings should be not less than either 1500 mm²/m run of external wall or 500 mm²/m² of floor area – whichever gives the greater opening area.	C4.14b
	Any pipes needed to carry ventilating air should have a diameter of at least 100 mm.	C4.14b
	Ventilation openings should incorporate suitable grilles to prevent the entry of vermin to the subfloor.	C4.14b
	If floor levels need to be nearer to the ground to provide level access, subfloor ventilation can be provided through offset (periscope) ventilators.	C4.14b
Damp-proof membrane	Damp-proof membranes should be of impervious sheet material, engineering brick or slates in cement mortar or other material which will prevent the passage of moisture.	C4.14c
Timber floor finish	In areas such as kitchens, utility rooms and bathrooms where water may be spilled, any board used as a flooring, irrespective of the storey, should be moisture resistant.	C4.15
	In the case of chipboard it should be of one of the grades with improved moisture resistance specified in BS 7331:1990 or BS EN 312 Part 5:1997.	C4.15
	Identification marks should be facing upwards.	C4.15
	Any softwood boarding should be at least 20 mm thick and from a durable species or treated with a suitable preservative.	C4.15

Suspended concrete ground floors exposed to moisture from the ground

Concrete suspended floors (including beam and block floors) that are next to the ground should:	C4.17

- adequately prevent the passage of moisture to the upper surface;
- be reinforced to protect against moisture.

There should be a facility for inspecting and clearing out the subfloor voids beneath suspended floors – particularly in localities where flooding is likely.	C4.20

Hardcore

Concrete	In situ concrete at least 100 mm thick containing at least 300 kg of cement for each m^3 of concrete; or precast concrete construction (with or without infilling slabs).	C4.1
	Reinforcing steel should be protected by a concrete cover of at least 40 mm (if the concrete is in situ) and at least the thickness required for a moderate exposure, if the concrete is precast.	C4.1
Ventilation	There should be a ventilated air space at least 150 mm clear from the ground to the underside of the floor (or insulation if provided).	C4.19b
	Two opposing external walls should have ventilation openings placed so that the ventilating air will have a free path between opposite sides and to all parts.	C4.19b
	Ventilating openings should be not less than either 1500 mm^2/m run of external wall or 500 mm^2/m^2 of floor area – whichever gives the greater opening area.	C4.19b
	Any pipes needed to carry ventilating air should have a diameter of at least 100 mm.	C4.19b
	Ventilation openings should incorporate suitable grilles to prevent the entry of vermin to the subfloor.	C4.19b
Damp-proof membrane	A suspended concrete floor should contain a damp-proof membrane (if the ground below the floor has been excavated below the lowest level of the surrounding ground and will not be effectively drained).	C4.19a

Ground floors and floors exposed from below (resistance to damage from interstitial condensation)

A ground floor (or floor exposed from below such as above an open parking space or passageway – see Figure 6.54) shall be designed in accordance with Clause 8.5 and Appendix D of BS 5250:2002.

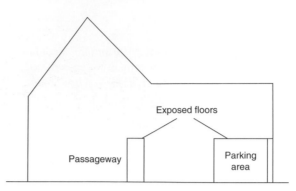

Figure 6.54 Typical floors exposed from below.

Floors (resistance to surface condensation and mould growth)

Ground floors should be designed and constructed so that the thermal transmittance (U-value) does not exceed $0.7\,\mathrm{W/m^2\,K}$ at any point.	C4.22a
Junctions between elements should be designed in accordance with robust construction recommendations.	C4.22b

Small single-storey non-residential buildings and annexes

The floor area of the building or annexe shall not exceed $36\,\mathrm{m^2}$.	A1/2 2C38(1)a
Where the floor area of the building or annex exceeds $10\,\mathrm{m^2}$, the walls shall have a mass of not less than $130\,\mathrm{kg/m^2}$.	A1/2 2C38(1)c

Tension straps

Tension straps (conforming to BS EN 845–1) should be used to strap walls to floors above ground level, at intervals not exceeding 2 m.	A1/2 2C35
For corrosion resistance purposes, the tension straps should be material reference 14 or 16.1 or 16.2 (galvanized steel) or other more resistant specifications including material references 1 or 3 (austenitic stainless steel).	A1/2 2C35

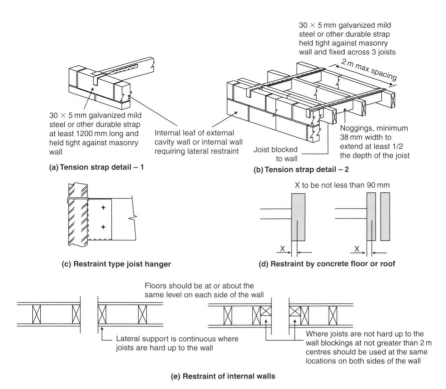

Figure 6.55 Lateral support by floors.

The declared tensile strength of tension straps should not be less than 8 kN.

Tension straps need **not** be provided:

- in the longitudinal direction of joists in houses of not more than 2 storeys if the joists: A1/2 2C35a
  - are at not more than 1.2 m centres;   A1/2 2C35b
  - have at least 90 mm bearing on the supported walls or 75 mm bearing on a timber wall plate at each end; or   A1/2 2C35c
  - are carried on the supported wall by joist hangers (in accordance with BS EN 845–1 and BS 5628 – see Figure 6.53(c));   A1/2 2C35d
  - and are incorporated at not more than 2 m centres;
- when a concrete floor has at least 90 mm bearing on the supported wall (see Figure 6.55(d)); and

- where floors are at or about the same level on each side of a supported wall, and contact between the floors and wall is either continuous or at intervals not exceeding 2 m. Where contact is intermittent, the points of contact should be in line or nearly in line on plan (see Figure 6.55(e)).

Interruption of lateral support

Where an opening in a floor or roof for a stairway or the like adjoins a supported wall and interrupts the continuity of lateral support:

- the maximum permitted length of the opening is to be 3 m, measured parallel to the supported wall;   A1/2 2C37a

- connections (if provided by means other than by anchor) should be throughout the length of each portion of the wall situated on each side of the opening;   A1/2 2C37b

- connections via mild steel anchors should be spaced closer than 2 m on each side of the opening to provide the same number of anchors as if there were no opening;   A1/2 2C37c

- there should be no other interruption of lateral support.   A1/2 2C37d

Lateral support by floors

Floors should: A1/2 2C33a

- act to transfer lateral forces from walls to buttressing walls, piers or chimneys;
- be secured to the supported wall by connections (see Figure 6.55).

Intermediate floors and roof shall be constructed so that they provide local support to the walls and act as horizontal diaphragms capable of transferring the wind forces to buttressing elements of the building.   A1/2 1A2d

A wall in each storey of a building should: A1/2 2C32

- extend to the full height of that storey;
- have horizontal lateral supports to restrict movement of the wall at right angles to its plane.

Walls should be strapped to floors above ground level, at intervals not exceeding 2 m and as shown in Figure 6.55 by tension straps conforming to BS EN 845–1.   A1/2 2C35

Where an opening in a floor for a stairway or the like adjoins a supported wall and interrupts the continuity of lateral support: A1/2 2C37

- the maximum permitted length of the opening is to be 3 m, measured parallel to the supported wall;

- connections (if provided by means other than by anchor) should be throughout the length of each portion of the wall situated on each side of the opening;
- connections via mild steel anchors should be spaced closer than 2 m on each side of the opening to provide the same number of anchors as if there were no opening;
- there should be no other interruption of lateral support.

The maximum span for any floor supported by a wall is 6 m where the span is measured centre to centre of bearing (see Figure 6.56).   A1/2 2C23

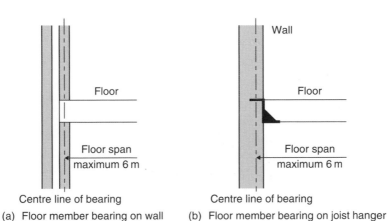

(a) Floor member bearing on wall   (b) Floor member bearing on joist hanger

Figure 6.56 Maximum span of floors.

Internal fire spread (structure)

Load-bearing elements of structure

All load-bearing elements of a structure shall have a minimum standard of fire resistance.   B3 4.1 (V1)

B3 7.1 (V2)

Structural frames, beams, floor structures and gallery structures should have at least the fire resistance given in Appendix A of Approved Document B.	B3 4.2 (V1) B3 7.2 (V2)
When altering an existing two-storey, single-family dwelling-house to provide additional storeys, the floor(s), both old and new, shall have the full 30 minute standard of fire resistance.	B3 4.7 (V1)

Fire resistance – compartmentation

To prevent the spread of fire within a building, whenever possible the building should be subdivided into compartments separated from one another by walls and/or floors of fire-resisting construction.	B3 5.1 (V1) B3 8.1 (V2)
Parts of a building that are occupied mainly for different purposes, should be separated from one another by compartment walls and/or compartment floors.	B3 5.3 (V1) B3 8.11 (V2)
The wall and any floor between the garage and the house shall have a 30-minute fire resistance.	B3
Any opening in the wall to be at least 100 mm above the garage floor level with an FD30 door.	B3

In buildings containing flats or maisonettes compartment   B3 8.13 (V2)
walls or compartment floors shall be constructed between:

- every floor (unless it is within a maisonette);
- one storey and another within one dwelling;
- every wall separating a flat or maisonette from any other part of the building;
- every wall enclosing a refuse storage chamber.

Every compartment floor should: B3 5.6 (V1)

- form a complete barrier to fire between the compart-   B3 8.20 (V2)
  ments they separate; and
- have the appropriate fire resistance as indicated in Appendix A of Approved Document B, Tables A1 and A2.

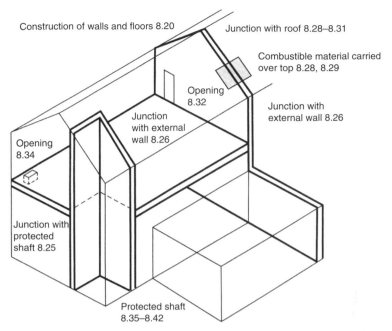

Figure 6.57 Compartment walls and compartment floors with reference to relevant paragraphs in Part B.

Where a compartment wall or compartment floor meets another compartment wall, or an external wall, the junction should maintain the fire resistance of the compartmentation.	B3 5.9 (V1) B3 8.25 (V2)
Junctions between a compartment floor and an external wall that has no fire resistance (e.g. a curtain wall) should be restrained at floor level to reduce the movement of the wall away from the floor when exposed to fire.	B2 8.26 (V2)
Compartment walls should be able to accommodate the predicted deflection of the floor above by either:	B2 8.27 (V2)

- having a suitable head detail between the wall and the floor, that can deform but maintain integrity when exposed to a fire; or
- the wall may be designed to resist the additional vertical load from the floor above as it sags under fire conditions and thus maintain integrity.

Underfloor voids

Extensive cavities in floor voids should be subdivided with cavity barriers.	B3

Concrete

With a concrete intermediate floor:

The ground floor may be a solid slab, laid on the ground, or a suspended concrete floor.	E2.51 E2.88 E2.126
A concrete slab floor on the ground may be continuous under a solid separating wall but may not be continuous under the cavity masonry core of the separating wall.	E2.51 E2.88 E2.127 E2.130
An internal concrete floor slab may only be carried through a separating wall if the floor base has a mass of at least 365 kg/m².	E2.46 E2.121

 Note: Internal concrete floors should generally be built into a separating wall and carried through to the cavity face of the leaf.

 The cavity should not be bridged (E2.85 and E2.122). Internal hollow-core concrete plank floors (and concrete beams with infilling block floors) should **not** be continuous through or under a separating wall (E2.47, E2.53 and E2.129).

A suspended concrete floor may only pass under a separating wall if the floor has a mass of at least 365 kg/m².	E2.52 E2.89
A suspended concrete floor should **not** be carried through to the cavity face of the leaf and the cavity should not be bridged.	

Floors – general

Floors that separate a dwelling from another dwelling (or part of the same building) shall resist the transmission of airborne sounds.	E
Floors above a dwelling that separate it from another dwelling (or another part of the same building) shall resist:	E

- the transmission of impact sound (such as speech, musical instruments and loudspeakers and impact sources such as footsteps and furniture moving);

- the flow of sound energy through walls and floors;
- the level of airborne sound.

Air paths, including those due to shrinkage, must be avoided – porous   E
materials and gaps at joints in the structure must be sealed.

The possibility of resonance in parts of the structure (e.g. a dry lining)   E
should be avoided.

Flanking transmission (i.e. the indirect transmission of sound from   E
one side of a wall or floor to the other side) should be minimized.

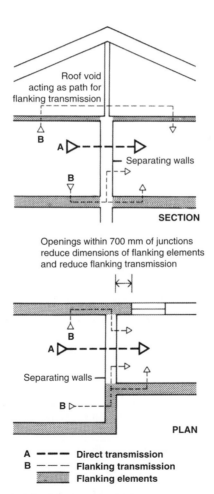

Figure 6.58 Direct and flanking transmission.

 Note: For clarity, not all flanking paths have been shown.

Requirement E1

Figure 6.59 illustrates the relevant parts of the building that should be protected from airborne and impact sound in order to satisfy requirement E1.

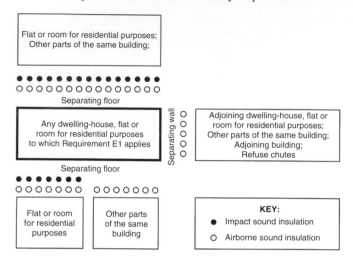

Figure 6.59 Requirement E1 – resistance to sound.

All new floors constructed within a dwelling-house (flat or room used for residential purposes) – whether purpose built or formed by a material change of use – shall meet the laboratory sound insulation values set out in Table 6.32.	E0.9
Floors: • for rooms for residential purposes; and • for dwelling houses and flats; that have a separating function, should achieve the sound insulation values as set out in Table 6.32.	E0.1

Table 6.32 Dwelling houses and flats – performance standards for separating floors and stairs that have a separating function

	Airborne sound insulation, $D_{nT,w}+C_{tr}$ (dB) (minimum values)	Impact sound insulation, $L_{nT,w}$ (dB) (maximum values)
Purpose-built rooms for residential purposes	45	62
Purpose-built dwelling houses and flats	45	62
Rooms for residential purposes formed by material change of use	43	64
Dwelling houses and flats formed by material change of use	43	64

Note:

(1) The sound insulation values in Table 6.32 include a built-in allowance for 'measurement uncertainty', and so, if any of these test values are not met, then that particular test will be considered as failed.

(2) Occasionally, a higher standard of sound insulation may be required between spaces used for normal domestic purposes and noise generated in and to an adjoining communal or non-domestic space. In these cases it would be best to seek specialist advice before committing yourself.

Flanking transmission from walls and floors connected to the separating wall shall be controlled.	E2
Tests should be carried out between rooms or spaces that share a common area formed by a separating wall or separating floor.	E1
Impact sound insulation tests should be carried out without a soft covering (e.g. carpet, foam backed vinyl etc.) on the floor.	E1
When a separating floor is used, a minimum mass per unit area of 120 kg/m^2 (excluding finish) shall always apply, irrespective of the presence or absence of openings.	E2
Care should be taken to correctly detail the junctions between the separating floor and other elements such as external walls, separating walls and floor penetrations.	E3
Spaces between floor joists should be sealed with full depth timber blocking.	E2
If the floor joists are to be supported on the separating wall then they should be supported on hangers and should not be built in.	E2
If the joists are at right angles to the wall, spaces between the floor joists should be sealed with full depth timber blocking.	E3
The floor base (excluding any screed) should be built into a cavity masonry external wall and carried through to the cavity face of the inner leaf.	E
The floor base should be continuous or above an internal masonry wall.	E3
All pipes and ducts that penetrate a floor separating habitable rooms in different flats should: • be enclosed for their full height in each flat; • have fire protection to satisfy Building Regulation Part B – Fire safety.	E3

Note:

(1) Where any building element functions as a separating element (e.g. a ground floor that is also a separating floor for a basement flat) then the separating element requirements should take precedence.

(2) In some circumstances (e.g. when a historic building is undergoing a material change of use) it may not be practical to improve the sound insulation to the standards set out in Approved Document E1, particularly if the special characteristics of such a building need to be recognized. In these circumstances the aim should be to improve sound insulation to the 'extent that it is practically possible'.

(3) BS 7913:1998 *The Principles of the Conservation of Historic Buildings* provides guidance on the principles that should be applied when proposing work on historic buildings.

Requirement E2

> Constructions for new floors within a dwelling-house (flat or room   E0.9
> for residential purposes) – whether purpose built or formed by a
> material change of use – shall meet the laboratory sound insulation
> values set out in Table 6.33.

Table 6.33 Laboratory values for new internal walls within dwelling-houses, flats and rooms for residential purposes – whether purpose built or formed by a material change of use

	Airborne sound insulation, R_W (dB) (minimum values)
Purpose-built dwelling houses and flats Floors	40

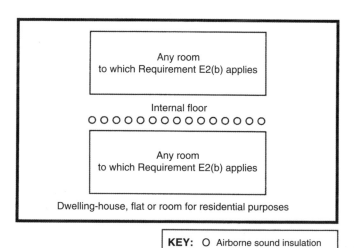

Figure 6.60 Requirement E2(b) – internal floors.

Requirement E3

Sound absorption measures described in Section 7 of Approved   E0.11
Document N shall be applied.

Requirement E4

The values for sound insulation, reverberation time and indoor   E0.12
ambient noise as described in Section 1 of Building Bulletin 93,
The Acoustic Design of Schools(produced by Department for Education and Skills (DFES) and published by The Stationery Office
(ISBN 0 11 271105 7)) shall be satisfied.

Separating floors and associated flanking constructions for new buildings

There are three groups of floor, as shown below:

Floor type 1
Concrete base
with ceiling
and soft floor
covering

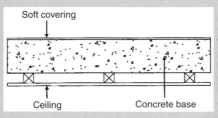

The resistance to
airborne sound
depends mainly on
the mass per unit
area of the concrete
base and partly on the
mass per unit area of
the ceiling. The soft
floor covering reduces
impact sound at
source.

Floor type 2
Concrete base
with ceiling and
floating floor
(three types of
floating floor
are available,
see p. 000)

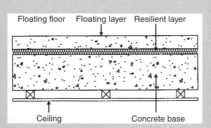

The resistance to
airborne and impact
sound depends on the
mass per unit area of
the concrete as well
as the mass per unit
area and isolation of
the floating layer and
ceiling. The floating
floor reduces impact
sound at source.

Floor type 3
Timber frame
base with
ceiling and
platform floor

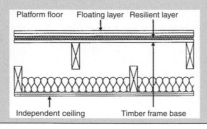

The resistance to
airborne and impact
sound depends on
the structural floor
base and the isolation
of the platform floor
and the ceiling. The
platform floor reduces
impact sound at
source.

General requirements

Floor types should be capable of achieving the performance standards shown in Table 6.32.

Care should be taken to correctly detail the junctions between the separating floor and other elements such as external walls, separating walls and floor penetrations.

E3.1

 Note: Where any building element functions as a separating element (e.g. a ground floor that is also a separating floor for a basement flat), then the separating element requirements should take precedence.

Ceiling treatments

Each floor type should use one of the following three ceiling treatments which are ranked in order of sound insulation performance from A to C.

E3.17 to E3.18

Note: Use of a better performing ceiling than that described in the guidance should improve the sound insulation of the floor, provided there is no significant flanking transmission.

Ceiling treatment A Independent ceiling with absorbent material	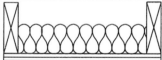	• At least two layers of plasterboard with staggered joints; • minimum total mass per unit area of plasterboard 20 kg/m²; • an absorbent layer of mineral wool (minimum thickness 100 mm, minimum density 10 kg/m³) laid in the cavity formed above the ceiling
The type of ceiling support depends on the floor type	**For floor types 1, 2 and 3** Use independent joists fixed only to the surrounding walls **For floor type 3** Use independent joists fixed to the surrounding walls with additional support provided by resilient hangers attached directly to the floor	Always ensure: • that you seal the perimeter of the independent ceiling with tape or sealant; • you do not create a rigid or direct connection between the independent ceiling and the floor base

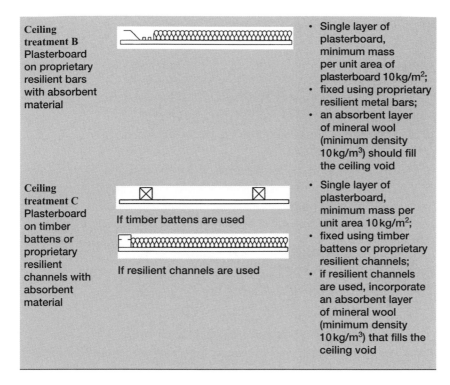

| Ceiling treatment B Plasterboard on proprietary resilient bars with absorbent material | | • Single layer of plasterboard, minimum mass per unit area of plasterboard 10 kg/m²; • fixed using proprietary resilient metal bars; • an absorbent layer of mineral wool (minimum density 10 kg/m³) should fill the ceiling void |
| Ceiling treatment C Plasterboard on timber battens or proprietary resilient channels with absorbent material | If timber battens are used If resilient channels are used | • Single layer of plasterboard, minimum mass per unit area 10 kg/m²; • fixed using timber battens or proprietary resilient channels; • if resilient channels are used, incorporate an absorbent layer of mineral wool (minimum density 10 kg/m³) that fills the ceiling void |

Note:
(1) Electrical cables give off heat when in use and special precautions may be required when they are covered by thermal insulating materials. See BRE BR 262, Thermal Insulation: avoiding risks, Section 2.3.
(2) Installing recessed light fittings in ceiling treatments A to C can reduce their resistance to the passage of airborne and impact sound.

Floor type 1: concrete base with ceiling and soft floor covering

A floor of type 1 consists of a concrete floor base with a soft floor covering and a ceiling. Its resistance to airborne sound mainly depends on:

• the mass per unit area of the concrete base;
• the mass per unit area of the ceiling;
• the soft floor covering (which helps to reduce the source of the impact sound).

General requirements

To allow for future replacements, the soft floor covering should not be fixed or glued to the floor.	E3.27a
To avoid air paths all joints between parts of the floor should be filled.	E3.27b

To reduce flanking transmission, air paths should be avoided at all points where a pipe or duct penetrates the floor.	E3.27c
A separating concrete floor should be built into the walls (around its entire perimeter) if the walls are masonry.	E3.27d
All gaps between the head of a masonry wall and the underside of the concrete floor should be filled with masonry.	E3.27e
Flanking transmission from walls connected to the separating floor should be controlled.	E3.27f
The floor base shall not bridge the cavity in a two-cavity masonry wall.	E3.27a
Non-resilient floor finishes (such as ceramic floor tiles and wood block floors that are rigidly connected to the floor base) shall not be used.	E3.27b2
Any soft floor covering that is used should be of resilient material with an overall uncompressed thickness of at least 4.5 mm (also see BS EN ISO 140–8:1998).	E3.28a

Two floor types (see below) will meet these requirements.

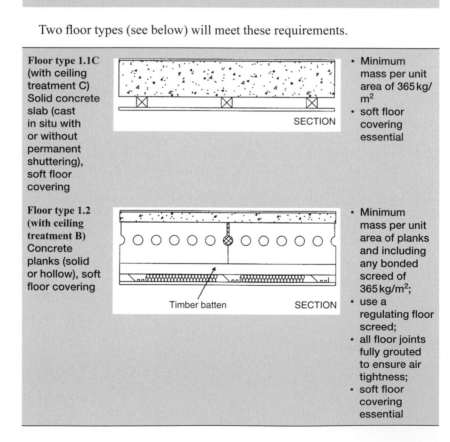

Floor type 1.1C (with ceiling treatment C) Solid concrete slab (cast in situ with or without permanent shuttering), soft floor covering	SECTION	• Minimum mass per unit area of 365 kg/m² • soft floor covering essential
Floor type 1.2 (with ceiling treatment B) Concrete planks (solid or hollow), soft floor covering	Timber batten SECTION	• Minimum mass per unit area of planks and including any bonded screed of 365 kg/m²; • use a regulating floor screed; • all floor joints fully grouted to ensure air tightness; • soft floor covering essential

Junction requirements for floor type 1

Junctions with an external cavity wall with masonry inner leaf

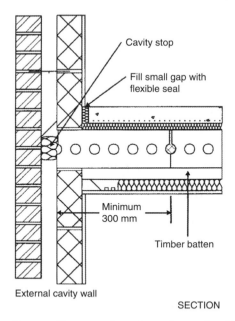

Cavity stop

Fill small gap with flexible seal

Minimum 300 mm

Timber batten

External cavity wall

SECTION

Figure 6.61 Junctions with an external cavity wall with masonry inner leaf.

If the external wall is a cavity wall:
   • the outer leaf of the wall may be of any construction;
   • the cavity should be stopped with a flexible closer to ensure adequate drainage.
	E3.31

The masonry inner leaf of an external cavity wall should have a mass per unit area of at least $120\,kg/m^2$ excluding finish.	E3.32
The floor base (excluding any screed) should be built into a cavity masonry external and carried through to the cavity face of the inner leaf.	E3.33

Junctions with an external cavity wall with timber frame inner leaf

Where the external wall is a cavity wall: • the outer leaf of the wall may be of any construction; • the cavity should be stopped with a flexible closer;	E3.36

- the wall finish of the inner leaf of the external wall should be two layers of plasterboard, each sheet of plasterboard to be a minimum mass per unit area $10\,kg/m^2$;
- all joints should be sealed with tape or caulked unenclosed.

Junctions with an external solid masonry wall

No official guidance currently available about junctions with a solid masonry wall. It is best to seek specialist advice.   E3.37

Junctions with internal framed walls

There are no restrictions on internal walls meeting a type 1 separating floor.   E3.38

Junctions with internal masonry walls

The floor base should be continuous through or above an internal masonry wall.   E3.39

The mass per unit area of any load-bearing internal wall (or any internal wall rigidly connected to a separating floor) should be at least $120\,kg/m^2$ excluding finish.   E3.40

Junctions with floor penetrations (excluding gas pipes)

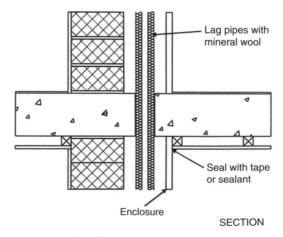

Figure 6.62 Floor type 1 – floor penetrations.

Pipes and ducts should be in an enclosure (both above and below the floor). In all cases:

The enclosure should be constructed of material having a mass per unit area of at least 15 kg/m². — E3.32

The enclosure should either be lined or the duct (or pipe) within the enclosure wrapped with 25 mm unfaced mineral fibre. — E3.42

Penetrations through a separating floor by ducts and pipes should have fire protection to satisfy Building Regulation Part B – Fire safety. — E3.43

Fire stopping should be flexible to prevent a rigid contact between the pipe and the floor. — E3.43

Gas pipes may be contained in a separate (ventilated) duct or can remain unenclosed. — E3.43

If a gas service is installed it shall comply with the Gas Safety (Installation and Use) Regulations 1998, S1 1998 No. 2451. — E3.43

If the pipes and ducts penetrate a floor separating habitable rooms in different flats, they should be enclosed for their full height in each flat. — E3.41

Junctions with separating wall type I – solid masonry

For floor types 1.1C and 1.2C, two possibilities exist:

A separating floor type 1.1C base (excluding any screed) should **not** pass through a separating wall type 1. — E3.44

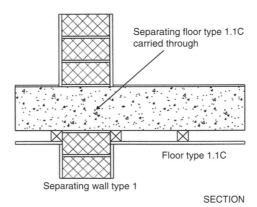

Separating floor type 1.1C carried through

Floor type 1.1C

Separating wall type 1

SECTION

Figure 6.63 Floor type 1.1C – wall type 1.

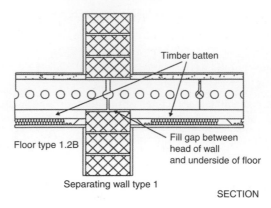

Figure 6.64 Floor type 1.1C – wall type 1.

A separating floor type 1.2B base (excluding any screed) should **not** pass through a separating wall type 1.	E3.44

Junctions with separating wall type 2 cavity masonry

The mass per unit area of any leaf that is supporting or adjoining the floor should be at least 15 kg/m² excluding finish.	E3.46
The floor base (excluding any screed) should be carried through to the cavity face of the leaf.	E3.47
The wall cavity should not be bridged.	E3.47

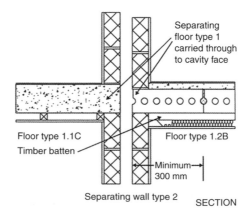

Figure 6.65 Floor types 1.1C and 1.2B – wall type 2.

Where floor type 1.2B is used (and the planks are parallel to the separating wall) the first joint should be a minimum of 300 mm from the inner face of the adjacent cavity leaf.   E3.48

Junctions with separating wall types 3.1 and 3.2 (solid masonry core)

A separating floor type 1.1C base (excluding any screed) should pass through separating wall types 3.1 and 3.2.   E3.49

A separating floor type 1.2B base (excluding any screed) should not be continuous through a separating wall type 3.   E3.50

Where a separating wall type 3.2 is used with floor type 1.2B (and the planks are parallel to the separating wall) the first joint should be a minimum of 300 mm from the centreline of the masonry core.   E3.51

Junctions with separating wall type 3.3 (cavity masonry core)

The mass per unit area of any leaf that is supporting or adjoining the floor should be at least 120 kg/m² excluding finish.   E3.52

The floor base (excluding any screed) should be carried through to the cavity face of the leaf of the core.   E3.53

The cavity should not be bridged.   E3.53

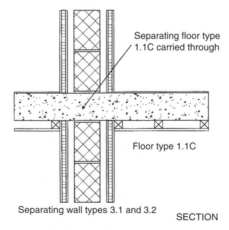

Separating floor type
1.1C carried through

Floor type 1.1C

Separating wall types 3.1 and 3.2   SECTION

Figure 6.66 Floor type 1.1C – wall types 3.1 and 3.2.

Where floor type 1.2B is used (and the planks are parallel to the   E3.54
separating wall) the first joint should be a minimum of 300 mm
from the inner face of the adjacent cavity leaf of the masonry core.

Junctions with separating wall type 4 timber frames with absorbent material

No official guidance is currently available. It is best to seek spe-   E3.55
cialist advice.

Floor type 2: concrete base with ceiling and floating floor

A floor of type 2 consists of a concrete floor base with a floating floor (which in turn consists of a floating layer and a resilient layer) and a ceiling. Its resistance to airborne and impact sound depends on:

- the mass per unit area of the concrete base;
- the mass per unit area and isolation of the floating layer and the ceiling;
- the floating floor (which reduces impact sound at source).

General requirements

All joints between parts of the floor should be filled to avoid air   E3.61a
paths.

To reduce flanking transmission, air paths should be avoided at all   E3.61b
points where a pipe or duct penetrates the floor.

A separating concrete floor should be built into the walls (around   E3.61c
its entire perimeter) if the walls are masonry.

All gaps between the head of a masonry wall and the underside of   E3.61d
the concrete floor should be filled with masonry.

Flanking transmission from walls connected to the separating   E3.61e
floor should be controlled.

The floor base shall not bridge a cavity in a cavity masonry wall.   E3.61

Two floor types (consisting of a floating layer and resilient layer – see below) will meet these requirements. A performance-based approach (type C) is also available.

**Floating floor
(a) Timber raft
floating layer
with resilient
layer**

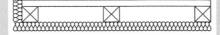

- Timber raft of board material (with bonded edges, e.g. tongued and grooved);
- minimum mass per unit area 12 kg/m^2;
- fixed to 45 mm × 45 mm battens laid loose on the resilient layer (but not along any joints in the resilient layer);
- resilient layer of mineral wool (which may be paper faced on the underside) with density 36 kg/m^{-3} and minimum thickness 25 mm

**Floating floor
(b) Sand
cement screed
floating layer
with resilient
layer**

Floating layer: of 65 mm sand–cement screed with a mass per unit area of at least 80 kg/m^2.

Resilient layer: protected while the screed is being laid (e.g. by a 20–50 mm wire mesh) and consisting of either:

- a layer of mineral wool of minimum thickness 25 mm with density 36 kg/m^3 (paper faced on the upper side);
- an alternative type of resilient layer with maximum dynamic stiffness of 15 kg/m^3;
- an alternative type of resilient layer with minimum thickness of 5 mm (see BS EN ISO 29052–1:1992)

Floating floor (c) Performance-based approach	Floating floors should meet the following specification: • Rigid boarding above a resilient and/or damping layer; • weighted reduction in impact sound pressure level of not less than 29 dB (see BS EN ISO 717–2:1997 and BS EN ISO 140–8:1998).

A small gap filled with a flexible sealant should be left between the floating layer and wall at all room edges.	E3.63a
A small gap (approx. 5 mm and filled with a flexible sealant) should be left between skirting and floating layer.	E3.63b
Resilient materials should be laid in rolls or sheets either with lapped joints or with joints tightly butted and taped.	E3.63c
Paper facing should be used on the upper side of fibrous materials to prevent screed entering the resilient layer.	E3.63d
The floating layer and the base or surrounding walls shall not be bridged (e.g. with services or fixings that penetrate the resilient layer).	E3.63a2
The floating screed shall create a bridge (e.g. through a gap in the resilient layer) to the concrete floor base or surrounding walls.	E3.63b2

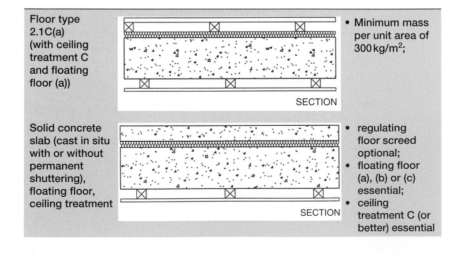

Floor type 2.1C(a) (with ceiling treatment C and floating floor (a))	SECTION	• Minimum mass per unit area of 300 kg/m²;
Solid concrete slab (cast in situ with or without permanent shuttering), floating floor, ceiling treatment	SECTION	• regulating floor screed optional; • floating floor (a), (b) or (c) essential; • ceiling treatment C (or better) essential

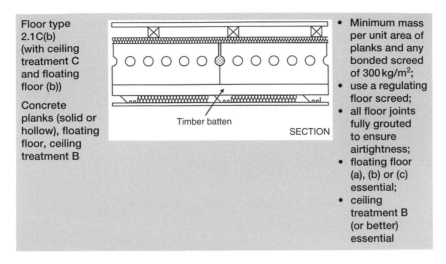

Floor type 2.1C(b) (with ceiling treatment C and floating floor (b))

Concrete planks (solid or hollow), floating floor, ceiling treatment B

Timber batten

SECTION

- Minimum mass per unit area of planks and any bonded screed of 300 kg/m²;
- use a regulating floor screed;
- all floor joints fully grouted to ensure airtightness;
- floating floor (a), (b) or (c) essential;
- ceiling treatment B (or better) essential

Junction requirements for floor type 2

Junctions with an external cavity wall with type 2 timber frame inner leaf

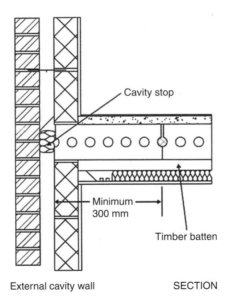

Cavity stop

Minimum 300 mm

Timber batten

External cavity wall   SECTION

Figure 6.67 Floor type 2 – external cavity wall with masonry internal leaf.

Where the external wall is a cavity wall: E3.69

- the outer leaf of the wall may be of any construction;
- the cavity should be stopped with a flexible closer.

The masonry inner leaf of an external cavity wall should have a mass per unit area of at least 120 kg/m^2.   E3.70

The floor base (excluding any screed) should be built into a cavity masonry external wall and carried through to the cavity face of the inner leaf.   E3.71

The cavity should not be bridged.   E3.71

If a floor 2.2B is used (and the planks through, or above, an internal masonry wall are parallel to the external wall) the first joint should be a minimum of 300 mm from the cavity face of the inner leaf.   E3.72

Junctions with an external cavity wall with timber frame inner leaf

Where the external wall is a cavity wall: E3.74
- the outlet leaf of the wall may be of any construction;
- the cavity should be stopped with a flexible closer;
- the wall face of the inner leaf of the external wall should be two layers of plasterboard;
- each sheet of plasterboard to be of minimum mass per unit area 10 kg/m^2;
- all joints should be sealed or caulked with sealant.

Junctions with an external solid masonry wall

No official guidance is currently available. It is best to seek specialist advice.   E3.75

Junctions with internal framed walls

There are no restrictions on internal walls meeting a type 4 separating wall.   E3.76

Junctions with internal masonry walls

The floor base should be continuous or above an internal masonry wall.   E3.77

The mass per unit area of any load bearing internal wall or any internal wall rigidly connected to a separating floor should be at least 120 kg/m^2 excluding finish.   E3.78

Junctions with floor penetrations (excluding gas pipes)

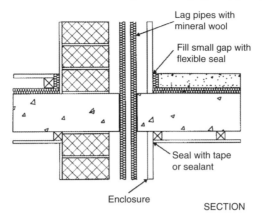

Lag pipes with mineral wool

Fill small gap with flexible seal

Seal with tape or sealant

Enclosure

SECTION

Figure 6.68 Floor type 2 – floor penetrations.

Pipes and ducts that penetrate a floor separating habitable rooms in different flats should be enclosed for their full height in each flat.	E3.79
The enclosure should be constructed of material having a mass per unit area of at least 15 kg/m².	E3.80
Either line the enclosure, or wrap the duct or pipe within the enclosure, with 25 mm unfaced mineral wool.	E3.80
A small gap (sealed with sealant or neoprene) of about 5 mm should be left between the enclosure and the floating layer.	E3.81
Where floating floor (a) or (b) is used the enclosure may go down to the floor base (provided that the enclosure is isolated from the floating layer).	E3.81
Penetrations through a separating floor by ducts and pipes should have fire protection to satisfy Building Regulation Part B – Fire safety.	E3.82
Gas pipes may be contained in a separate (ventilated) duct or can remain unenclosed.	E3.82
If a gas service is installed it shall comply with the Gas Safety (Installation and Use) Regulations 1998, S1 1998 No. 2451.	E3.82

Junctions with a separating wall type 1 – solid masonry

A separating floor type 2.1C base (excluding any screed) should pass through a separating wall type 1.	E3.84

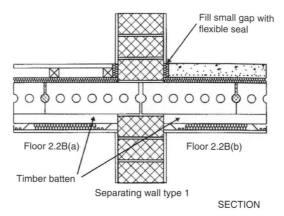

Figure 6.69 Floor type 2.1C – wall types 3.1 and 3.2.

A separating floor type 2.2B base (excluding any screed) should **not** be continuous through a separating wall type 1.	E3.84

Junctions with a separating wall type 2 cavity masonry

The floor base (excluding any screed) should be carried through to the cavity face of the leaf.	E3.85
The cavity should not be bridged.	E3.85
If a floor type 2.2B is used (and the planks are parallel to the separating wall) the first joint should be a minimum of 300 mm from the cavity face of the leaf.	E3.86

Junctions with separating wall type 3.1 and 3.2 (solid masonry core)

A separating floor type 2.1C base (excluding any screed) should pass through separating wall types 3.1 and 3.2.	E3.87
A separating floor type 2.2B base (excluding any screed) should not be continuous through a separating wall type 3.	E3.88
If a separating wall type 3.2 is used with floor type 2.2B (and the planks are parallel to the separating wall) the first joint should be a minimum of 300 mm from the centreline of the masonry core.	E3.89

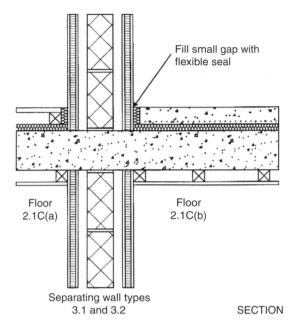

Fill small gap with
flexible seal

Floor
2.1C(a)

Floor
2.1C(b)

Separating wall types
3.1 and 3.2

SECTION

Figure 6.70 Floor type 2.1C – wall types 3.1 and 3.2.

Junctions with separating wall type 3.3 (cavity masonry core)

The mass per unit area of any leaf that is supporting or adjoining the floor should be at least 120 kg/m² excluding finish.	E3.90
The floor base (excluding any screed) should be carried through to the cavity face of the leaf of the core.	E3.91
The cavity should not be bridged.	E3.91
If a floor type 2.2B is used (and the planks are parallel to the separating wall) the first joint should be a minimum of 300 mm from the inner face of the adjacent cavity leaf of the masonry core.	E3.92

Junctions with separating wall type 4 timber frames with absorbent material

No official guidance is currently available. It is best to seek specialist advice.	E3.93

Floor type 3: timber frame base with ceiling and platform floor

A floor of type 3 consists of a timber frame structural floor base with a deck, platform floor (consisting of a floating layer and a resilient layer) and ceiling treatment. Its resistance to airborne and impact sound depends on:

- the structural floor base;
- the isolation of the platform floor and the ceiling;
- the platform floor (which reduces impact sound at source).

General requirements

To reduce flanking transmission, air paths should be avoided at all points where the floor is penetrated.	E3.99a
Flanking transmission from walls connected to the separating floor should be as described in the following junction requirements for floor type 3.	E3.99b
There should be no bridge (e.g. formed by services or fixings that penetrate the resilient layer) between the floating layer and the base or surrounding walls.	E3.99
For the platform floor, ensure that: • the correct density of resilient layer is used; • the layer can carry the anticipated load; • during construction a gap is maintained between the wall and the floating layer (filled with a flexible sealant, expanded or extruded polystyrene strip); • resilient materials are laid in sheets with all joints tightly butted and taped.	E3.99

The following floor type (floor type 3.1A) will meet these requirements.

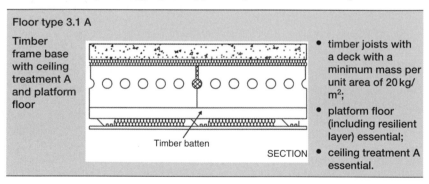

Floor type 3.1 A

Timber frame base with ceiling treatment A and platform floor

Timber batten

SECTION

- timber joists with a deck with a minimum mass per unit area of 20 kg/m²;
- platform floor (including resilient layer) essential;
- ceiling treatment A essential.

Platform floor

The floating layer should: • be a minimum of two layers of board material; • be minimum total mass per unit area 25 kg/m²; • have layers of minimum thickness 8 mm; • be fixed together with joints staggered; • be laid loose on a resilient layer.	E3.101

Resilient layer

The resilient layer should be of mineral wool: E3.102
- minimum thickness 25 mm;
- density 60 kg/m^3 to 100 kg/m^3;
- paper faced on the underside.

Junction requirements for floor type 3

Junctions with an external cavity wall with masonry inner leaf

Where the external wall is a cavity wall: E3.103
- the outer leaf of the wall may be of any construction;
- the cavity should be stopped with a flexible closer.

The masonry inner leaf of a cavity wall should be lined with an   E3.104
independent panel.

The ceiling should be taken through to the masonry.   E3.105

The junction between the ceiling and the independent panel   E3.105
should be sealed with tape or caulked with sealant.

Air paths between floor and wall cavities should be blocked.   E3.106

Note:
(1) Any normal method of connecting floor base to wall may be used.
(2) Independent panels are not required if the mass per unit area of the inner
   leaf is greater than 375 kg/m^2.

Junctions with an external cavity wall with timber frame inner leaf

Where the external wall is a cavity wall: E3.109
- the outer leaf of the wall may be of any construction;
- the cavity should be stopped with a flexible closer.

The wall finish of the inner leaf of the external wall should: E3.110
- be two layers of plasterboard;
- be each sheet of plasterboard of minimum mass per unit area
   10 kg/m^2;
- have all joints sealed with tape or caulked with sealant.

Any normal method of connecting floor base to wall may be used. E3.111

If the joists are at right angles to the wall, spaces between the   E3.112
floor joists should be sealed with full depth timber blocking.

The junction between the ceiling and wall lining should be   E3.113
sealed with tape or caulked with sealant.

Junctions with an external solid masonry wall

No official guidance is currently available. It is best to seek specialist advice. E3.113

Junctions with internal framed walls

The spaces between joists are at right angles and should be sealed with full depth timber blocking. E3.114

The junction between the ceiling and the internal framed wall should be sealed with tape or caulked with sealant. E3.115

Junctions with internal masonry walls

No official guidance currently available. It is best to seek specialist advice. E3.116

Junctions with floor penetrations (excluding gas pipes)

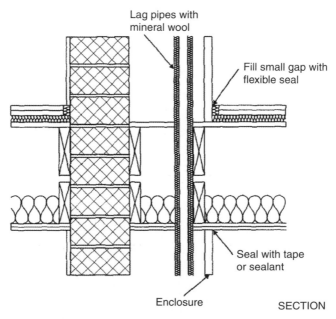

Figure 6.71 Floor type 3 – floor penetrations.

Pipes and ducts that penetrate a floor separating habitable rooms   E3.117
in different flats should be enclosed for their full height in each
flat.

The enclosure should:

- be constructed of material having a mass per unit area of at   E3.118
  least 15 kg/m^2;

- have a small, sealed (with sealant or neoprene) 5 mm gap   E3.119
  between the enclosure and floating layer;

- go down to the floor base;   E3.119

- be isolated from the floating layer.   E3.119

The duct or pipe within the enclosure should be lined or wrapped   E3.118
with 25 mm unfaced mineral wool.

Penetrations through a separating floor by ducts and pipes should   E3.120
have fire protection to satisfy Building Regulation Part B – Fire
safety.

Fire stopping should be flexible and also prevent rigid contact   E3.121
between the pipe and floor.

Gas pipes may be contained in a separate (ventilated) duct or can   E3.120
remain unenclosed.

If a gas service is installed it shall comply with the Gas Safety   E3.120
(Installation and Use) Regulations 1998, S1 1998 No. 2451.

Junctions with a separating wall type 1 solid masonry

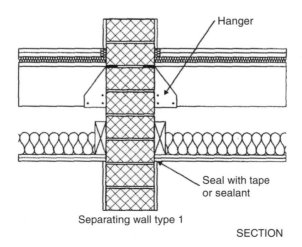

Figure 6.72 Floor type 3 – wall type 1.

Floor joists supported on a separating wall should be supported   E3.121
on hangers as opposed to being built in.

The junction between the ceiling and wall should be sealed with   E3.122
tape or caulked with sealant.

 Note: The above is particularly relevant for flats where there are separating walls.

Junctions with a separating wall type 2 – cavity masonry

Floor joists that are supported on a separating wall should be sup-   E3.123
ported on hangers and not built in.

The adjacent leaf of a cavity separating wall should be lined with   E3.124
an independent panel.

The ceiling should be taken through to the masonry.   E3.125

The junction between the ceiling and the independent panel   E3.125
should be sealed with tape or caulked with sealant.

 Note: Independent panels are not required if the mass per unit area of the inner leaf is greater than $375\,kg/m^2$.

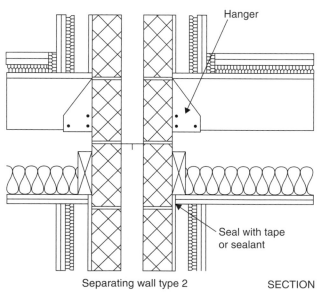

Figure 6.73 Floor type 3 – wall type 2.

Junctions with a separating wall type 3 – masonry between independent panels

Floor joists that are supported on a separating wall should be sup-   E3.127
ported on hangers and not built in.

The ceiling should be taken through to the masonry.   E3.128

The junction between the ceiling and the independent panel   E3.128
should be sealed with tape or caulked with sealant.

Junctions with a separating wall type 4 – timber frames with absorbent material

Spaces between the floor joists that are at right angles to the wall   E3.129
should be sealed with full depth timber blocking.

The junction of the ceiling and wall lining should be sealed with   E3.130
tape or caulked with sealant.

Conservation of fuel and power

If work being undertaken on an existing buildings which has a total useful floor area of over $1000\,m^2$, then, in addition to the principal works (which must still comply with the energy efficiency requirements detailed in Part L:2010 in the normal way), consequential improvements (where technically, functionally and economically feasible) will have to be completed, and

If upgrades are proposed to the existing dwelling, then such   L1B 4.7
upgrades should be implemented to a standard that is no worse
than that shown in column (b) of Table 6.34.

Table 6.34 Upgrading retained thermal elements

Element	(a) Threshold U-value (W/m²K)	(b) Improved U-value (W/m²K)
Floor	0.70	0.25

In addition, the standards for new thermal elements should be no worse than those shown in Table 6.35.

Table 6.35 Upgrading required for new thermal elements

Location	Element	Standard (W/m²K)
Conservatories and porches	Floor	0.262
Fabric standards	Floor	0.20
Buildings (other than dwellings)	Floor	0.25

6.8 Walls

In a brick-built house, the external walls are load-bearing elements that support the roof, floors and internal walls. These walls are normally cavity walls comprising two leaves braced with metal ties, but older houses will have solid walls, at least 225 mm (9 in.) thick. Bricks are laid with mortar in overlapping bonding patterns to give the wall rigidity and a damp-proof course is laid just above ground level to prevent the moisture rising. Window and door openings are spanned above with rigid supporting beams called 'lintels'. The internal walls of a brick-built house are either non-load-bearing divisions (made from lightweight blocks, manufactured boards or timber studding) or load-bearing structures made of brick or block.

Modern timber-framed house walls are constructed of vertical timber studs with horizontal top and bottom plates nailed to them. The frames, which are erected on a concrete slab or a suspended timber platform supported by cavity brick walls, are faced on the outside with plywood sheathing to stiffen the structure. Breather paper is fixed over the top to act as a moisture barrier. Insulation quilt is used between studs. Rigid timber lintels at openings carry the weight of the upper floor and roof. Brick cladding is typically used to cover the exterior of the frame. It is attached to the frame with metal ties. Weatherboarding often replaces the brick cladding on upper floors.

 Note: When reading this section, you will probably notice that a few of the requirements have already been covered in Sections 6.7 Floors and Section 6.9 Ceilings. This has been done in order to save the reader having to constantly turn back and re-read a previous page.

6.8.1 Requirements

Fire precautions

Materials and/or products used for the internal linings of walls shall restrict:

- *the spread of flame;*
- *the amount of heat released.*

(Approved Document B2)

Internal fire spread (structure)

A wall common to two or more buildings shall be designed and constructed so that it adequately resists the spread of fire between those buildings.

(Approved Document B3)

External walls shall be constructed so as to have a low rate of heat release and thereby be capable of reducing the risk of ignition from an external source and the spread of fire over their surfaces.

The amount of unprotected area in the sides of the building shall be restricted so as to limit the amount of thermal radiation that can pass through the wall.

(Approved Document B4)

The walls of the building shall adequately protect the building and people who use the building from harmful effects caused by:

- *ground moisture;*
- *precipitation and wind-driven spray;*
- *interstitial and surface condensation; and*
- *spillage of water from or associated with sanitary fittings or fixed appliances.*

(Approved Document C2)

Cavity insulation

Fumes given off by insulating materials such as by Urea Formaldehyde (UF) foams should not be allowed to penetrate occupied parts of buildings to an extent where it could become a health risk to persons in the building by becoming an irritant concentration.

(Approved Document D)

Airborne and impact sound

Dwellings shall be designed so that the noise from domestic activity in an adjoining dwelling (or other parts of the building) is kept to a level that:

- *does not affect the health of the occupants of the dwelling;*
- *will allow them to sleep, rest and engage in their normal activities in satisfactory conditions.*

(Approved Document E1)

Dwellings shall be designed so that any domestic noise that is generated internally does not interfere with the occupants' ability to sleep, rest and engage in their normal activities in satisfactory conditions.

(Approved Document E2)

Domestic buildings shall be designed and constructed so as to restrict the transmission of echoes.

(Approved Document E3)

Schools shall be designed and constructed so as to reduce the level of ambient noise (particularly echoing in corridors).

(Approved Document E4)

Conservation of fuel and power

Reasonable provision shall be made for the conservation of fuel and power in buildings by:

(a) *limiting heat gains and losses –*
  (i) *through thermal elements and other parts of the building fabric; and*
  (ii) *from pipes, ducts and vessels used for space heating, space cooling and hot water services;*
(b) *providing fixed building services which –*
  (i) *are energy efficient;*
  (ii) *have effective controls; and*
  (iii) *are commissioned by testing and adjusting as necessary to ensure they use no more fuel and power than is reasonable in the circumstances; and*
(c) *providing to the owner sufficient information about the building, the fixed building services and their maintenance requirements so that the building can be operated in such a manner as to use no more fuel and power than is reasonable in the circumstances*

(Approved Document L)

A new Part L came into force on 1 October 2010.

6.8.2 Meeting the requirements

General

Walls should comply with the relevant requirements of BS 5628: Part 3:2001.	A1/2 2C2c

Basic requirements for stability

The layout of walls (both internal and external) shall: • form a robust three-dimensional box structure in plan; • be constructed according to the specific guidance for each form of construction.	A1/2 1A2b
Internal and external walls shall be adequately connected by either masonry bonding or by using mechanical connections.	A1/2 1A2c

Building height

> For residential buildings, the maximum height of the building measured from the lowest finished ground level adjoining the building to the highest point of any wall or roof should not be greater than 15 m.   A1/2 2C4i
>
> Types of wall shown in Table 6.36 must extend to the full storey height.   A1/2 2C2

Table 6.36 Wall types considered in this section

Residential buildings of up to three storeys	Small, single-storey, non-residential buildings and annexes
External walls	External walls
Internal load-bearing walls	Internal load-bearing walls
Compartment walls	
Separating walls	

Thickness of walls

The thickness of the wall depends on the general conditions relating to the building of which the wall forms a part (e.g. floor area, roof loading, wind speed etc.) and the design conditions relating to the wall (e.g. type of materials, loading, end restraints, openings, recesses, overhangs and lateral floor support requirements etc.).

Note: Where walls are constructed of bricks or blocks, they shall be in accordance with BS 6649:1985.

Masonry units

> Walls should be properly bonded and solidly put together with mortar and constructed of masonry units conforming to:
>
> - clay bricks or blocks conforming to BS 3921:1985 or BS 6649:1985 or BS EN 771-1;   A1/2 2C20a
> - calcium silicate bricks conforming to BS 187:1978 or BS 6649:1985 or BS EN 771-2;   A1/2 2C20b
> - concrete bricks or blocks conforming to BS 6073: Part 1:1981 or BS EN 771-3 or 4;   A1/2 2C20c
> - square dressed natural stone conforming to the appropriate requirements described in BS EN 771-6 or BS 5628: Part 3:2001;   A1/2 2C20d
> - manufactured stone complying with BS 6457:1984 and BS EN 771-5.   A1/2 2C20e

Note: See BS 3921, BS 6073-1, BS 187, BS 5390 and BS 6649 for further details about the minimum compressive strength requirements for masonry units.

Mortar

Mortar should be equivalent to (or of greater strength and durability) than:

- mortar designation (iii) according to BS 5628: Part 3:2001;   A1/2 2C22
- strength class M4 according to BS EN 998-2;   A1/2 2C22b
- 1:1:5 or 6 CEM 1, lime and fine aggregate measured by volume of dry materials.   A1/2 2C22

Tension straps

Tension straps (conforming to BS EN 845-1) should be used to strap walls to floors above ground level, at intervals not exceeding 2 m and as shown in Figure 6.74.   A1/2 2C35

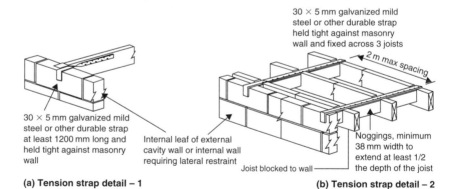

(a) Tension strap detail – 1   (b) Tension strap detail – 2

Figure 6.74 Lateral support by floors.

Gable walls should be strapped to roofs as shown in Figure 6. 75(a) and (b) by tension straps.   A1/2 2C36

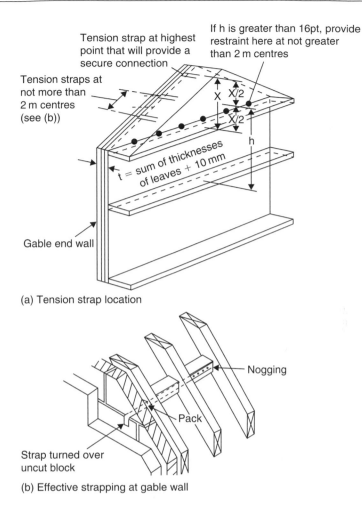

(a) Tension strap location

(b) Effective strapping at gable wall

Figure 6.75 Lateral support at roof level.

For corrosion resistance purposes, tension straps should be material reference 14 or 16.1 or 16.2 (galvanized steel) or other more resistant specifications including material references 1 or 3 (austenitic stainless steel).   A1/2 2C35

The declared tensile strength of tension straps should not be less than 8 kN.

Tension straps need **not** be provided: A1/2 2C35

- in the longitudinal direction of joists in houses of not more than two storeys if the joists:

– are at not more than 1.2 m centres;	A1/2 2C35a
– have at least 90 mm bearing on the supported walls or 75 mm bearing on a timber wall plate at each end; or	A1/2 2C35a
– are carried on the supported wall by joist hangers (in accordance with BS EN 845-1 and BS 5628 – see Figure 6.76); and	A1/2 2C35b
– are incorporated at not more than 2 m centres;	A1/2 2C35b
• when a concrete floor has at least 90 mm bearing on the supported wall (see Figure 6.77); and	A1/2 2C35c
• where floors are at or about the same level on each side of a supported wall, and contact between the floors and wall is either continuous or at intervals not exceeding 2 m. Where contact is intermittent, the points of contact should be in line or nearly in line on plan (Figure 6.78).	A1/2 2C35d

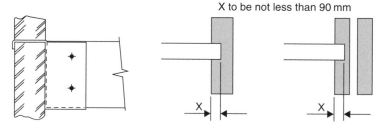

X to be not less than 90 mm

Figure 6.76 Restraint type joist hanger.

Figure 6.77 Restraint by concrete floor or roof.

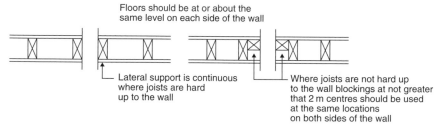

Floors should be at or about the same level on each side of the wall

Lateral support is continuous where joists are hard up to the wall

Where joists are not hard up to the wall blockings at not greater that 2 m centres should be used at the same locations on both sides of the wall

Figure 6.78 Restraint of internal walls.

Internal load-bearing walls in brickwork or blockwork

The maximum span for any floor supported by a wall is 6 m, where the span is measured centre to centre of bearing (Figure 6.79).	A1/2 2C23

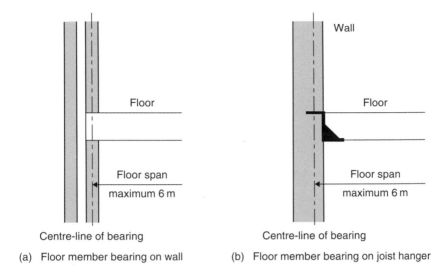

(a) Floor member bearing on wall   (b) Floor member bearing on joist hanger

Figure 6.79 Maximum span of floors.

Vertical loading on walls should be distributed.	A1/2 2C23a
Differences in level of ground or other solid construction between one side of the wall and the other should be less than four times the thickness of the wall, as shown in Figure 6.80.	A1/2 2C23b

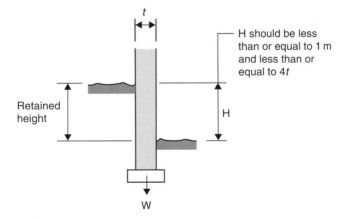

Figure 6.80 Maximum permitted difference in level.

Dead load, imposed load and wind load should be in accordance with current codes of practice.

A1/2 0.2b

Loads used in calculations should allow for possible dynamic, concentrated and peak load effects that may occur.

A1/2 0.2

All walls (except compartment and/or separating walls) should have a thickness not less than:

A1/2 2C10

$$\frac{\text{Specified thickness from Table 6.33}}{2} - 5\,\text{mm}$$

Note: Except for a wall in the lowest storey of a three-storey building, carrying load from both upper storeys, walls should have a thickness as determined by the equation or 140 mm whichever is the greatest.

Solid external walls, compartment walls and separating walls in coursed brickwork or blockwork

Solid walls constructed of coursed brickwork or blockwork should be at least as thick as 1/16 of the storey height.

A1/2 2C6

Solid external walls, compartment walls and separating walls in uncoursed stone, flints, etc.

The thickness of walls constructed in uncoursed stone, flints, clunches, or bricks or other burnt or vitrified material should not be less than 1.33 times the thickness of the storey height.

A1/2 2C7

Cavity walls in coursed brickwork or blockwork

All cavity walls should have leaves at least 90 mm thick and cavities at least 50 mm wide.

A1/2 2C8

The combined thickness of the two leaves plus 10 mm should be at least 1/16 of the storey height (Table 6.37).

A1/2 2C8

Table 6.37 Minimum thickness of certain external walls, compartment walls and separating walls

Height of wall	Length of wall	Minimum thickness of wall
Not exceeding 3.5 m	Not exceeding 12 m	190 mm for whole of its height
Exceeding 3.5 m but not exceeding 9 m	Not exceeding 9 m	190 mm for whole of its height
	Exceeding 9 m	290 mm from the base for the height of one storey and 190 mm for the rest of its height
Exceeding 9 m but not exceeding 12 m	Not exceeding 9 m	290 mm from the base for the height of one storey and 190 mm for the rest of its height
	Exceeding 9 m but not exceeding 12 m	290 mm from the base for the height of two storeys and 190 mm for the rest of its height

Wall ties should either comply with BS 1243, DD 140 or BS EN 845-1.	A1/2 2C19
Wall ties should have a horizontal spacing of 900 mm and a vertical spacing of 450 mm.	A1/2 2C8
Equivalent to 2.5 ties per square metre.	
Wall ties should also be provided, spaced not more than 300 mm apart vertically, within a distance of 225 mm from the vertical edges of all openings, movement joints and roof verges.	A1/2 2C8
For external walls, compartment walls and separating walls in cavity construction, the combined thickness of the two leaves plus 10 mm should be at least as thick as 1/16 of the storey height.	A1/2 2C8

Walls providing vertical support to other walls

Irrespective of the material used in the construction, a wall should not be less than the thickness of any part of the wall to which it gives vertical support.	A1/2 2C9

Parapet walls

The minimum thickness and maximum height of parapet walls should be as shown in Figure 6.81.	A1/2 2C11

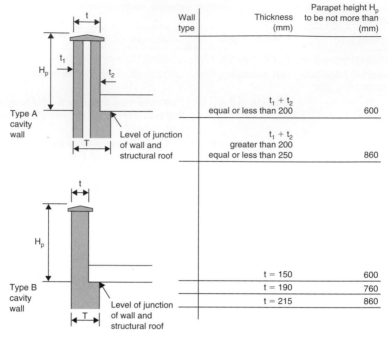

Wall type	Thickness (mm)	Parapet height H_p to be not more than (mm)
Type A cavity wall	$t_1 + t_2$ equal or less than 200	600
	$t_1 + t_2$ greater than 200 equal or less than 250	860
Type B cavity wall	t = 150	600
	t = 190	760
	t = 215	860

Note: t should be less than or equal to T

Figure 6.81 Height of parapet walls.

Single leaves of certain external walls

> The single leaf of external walls of small, single-storey, non-residential buildings and of annexes need be only 90 mm thick.   A1/2 2C12

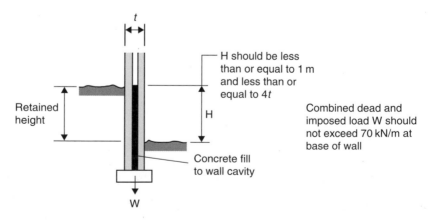

Figure 6.82 Combined and imposed dead load.

Vertical lateral restraint to walls

The ends of every wall should be bonded or otherwise securely tied throughout their full height to a buttressing wall, pier or chimney.

A1/2 2C25

Long walls may be provided with intermediate buttressing walls, piers or chimneys dividing the wall into distinct lengths within each storey.

A1/2 2C25

Note: Each distinct length is considered to be a supported wall for the purposes of the Building Regulations.

Intermediate buttressing walls, piers or chimneys should provide lateral restraint to the full height of the supported wall.

A1/2 2C25

They may be staggered at each storey.

A wall in each storey of a building should:

A1/2 2C32

- extend to the full height of that storey;
- have horizontal lateral supports to restrict movement of the wall at right angles to its plane.

The requirements for lateral restraint are shown in Table 6.38.

A1/2 2C34

Table 6.38 Lateral support for walls

Wall type	Wall length	Lateral support required
Solid or cavity: external compartment separating	Any length	Roof lateral support by every roof forming a junction with the supported wall
	Greater than 3 m	Floor lateral support by every floor forming a junction with the supported wall
Internal load-bearing wall (not being a compartment or separating wall)	Any length	Roof or floor lateral support at the top of each storey

Walls should be strapped to floors above ground level, at intervals not exceeding 2 m and, as shown in Figure 6.83, by tension straps conforming to BS EN 845-1.

A1/2 2C35

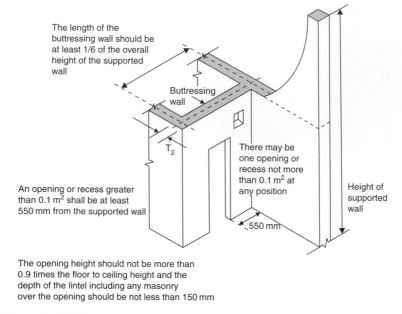

The length of the buttressing wall should be at least 1/6 of the overall height of the supported wall

Buttressing wall

T_2

There may be one opening or recess not more than 0.1 m² at any position

An opening or recess greater than 0.1 m² shall be at least 550 mm from the supported wall

Height of supported wall

550 mm

The opening height should not be more than 0.9 times the floor to ceiling height and the depth of the lintel including any masonry over the opening should be not less than 150 mm

Figure 6.83 Openings in a buttressing wall.

Buttressing walls

If the buttressing wall is not itself a supported wall, its thickness T_2 should not be less than:

- half the thickness required for an external or separating wall of similar height and length less 5 mm; or   A1/2 2C26a
- 75 mm if the wall forms part of a dwelling-house and does not exceed 6 m in total height and 10 m in length; and   A1/2 2C26b
- 90 mm in other cases.   A1/2 2C26c

The length of the buttressing wall should be:

- at least 1/6 of the overall height of the supported wall;
- bonded or securely tied to the supporting wall and at the other end to a buttressing wall, pier or chimney.

The size of any opening in the buttressing wall should be restricted as shown in Figure 6.83.   A1/2 2C26c

Gable walls

> Gable walls should be strapped to roofs as shown in *Figure 6.84*(a) and (b) by tension straps.
>
> A1/2 2C36
>
> Vertical strapping at least 1 m in length should be provided at eaves level at intervals not exceeding 2 m as shown in *Figure 6.84*(c) and (d).
>
> A1/2 2C36
>
> Vertical strapping may be omitted if the roof:
>
> A1/2 2C36a–d
>
> - has a pitch of 15° or more; and
> - is tiled or slated; and
> - is of a type known by local experience to be resistant to wind gusts; and
> - has main timber members spanning onto the supported wall at not more than 1.2 m centres.

Piers

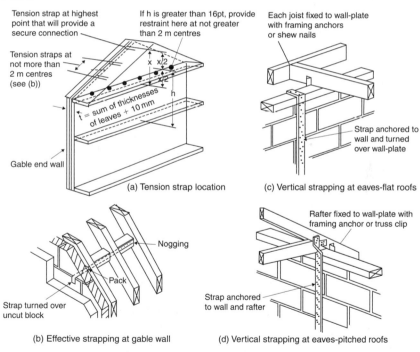

(a) Tension strap location

(b) Effective strapping at gable wall

(c) Vertical strapping at eaves-flat roofs

(d) Vertical strapping at eaves-pitched roofs

Figure 6.84 Lateral support at roof level.

Piers should have a minimum width of 190 mm (Figure 6.85). A1/2 2C27a

Piers should measure at least three times the thickness of the supported wall. A1/2 2C27a

Chimneys

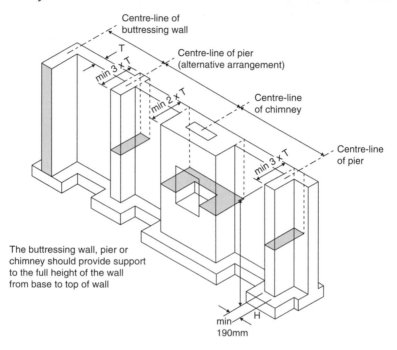

Figure 6.85 Buttressing.

Chimneys should measure at least twice the thickness, measured at right angles to the wall (see Figure 6.85). A1/2 2C27a

The sectional area on plan of chimneys (excluding openings for fireplaces and flues): A1/2 2C27b

- should be not less than the area required for a pier in the same wall; and
- the overall thickness should not be less than twice the required thickness of the supported wall (Figure 6.85).

Openings and recesses

The number, size and position of openings and recesses should not impair the stability of a wall or the lateral restraint afforded by a buttressing wall to a supported wall.   A1/2 2C28

Construction over openings and recesses should be adequately supported.   A1/2 2C28

No openings should be provided in walls below ground floor except for small holes for services and ventilation etc., which should be limited to a maximum area of $0.1\,m^2$ at not less than 2 m centres (Figure 6.86 and Table 6.39).   A1/2 2C29

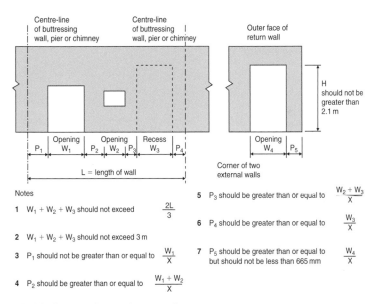

Notes

1. $W_1 + W_2 + W_3$ should not exceed $\dfrac{2L}{3}$

2. $W_1 + W_2 + W_3$ should not exceed 3 m

3. P_1 should not be greater than or equal to $\dfrac{W_1}{X}$

4. P_2 should be greater than or equal to $\dfrac{W_1 + W_2}{X}$

5. P_3 should be greater than or equal to $\dfrac{W_2 + W_3}{X}$

6. P_4 should be greater than or equal to $\dfrac{W_3}{X}$

7. P_5 should be greater than or equal to $\dfrac{W_4}{X}$ but should not be less than 665 mm

Figure 6.86 Sizes of openings and recesses.

Table 6.39 Value of X factor for Figure 6.86

Nature of roof span	Maximum roof span (m)	Minimum thickness of wall inner (mm)	Span of floor is parallel to wall	Span of timber floor into wall Max. 4.5 m	Max. 6.0 m	Span of concrete floor into wall Max. 4.5 m	Max. 6.0 m
			Value of factor X				
Roof spans parallel to wall	Non-applicable	100	6	6	6	6	6
		90	6	6	6	6	5
Timber roof spans into wall	9	100	6	6	5	4	3
		90	6	4	4	3	3

Overhangs

The amount of any projection should not impair the stability of the wall.	A1/2 2C31

Chases

Vertical chases should not be deeper than 1/3 of the wall thickness.	A1/2 2C30a

 Note: Or, in cavity walls, 1/3 of the thickness of the leaf.

Horizontal chases should not be deeper than 1/6 of the thickness of the leaf of the wall.	A1/2 2C30b
Chases should not be so positioned as to impair the stability of the wall (particularly where hollow blocks are used).	A1/2 2C30c

Small, single-storey, non-residential buildings and annexes

General

The walls shall be solidly constructed in brickwork or blockwork.	A1/2 2C38(i)b
Where the floor area of the building or annexe exceeds $10\,m^2$, the walls shall have a mass of not less than $130\,kg/m^2$.	A1/2 2C38(i)c
The only lateral loads are wind loads.	A1/2 2C38(i)e
The maximum length or width of the building or annexe shall not exceed $9\,m$.	A1/2 2C38(i)
The height of the building or annexe shall not exceed the lower value derived from Figure 6.87.	A1/2 2C38(i)

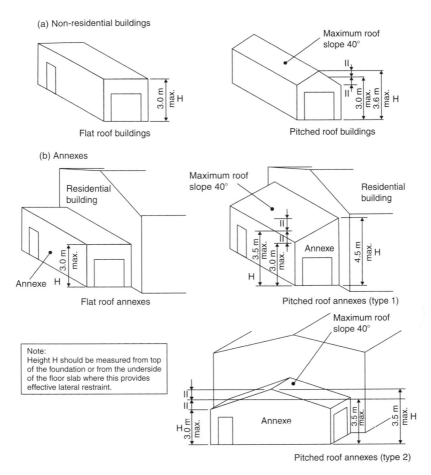

Figure 6.87 Size and proportions of non-residential buildings and annexes.

Walls shall be tied to the roof structure vertically and horizontally and have a horizontal lateral restraint at roof level.	A1/2 2C38(i)i

Size and location of openings

One or two major openings not more than 2.1 m in height are permitted in one wall of the building or annexe only.	A1/2 2C38(ii)
The width of a single opening or the combined width of two openings should not exceed 5 m.	A1/2 2C38(ii)

The only other openings permitted in a building or annexe are for windows and a single leaf door.

A1/2 2C38(ii)

The size and location of these openings should be in accordance with Figure 6.88.

A1/2 2C38(ii)

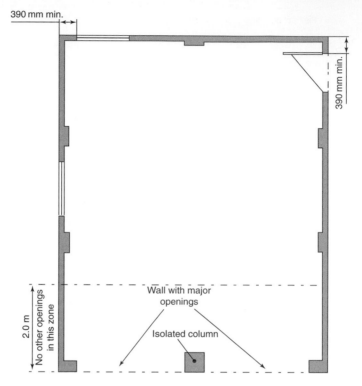

Figure 6.88 Size and location of openings.

Major openings should be restricted to one wall only. Their aggregate width should not exceed 0.5 m and their height should not be greater than 2.1 m.

A1/2 2C38(ii)

There should be no openings within 2.0 m of a wall containing a major opening.

A1/2 2C38(ii)

The aggregate size of the openings in a wall not containing a major opening should not exceed 2.4 m.

A1/2 2C38(ii)

There should not be more than one opening between piers.

A1/2 2C38(ii)

Unless there is a corner pier, the distance from a window or a door to a corner should not be less than 390 mm.

A1/2 2C38(ii)

Wall thicknesses and recommendations for piers

The walls should have a minimum thickness of 90 mm.

A1/2 2C38(iii)

Walls that do not contain a major opening but exceed 2.5 m in length or height should be bonded or tied to piers for their full height at not more than 3 m centres as shown in Figure 6.89.

A1/2 2C38(iii)

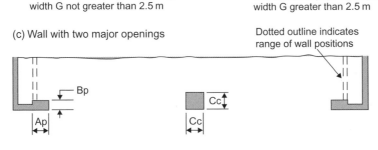

(a) Wall without a major opening

(b) Wall with a single major opening

Orientation of piers with opening width G not greater than 2.5 m

Orientation of piers with opening width G greater than 2.5 m

(c) Wall with two major openings

Figure 6.89 Wall thicknesses.

Walls that contain one or two major openings should in addition have piers as shown in Figure 6.87(b) and (c).

A1/2 2C38(iii)

Where ties are used to connect piers to walls they should be: A1/2 2C38(iii)

- flat;
- 20 mm × 3 mm in cross-section;
- stainless steel;
- placed in pairs;
- spaced at not more than 300 mm centres vertically.

Walls should be tied horizontally at no more than 2 m centres to the roof structure at eaves level, base of gables and along roof slopes (as shown in Figure 6.90) with straps.   A1/2 2C38(iv)

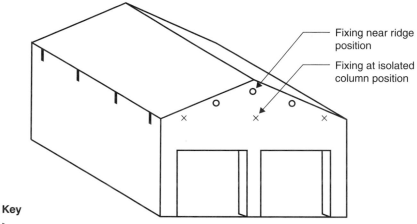

Key

| denotes fixings at eaves level.   × denotes fixings at base of gable.
o denotes fixings along roof slope.

Figure 6.90 Lateral restraint at roof level.

Where straps cannot pass through a wall, they should be adequately secured to the masonry using suitable fixings.   A1/2 2C38(iv)

Isolated columns should also be tied to the roof structure (Figure 6.90).   A1/2 2C38(iv)

Foundations

> A wall shall be erected to prevent undue moisture from the ground reaching the inside of the building and (if it is an outside wall) adequately resisting the penetration of rain and snow to the inside of the building (Figure 6.91).   C3

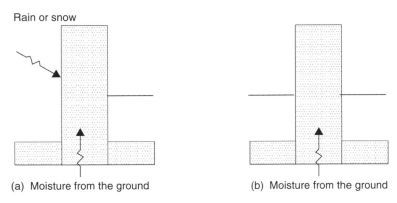

Rain or snow

(a) Moisture from the ground   (b) Moisture from the ground

Figure 6.91 Resistance to moisture. (a) External wall. (b) Internal wall.

Resistance to the passage of moisture

> Walls should:
>
> - resist the passage of moisture from the ground to the inside of the building;   C5.2
> - not be damaged by moisture from the ground;   C5.2b
> - not carry moisture from the ground to any part that would be damaged by moisture.   C5.2b
>
> External walls should:
>
> - resist rain penetrating components of the structure that might be damaged by moisture;   C5.2c
> - resist rain penetrating to the inside of the building;   C5.2d
> - be designed and constructed so that their structural and thermal performance are not adversely affected by interstitial condensation;   C5.2e
> - not promote surface condensation or mould growth.   C5.2f
>
> For buildings used wholly for storing goods or where provisions put in place do not increase the health and safety of persons employed in that building, this requirement may not apply.   C5.3

Internal and external walls exposed to moisture from the ground

Internal and external walls (subject to moisture from the ground) shall have a damp-proof course of bituminous material, polyethylene, engineering bricks or slates in cement mortar, or any other material that will prevent the passage of moisture.   C5.5a

The damp-proof course should be continuous with any damp-proof membrane in the floors.   C5.5a

If the wall is an external wall, the damp-proof course should be at least 150 mm above the level of the adjoining ground (Figure 6.92) unless the design is such that a part of the building will protect the wall.   C5.5b

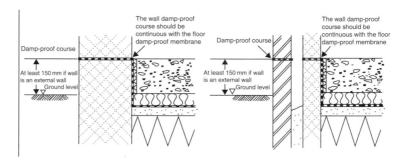

Figure 6.92 Damp-proof courses.

If the wall is an external cavity wall (Figure 6.93) the cavity should either:

- be taken down at least 225 mm below the level of the low-est damp-proof course; or   C5.5c
- a damp-proof tray should be provided so as to prevent pre-cipitation passing into the inner leaf (Figure 6.94), with weep holes every 900 mm to assist in the transfer of moisture through the external leaf.   C5.5c

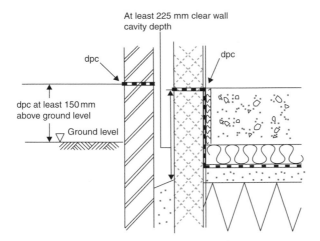

Figure 6.93 Cavity carried down.

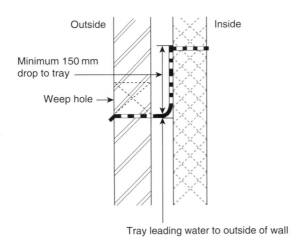

Figure 6.94 Damp-proof (cavity) tray.

Where the damp-proof tray does not extend the full length of the exposed wall (i.e. above an opening), stop ends and at least two weep holes should be provided.	C5.5c
As well as giving protection against moisture from the ground, an external wall should give protection against precipitation.	C5.7

Solid external walls

Solid walls shall hold moisture arising from rain and snow until it can be released in a dry period without penetrating to the inside of the building, or causing damage to the building.	C5.8

Solid external walls exposed to **very severe** conditions should   C5.9
be protected by external impervious cladding.

Solid external walls exposed to **severe** conditions may be built   C5.9a
with:

- brickwork (or stonework) at least 328 mm thick;
- dense aggregate concrete blockwork at least 250 mm thick; or
- lightweight aggregate (aerated autoclaved concrete block-work) at least 215 mm thick.

Solid external walls exposed to **severe** conditions may be   C5.9b
built, providing:

- the rendering is in two coats with a total thickness of at least 20 mm and has a scraped or textured finish;
- the strength of the mortar is compatible with the strength of the bricks or blocks;
- the joints (if the wall is to be rendered) are raked out to a depth of at least 10 mm;
- the rendering mix is 1 part of cement, 1 part of lime and 6 parts of well-graded sharp sand (nominal mix 1:1:6) unless the blocks are of dense concrete aggregate, in which case the mix may be 1:½.

Adequate protection should be provided at the top of walls,   C5.9c
etc. (Figure 6.95).

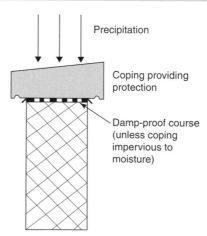

Figure 6.95 Projection of wall head from precipitation.

Unless the protection and joints are a complete barrier to moisture, a damp-proof course should also be provided. — C5.9c

Damp-proof courses, cavity trays and closers should be provided and designed to ensure that water drains outwards:

- where the downward flow will be interrupted by an obstruction (e.g. from some types of lintel); — C5.9d(i)
- under openings – unless there is a sill and the sill and its joints will form a complete barrier; — C5.9d(ii)
- at abutments between walls and roofs. — C5.9d(iii)

A solid external wall may be insulated on the inside or on the outside. — C5.10

Where the insulation is on the inside, a cavity should be provided to give a break in the path for moisture. — C5.10

Where the insulation is on the outside, it should provide some resistance to the ingress of moisture to ensure the wall remains relatively dry (see Figure 6.96). — C5.10

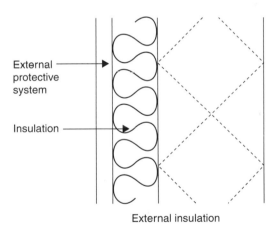

External protective system

Insulation

External insulation

Figure 6.96 Insulated (solid) external wall.

Cavity external walls

The outer leaf shall be separated from the inner leaf by a drained air space (or in any other way which will prevent precipitation from being carried to the inner leaf). — C5.12

The construction of a cavity external wall could include:

- outer leaf masonry (bricks, blocks, stone or a manufac-   C5.13a
  tured stone);
- a cavity at least 50 mm wide;   C5.13b
- inner leaf masonry or frame with lining.   C5.13b

Masonry units should be laid on a full bed of mortar with the   C5.13c
cross joints substantially and continuously filled to ensure
structural robustness and weather resistance.

Where a cavity is to be partially filled, the residual cavity   C5.13c
should not be less than 50 mm wide (Figure 6.97).

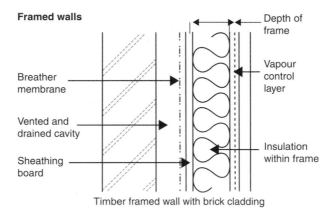

Figure 6.97 Insulated framed wall.

Cavity insulation

The suitability of the wall for installing insulation   C5.15a
material(s) is to be assessed before the work is carried out.   and d

When the cavity of an existing house is being filled,   C5.15e
attention should be given to the condition of the external leaf
of the wall, e.g. its state of repair and type of pointing.

A full or partial fill insulating material may be placed in the   C5.15a
cavity between the outer leaf and an inner leaf of masonry,
subject to the suitability of a wall for installing insulation
into the cavity (Table 6.40).

Table 6.40 Maximum recommended exposure zones for insulated masonry walls

Wall construction	Min. width of filled or clear cavity (mm)	Maximum recommended exposure zone for each construction						
Insulation method		Impervious cladding		Rendered finish		Facing masonry		Flush sills and copings
		Full height of wall	Above facing masonry	Full height of wall	Above facing masonry	Tooled flush joints	Recessed mortar joints	
Built-in full fill	50	4	3	3	3	2	1	1
	75	4	3	4	3	3	1	1
	100	4	4	4	3	3	1	2
	125	4	4	4	3	3	1	2
	150	4	4	4	4	4	1	2
Injected fill, not UF foam	50	4	2	3	2	2	1	1
	75	4	3	4	3	3	1	1
	100	4	3	4	3	3	1	1
	125	4	4	4	3	3	1	2
	150	4	4	4	4	4	1	2
Injected fill, UF foam	50	4	2	3	2	1	1	1
	75	4	2	3	2	2	1	1
	100	4	2	3	2	2	1	1
Partial fill								
Residual 50 mm cavity	50	4	4	4	4	3	1	1
Residual 75 mm cavity	75	4	4	4	4	4	1	1
Residual 100 mm cavity	100	4	4	4	4	4	2	1
Internal insulation								
Clear cavity 50 mm	50	4	3	4	3	3	1	1
Clear cavity 100 mm	100	4	4	4	4	4	2	2
Fully filled								
Cavity 50 mm	50	4	3	3	3	2	1	1
Cavity 100 mm	100	4	4	4	3	3	1	2

UF, urea–formaldehyde.

The insulating material should be the subject of current certification from an appropriate body or a European Technical Approval.   C5.15c

When partial fill materials are used, the residual cavity should not be less than 50 mm nominal.   C5.15b

Rigid (board or batt) thermal insulating material built into the wall must be certified as being in conformance by an approved installer.   C5.15b

Urea–formaldehyde foam inserted into the cavity should be: C5.15d

- in accordance with BS 5617:1985;
- installed in accordance with BS 5618:1985.

The person undertaking installation work should operate under an Approved Installer Scheme.   C5.15c

Framed external walls

The cladding shall be separated from the insulation or sheathing by a vented and drained cavity with a membrane that is vapour open, but resists the passage of liquid water, on the inside of the cavity (Figure 6.97).   C5.17

Cracking of external walls

The possibility of severe rain penetration occurring through cracks in masonry external walls should be taken into account when designing a building.   C5.18

Impervious cladding systems for walls

Cladding systems for walls should:

- resist the penetration of precipitation to the inside of the building;   C5.19a
- not be damaged by precipitation;   C5.19b
- not carry precipitation to any part of the building that would be damaged by it.   C5.19b

Cladding that is designed to protect a building from precipitation shall be:

- jointless or have sealed joints;   C5.21a
- impervious to moisture.   C5.21a

If the cladding has overlapping dry joints it shall be:

- impervious or weather resisting;   C5.21b
- backed by a material which will direct precipitation that   C5.21b
  enters the cladding towards the outer face.

Materials that can deteriorate rapidly without special care   C5.22
should only be used as the weather-resisting part of a
cladding system.

Cladding may be:

- impervious (e.g. metal, plastic, glass and bituminous   C5.23a
  products);
- weather-resisting (e.g. natural stone or slate, cement-   C5.23b
  based products, fired clay and wood);
- moisture-resisting – (e.g. bituminous and plastic prod-   C5.23c
  ucts lapped at the joints);
- jointless materials and sealed joints – (i.e. to allow for   C5.23d
  structural and thermal movement).

Dry joints between cladding units should be designed so
that:

- precipitation will not pass through them;   C5.24
- precipitation which enters the joints will be directed   C5.24
  towards the exposed face without it penetrating beyond
  the back of the cladding.

 Note: Whether dry joints are suitable will depend on the design of the joint or the design of the cladding and the severity of the exposure to wind and rain.

Each sheet, tile and section of cladding should be securely   C5.25
fixed (as per guidance contained in BS 8000-6:1990).

Particular care should be taken with detailing and   C5.25
workmanship at the junctions between cladding and window
and door openings as they are vulnerable to moisture ingress.

Insulation may be incorporated into the construction   C5.26
provided it is either protected from moisture or is unaffected
by it.

Where cladding is supported by timber components (or is on   C5.27
the facade of a timber framed building) the space between
the cladding and the building should be ventilated to ensure
rapid drying of any water that penetrates the cladding.

Joints between walls and doors/window frames

The joint between walls and doors and window frames should:
- resist the penetration of precipitation to the inside of the building;   C5.29a
- not be damaged by precipitation;   C5.29b
- not permit precipitation to reach any part of the building that would be damaged by it.   C5.29

Damp-proof courses should be provided to direct moisture towards the outside, particularly:
- where the downward flow of moisture would be interrupted at an obstruction, e.g. at a lintel;   C5.30a
- where sill elements (including joints) do not form a complete barrier to the transfer of precipitation (e.g. under openings, windows and doors);   C5.30b
- where reveal elements, including joints, do not form a complete barrier to the transfer of rain and snow (e.g. at openings, windows and doors).   C5.30c

Direct plastering of the internal reveal of any window frame should only be used with a backing of expanded metal lathing or similar.   C5.31

In areas of the country that are exposed to very severe driving rain:
- checked rebates should be used in all window and door reveals;   C5.32
- the frame should be set back behind the outer leaf of masonry as shown in Figure 6.98;   C5.32
- alternatively an insulated finned cavity closer may be used.   C5.32

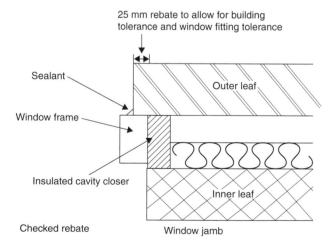

Figure 6.98 Window reveals for use in areas subject to very severe driving rain.

Door thresholds

Where an accessible threshold is provided to allow
unimpeded access (as specified in Part M):

- the external landing (Figure 6.99) should be laid to a   C5.33a
  fall between 1 in 40 and 1 in 60 in a single direction
  away from the doorway;
- the sill leading up to the door threshold has a maxi-   C5.33b
  mum slope of 15°.

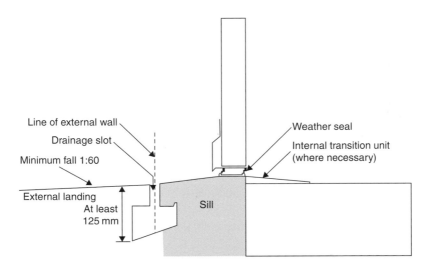

Figure 6.99 Accessible threshold for use in exposed areas.

Interstitial condensation (external walls)

External walls shall be designed and constructed in   C5.34
accordance with Clause 8.3 of BS 5250:2002.

Specialist advice should be sought when designing   C5.35
swimming pools and other buildings where
interstitial condensation in the walls (caused by high
internal temperatures and humidities) can cause high levels
of moisture being generated.

Surface condensation and mould growth (external walls)

External walls shall be designed and constructed so that the:

- thermal transmittance (U-value) does not exceed 0.7 W/   C5.36a
  m^{2}K at any point;
- junctions between elements and details of openings (such   C5.36b
  as doors and windows) meet with the recommendations in
  the report on robust construction details.

Wall cladding

Wall cladding presents a hazard if it becomes detached from the building. An acceptable level of safety can be achieved depending on the type and location of the cladding.

 The guidance given below relates to all forms of cladding, including curtain walling and glass façades.

Cladding shall be capable of safely sustaining and   A1/2 3.2a
transmitting (to the supporting structure of the building)
all dead, imposed and wind loads.

Provision shall be made, where necessary, to accommodate   A1/2 3.2
differential movement of the cladding and the supporting
structure of the building.

Wind loading on the cladding should be derived from BS   A1/2 3.3
6399: Part 2:2001.

Due consideration shall be given to local increases in wind   A1/2 3.3
suction arising from funnelling of the wind through gaps
between buildings.

 Note: Guidance on funnelling effects is given in BRE Digest 436, *Wind Loading on Buildings – Brief Guidance for Using BS 6399-2:1997*, available from BRE, Bucknalls Lane, Garston, Watford, Herts WD2 7JR.

The cladding shall be securely fixed to, and supported   A1/2 3.2b
by, the structure of the building using both vertical
support and horizontal restraint.

The cladding and its fixings (including any support   A1/2 3.2d
components) shall be of durable materials.

The design life of the fixings shall not be less than that   A1/2 3.2d
of the cladding.

Fixings shall be corrosion resistant and of a material   A1/2 3.2d
type appropriate for the local environment.

Fixings for supporting cladding should be determined from a consideration of the proven performance of the fixing and the risks associated with the particular application.	A1/2 3.7
The strength of fixings should be derived from tests using materials representative of the material into which the fixing is to be anchored, taking account of any inherent weaknesses that may affect the strength of the fixing (e.g. cracks in concrete due to shrinkage and flexure, or voids in masonry construction).	A1/2 3.8
Where the cladding is required to support other fixtures (e.g. handrails or fittings such as antennae and signboards) account should be taken of the loads and forces arising from such fixtures and fittings.	A1/2 3.4
Where the wall cladding is required to function as pedestrian guarding to stairs, ramps, vertical drops of 600 mm or greater or as a vehicle barrier, account should be taken of the additional imposed loading as stipulated in Part K.	A1/2 3.5
Where wall cladding is required to safely withstand lateral pressures from crowds, an appropriate design loading is given in BS 6399: Part 1 and the *Guide to Safety at Sports Grounds* (4th edition, 1997).	
Applications should be designated as being either non-redundant (where the failure of a single fixing could lead to the detachment of the cladding) or redundant (where failure or excessive movement of one fixing results in load sharing by adjacent fixings) and the required reliability of the fixing determined accordingly.	A1/2 3.7
All cladding (used to protect the building from rain or snow) shall be jointless or have sealed joints.	C4 (5.1–5.6)

Note: Large glass panels in cladding of walls and roofs (where the cladding is not divided into small areas by load-bearing framing) needs special consideration. Guidance is given in the following documents:

The Institution of Structural Engineers' report on *Structural use of Glass in Buildings*, 1999, available from 11 Upper Belgrave Street, London SW1X 8BH. *Nickel Sulfide in Toughened Glass*, published by the Centre for Window Cladding and Technology, 2000.

Further guidance on cladding is given in the following documents:

- The Institution of Structural Engineers' report on *Aspects of Cladding*, 1995.
- The Institution of Structural Engineers' report on *Guide to the Structural Use of Adhesives*, 1999.

- BS 8297 *Code of Practice for the Design and Installation of Non-load-bearing Precast Concrete Cladding.*
- BS 8298 *Code of Practice for the Design and Installation of Natural Stone Cladding and Lining.*

Internal fire spread (linings)

The choice of materials for walls and ceilings can significantly affect the spread of a fire and its rate of growth, even though they are not likely to be the materials first ignited. Although furniture and fittings can have a major effect on fire spread it is not possible to control them through Building Regulations.

The surface linings of walls should meet the classifications shown in Table 6.41.

Table 6.41 Classification of linings

Location	Class*
Small rooms with an area of not more than 4 m² (in residential accommodation) or 30 m² (in non-residential accommodation)	3
Domestic garages not more than 40 m²	3
Other rooms (including garages)	1
Circulation spaces within buildings	1
Other circulation spaces (including the common area of flats and maisonettes)	0

* Classifications are based on tests as per BS 476 and as described in Appendix A of Approved Document B.

Any flexible membrane covering a structure (other than an air supported structure) should comply with the recommendations given in Appendix A of BS 7157.	B2 6.8 (V2)
The wall and any floor between the garage and the house shall have a 30 minute standard of fire resistance.	B2

Load-bearing elements of structure

All load-bearing elements of a structure shall have a minimum standard of fire resistance.	B3 4.1 (V1) B3 7.1 (V2)
Structural frames, beams, columns, load-bearing walls (internal and external), floor structures and gallery structures, should have at least the fire resistance given in Appendix A of Part B.	B3 4.2 (V1) B3 7.2 (V2)

Compartmentation

To prevent rapid fire spread and to reduce the chance of fires becoming large, the spread of fire within a building can be restricted by subdividing that building into compartments that are separated from one another by walls and/or floors of fire-resisting construction.

The appropriate degree of subdivision depends on:

- the use of and fire load in the building;
- the height to the floor of the top storey in the building; and
- the availability of a sprinkler system.

General

To prevent the spread of fire within a building, whenever possible, the building should be subdivided into compartments separated from one another by walls and/or floors of fire-resisting construction.	B3 5.1 (V1) B3 8.1 (V2)
Walls separating semi-detached houses, or houses in terraces, should be constructed as a compartment wall and the houses should be considered as separate buildings.	B3 5.3 (V1)
Compartment walls that are common to two or more buildings should: • be constructed as a compartment wall; • run the full height of the building in a continuous vertical plane.	B3 8.10 (V2) B3 5.7 (V1)

The lowest floor in a building does not need to be constructed as a compartment floor.

Compartment walls should: • form a complete barrier to fire between the compartments they separate; and • have the appropriate fire resistance as indicated in Appendix A, Tables A1 and A2.	B3 5.6 (V1) B2 8.20a (V2) B3 5.6 (V1) B2 8.20b (V2)

 Note: Adjoining buildings should only be separated by walls, not floors.

Junction of compartment wall with other walls

Where a compartment wall meets another compartment wall, the junction should maintain the fire resistance of the compartmentation.	B3 5.9 (V1) B3 8.25 (V2)

At the junction of a compartment floor with an external wall that has no fire resistance (such as a curtain wall) the external wall should be restrained at floor level to reduce the movement of the wall away from the floor when exposed to fire.

B3 5.10 (V1)

Junction of compartment wall with roof

If a fire penetrates a roof near a compartment wall there is a risk that it will spread over the roof to the adjoining compartment. To reduce this risk the wall should be:

B3 5.11 (V1)

- taken up to meet the underside of the roof;
- covered with fire-stopping (where necessary) at the wall/roof junction to maintain the continuity of fire resistance;

B3 8.28 (V2)

- continued across any eaves;
- either extended up through the roof for a height of at least 375 mm above the top surface of the adjoining roof covering (Figure 6.100); or a 1500 mm wide zone on either side of the wall should have a suitable covering (Figure 6.100).

B3 5.12 (V1)
B3 8.29 (V2)
B3 8.30 (V2)

Compartment walls in a top storey beneath a roof should be continued through the roof space.

B3 5.8 (V1)
B3 8.24 (V2)

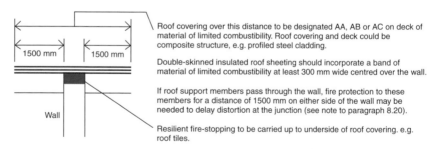

Figure 6.100 Junction of compartment wall with roof.

Roof covering over this distance to be designated AA, AB or AC on deck of material of limited combustibility. Roof covering and deck could be composite structure, e.g. profiled steel cladding.

Double-skinned insulated roof sheeting should incorporate a band of material of limited combustibility at least 300 mm wide centred over the wall.

If roof support members pass through the wall, fire protection to these members for a distance of 1500 mm on either side of the wall may be needed to delay distortion at the junction (see note to paragraph 8.20).

Resilient fire-stopping to be carried up to underside of roof covering. e.g. roof tiles.

1500 mm   1500 mm

Wall

Openings in compartment walls separating buildings or occupancies

Any openings in a compartment wall that is common to two or more buildings should be limited to those for a door which is providing 'means of access' in case of fire (and which has the same fire resistance as that required for the wall).	B3 5.13 (V1) B3 8.32 (V2)
All other openings in compartment walls or compartment floors should be limited to those for:	B3 8.34 (V2)

- doors that have the appropriate fire resistance;
- the passage of pipes, ventilation ducts, service cables, chimneys, appliance ventilation ducts or ducts encasing one or more flue pipes;
- refuse chutes of non-combustible construction;
- atria designed in accordance with BS 5588-7:1997; and
- protected shafts (see B3 8.35V2 for details of the relevant requirements).

All purpose groups

Parts of a building that are used and/or occupied for different purposes should be separated from one another by compartment walls and/or compartment floors (also see Appendix D to Part B (V2).	B2 8.11 (V2)
Walls that are common to two or more buildings should be constructed as a compartment wall.	B2 8.10 (V2)

Flats

In buildings containing flats, the following should be constructed as compartment walls:	B2 8.13 (V2)

- every wall separating a flat from any other part of the building; and
- every wall enclosing a refuse storage chamber.

Non-residential buildings

In non-residential, purpose group buildings (such as office, shop and commercial, assembly and recreation, industrial, storage etc.), the following walls should be constructed as compartment walls:	B2 8.18a (V2)
• walls that are required to subdivide buildings in order to meet the size limits on compartments given in Table 12 of Volume 2;	
• the walls of a building that form part of a shopping complex;	B2 8.18e (V2)
• walls that divide a building into separate occupancies, (i.e. spaces used by different organizations whether they fall within the same purpose group or not).	B2 8.18f (V2)

Construction of compartment walls

Adjoining buildings should only be separated by walls, not floors.	B2 8.21 (V2)
Compartment walls should be able to accommodate the predicted deflection of the floor above by either:	B2 8.27 (V2)
• having a suitable head detail between the wall and the floor, that can deform (but still maintain its integrity) when exposed to a fire; or	
• the wall may be designed to resist the additional vertical load from the floor above as it sags under fire conditions and thus maintain integrity.	
Compartment walls that are common to two or more buildings should run the full height of the building in a continuous vertical plane.	B2 8.21 (V2)
Compartment walls used to form a separated part of a building should run the full height of the building in a continuous vertical plane.	B2 8.22 (V2)
If trussed rafters bridge the wall, they should be designed so that failure of any part of the truss due to a fire in one compartment will not cause failure of any part of the truss in another compartment.	B3 5.6 (V1) B3 8.20 (V2)
Junctions between a compartment floor and an external wall that has no fire resistance (such as a curtain wall) should be restrained at floor level to reduce the movement of the wall away from the floor when exposed to fire.	B2 8.26 (V2)

Load-bearing walls (internal and external) should have at least the fire resistance given in Appendix A Table A1 of Part B.   B3 7.2 (V2)

Timber beams, joists, purlins and rafters may be built into or carried through a masonry or concrete compartment wall if the openings for them are kept as small as practicable and then fire-stopped.   B3 5.6 (V1)
B3 8.20 (V2)

There should be continuity at the junctions of the fire-resisting elements enclosing a compartment.   B3 8.6 (V2)

Note: Generally speaking, an external wall of a protected shaft does not need to have fire resistance (but see BS 5588-5:2004 for fire resistance of external walls of fire-fighting shafts).

Garages

If a domestic garage is attached to (or forms an integral part of) a house: B3 5.4 (V1)

- the wall between the garage and the house shall have a 30 minute standard of fire resistance;
- any opening in the wall should be at least 100 mm above the garage floor level with an FD30 door.

Protection of openings for pipes

Openings in compartment walls should be limited to those for: B3 8.34 (V2)

- doors that have the appropriate fire resistance;
- the passage of pipes, ventilation ducts, service cables, chimneys, appliance ventilation ducts or ducts encasing one or more flue pipes;
- refuse chutes of non-combustible construction;
- atria designed in accordance with BS 5588-7:1997; and
- protected shafts (see B3 8.35 V2 for details of the relevant requirements).

Ventilation ducts and flues etc.

Air-circulation-system transfer grilles should not be fitted in any wall enclosing a protected stairway.   B3 7.10 (V1)
B1 2.17 (V1)

If a flue (or a duct containing flues and/or ventilation duct(s)), passes through a compartment wall, or is built into a compartment wall, each wall of the flue or duct should have a fire resistance of at least half that of the wall or floor (Figure 6.101).   B1 7.11 (V1)

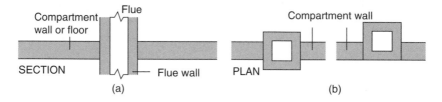

Figure 6.101 Flues penetrating compartment walls or floors.

Fire resistance

Proprietary fire-stopping and sealing systems (including those designed for service penetrations) that have been shown by test to maintain the fire resistance of the wall or other element, are available and may be used. Other fire-stopping materials include:

- cement mortar;
- gypsum-based plaster;
- cement or gypsum-based vermiculite/perlite mixes;
- glass fibre, crushed rock, blast furnace slag or ceramic-based products (with or without resin binders); and
- intumescent mastics (B3 11.14).

Joints between fire-separating elements should be fire-stopped.   B3 7.12a (V1)
B3 10.17a (V2)

All openings for pipes, ducts, conduits or cables to pass through any part of a fire-separating element should be: B3 7.12b (V1)
B3 10.17b (V2)

- kept as few in number as possible;
- kept as small as practicable;
- fire-stopped (which in the case of a pipe or duct, should allow thermal movement).

To prevent displacement, materials used for fire-stopping should be reinforced with (or supported by) materials of limited combustibility.   B3 7.13 (V1)
B3 10.18 (V2)

Construction of an external wall

Where a portal framed building is near a relevant boundary, the external wall near the boundary may need fire resistance to restrict the spread of fire between buildings.	B4 12.4 (V2)
In cases where the external wall of the building cannot be wholly unprotected, the rafter members of the frame, as well as the column members, may need to be fire protected.	B4 12.4 (V2)
The external surfaces of walls should meet the provisions shown in Figure 6.102.	B4 8.4 (V1) B4 12.4 (V2)

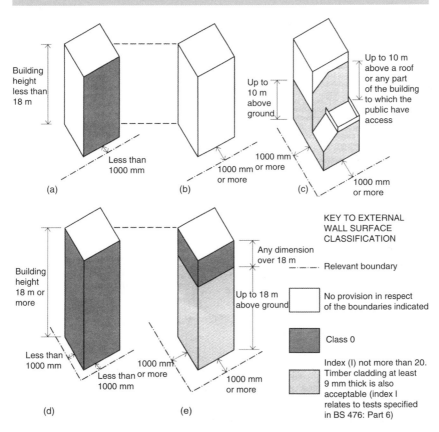

Figure 6.102 Provisions for external surfaces of walls. (a), (d), (e) Any building. (b) Any building other than (c). (c) Assembly or recreation building of more than one storey.

It should be noted that the use of combustible materials for cladding framework, or the use of combustible thermal insulation as an overcladding may be risky in tall buildings, even though the provisions for external surfaces in Figure 6.102 may have been satisfied.

The external envelope of a building should not provide   B4 12.5 (V2)
a medium for fire spread if it is likely to be a risk to
health or safety.

In a building with a storey 18 m or more above ground   B4 12.5 (V2)
level, insulation material used in ventilated cavities
in the external wall construction should be of limited
combustibility (this restriction does not apply to
masonry cavity wall construction).

Combustible material should not be placed in or exposed to the cavity, except for:

*   timber lintels, window or door frames, or the end stairway of timber joists;
*   pipes, conduits or cables;
*   damp-proof course, flashing, cavity closer or wall ties;
*   fire-resisting thermal insulating material;
*   a domestic meter cupboard.

Masonry wall construction

SECTION THROUGH CAVITY WALL

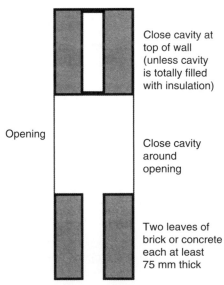

Opening

Close cavity at
top of wall
(unless cavity
is totally filled
with insulation)

Close cavity
around
opening

Two leaves of
brick or concrete
each at least
75 mm thick

Figure 6.103 Masonry cavity walls excluded from the previous for cavity barriers.

Cavity insulation

The outer leaf of the wall should be built of masonry or concrete.	D1 (1.1–1.2)
The inner leaf of the wall should be built of masonry (bricks or blocks).	D1 (1.1–1.2)
The wall being insulated with UF (urea formaldehyde) shall be assessed (in accordance with BS 8208) for suitability before any work commences.	D1 (1.1–1.2)
The person carrying out the work needs to hold (or operate under) a current BSI Certificate of Registration of Assessed Capability for the work he is doing.	D1 (1.1–1.2)
The installation shall be in accordance with BS 5618:1985.	D1 (1.1–1.2)
The material shall be in accordance with the relevant recommendations of BS 5617:1985.	D1 (1.1–1.2)

Airborne sound

The flow of sound energy through walls should be restricted.

Walls should reduce the level of airborne sound.

Walls that separate a dwelling from another building (or another dwelling) shall resist the transmission of airborne sound.

Habitable rooms (or kitchens) within a dwelling shall resist the transmission of airborne sound.

Air paths, including those due to shrinkage, must be avoided.

Porous materials and gaps at joints in the structure must be sealed.

Flanking transmission (i.e. the indirect transmission of sound from one side of a wall to the other side) should be minimized.

The possibility of resonance in parts of the structure (such as a dry lining) should be avoided.

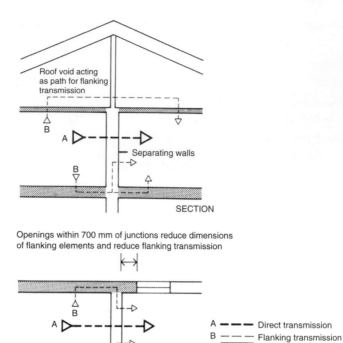

Figure 6.104 Direct and flanking transmission.

Separating walls (new buildings)

All new walls constructed within a dwelling-house (flat   E0.9
or room used for residential purposes) – whether purpose
built or formed by a material change of use – shall meet the
laboratory sound insulation values set out in Table 6.42.

Walls for: E0.1

• rooms for residential purposes; and
• dwelling-houses and flats;

that have a separating function should achieve the sound insulation values, as set out in Table 6.42.

Table 6.42 Dwelling houses and flats – performance standards for walls that have a separating function

	Airborne sound insulation $D_{nT,W} + C_{tr}$ (dB) (minimum values)
Purpose built rooms for residential purposes	43
Purpose built dwelling-houses and flats	45
Rooms for residential purposes formed by material change of use	43
Dwelling-houses and flats formed by material change of use	43

Note:

(1) The sound insulation values in Table 6.42 include a built-in allowance for 'measurement uncertainty', and so, if any of these test values are not met, that particular test will be considered as failed.

(2) Occasionally, a higher standard of sound insulation may be required between spaces used for normal domestic purposes and noise generated in and to an adjoining communal or non-domestic space. In these cases it would be best to seek specialist advice before committing yourself.

Flanking transmission from walls connected to the separating wall shall be controlled.	E2
Tests should be carried out between rooms or spaces that share a common area formed by a separating wall or separating floor.	E1
Impact sound insulation tests should be carried out without a soft covering (e.g. carpet, foam backed vinyl etc.) on the floor.	E1
If the floor joists are to be supported on the separating wall then they should be supported on hangers and should not be built in.	E2
If the joists are at right angles to the wall, spaces between the floor joists should be sealed with full depth timber blocking.	E3
The floor base (excluding any screed) should be built into a cavity masonry external wall and carried through to the cavity face of the inner leaf.	E
Walls that separate a dwelling from another dwelling (or part of the same building) shall resist:	E

- the level (and transmission) of airborne sounds;
- the transmission of impact sound (such as speech, musical instruments and loudspeakers and impact sources such as footsteps and furniture moving);
- the flow of sound energy through walls and floors.

Requirements

Requirement E1

Figure 6.105 illustrates the relevant parts of the building that should be protected from airborne and impact sound in order to satisfy Requirement E1.

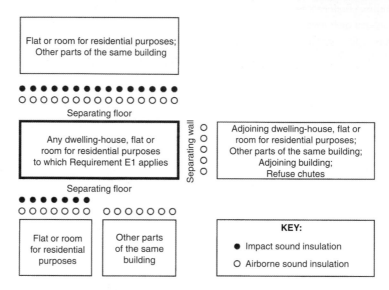

Figure 6.105 Requirement E1 – resistance to sound.

In some circumstances (for example, when a historic building is undergoing a material change of use) it may not be practical to improve the sound insulation to the standards set out in Approved Document E1, particularly if the special characteristics of such a building need to be recognized. In these circumstances the aim should be to improve sound insulation to the '*extent that it is practically possible*'.

 Note: BS 7913:1998 *The Principles of the Conservation of Historic Buildings* provides guidance on the principles that should be applied when proposing work on historic buildings.

Requirement E2

> Constructions for new walls within a dwelling-house (flat   E0.9
> or room for residential purposes) – whether purpose built or
> formed by a material change of use – shall meet the laboratory
> sound insulation values set out in Table 6.43.

Table 6.43 Laboratory values for new internal walls within dwelling-houses, flats and rooms for residential purposes – whether purpose built or formed by a material change of use

	Airborne sound insulation R_W (dB) (minimum values)
Purpose-built dwelling-houses and flats	40

Figure 6.106 illustrates the relevant parts of the building that should be protected from airborne and impact sound in order to satisfy Requirement E2a.

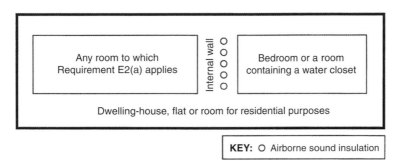

Figure 6.106 Requirement E2a – internal walls.

Requirement E3

Sound absorption measures described in Section 7 of Approved Document N shall be applied.	E0.11

Requirement E4

The values for sound insulation, reverberation time and indoor ambient noise as described in Section 1 of Building Bulletin 93, *The Acoustic Design of Schools* (produced by DFES and published by The Stationery Office (ISBN 0 11 271105 7)) shall be satisfied.	E0.12

Types of Wall

As shown in Figure 6.107 there are four main types of separating walls that can be used in order to achieve the required performance standards shown in Table 6.44.

Other designs, materials and/or products may also be available and so it is always worthwhile talking to the manufacturers and/or suppliers first.

The resistance to airborne sound depends mainly on the mass of the wall.

Solid masonry
(Wall type 1)

The resistance to airborne sound depends mainly on the mass per unit area of the wall.

Plaster

Masonry

Wall type 1

Cavity masonry
(Wall type 2)

The resistance to airborne sound depends on the mass per unit area of the leaves and on the degree of isolation achieved. The isolation is affected by connections (such as wall ties and foundations) between the wall leaves and by the cavity width.

Plaster

Masonry

Wall type 2

Cavity

Masonry between independent panels
(Wall type 3)

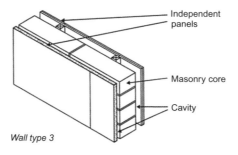

Independent panels

Masonry core

Cavity

Wall type 3

The resistance to airborne sound depends partly on the type and mass per unit area of the core, and partly on the isolation and mass per unit area of the independent panels.

Framed wall absorbent with material
(Wall type 4)

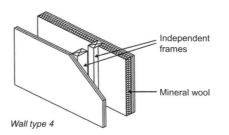

Independent frames

Mineral wool

Wall type 4

The resistance to airborne sound depends on the mass per unit area of the leaves, the isolation of the frames, and the absorption in the cavity between the frames.

Figure 6.107 Types of separating walls.

Table 6.44 Dwelling-houses and flats – performance standards for separating walls, separating floors, and stairs that have a separating function

	Airborne sound insulation $D_{nT,w} C_{tr}$ dB (minimum values)	Impact sound insulation $L^9 {}_{nT,w}$ dB (maximum values)
Purpose built dwelling-houses and flats		
Walls	45	–
Floors and stairs	45	62
Dwelling-houses and flats formed by material change of use		
Walls	43	–
Floors and stairs	43	64

Junctions between separating walls and other building elements

> Care should be taken to correctly detail the junctions between the separating wall and other elements, such as floors, roofs, external walls and internal walls.   E2.9

Note: Where any building element functions as a separating element (e.g. a ground floor that is also a separating floor for a basement flat) then the separating-element requirements should take precedence.

Mass per unit area of walls

The mass per unit area of a wall is expressed in kilograms per square metre (kg/m^2) and is equivalent to:

$$\text{mass per unit area of wall} = \frac{\text{mass of co-ordinating area}}{\text{co-ordinating area}} \qquad (6.1)$$

Mass per unit area of a wall can be calculated as follows:

$$\text{mass per unit area of wall} = \frac{M_B + \rho_m [Td(1 + h - d) + V]}{LH} \, kg/m^2 \qquad (6.2)$$

Where:
M_B = brick/block mass (kg) at appropriate moisture content
ρ_m = density of mortar (kg/m^3) at appropriate mortar content
T = the brick/block finish without surface finish (m)
d = mortar thickness (m)
L = co-ordinating length (m)
H = co-ordinating height (m)
V = volume of any frog/void filled with mortar (m^3).

Note: The method for calculating mass per unit area is provided in Annex A to Part E of the Regulations, together with some examples.

Density of the materials

The density of the materials used (and on which the mass per unit area of the wall depends) is expressed in kilograms per cubic metre (kg/m^3).

Plasterboard linings on separating and external masonry walls

Wherever plasterboard is recommended (or the finish is not specified) a dry lining laminate of plasterboard with mineral wool may be used.	E2.15
Plasterboard linings should be fixed according to manufacturer's instructions.	E2.16

Note: Recommended cavity widths in separating cavity masonry walls are minimum values.

Wall ties in separating and external cavity masonry walls

There are two types of wall tie that can be used in masonry cavity walls: type A (butterfly ties), which are normal; and type B (double-triangle ties), which are used only in external masonry cavity walls where tie type A does not satisfy the requirements of Building Regulation Part A – Structure.

Note:
(1) Recommended cavity widths in separating cavity masonry walls are minimum values.
(2) In external cavity masonry walls, tie type B may decrease the airborne sound insulation due to flanking transmission via the external wall leaf compared to tie type A.

Stainless steel cavity wall ties are specified for all houses regardless of their location.	A1/2
Wall ties should have a horizontal spacing of 900 mm and a vertical spacing of 450 mm.	A1/2 2C8

Equivalent to 2.5 ties per square metre.

Wall ties should be spaced not more than 300 mm apart vertically, within a distance of 225 mm from the vertical edges of all openings, movement joints and roof verges.	A1/2 2C8
Wall ties should either comply with BS 1243, DD 140, or BS EN 845-1.	A1/2 2C19
Wall ties should be selected in accordance with Table 6.45.	A1/2 2C19

Table 6.45 Cavity wall ties

Normal cavity width (mm)	Tie length (mm)	Tie shape in accordance with BS 1243
50–75	200	Butterfly, double triangle or vertical twist
76–90	225	Double triangle or vertical twist
91–100	225	Double triangle or vertical twist
101–125	250	Vertical twist
126–150	275	Vertical twist
151–175	300	Vertical twist
176–300	Where face insulated blocks are used the cavity width should be measured from the face of the masonry unit	Vertical twist

The leaves of a cavity masonry wall construction should be connected by either butterfly ties or double-triangle ties spaced as per BS 5628-3:2001, which limits this tie type and spacing to cavity widths of 50 mm to 75 mm with a minimum masonry leaf thickness of 90 mm.   E2.19

Note: Wall ties may be used provided that they have the measured dynamic stiffness for the cavity width (see E2.20 and E2.21 for details of the relevant formula for measuring the dynamic stiffness).

In conditions of severe exposure, austenitic stainless steel or suitable non-ferrous ties should be used.   A1/2 (1C20)

The number of ties per square metre, n, shall be calculated from the horizontal (S_x) and vertical (S_y) tie spacing distances (in metres) using the formula $n = 1/(S_x \cdot S_y)$.   E2.22

All wall ties and spacings specified using the dynamic stiffness parameter should also satisfy the requirements of Building Regulation Part A – Structure.   E2.24

Corridor walls and doors

Separating walls should be used between corridors and rooms in flats, in order to control flanking transmission and to provide the required sound insulation.   E2.25

Note: It is highly likely that the amount of sound insulation gained by using a separating wall will be reduced by the presence of a door.

Noisy parts of the building should preferably have a lobby, double door or high performance doorset to contain the noise.	E2.27
All corridor doors shall have a good perimeter sealing (including the threshold where practical).	E2.26
All corridor doors shall have a minimum mass per unit area of 25 kg/m².	E2.26
All corridor doors shall have a minimum sound reduction index of 29 dB Rw (measured according to BS EN ISO 140-3:1995 and rated according to BS EN ISO 717-1:1997).	E2.26
All corridor doors shall meet the requirements for fire safety (see Building Regulations Part B – Fire Safety).	E2.26

Refuse chutes

A wall separating a habitable room (or kitchen) from a refuse chute should have a mass per unit area (including any finishes) of at least 1320 kg/m².	E2.28
A wall separating a non-habitable room from a refuse chute should have a mass per unit area (including any finishes) of at least 220 kg/m².	E2.28

Wall type 1 (solid masonry)

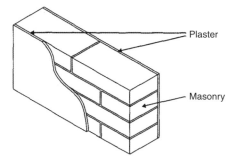

Plaster

Masonry

When using a solid masonry wall, the resistance to airborne sound depends mainly on the mass per unit area of the wall. As shown below, there are three different categories of solid masonry walls:

Table 6.46 Wall type 1 – categories

Wall type 1

Category 1.1 Solid masonry

Dense aggregate concrete block, plaster on both room faces

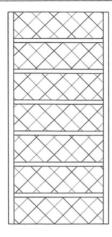

Minimum mass per unit area (including plaster) 415 kg/m²

Plaster on both room faces

Blocks laid flat to the full thickness of the wall

For example:

Size	215 mm laid flat
Density	1840 kg/m³
Coursing	110 mm
Plaster	13 mm lightweight

Wall type 1

Category 1.2 Dense aggregate concrete

Dense aggregate concrete, cast in situ, plaster on both room faces

Minimum mass per unit area (including plaster) 415 kg/m²

Plaster on both room faces

For example:

Concrete	190 mm
Density	2200 kg/m³
Plaster	13 mm lightweight

Wall type 1

Category 1.3 Brick

Brick, plaster on both room faces

Minimum mass per unit area (including plaster) 375 kg/m²

Bricks to be laid frog up, coursed with headers

For example:

Size	215 mm laid flat
Density	1610 kg/m³
Coursing	75 mm
Plaster	13 mm lightweight

Wall type 1 – general requirements

Fill and seal all masonry joints with mortar.	E2.32a
Lay bricks frog up to achieve the required mass per unit area and avoid air paths.	E2.32b
Use bricks/blocks that extend to the full thickness of the wall.	E2.32c
Ensure that an external cavity wall is stopped with a flexible closer at the junction with a separating wall.	E2.32d

 Unless the cavity is fully filled with mineral wool or expanded polystyrene beads.

Control flanking transmission from walls and floors connected to the separating wall (see guidance on junctions).	
Deep sockets and chases should **not** be used in separating walls.	E2.32
Stagger the position of sockets on opposite sides of the separating wall.	
Ensure flue blocks: • will not adversely affect the sound insulation; • use a suitable finish.	
A cavity separating wall may **not** be changed into a solid masonry (i.e. type 1) wall by filling in the cavity with mortar and/or concrete.	E2.32
When the cavity wall is bridged by the solid wall, ensure that there is no junction between the solid masonry wall and a cavity wall.	E2.32

Wall type 1 – junction requirements

Junctions with an external cavity wall with masonry inner leaf

Where the external wall is a cavity wall: • the outer leaf of the wall may be of any construction; • the cavity should be stopped with a flexible closer.	E2.36a E2.36b

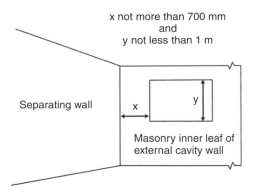

x not more than 700 mm
and
y not less than 1 m

Separating wall

x

y

Masonry inner leaf of
external cavity wall

Figure 6.108 Wall type 1 – position of openings in a masonry inner leaf of an external cavity wall.

The masonry inner leaf should have a mass per unit area of at least 120 kg/m² excluding finish unless there are openings in the external wall (Figure 6.108) that are:	E2.38a
• not less than 1 m high;	E2.38b
• on both sides of the separating wall at every storey;	E2.38c
• not more than 700 mm from the face of the separating wall on both sides.	E2.39

 Note: If there is also a separating floor, then the minimum mass per unit area of 120 kg/m² (excluding finish) will always apply, irrespective of the presence or absence of openings.

The separating wall should be joined to the inner leaf of the external cavity wall by one of the following methods shown in Figure 6.109.

Junctions with an external cavity wall with timber frame inner leaf

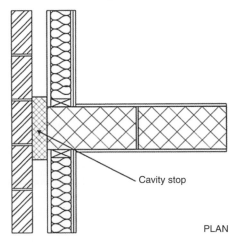

Cavity stop

PLAN

Bonded junction

Masonry inner leaf of an external cavity wall with a solid separating wall

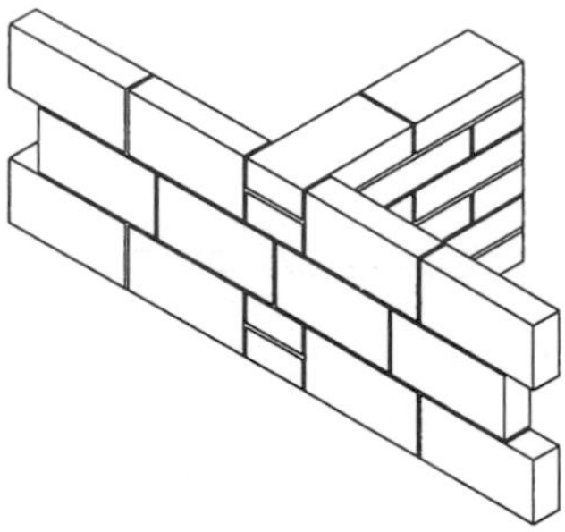

The separating wall should be bonded to the external wall in such a way that the separating wall contributes at least 50% of the bond at the junction

E2.37a

Tied junction

External cavity wall with an internal masonry wall

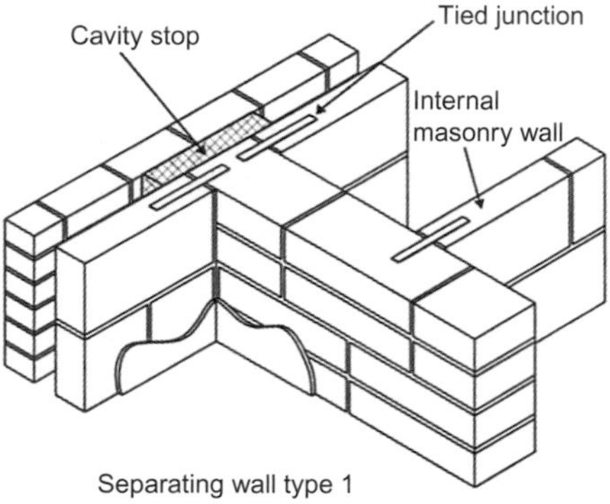

Cavity stop

Tied junction

Internal masonry wall

Separating wall type 1

The external wall should abut the separating wall and be tied to it

E2.37b

Figure 6.109 Separating wall junctions for a type-1 wall.

Where the external wall is a cavity wall:	E2.40a
• the outer leaf of the wall may be of any construction;	E2.40b
• the cavity should be stopped with a flexible closer.	
Where the inner leaf of an external cavity wall is of framed construction, the framed inner leaf should:	E2.41a1 E2.41b1
• abut the separating wall;	
• be tied to it with ties at no more than 300 mm centres vertically.	
The wall finish of the framed inner leaf of the external wall should be:	E2.41a2 E2.41b2 E2.41c E2.41d
• one layer of plasterboard; or	
• two layers of plasterboard where there is a separating floor;	
• each sheet of plasterboard should be of minimum mass per unit area 10 kg/m²;	
• all joints should be sealed with tape or caulked with sealant.	

Junctions with internal timber floors

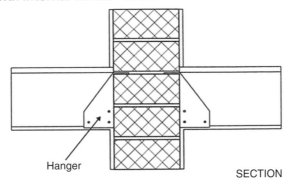

Hanger

SECTION

If the floor joists are to be supported on a Type-1 separating wall then they should be supported on hangers as opposed to being built in.

E2.45

Junctions with internal concrete floors

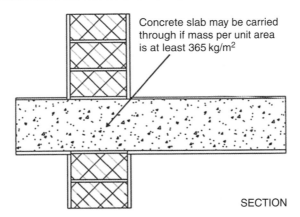

Concrete slab may be carried through if mass per unit area is at least 365 kg/m^2

SECTION

An internal concrete floor slab may only be carried through a type-1 separating wall if the floor base has a mass per unit area of at least 365 kg/m^2.

E2.46

Internal hollow-core concrete plank floors and concrete beams with infilling block floors should not be continuous through a type-1 separating wall.

E2.47

 Note: For internal floors of concrete beams with infilling blocks, avoid beams built into the separating wall unless the blocks in the floor fill the space between the beams where they penetrate the wall.

Junctions with concrete ground floors

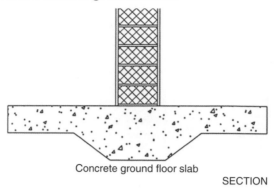

Concrete ground floor slab

SECTION

The ground floor may be a solid slab, laid on the ground, or a suspended concrete floor.	E2.51
A concrete-slab floor on the ground may be continuous under a type-1 separating wall.	E2.51
A suspended concrete floor may only pass under a type-1 separating wall if the floor has a mass of at least 365 kg/m².	E2.52
Hollow core concrete plank and concrete beams within filling block floors should not be continuous under a type-1 separating wall.	E2.53

 Note: See also Building Regulation Part C (*Site Preparation and Resistance to Moisture*) and Building Regulation Part L (*Conservation of Fuel and Power*).

Junctions with ceiling and roof

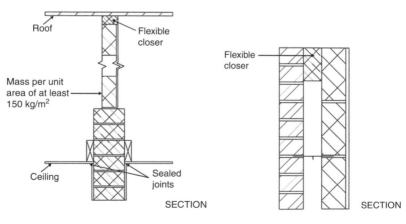

Ceiling and roof junction   External cavity at roof level

Where a type-1 separating wall is used it should be continuous to the underside of the roof.   E2.55

The junction between the separating wall and the roof should be filled with a flexible closer which is also suitable as a fire stop.   E2.56

Where the roof or loft space is not a habitable room (and there is a ceiling with a minimum mass per unit area of $10\,kg/m^2$ with sealed joints), the mass per unit area of the separating wall above the ceiling may be reduced to $150\,kg/m^2$.   E2.57

If lightweight aggregate blocks of density less than $1200\,kg/m^3$ are used above ceiling level, one side should be sealed with cement paint or plaster skim.   E2.58

Where there is an external cavity wall, the cavity should be closed at eaves level with a suitable flexible material (e.g. mineral wool).   E2.59

A rigid connection between the inner and external wall leaves should be avoided.   Ep23

If a rigid material is used, then it should only be rigidly bonded to one leaf.   Ep23

Guidance for other types of wall type 1 junctions

Junctions with an external solid masonry wall	No guidance available (seek specialist advice).	E2.42
Junctions with internal framed walls	There are no restrictions on internal framed walls meeting a type-1 separating wall.	E2.43
Junctions with internal masonry walls	Internal masonry walls that about a type-1 separating wall should have a mass per unit area of at least $120\,kg/m^2$ excluding finish.	E2.44
Junctions with timber ground floors	If the floor joists are to be supported on a type-1 separating wall then they should be supported on hangers and should not be built in.	E2.49

Wall type 2 (cavity masonry)

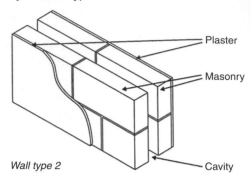

Wall type 2

When using a cavity masonry wall, the resistance to airborne sound depends on the mass per unit area of the leaves and on the degree of isolation achieved. The isolation is affected by connections (e.g. wall ties and foundations) between the wall leaves and by the cavity width.

As shown in Table 6.47, there are four different categories of cavity masonry wall.

Table 6.47 Wall type 2 – categories

Wall type 2		
Wall type 2 **Category 2.1** **Two leaves of dense aggregate concrete block with 50 mm cavity incorporated at the separating wall**	SECTION	Minimum mass per unit area (including plaster) 415 kg/m^2 Plaster on both room faces Minimum cavity width 50 mm For example: Block leaves 100 mm Density 1990 kg/m^3 Coursing 225 mm Plaster 13 mm lightweight
Wall type 2 **Category 2.2** **Two leaves of lightweight aggregate block with 75 mm cavity, plaster on both room faces**	SECTION	Minimum mass per unit area (including plaster) 300 kg/m^2 Plaster on both room faces Minimum cavity width of 75 mm For example: Block leaves 100 mm Density 1375 kg/m^3 Coursing 225 mm Plaster 13 mm lightweight

Table 6.47 Wall type 2 – categories (Continued)

Wall type 2

Category 2.3

Two leaves of lightweight aggregate block with 75 mm cavity and step/ stagger plasterboard on both room faces

Wall type 2.3 should only be used where there is a step and/or stagger of at least 300 mm

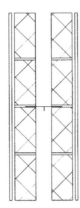

SECTION

Increasing the size of the step or stagger in the separating wall tends to increase the airborne sound insulation

Minimum mass per unit area (including plaster) 290 kg/m²

Lightweight aggregate blocks should have a density in the range 1350 to 1600 m³

Minimum cavity width of 75 mm

Plasterboard (lightweight) each sheet of minimum mass per unit area 10 kg/m² on both room faces

For example:

Block leaves   100 mm
Density   1375 kg/m³
Coursing   225 mm

Lightweight plasterboard (minimum mass per unit area 10 kg/m²) on both room faces

Wall type 2

Category 2.4

Two leaves of Aircrete block with 75 mm cavity and step/ stagger plasterboard or plaster on both room faces

Wall type 2.4 should only be used where there is a step and/or stagger of at least 300 mm

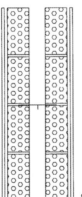

SECTION

Increasing the size of the step or stagger in the separating wall tends to increase the airborne sound insulation.

Minimum mass per unit area (including plaster) 150 kg/m²

Lightweight aggregate blocks should have a density in the range 1350 to 1600 kg/m³

Minimum cavity width of 75 mm

Plasterboard (lightweight) minimum mass per unit area 10 kg/m² on both room faces or 13 mm plasterboard on both faces

For example:

Aircrete block 100 mm
leaves   650 kg/m³
Density   225 mm
Coursing

Plaster (lightweight) minimum mass per unit area 10 kg/m² on both room faces

Wall type 2 – general requirements

Fill and seal all masonry joints with mortar.	E2.65a
Keep the cavity leaves separate below ground floor level.	E2.65b
Ensure that any external cavity wall is stopped with a flexible closer at the junction with the separating wall.	E2.65c

Control flanking transmission from walls and floors connected to the separating wall.	E2.65d
Stagger the position of sockets on opposite sides of the separating wall.	E2.65e
Ensure that flue blocks will not adversely affect the sound insulation and that a suitable finish is used over the flue blocks.	E2.65f
The cavity separating wall should **not** be converted to a type-1 (solid masonry) separating wall by inserting mortar or concrete into the cavity between the two leaves.	E2.65a2
A solid wall construction in the roof space should **not** be changed.	E2.65b2
Cavity walls should **not** be built off a continuous solid concrete slab floor.	E2.65c2
Deep sockets and chases should **not** be used in a separating wall.	E2.65d2
Deep sockets and chases in a separating wall should **not** be placed back to back.	E2.65d2
Wall ties used to connect the leaves of a cavity masonry wall should be tie type A.	E2.66

Wall type 2 – junction requirements

Junctions with an external cavity wall with masonry inner leaf

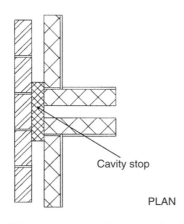

Wall types 2.1 and 2.2 – external cavity wall with masonry inner leaf

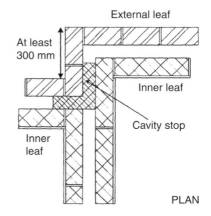

Wall types 2.3 and 2.4 – external cavity wall with masonry inner leaf – stagger

Where the external wall is a cavity wall:

- the outer leaf of the wall may be of any construction;   E2.73a
- the cavity should be stopped with a flexible closer.   E2.73b

The separating wall should be joined to the inner leaf of   E2.74
the external cavity wall.

The masonry inner leaf should have a mass per unit area   E2.75
of at least $120\,\text{kg/m}^2$ excluding finish.

There is no minimum mass requirement where separating   E2.76
wall type 2.1, 2.3 or 2.4 is used **unless** there is also a
separating floor.

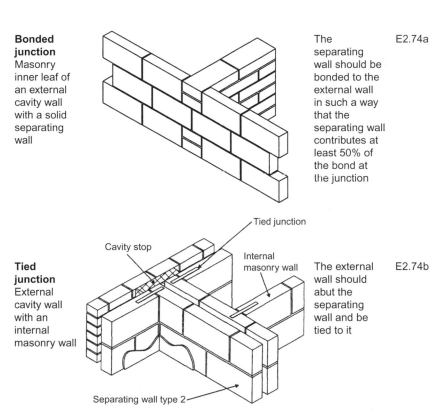

Bonded junction
Masonry inner leaf of an external cavity wall with a solid separating wall

The separating wall should be bonded to the external wall in such a way that the separating wall contributes at least 50% of the bond at the junction   E2.74a

Tied junction
External cavity wall with an internal masonry wall

The external wall should abut the separating wall and be tied to it   E2.74b

Tied junction

Cavity stop

Internal masonry wall

Separating wall type 2

Figure 6.110 Separating wall junctions for a type-2 wall.

Junctions with an external cavity wall with timber frame inner leaf

Where the external wall is a cavity wall:

- the outer leaf of the wall may be of any construction;   E2.77a
- the cavity should be stopped with a flexible closer   E2.77b
  (Figure 6.111).

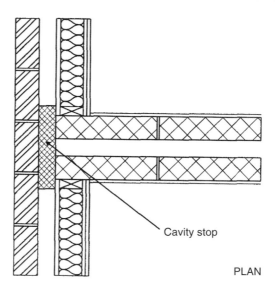

Cavity stop

PLAN

Figure 6.111 Wall type 2 – external cavity wall with timber frame inner leaf.

Where the inner leaf of an external cavity wall is of framed
construction, the framed inner leaf should:

- abut the separating wall;   E2.78a
- be tied to it with ties at no more than 300 mm centres   E2.78b
  vertically.

The wall finish of the inner leaf of the external wall should   E2.98a2
be: E2.78b2

- one layer of plasterboard;   E2.78c2
- two layers of plasterboard where there is a separating   E2.78d2
  floor;
- each sheet of plasterboard to be of minimum mass per
  unit area $10 \, \text{kg/m}^2$;
- all joints should be sealed with tape or caulked with
  sealant.

Junctions with internal timber floors

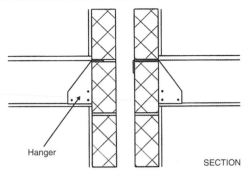

Hanger

SECTION

Figure 6.112 Wall type 2 – internal timber floor.

> If the floor joists are to be supported on the separating wall, they   E2.84
> should be supported on hangers as opposed to being built in.

Junctions with internal concrete floors

> Internal concrete floors should generally be built into a type-2   E2.85
> separating wall and carried through to the cavity face of the leaf.

 The cavity should not be bridged.

Junctions with concrete ground floors

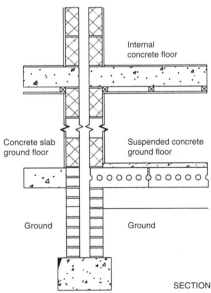

Internal
concrete floor

Concrete slab
ground floor

Suspended concrete
ground floor

Ground

Ground

SECTION

Figure 6.113 Wall type 2 – internal concrete floor and concrete ground floor.

The ground floor may be a solid slab, laid on the ground, or a suspended concrete floor.	E2.88
A concrete-slab floor on the ground should not be continuous under a Type-2 separating wall.	E2.88
A suspended concrete floor should not be continuous under a Type-2 separating wall.	E2.89
A suspended concrete floor should be carried through to the cavity face of the leaf.	E2.89
The cavity should not be bridged.	

Junctions with ceiling and roof space

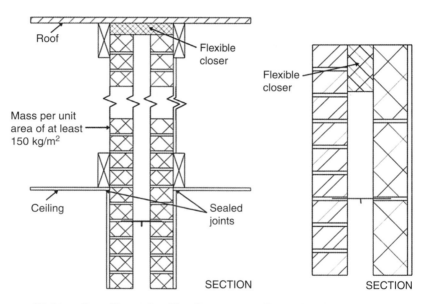

Wall type 2 – ceiling and roof junction

External cavity wall at eaves level

A type-2 separating wall should be continuous to the underside of the roof.	E2.91
The junction between the separating wall and the roof should be filled with a flexible closer that is also suitable as a fire stop.	E2.92

If lightweight aggregate blocks (with a density less than $1200\,kg/m^3$) are used above ceiling level, then one side should be sealed with cement paint or plaster skim.	E2.94
The cavity of an external cavity wall should be closed at eaves level with a suitable flexible material (e.g. mineral wool).	E2.95
A rigid connection between the inner and external wall leaves should be avoided.	E2.95
If a rigid material has to be used, then it should only be rigidly bonded to one leaf.	E2.95

 Note: If the roof or loft space is not a habitable room (and there is a ceiling with a minimum mass per unit area of $10\,kg/m^2$ with sealed joints), then the mass per unit area of the separating wall above the ceiling may be reduced to $150\,kg/m^2$ – **but** it should still be a cavity wall.

Guidance for other wall type 2 junctions

Junctions with internal masonry walls	Internal masonry walls that abut a type 2 separating wall should have a mass per unit area of at least $120\,kg/m^2$ excluding finish.	E2.81
	When there is a separating floor, the internal masonry walls should have a mass per unit area of at least $120\,kg/m^2$ excluding finish.	E2.82
Junctions with internal framed walls	There are no restrictions on internal framed walls meeting a type-2 separating wall.	E2.80
Junctions with an external solid masonry wall	No guidance available (seek specialist advice).	E2.79
Junctions with timber ground floors	If the floor joists are to be supported on a type-1 separating wall, they should be supported on hangers and should not be built in.	E2.49

Wall type 3 (masonry between independent panels)

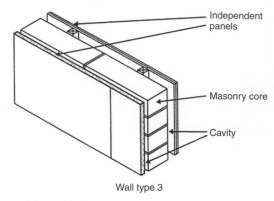

Wall type 3

Wall type 3 provides a high resistance to the transmission of both airborne sound and impact sound on the wall. As shown below, there are three different categories of wall type 3, which comprise either a solid or a cavity masonry core wall with independent panels on both sides. Their resistance to sound depends partly on the type (and mass) of the core and partly on the isolation and mass of the panels.

Wall type 3 – general requirements

Fill and seal all masonry joints with mortar.	E2.101a
Control flanking transmission from walls and floors connected to the separating wall.	E2.101b
The panels and any frame should not be in contact with the core wall.	E2.99
The panels and/or supporting frames should be fixed to the ceiling and floor only.	E101c
All joints should be taped and sealed.	E2.101d
Flue blocks shall not adversely affect the sound insulation.	E2.101e
A suitable finish is used over the flue blocks (see BS 1289-1:1986).	E2.101e
Free-standing panels and/or the frame should not be fixed, tied or connected to the masonry core.	E2.101
Wall ties in cavity masonry cores, used to connect the leaves of a cavity masonry core together, should be tie type A.	E2.102
The minimum mass per unit area of independent panels (excluding any supporting framework) should be $20\,kg/m^2$.	E2.104
Panels should be either at least two layers of plasterboard with staggered joints or a composite panel consisting of two sheets of plasterboard separated by a cellular core.	E2.104

Panels that are not supported on a frame should be at least 35 mm from the masonry core.   E2.104

Panels that are supported on a frame should have a gap of at least 10 mm between the frame and the masonry core.   E2.104

Table 6.48 Wall type 3 – categories

Wall type 3	
Category 3.1 Solid masonry core (dense aggregate concrete block), independent panels on both room faces	Minimum mass per unit area (including plaster) 300 kg/m²; Independent panels on both room faces Minimum core width determined by structural requirements. For example: Size 140 mm core block Density 2200 kg/m³ Coursing 110 mm Independent panels – each panel of mass per unit area 20 kg/m², to be two sheets of plasterboard with joints staggered
Category 3.2 Solid masonry core (lightweight concrete block), independent panels on both room faces 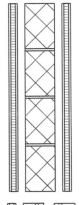 SECTION	Minimum mass per unit area (including plaster) 150 kg/m²; Independent panels on both room faces; Minimum core width determined by structural requirements For example: Size 140 mm core block Density 1400 kg/m³ Coursing 225 mm Independent panels – each panel of mass per unit area 20 kg/m², to be two sheets of plasterboard with joints staggered
Category 3.3 Cavity masonry core (brickwork or block work), 50 mm cavity, independent panels on both room faces 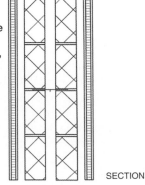 SECTION	Core mass – unrestricted Minimum cavity width of 50 mm Independent panels on both room faces Minimum core width determined by structural requirements For example: Concrete block – two leaves (each leaf at least 100 mm thick) Minimum cavity width – 50 mm Independent panels – each panel of mass per unit area 20 kg/m², to be two sheets of plasterboard with joints staggered

Junction requirements for wall type 3

Junctions with an external cavity wall with masonry inner leaf

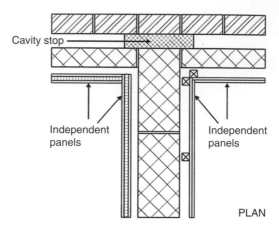

Figure 6.114 Wall type 3 – external cavity wall with masonry inner leaf.

If the external wall is a cavity wall: E2.108

- the outer leaf of the wall may be of any construction;
- the cavity should be stopped with a flexible closer.

If the inner leaf of an external cavity wall is masonry: E2.109

- the inner leaf of the external wall should be bonded or tied to the masonry core;
- the inner leaf of the external wall should be lined with independent panels.

If there is a separating floor, the masonry inner leaf (of the external wall) should have a minimum mass per unit area of at least 120 kg/m^2 excluding finish.   E2.110

If there is no separating floor:

- the external wall may be finished with plaster orplasterboard of minimum mass per unit area 10 kg/m^2 (provided the masonry inner leaf of the external wall has a mass per unit area of at least 120 kg/m^2 excluding finish);   E2.111
- there is no minimum mass requirement on the masonry inner leaf (provided that the masonry inner leaf of the external wall is lined with independent panels in the same manner as the separating walls).   E2.112

Junctions with internal framed walls

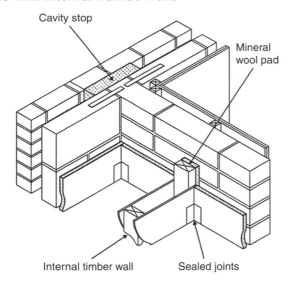

Figure 6.115 Wall type 3 – external cavity wall with internal timber wall.

Load-bearing (framed) internal walls should be fixed to the masonry core through a continuous pad of mineral wool.	E2.115
Non-load-bearing internal walls should be butted to the independent panels.	E2.116
All joints between internal walls and panels should be sealed with tape or caulked with sealant.	E2.117

Junctions with internal timber floors

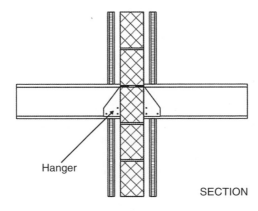

Figure 6.116 Wall type 3 – internal timber floor.

Junctions with internal masonry walls

If the floor joists are to be supported on the separating wall then they should be supported on hangers as opposed to being built in.	E2.119
Spaces between the floor joists should be sealed with full-depth timber blocking.	E2.120

Junctions with internal concrete floors

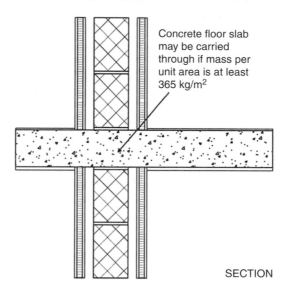

Concrete floor slab may be carried through if mass per unit area is at least 365 kg/m²

SECTION

Figure 6.117 Wall types 3.1 and 3.2 – internal concrete floor.

For wall types 3.1 and 3.2 (i.e. those with solid masonry cores) internal concrete floor slabs may only be carried through a solid masonry core if the floor base has a mass per unit area of at least 365 kg/m².	E2.121
For wall type 3.3 (cavity masonry core): • internal concrete floors should generally be built into a cavity masonry core and carried through to the cavity face of the leaf; • the cavity should not be bridged.	E2.122

Junctions with ceiling and roof space

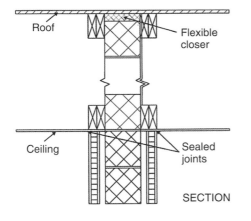

Roof

Flexible closer

Ceiling

Sealed joints

SECTION

Wall types 3.1 and 3.2 – ceiling and roof junction

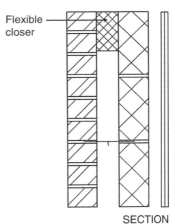

Flexible closer

SECTION

External cavity wall at eaves level

The masonry core should be continuous to the underside of the roof.	E2.133
The junction between the separating wall and the roof should be filled with a flexible closer that is also suitable as a fire stop.	E2.134
The junction between the ceiling and independent panels should be sealed with tape or caulked with sealant.	E2.135
If there is an external cavity wall, the cavity should be closed at eaves level with a suitable flexible material (e.g. mineral wool).	E2.136
Rigid connections between the inner and external wall leaves should be avoided where possible.	E2.136
If a rigid material is used, then it should only be rigidly bonded to one leaf.	E2.136

For wall types 3.1 and 3.2 (solid masonry core):

- if the roof or loft space is not a habitable room (and there is a ceiling with a minimum mass per unit area of $10\,kg/m^2$ and it has sealed joints), the independent panels may be omitted in the roof space and the mass per unit area of the separating wall above the ceiling may be a minimum of $150\,kg/m^2$;   E2.137
- if lightweight aggregate blocks with a density less than $1200\,kg/m^3$ are used above ceiling level, one side should be sealed with cement paint or plaster skim.   E2.138

For wall type 3.3 (cavity masonry core) if the roof or loft space is not a habitable room (and there is a ceiling with a minimum mass per unit area of $10\,kg/m^2$ and it has sealed joints), the independent panels may be omitted in the roof space, but the cavity masonry core should be maintained to the underside of the roof.	E2.139

Junctions with internal masonry floors

Internal walls that abut a type-2 separating wall should not be of masonry construction.	E2.118

Junctions with timber ground floors

Floor joists supported on a separating wall should be supported on hangers as opposed to being built in.	E2.123
The spaces between floor joists should be sealed with full depth timber blocking.	E2.124

Junctions with an external cavity wall with timber frame inner leaf

No official guidance is currently available. It is best to seek specialist advice.	E2.113

Junctions with an external solid masonry wall

No official guidance is currently available. It is best to seek specialist advice	E2.114

Junctions with concrete ground floors

The ground floor may be a solid slab, laid on the ground, or a suspended concrete floor.	E2.126
For wall types 3.1 and 3.2 (solid masonry core):	
• a concrete slab floor on the ground may be continuous under the solid masonry core of the separating wall;	E2.127
• a suspended concrete floor may only pass under the solid masonry core if the floor has a mass per unit area of at least $365\,kg/m^2$;	E2.128
• hollow core concrete plank (and concrete beams with infilling block floors) should **not** be continuous under the solid masonry core of the separating wall.	E2.129

For wall type 3.3 (cavity masonry core):

• a concrete–masonry core of a slab floor on the ground should **not** be continuous under the cavity masonry core of the separating wall;	E2.130
• a suspended concrete floor should **not** be continuous under the cavity type-3.3 separating wall;	E2.131
• a suspended concrete floor **should** be carried through to the cavity face of the leaf but the cavity should not be bridged.	E2.132

Junctions with internal masonry walls

Internal walls that abut a type-3 separating wall should not be of masonry construction.	E2.118

Wall type 4 (framed walls with absorbent material)

A wall type 4 consists of a timber frame with a plasterboard lining on the room surface with an absorbent material between the frames. Its resistance to airborne sound depends on:

- the mass per unit area of the leaves;
- the isolation of the frames;
- the absorption in the cavity between the frames.

Wall type 4 – general requirements

If a fire stop is required in the cavity between frames, then it should either be flexible or only be fixed to one frame.	E2.146
Layers of plasterboard should:	
• be independently fixed to the stud frame;	E2.146c
• not be chased.	E2.146b2
If two leaves have to be connected together for structural reasons, then:	
• the cross-section of the ties shall be less than 40 mm × 3 mm;	E2.146a2
• ties should be fixed to the studwork at or just below ceiling level;	E2.146a2
• ties should not be set closer than 1.2 m centres.	E2.146a2
Sockets should:	
• be positioned on opposite sides of a separating wall;	E2.146b
• not be connected back to back;	E2.146b2
• be staggered a minimum of 150 mm edge to edge.	E2.146b2

The flanking transmission from walls and floors connected   E2.146d
to a separating wall should be controlled (see guidance on
junctions).

Wall type 4.1 (double-leaf frames with absorbent material)

General requirements

The lining shall be two or more layers of plasterboard with   E2.147
a minimum sheet mass per unit area $10\,kg/m^2$ and with
staggered joints.

If a masonry core is used for structural purposes, then the   E2.147
core should only be connected to one frame.

The minimum distance between inside lining faces shall be   E2.147
200 mm.

Plywood sheathing may be used in the cavity if required for   E2.147
structural reasons.

Absorbent material: E2.147

- shall have a minimum density of $10\,kg/m^3$;
- shall be unlaced mineral wool batts (or quilt);
- may be wire reinforced;
- shall have a minimum thickness of between 25 and
  50 mm as shown in Figure 6.118.

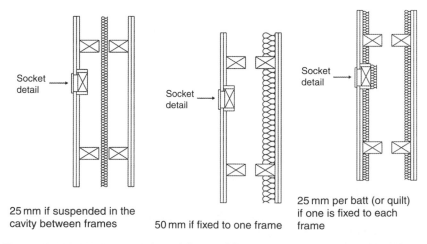

25 mm if suspended in the
cavity between frames

50 mm if fixed to one frame

25 mm per batt (or quilt)
if one is fixed to each
frame

Socket
detail

Socket
detail

Socket
detail

Figure 6.118 Wall type 4.1 – minimum thickness of absorbent material.

Junction requirements for wall type 4

Junctions with an external cavity wall with timber-frame inner leaf

If the external wall is a cavity wall:	
• the outer leaf of the wall may be of any construction;	E2.149
• the cavity should be stopped between the ends of the separating wall and the outer leaf with a flexible closer.	E2.149
The wall finish of the inner leaf of the external wall should be one layer of plasterboard (or two layers of plasterboard if there is a separating floor).	E2.150a E2.150b
Each sheet of plasterboard to be of minimum mass per unit area 10 kg/m².	E2.150c
All joints should be sealed with tape or caulked with sealant.	E2.150

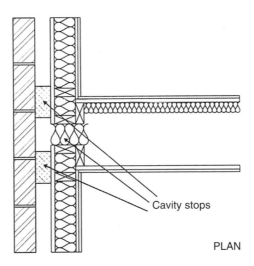

Cavity stops

PLAN

Figure 6.119 Junctions with an external solid masonry wall.

Junction with ceiling and roof space

The wall should preferably be continuous to the underside of the roof.	E2.160
The junction between the separating wall and the roof should be filled with a flexible closer.	E2.161

The junction between the ceiling and the wall linings should be sealed with tape or caulked with sealant.   E2.162

If the roof or loft space is not a habitable room (and there is a ceiling with a minimum mass per unit area of $10\,kg/m^2$ with sealed joints), then:

- either the linings on each frame may be reduced to two   E2.162a
  layers of plasterboard, each sheet with a minimum mass
  per unit area of $10\,kg/m^2$; or
- the cavity may be closed at ceiling level without con-   E2.162b
  necting the two frames rigidly together.

Note: In which case there need only be one frame in the roof space provided there is a lining of two layers of plasterboard, each sheet of minimum mass per unit area of $10\,kg/m^2$, on both sides of the frame.

External wall cavities should be closed at eaves level with a   E2.163
suitable material.

Junctions with timber ground floors

Air paths through the wall into the cavity shall be blocked   E2.156
using solid timber blockings, continuous ring beam or joists.

See also Building Regulation Part C (*Site Preparation and Resistance to Moisture*) and Building Regulation Part L (*Conservation of Fuel and Power*).

Junctions with concrete ground floors

If the ground floor is a concrete slab laid on the ground, it   E2.158
may be continuous under a type-4 separating wall.

If the ground floor is a suspended concrete floor, it may
only pass under a wall type 4 if the floor has a mass per
unit area of at least $365\,kg/m^2$.

See also Building Regulation Part C (*Site Preparation and Resistance to Moisture*) and Building Regulation Part L (*Conservation of Fuel and Power*).

Junctions with internal timber floors

Air paths through the wall into the cavity shall be blocked using solid timber blockings, continuous ring beam or joists.	E2.154

Junctions with internal concrete floors

No official guidance is currently available. It is best to seek specialist advice.	E2.155

Junctions with internal framed walls

There are no restrictions on internal framed walls meeting a type-4 separating wall.	E2.152

Junctions with internal masonry walls

There are no restrictions on internal masonry walls meeting a type-4 separating wall.	E2.153

Junctions with an external solid masonry wall

No official guidance is currently available. It is best to seek specialist advice.	E2.151

Junctions with an external cavity wall with masonry inner leaf

No official guidance is currently available. It is best to seek specialist advice.	E2.148

Walls adjacent to hearths

Walls that are not part of a fireplace recess or a prefabricated appliance chamber but are adjacent to hearths or appliances also need to protect the building from catching fire. A way of achieving the requirement is shown in Figure 6.120.	J (2.31)

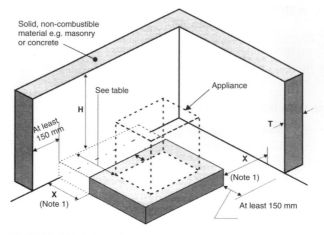

Location of hearth or appliance	Solid, non-combustible material	
	Thickness (T)	Height (H)
Where the hearth abuts a wall and the appliance is not more than 50 mm from the wall	200 mm	At least 300 mm above the appliance and 1.2 m above the hearth
Where the hearth abuts a wall and the appliance is more than 50 mm but not more than 300 mm from the wall	75 mm	At least 300 mm above the appliance and 1.2 m above the hearth
Where the hearth does not abut a wall and is no more than 150 mm from the wall (see Note 1)	75 mm	At least 1.2 m above the hearth
Note 1: There is no requirement for protection of the wall where X is more than 150 mm		

Figure 6.120 Walls adjacent to hearths.

6.8.3 Conservation of energy and power

Conservatories and porches in dwellings

Regulation 9 of the Building Regulations exempts some conservatory and porch extensions from the energy efficiency requirements. These consist of conservatories or porches:

- which are at ground level;
- where the floor area is less than 30 m²;
- where the existing walls, doors and windows in the part of the dwelling which separates the conservatory are retained or, if removed, replaced by walls, windows and doors which meet the energy efficiency requirements;
- where the heating system of the dwelling is not extended into the conservatory or porch;
- where the worst standards for fabric properties does not exceed the values given in Table 6.49.

Table 6.49 Limiting fabric parameters

Element	Standard (W/m²K)
Roof	0.20
Wall	0.30
Floor	0.25
Party wall	0.20
Windows, roof windows, glazed rooflights, curtain walling and pedestrian doors	2.00
Opaque elements	0.262

If a new conservatory or porch does not meet all these requirements then it is not exempt and must comply with the relevant energy efficiency requirements of Part L:2010.

If a building extension is a conservatory or porch that is not exempt from the energy efficiency requirements, then:	L1A 4.11 L2B 4.12

- the effective thermal separation between the heated area in the existing building (i.e. the walls, doors and windows between the building and the extension) should be insulated and draught-proofed to at least the same extent as in the existing building; and;
- Curtain walling should be less than 1.8 W/m²K.

Removing and not replacing any of the thermal separation between the building and an existing exempt extension, or extending the building's heating system into the extension, means that the extension ceases to be exempt!

Dwellings

Extensions to dwellings should either use newly constructed thermal elements (that meet the requirements for the conservation of fuel and power) or the use of existing (or new) windows, roof windows and rooflights that meet the standards shown in Table 6.50.

> If upgrades are proposed to the existing dwelling, such   L1B 4.7
> upgrades should be implemented to a standard that is no
> worse than that shown in column (b) of Table 6.50.

Table 6.50 Upgrading retained thermal elements

Element	(a) Threshold U-value (W/m²K)	(b) Improved U-value (W/m²K)
Wall – cavity insulation	0.70	0.55
Wall – external or cavity insulation	0.70	0.30

Building fabric

In accordance with Part L:2010 and Regulation 7:

> The building fabric of all dwelling and buildings other than   L1A 5.2
> dwellings should be constructed to a reasonable standard so   L2A 5.2
> that:
>
> - the insulation is reasonably continuous over the whole
>   building envelope; and
> - the air permeability is within reasonable limits.
>
> Reductions in thermal performance can occur where   L2A 5.4
> the air barrier and the insulation layer are not contiguous
> and the cavity between them is subject to air movement.
>
> To avoid this problem, either the insulation layer should
> be contiguous with the air barrier at all points in the
> building envelope, or the space between them should
> be filled with solid material such as in a masonry wall

Thermal bridges

> The heat loss through a party wall can be reduced by fully   L1A 5.4
> filling the cavity and/or by providing effective sealing
> around the perimeter so as to restrict air movement through
> the cavity

6.9 Ceilings

6.9.1 The requirement

Structure

As a fire precaution, all materials used for internal linings of a building should have a low rate of surface flame spread and (in some cases) a low rate of heat release.

(Approved Document B2)

Dwellings shall be designed so that the noise from domestic activity in an adjoining dwelling (or other parts of the building) is kept to a level that:

* *does not affect the health of the occupants of the dwelling;*
* *will allow them to sleep, rest and engage in their normal activities in satisfactory conditions.*

(Approved Document E1)

Dwellings shall be designed so that any domestic noise that is generated internally does not interfere with the occupants' ability to sleep, rest and engage in their normal activities in satisfactory conditions.

(Approved Document E2)

Domestic buildings shall be designed and constructed so as to restrict the transmission of echoes.

(Approved Document E3)

Schools shall be designed and constructed so as to reduce the level of ambient noise (particularly echoing in corridors).

(Approved Document E4)

6.9.2 Meeting the requirements

Suspended ceilings

Table 6.51 sets out criteria appropriate to suspended ceilings that can be accepted as contributing to the fire resistance of a floor.

Table 6.51 Limitations on fire-protected suspended ceilings

Height of building or separated part	Type of floor	Provision for fire resistance of floor	Description of suspended ceiling
≤18 m	Not compartment	60 min or less	Type A, B, C or D
	Compartment	≤60 min	Type A, B, C or D
		60 min	Type B, C or D
18 m or more	Any	60 min or less	Type C or D
No limit	Any	60 min	Type D

Ceilings – general

The resistance to airborne and impact sound depends on the independence and isolation of the ceiling and the type of material used.   E

Three ceiling treatments (which are ranked in order of sound insulation) may be used: E3

- ceiling treatment A – independent ceiling with absorbent material;
- ceiling treatment B – plasterboard on proprietary resilient bars with absorbent material;
- ceiling treatment C – plasterboard on timber battens (or proprietary resilient channels) with absorbent material.

If the roof or loft space is not a habitable room (and provided that there is a ceiling with a minimum mass per unit area of $10\,kg/m^2$ with sealed joints and the cavity masonry core is maintained to the underside of the roof) then: E

- the mass per unit area of the separating wall above the ceiling may be reduced to $150\,kg/m^2$;
- the independent panels may be omitted in the roof space;
- the linings on each frame may be reduced to two layers of plasterboard or the cavity may be closed at ceiling level without connecting the two frames rigidly together.

All junctions between ceilings and independent panels (and joints between casings and ceiling) should be sealed with tape or caulked with sealant.   E

At junctions with external cavity walls (with masonry inner leaf) the ceiling should be taken through to the masonry.   E3

The ceiling void and roof space detail can only be used where the requirements of Building Regulation Part B – Fire safety can also be satisfied.   E3

If there is an existing lath and plaster ceiling it should be retained as long as it satisfies Building Regulation Part B – Fire safety.   E3

If the existing ceiling is not lath and plaster, it should be upgraded to provide: E4

- at least two layers of plasterboard with staggered joints;
- a minimum total mass per unit area of $20\,kg/m^2$;
- an absorbent layer of mineral wool laid on the ceiling (minimum thickness 100 mm, minimum density $10\,kg/m^3$);

- plasterboard with joints staggered, total mass per unit area $20\,kg/m^2$.

Care should be taken at the design stage to ensure that adequate ceiling height is available in all rooms to be treated.

The ceiling should be supported by either: E2

- independent joists fixed only to the surrounding walls; or
- independent joists fixed to the surrounding walls with additional support provided by resilient hangers attached directly to the existing floor base.

Note: A clearance of at least 25 mm should be left between the top of the independent ceiling joists and the underside of the existing floor construction.

Where a window head is near to the existing ceiling, the new   E4
independent ceiling may be raised to form a pelmet recess.

A rigid or direct connection should not be created between an   E4
independent ceiling and the floor base.

Where the roof or loft space is not a habitable room (and there   E2.57
is a ceiling with a minimum mass per unit area of $10\,kg/m^2$
with sealed joints) then the mass per unit area of the separating
wall above the ceiling may be reduced to $150\,kg/m^2$.

If lightweight aggregate blocks of density less than $1200\,kg/$ E2.58
m^3 are used above ceiling level, then one side should be   E2.94
sealed with cement paint or plaster skim.   E2.138

Where the external wall is a cavity wall with a masonry inner   E3.105
leaf (or a simple cavity masonry wall or masonry between   E3.125
independent panels), the ceiling should be taken through to   E3.125
the masonry.

Where a window head is near to the existing ceiling, the new   E4.29
independent ceiling may be raised to form a pelmet recess.

A rigid or direct connection should not be created between
the independent ceiling and the floor base.

Ceiling joists

Softwood timber used for roof construction or fixed in   A1/2 2B2
the roof space (including ceiling joists within the void
spaces of the roof) should be adequately treated to prevent
infestation by the house longhorn beetle (*Hylotrupes
bajulus* L.), particularly in the following areas:

- the Borough of Bracknell Forest, in the parishes of Sandhurst and Crowthorne;
- the Borough of Elmbridge;
- the District of Hart, in the parishes of Hawley and Yateley;
- the District of Runnymede;
- the Borough of Spelthorne;
- the Borough of Surrey Heath;
- the Borough of Rushmoor, in the area of the former district of Farnborough;
- the Borough of Woking.

Note:

(1)   Guidance on suitable preservative treatments is given within the *British Wood Preserving and Damp-Proofing Association's Manual* (2000 revision), available from 1 Gleneagles House, Vernongate, South Street, Derby DE1 1UP.

(2)   Guidance on the sizing of certain members in floors and roofs is given in BS 5268: Part 2:2002 and Part 3:1998 as *Span Tables for Solid Timber Members in Floors, Ceilings and Roofs (Excluding Trussed Rafter Roofs) for Dwellings*, published by TRADA, available from Chiltern House, Stocking Lane, Hughenden Valley, High Wycombe, Bucks HP14 4ND.

Air-circulation systems

Transfer grilles of air-circulation systems should **not** be fitted in any ceiling enclosing a protected stairway.	B1 2.17 (V1)
	B1 2.18 (V2)
	B3 7.10 (V1)
	B3 10.2 (V2)

Ceiling linings

To inhibit the spread of fire within the building, ceiling internal linings shall:

- adequately resist the spread of flame over their surfaces; and
- have, if ignited, a rate of heat release or a rate of fire growth that is reasonable in the circumstances.

Note: Flame spread over wall or ceiling surfaces is controlled by ensuring that the lining materials or products meet given performance levels that are measured in terms of performance with reference to Tables A1 and A3 of Part B of the Regulations.

For the purpose of this requirement, the content of ceilings is as described in Table 6.52.

Table 6.52 Content of ceiling linings

Include	Do not include
• The surface of glazing • Any part of a wall which slopes at an angle of 70° or less to the horizontal • The underside of a gallery • The underside of a roof exposed to the room below	• Trap doors and their frames • The frames of windows or rooflights and frames in which glazing is fitted • Architraves, cover moulds, picture rails • Exposed beams and similar narrow members

Classification of linings

In general terms (but see paragraphs 3.2 to 3.14 of Part B V1 and 6.2 to 6.14 of Part B V2 for more details), the surface linings for ceilings should meet the classifications given in Table 6.53.

Table 6.53 Classification of linings

Location	National Class	European Class
Small rooms less than 4 m²	3	D-s3, d2
Domestic garages less than 40 m²	3	D-s3, d2
Other rooms (including garages)	1	C-s3, d2
Circulation spaces within dwelling houses	1	C-s3, d2

Fire-resisting ceilings

The need for cavity barriers in some concealed floor or roof spaces can be reduced by using a fire-resisting ceiling below the cavity.	B2 3.6 (V1) B2 6.6 (V2)

Heat alarms

Heat detectors and heat alarms should:

• be designed and installed in accordance with BS 5839–6:2004;	B1 1.10
• be sited so that the sensor in ceiling-mounted devices is between 25 mm and 150 mm below the ceiling;	B1 1.15c
• be mains-operated and conform to BS 5446–2:2003; and	B1 1.4 (V1)
• have a standby power supply such as a rechargeable (or non-rechargeable) battery.	B1 1.5 (V2)

Lighting diffusers

Thermoplastic lighting diffusers (i.e. translucent or open-structured elements that allow light to pass through) should not be used in fire-protecting or fire-resisting ceilings, unless they have been satisfactorily tested as part of the ceiling system that is to be used to provide the appropriate fire protection.	B2 3.12 (V1) B2 6.14 (V2)

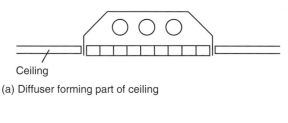

Ceiling

(a) Diffuser forming part of ceiling

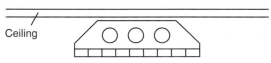

Ceiling

(b) Diffuser in fitting below and not forming part of ceiling

Figure 6.121 Lighting diffuser in relation to ceiling.

Rooflights

Rooflights should meet the relevant classification in Table 6.53.	B2 3.7 (V1) B2 6.7 (V2)
Rooflights may be constructed of a thermoplastic material if:	B2 3.10 (V1)
• the lower surface has a TPI(a) (rigid) or a TP(b) classification; • the size and disposition of the rooflights accords with the limits in Table 6.54.	B2 6.12 (V2)

Smoke alarms

Smoke alarm systems should be ceiling-mounted and at least 300 mm from walls and light fittings.	B1 1.15b (V1) B1 1.14b (V2)

> The sensor in ceiling-mounted devices shall be
> between 25 mm and 600 mm below the ceiling (25–
> 150 mm in the case of heat detectors or heat alarms).
>
> B1 1.15c (V1)
> B1 1.14c (V2)

Table 6.54 Limitations applied to thermoplastic rooflights and lighting diffusers in suspended ceilings and class 3 plastic rooflights

Minimum classification of lower surface	Use of space below the diffusers or rooflight	Maximum area of each diffuser panel or rooflight	Maximum total area of diffuser panels and rooflights as % of floor area of the space in which the ceiling is located	Minimum separation distance between diffuser panels or rooflights
TP(a)	Any (except a protected stairway)	No limit	No limit	No limit
Class 3 or TP(b)	Rooms	5	50	3
	Circulation spaces (except protected stairways)	5	15	3

Suspended ceilings

A suspended, fire-resisting ceiling should meet the requirements given in Table 6.55.

Table 6.55 Limitations on fire-resisting suspended ceilings

Height of building or separated part	Type of floor	Provision for fire resistance of floor (minutes)
Less than 18	Not compartment	60 or less
	Compartment	Less than 60
18 or more	Any	60 or less
No limit	Any	More than 60

for further details see Part B, Appendix A, Table A3.

Suspended or stretched-skin ceilings

> The ceiling of a room may be constructed from panels
> of a thermoplastic material of the TP(d) flexible
> classification, provided that it is not part of a fire-
> resisting ceiling.
>
> B2 3.14 (V1)
> B2 6.16 (V2)

Each panel should not exceed $5\,m^2$ in area and should
be supported on all of its sides.

Thermoplastic materials

Thermoplastic materials may be used in windows, rooflights and lighting diffusers in suspended ceilings.	B2 3.8 (V1) B2 6.10 (V2)
Flexible thermoplastic material may be used in panels to form a suspended ceiling.	B2 3.8 (V1) B2 6.10 (V2)
Flexible membranes covering a structure shall be in accordance with Appendix A of BS 7157:1989.	B2 6.8 (V2)

Venting of heat and smoke from basements

Smoke outlets should be sited at high level, either in the ceiling or in the wall of the space they serve.	B5 18.3 (V2) B5 18.5 (V2) B5 18.7 (V2)

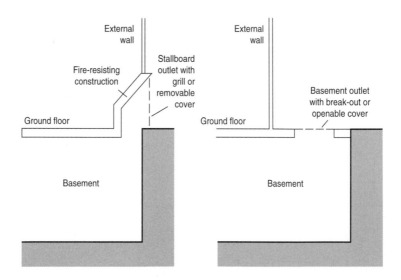

Figure 6.122 Fire-resisting construction for smoke outlet shafts.

6.10 Roofs

The roof of a brick-built house is normally an aitched (sloping) roof comprising rafters fixed to a ridge board, braced by purlins, struts and ties and fixed to
wall plates bedded on top of the walls. They are then usually clad with slates or
tiles to keep the rain out.

Timber-framed houses usually have trussed roofs – prefabricated triangu-
lated frames that combine the rafters and ceiling joists – which are lifted into
place and supported by the rails. The trusses are joined together with horizontal
and diagonal ties. A ridge board is not fitted, nor are purlins required. Roofing-
felt battens and tiling are applied in the usual way.

6.10.1 Requirements

Structure

*The building shall be constructed so that the combined dead, imposed and wind
loads are sustained and transmitted by it to the ground.*

(Approved Document A1)

*As a fire precaution, all materials used for internal linings of a building should
have a low rate of surface flame spread and (in some cases) a low rate of heat
release.*

(Approved Document B2)

*The roof of the building shall be resistant to the penetration of moisture from
rain or snow to the inside of the building.*

*All floors next to the ground, walls and roof shall not be damaged by moisture
from the ground, rain or snow and shall not carry that moisture to any part of
the building that it would damage.*

(Approved Document C2)

*Rainwater from roofs shall be carried away from the surface either by a drain-
age system or by other means.*

*The rainwater drainage system shall carry the flow of rainwater from the roof
to an outfall (e.g. a soakaway, a watercourse, a surface water or combined
sewer).*

(Approved Document H3)

External fire spread

- *The roof shall be constructed so that the risk of spread of flame and/or fire
  penetration from an external fire source is restricted.*
- *The risk of a fire spreading from the building to a building beyond the
  boundary, or vice versa shall be limited.*

(Approved Document B4)

Internal fire spread (structure)

Ideally the building should be sub-divided by elements of fire-resisting construction into compartments.

- *All openings in fire-separating elements shall be suitably protected in order to maintain the integrity of the continuity of the fire separation.*
- *Any hidden voids in the construction shall be sealed and sub-divided to inhibit the unseen spread of fire and products of combustion, in order to reduce the risk of structural failure, and the spread of fire.*

(Approved Document B3)

Ventilation

There shall be adequate means of ventilation provided for people in the building.

(Approved Document F)

 A new Part F *Conservation of Fuel and Power* came into force in October 2010.

Conservation of fuel and power

Reasonable provision shall be made for the conservation of fuel and power in buildings by:

(a) *limiting heat gains and losses-*

  (i) *through thermal elements and other parts of the building fabric; and*
  (ii) *from pipes, ducts and vessels used for space heating, space cooling and hot water services;*

(b) *providing fixed building services which-*

  (i) *are energy efficient;*
  (ii) *have effective controls; and*
  (iii) *are commissioned by testing and adjusting as necessary to ensure they use no more fuel and power than is reasonable in the circumstances; and*

(c) *providing to the owner sufficient information about the building, the fixed building services and their maintenance requirements so that the building can be operated in such a manner as to use no more fuel and power than is reasonable in the circumstances*

(Approved Document L)

 A new Part L came into force on 1 October 2010.

Safety

Pedestrian guarding should be provided for any roof to which people have access.

(Approved Document K2)

6.10.2 Meeting the requirements

Precipitation

Roofs should:	C6.2a
• resist the penetration of precipitation to the inside of the building;	
• not be damaged by precipitation;	C6.2b
• not carry precipitation to any part of the building that would be damaged by it;	C6.2b
• be designed and constructed so that their structural and thermal performance are not adversely affected by interstitial condensation.	C6.2c

Resistance to moisture from the outside

Roofs should be designed so as to protect the building from precipitation either by holding the precipitation at the face of the roof or by stopping it from penetrating beyond the back of the roofing system.	C6.3
Roofs that are jointless or have sealed joints should be impervious to moisture.	C6.4a
Roofs that have overlapping dry joints should be weather resistant and backed by a material (such as roofing felt) to direct any precipitation that does enter the roof towards the outer face.	C6.4b
Materials that can deteriorate rapidly without special care should only be used as the weather-resisting part of a roof.	C6.5
Weather-resisting parts of a roofing system shall not include paint or include any coating, surfacing or rendering that will not itself provide all the weather resistance.	C6.5
Roofing systems may be:	C6.6a
• impervious – such as metal, plastic and bituminous products;	

- weather-resistant – such as natural stone or slate, cement-based products, fired clay and wood;   C6.6b

- moisture-resisting – such as bituminous and plastic products lapped at the joints;   C6.6c

- jointless materials and sealed joints – that would allow for structural and thermal movement.   C6.6d

Dry joints between roofing sheets should be designed so that precipitation will not pass through them.   C6.7

Any precipitation that does enter a joint shall be drained away without penetrating beyond the back of the roofing system.   C6.7

Each sheet, tile and section of roof should be fixed in accordance with the guidance contained in BS 8000–6:1990.   C6.8

Resistance to damage from interstitial condensation

Roofs shall be designed and constructed in accordance with Clause 8.4 of BS 5250:2002 and BS EN ISO 13788:2001.   C6.10

Cold deck roofs (i.e. those roofs where the moisture from the building can permeate the insulation) shall be ventilated.   C6.11

Any parts of a roof that have a pitch of 70° or more shall be insulated as though it were a wall.   C6.11

All gaps and penetrations for pipes and electrical wiring should be filled and sealed to avoid excessive moisture transfer to roof voids.   C6.12

An effective draught seal should be provided to loft hatches to reduce inflow of warm air and moisture.   C6.12

Specialist advice should be sought when designing swimming pools and other buildings where interstitial condensation in the walls (caused by high internal temperatures and humidities) can cause high levels of moisture being generated.   C6.13

Resistance surface condensation and mould growth

Roofs shall be designed and constructed so that the: C6.14a

- thermal transmittance (U-value) does not exceed $0.35 \, \text{W/m}^2 \, \text{K}$ at any point;

- junctions between elements and the details of openings (such as windows) are in accordance with the recommendations in the report on robust construction details. C6.14b

Building height

For residential buildings, the maximum height of the building measured from the lowest finished ground level adjoining the building to the highest point of any roof should not be greater than 15 m. A1/2 2C4i

General

Roofs shall be constructed so that they: A1/2 1A2d

- provide local support to the walls;
- act as horizontal diaphragms capable of transferring the wind forces to buttressing elements of the building.

Note: A traditional cut timber roof (i.e. using rafters, purlins and ceiling joists) generally has sufficient built-in resistance to instability and wind forces. However, the need for diagonal rafter bracing equivalent to that recommended in BS 5268: Part 3:1998 or Annex H of BS 8103: Part 3:1996 for trussed rafter roofs, should be considered especially for single-hipped and non-hipped roofs of greater than 40° pitch to detached houses.

Roofs should: A1/2 2C33a

- act to transfer lateral forces from walls to buttressing walls, piers or chimneys;
- be secured to the supported wall. A1/2 2C33b

The roof shall be braced (in accordance with BS 5268: Part 3): A1/2 2C38(i)h

- at rafter level;
- horizontally at eaves level;
- at the base of any gable by roof decking, rigid sarking or diagonal timber bracing (as appropriate).

Vertical strapping may be omitted if the roof: A1/2 2C36a–d

- has a pitch of 15° or more; and
- is tiled or slated; and
- is of a type known by local experience to be resistant to wind gusts; and
- has main timber members spanning onto the supported wall at not more than 1.2 m centres.

Gable walls should be strapped to roofs as shown in A1/2 2C36
Figure 6.123(a) and (b) by tension straps.

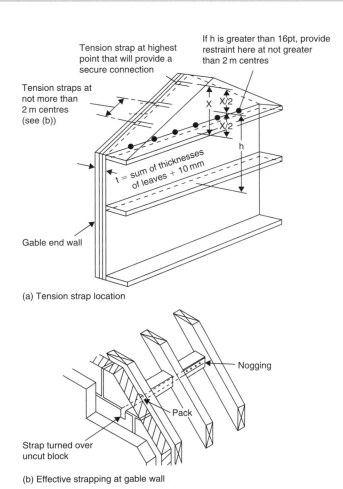

(a) Tension strap location

(b) Effective strapping at gable wall

Figure 6.123 Lateral support at roof level.

Walls shall be tied to the roof structure vertically and horizontally and have a horizontal lateral restraint at roof level.   A1/2 2C38(i)i

Wall ties should also be provided, spaced not more than 300 mm apart vertically, within a distance of 225 mm from the vertical edges of all roof verges.   A1/2 2C8

Walls shall be tied to the roof structure vertically and horizontally and have a horizontal lateral restraint at roof level.   A1/2 2C38(i)i

Walls should be tied horizontally at no more than 2 m centres to the roof structure at eaves level, base of gables and along roof slopes (as shown in Figure 6.124) with straps.   A1/2 2C38(iv)

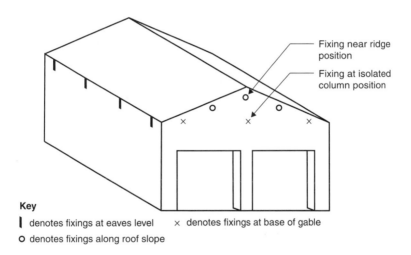

Fixing near ridge position

Fixing at isolated column position

Key

❙ denotes fixings at eaves level   × denotes fixings at base of gable

○ denotes fixings along roof slope

Figure 6.124 Lateral restraint at roof level.

Isolated columns should also be tied to the roof structure (see Figure 6.124).   A1/2 2C38(iv)

The roof structure of an annexe shall be secured to the structure of the main building at both rafter and eaves level.   A1/2 2C38(1)j

Access to the roof shall only be for the purposes of maintenance and repair.   A1/2 2C38(1)d

Timber

Softwood timber used for roof construction or fixed in the   A1/2 2B2
roof space (including ceiling joists within the void spaces
of the roof), should be adequately treated to prevent
infestation by the house longhorn beetle (*Hylotrupes
bajulus* L.), particularly in the following areas:

- the Borough of Bracknell Forest, in the parishes of
  Sandhurst and Crowthorne;
- the Borough of Elmbridge;
- the District of Hart, in the parishes of Hawley and
  Yateley;
- the District of Runnymede;
- the Borough of Spelthorne;
- the Borough of Surrey Heath;
- the Borough of Rushmoor, in the area of the former
  district of Farnborough;
- the Borough of Woking.

 Note:

(1)   Guidance on suitable preservative treatments is given in the British
      Wood Preserving and Damp-Proofing Association's Manual (2000
      revision), available from 1 Gleneagles House, Vernongate, South
      Street, Derby DE1 1UP.
(2)   Guidance on the sizing of roof members is given in BS 5268: Part
      2:2002 and Part 3:1998, as well as *Span Tables for Solid Timber
      Members in Floors, Ceilings and Roofs (Excluding Trussed Rafter
      Roofs) for Dwellings*, published by TRADA (available from Chiltern
      House, Stocking Lane, Hughenden Valley, High Wycombe, Bucks
      HP14 4ND).

Openings

Where an opening in a roof for a stairway adjoins a   A1/2 2C37a
supported wall and interrupts the continuity of lateral
support:

- the maximum permitted length of the opening
  is to be 3 m, measured parallel to the supported
  wall;

- connections (if provided by means other than by anchor) should be throughout the length of each portion of the wall situated on each side of the opening; A1/2 2C37b

- connections via mild steel anchors should be spaced closer than 2 m on each side of the opening to provide the same number of anchors as if there were no opening; A1/2 2C37c

- there should be no other interruption of lateral support. A1/2 2C37d

Means of escape

A flat roof being used as a means of escape should: C1

- be part of the same building from which escape is being made;
- lead to a storey exit or external escape route;
- provide 30 minutes of fire resistance.

Where a balcony or flat roof is provided for escape purposes, guarding may be needed (see also Approved Document K, *Protection from Falling, Collision and Impact*).

Construction of escape stairs

If an escape route is over a flat roof: B1 2.10 (V1)

- the roof should be part of the same building from which escape is being made; B1 2.7 (V2)

- the route across the roof should lead to a storey exit or external escape route; B1 5.35 (V2)

- the part of the roof forming the escape route, its supporting structure, together with any opening within 3 m of the escape route, should provide 30 minutes of fire resistance (see Appendix A, Table A1 of Approved Document B for fire resistance figures for elements of structure, etc.);

- the route should be adequately defined and guarded by walls and/or protective barriers (that meet the provisions of Part K). B1 2.11 (V1)

Note: Part K (Protection from falling, collision and impact) specifies a minimum guarding height of 800 mm, except in the case of a window in a roof where the bottom of the opening may be 600 mm above the floor.

An escape over a flat roof is permissible if:	B1 2.31 (V2)
• more than one escape route is available from a storey, or part of a building;	
• the route does not serve as an institutional building;	
• part of a building is intended for use by members of the public;	
• the roof is fire-resistant in accordance with Tables A1 and A2 of Appendix A to Part B;	B1 5.3 (V2)
• the exit from a flat etc. is remote from the main entrance door to that flat.	B1 2.17 (V2)

Provision of refuges

Refuges are relatively safe waiting areas for short periods. They are **not** areas where disabled people should be left alone indefinitely until rescued by the fire and rescue service, or until the fire is extinguished.

A refuge such as a flat roof (balcony, podium or similar compartment, protected lobby, protected corridor or protected stairway) should be provided for each protected stairway.	B1 4.8 (V2)

Note: The number of refuge spaces need not necessarily equal the number of wheelchair users who can be present in the building.

Compartmentation

To prevent the spread of fire within a building, whenever possible the building should be subdivided into compartments separated from one another by walls and/or floors of fire-resisting construction.	B3 5.1 (V1) B3 8.1 (V2)
Compartment walls in a top storey beneath a roof should be continued through the roof space.	B3 5.8 (V1) B3 8.24 (V2)
When a compartment wall meets the underside of the roof covering or deck, the wall/roof junction shall maintain continuity of fire resistance.	B3 8.28 (V2)

Double-skinned insulated roof sheeting should
incorporate a band of material of limited
combustibility.

B3 8.29 (V2)

Compartment walls between buildings

If a fire penetrates a roof near a compartment wall
there is a risk that it will spread over the roof to the
adjoining compartment. To reduce this risk either:

B3 8.29 (V2)
B3 8.30 (V2)

- the wall should be extended up through the roof for
  a height of at least 375 mm above the top surface of
  the adjoining roof covering; or
- a 1500 mm wide zone on either side of the wall
  should have a suitable covering (Figure 6.125).

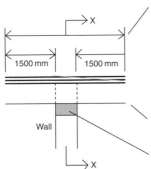

Roof covering to be designated AA, AB or AC for at least this distance.

Boarding (used as a substrate), wood-wool slabs or timber tiling battens may be carried over the wall provided that they are fully bedded in mortar (or other suitable material).

Thermoplastic insulation materials should not be carried over the wall.

Double-skinned insulated roof sheeting with a thermoplastic core should incorporate a band of material of limited combustibility at least 300 mm wide centred over the wall.

Sarking felt may also be carried over the wall.

If roof support members pass through the wall, fire protection to these members for a distance of 1500 mm on either side of the wall may be needed to delay distortion at the junction (see note to paragraph 8.20).

Fire-stopping to be carried up to underside of roof covering, boarding or slab.

Figure 6.125 Junction of compartment wall with roof.

Concealed spaces (cavities)

Concealed spaces or cavities in roofs will provide an easy route for smoke and
flame spread, which, because it is obscured, will present a greater danger than
would a more obvious weakness in the fabric of the building. For this reason:

Cavity barriers need not be provided between double-
skinned corrugated or profiled insulated roof sheeting,
if the sheeting is a material of limited combustibility.

B3 6.4 (V1)
B3 9.5 (V2)

Note: Separate rules exist for bedrooms in institutional and other residential
buildings (see B3 9.7 (V2)).

Roof covering

The re-covering of roofs is commonly undertaken to extend the useful life of buildings; however, roof structures may be required to carry under drawing or insulation provided at a time later than their initial construction.

All materials used to cover roofs (including transparent or translucent materials, but excluding windows of glass in residential buildings with roof pitches of not less than 15°) shall be capable of safely withstanding the concentrated imposed loads upon roofs specified in BS 6399: Part 3.	A1/2 4.1
Where the work involves a significant change in the applied loading, the structural integrity of the roof structure and the supporting structure should be checked to ensure that upon completion of the work the building is not less compliant.	A1/2 4.3

Note: Re-covering roofs is commonly undertaken to extend the useful life of buildings (e.g. roof structures may be required to be insulated at a later date).

Where such checking of the existing roof structure indicates that the construction is unable to sustain any proposed increase in loading (e.g. due to overstressed members or unacceptable deflection leading to ponding), appropriate strengthening work or replacement of roofing members should be undertaken.	A1/2 4.5

This is classified as a *material alteration*.

Where work will significantly decrease the roof dead loading, the roof structure and its anchorage to the supporting structure should be checked to ensure that an adequate factor of safety is maintained against uplift of the roof under imposed wind loading.	A1/2 4.7

Note: A significant change in roof loading is when the loading upon the roof is increased by more than 15 per cent.

Plastic rooflights should have a minimum of class 3 lower surface.	B4 10.6 (V1) B4 14.6 (V2)
When used in rooflights, unwired glass shall be at least 4 mm thick and shall be AA designated (see Table 6.56).	B4 10.8 (V1) B4 14.8 (V2)

Thatch and wood shingles should be regarded as having an AD/BD/CD designation (see Table 6.56).

B4 10.9 (V1)
B4 14.9 (V2)

Table 6.56 Limitations on roof coverings

Designation of roof covering	Minimum distance from any point on relevant boundary			
	Less than 6 m	At least 6 m	At least 12 m	At least 20 m
AA, AB or AC	Acceptable	Acceptable	Acceptable	Acceptable
BA, BB or BC	Not acceptable	Acceptable	Acceptable	Acceptable
CA, CB or CC	Not acceptable	Acceptable	Acceptable	Acceptable
AD, BD or CD	Not acceptable	Acceptable	Acceptable	Acceptable
DA, DB, DC or DD	Not acceptable	Not acceptable	Not acceptable	Acceptable

Separation distances (i.e. the minimum distance from the roof, or part of the roof, to the relevant or notional boundary) shall be in accordance with Part B, Volume 1, Table 5 according to the type of roof covering and the size and use of the building.

B4 10.5 (V1)
B4 14.5 (V2)

A rigid thermoplastic sheet product made from polycarbonate or from unplasticized PVC may be used in rooflights.

B4 10.7 (V1)
B4 14.7 (V2)

Unwired glass at least 4 mm thick may be used in rooflights.

B4 10.8 (V1)
B4 14.8 (V2)

Note: See Part B, Volume 1, Table 5 for limitations on roof coverings and BSEN 13501–5:2005 for guidance on roof coverings (which has been updated to incorporate the new European classification system).

Rooflights

Rooflights should meet the relevant classification in Table 6.57.

B2 3.7 (V1)
B2 6.7 (V2)

Table 6.57 Classification of linings

Location	National Class	European Class
Small rooms less than 4 m²	3	D-s3, d2
Domestic garages less than 40 m²	3	D-s3, d2
Other rooms (including garages)	1	C-s3, d2
Circulation spaces within dwelling houses	1	C-s3, d2

Thermoplastic materials may be used in windows, rooflights and lighting diffusers in suspended ceilings.	B2 3.8 (V1) B2 6.10 (V2)
Rooflights may be constructed of a thermoplastic material if:	B2 3.10 (V1)
• the lower surface has a TPI(a) (rigid) or a TP(b) classification;	B2 6.12 (V2)
• the size and disposition of the rooflights accords with the limits in Table 6.58.	

Table 6.58 Limitations applied to thermoplastic rooflights and lighting diffusers in suspended ceilings and class 3 plastic rooflights

Minimum classification of lower surface	Use of space below the diffusers or rooflight	Maximum area of each diffuser panel or rooflight	Maximum total area of diffuser panels and rooflights as % of floor area of the space in which the ceiling is located	Minimum separation distance between diffuser panels or rooflights
TP(a)	Any (except a protected stairway)	No limit	No limit	No limit
Class 3 or TP(b)	Rooms	5	50	3
	Circulation spaces (except protected stairways)	5	15	3

Thatched roofs

Thatched roofs can sometimes be vulnerable to spontaneous combustion caused by heat transferred from flues building up in thick layers of thatch in contact with the chimney. To reduce the risk it is recommended that rigid twin-walled insulated metal flue liners be used within a ventilated (top and bottom) masonry chimney void provided they are adequately supported and not in direct contact with the masonry. Non-metallic chimneys and cast in situ flue liners can also be used provided the heat transfer to the thatch is assessed in relation to the depth of thatch and risk of spontaneous combustion.

Spark arrestors are not generally recommended as they can be difficult to maintain and may increase the risk of flue blockage and flue fires.

 Further information and recommendations are contained in HETAS Information Paper 1/007 *Chimneys in Thatched Properties.*

In thatched roofs:

• the rafters should be overdrawn with construction having not less than 30 minutes of fire resistance;	B4 10.9 (V1)
• smoke alarms should be installed in the roof space.	B4 14.9 (V2)

Fire resistance

The need for cavity barriers in some concealed roof spaces can be reduced by using a fire-resisting ceiling below the cavity.	B2 3.6 (V1) B2 6.6 (V2)
Where the conversion of an existing roof space (such as a loft conversion to a two-storey house) means that a new storey is going to be added, then the stairway will need to be protected with fire-resisting doors and partitions.	B1 2.20b

Pitched roofs covered with slates or tiles

Table 6.59 Pitched roofs – slates or tiles

Covering material	Supporting structure	Designation
1. Natural slates 2. Fibre reinforced 3. Clay tiles 4. Concrete tiles	Timber rafters with or without underfelt, sarking, boarding, wood-wool slabs, compressed straw slabs, plywood, wood chipboard, or fibre cement slates insulating board	AA

 Note: Although the table does not include guidance for pitched roofs covered with bitumen felt, it should be noted that there is a wide range of materials on the market, and information on specific products is readily available from manufacturers.

Pitched roofs covered with self-supporting sheet

Flat roofs with bitumen felt

A flat roof consisting of bitumen felt should (irrespective of the felt specification) be deemed to be of designation AA if the felt is laid on a deck constructed of 6 mm plywood, 12.5 mm wood chipboard, 16 mm (finished) plain-edged timber boarding, compressed straw slab, screeded wood-wool slab, profiled fibre-reinforced cement or steel deck (single or double skin) with or without fibre insulating board overlay, profiled aluminium deck (single or double skin) with or without fibre insulating board overlay, or concrete or clay-pot slab (in situ or pre-cast), and has a surface finish of:

Table 6.60 Pitched roofs – self-supporting sheet

Covering material	Construction	Supporting structure	Designation
Profiled sheet of galvanized steel, aluminium, fibre-reinforced cement, or pre-painted (coil-coated) steel or aluminium with a PVC or PVF2 coating	Single skin without underlay, or with underlay or plasterboard, or wood-wool slab	Structure of timber, steel or concrete	AA
Profiled sheet of galvanized steel, aluminium, fibre-reinforced cement, or pre-painted (coil-coated) steel or aluminium with a PVC or PVF2 coating	Double skin without interlayer, or with interlayer of resin bonded or concrete glass fibre, mineral-wool slab, polystyrene, or polyurethane	Structure of timber, steel or concrete	AA

- bitumen-bedded stone chippings covering the whole surface to a depth of at least 12.5 mm;
- bitumen-bedded tiles of a non-combustible material;
- sand and cement screed; or
- tarmacadam.

Pitched or flat roofs covered with fully supported material

Table 6.61 Pitched roofs – fully supported material

Covering material	Supporting structure	Designation
1. Aluminium sheet 2. Copper sheet 3. Zinc sheet 4. Lead sheet 5. Mastic asphalt 6. Vitreous enamelled steel 7. Lead/tin alloy-coated steel sheet 8. Zinc/aluminium alloy-coated steel sheet 9. Pre-painted (coil-coated) steel sheet including liquid-applied PVC coatings	Timber joists and tongued and grooved boarding, or plain-edged boarding	AA
	Steel or timber joists with deck of wood-wool slabs, compressed-straw slab, wood chipboard, fibre insulating board, or 9.5 mm plywood	AA
	Concrete or clay-pot slab (in situ or pre-cast) or non-combustible deck of steel, aluminium, or fibre cement (with or without insulation)	AA

 Note: Lead sheet supported by timber joists and plain edged boarding should be regarded as having a BA designation.

Rating of material and products

Table 6.62 Typical performance ratings of some generic materials and products

Rating	Material or product
Class 0	1. Any non-combustible material or material of limited combustibility
	2. Brickwork, blockwork, concrete and ceramic tiles
	3. Plasterboard (painted or not with a PVC facing not more than 0.5 mm thick) with or without an air gap or fibrous or cellular insulating material behind
	4. Wood-wool cement slabs
	5. Mineral-fibre tiles or sheets with cement or resin binding
Class 3	6. Timber or plywood with a density more than 400 kg/ml, painted or unpainted
	7. Wood-particle board or hardboard, either untreated or painted
	8. Standard glass reinforced polyesters

Roof with a pitch of 15° or more

- Pitched roof spaces should have ventilation openings at least 10 mm wide at eaves level to promote cross-ventilation.
- A pitched roof that has a single slope and abuts a wall should have ventilation openings at eaves level at least 10 mm wide and at high level (i.e. at the junction of the roof and the wall) at least 5 mm wide.

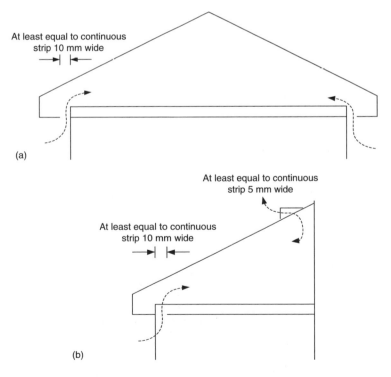

Figure 6.126 Ventilating roof voids. (a) Pitched roof. (b) Lean-to roof.

Roof with a pitch of less than 15°

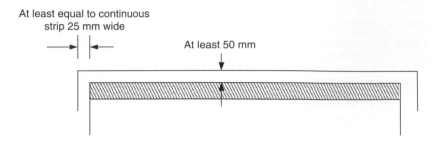

Figure 6.127 Ventilating roof void – flat roof.

- Roof spaces should have ventilation openings at least 25 mm wide in two opposite sides to promote cross-ventilation.
- The void should have a free air space of at least 50 mm between the roof deck and the insulation.
- Pitched roofs where the insulation follows the pitch of the roof need ventilation at the ridge at least 5 mm wide.
- Where the edges of the roof abut a wall or other obstruction in such a way that free air paths cannot be formed to promote cross-ventilation or the movement of air outside any ventilation openings would be restricted, an alternative form of roof construction should be adopted.

 Roofs with a span exceeding 10 m may require more ventilation, totalling 0.6 per cent of the roof area.

Ventilation openings may be continuous or distributed along the full length and may be fitted with a screen, facia, baffle, etc.

Where necessary (i.e. for the purposes of health and safety), ventilation to small roofs such as those over porches and bay windows, should always be provided and a roof that has a pitch of 70° or more shall be insulated as though it were a wall.

 If the ceiling of a room follows the pitch of the roof, ventilation should be provided as if it were a flat roof.

Passive stack ventilation

In roof spaces: F Table 5.2a to 5.2d

- ducts should, ideally, be secured to a wooden strut that is securely fixed at both ends;
- flexible ducts should be allowed to curve gently at each end of the strut;

- ducts should be insulated with at least 25 mm
  of a material having a thermal conductivity
  of 0.04 W/mK.

For stability, rigid ducts should be used for any outside part of the PSV system that is above the roof slope and to provide stability, it should project down into the roof space far enough to allow firm support.	F Table 5.2a to 5.2d
If a duct penetrates the roof more than 0.5 m from the roof ridge, it must extend above the roof slope to at least the height of the roof ridge.	F Table 5.2a to 5.2d
If tile ventilators are used on the roof slope they must be positioned no more than 0.5 m from the roof ridge.	F Table 5.2a to 5.2d
If a duct extends above the roof level, then that section of the duct should be insulated or be fitted with a condensation trap just below roof level.	F Table 5.2a to 5.2d
Ducts should be securely fixed to the roof outlet terminal so that it cannot sag or become detached.	F Table 5.2a to 5.2d
Separate ducts shall be taken from the ceilings of wet rooms to separate terminals on the roof.	F Table 5.2a to 5.2d
Terminals should be designed such that any condensation forming inside it cannot run down into the dwelling but will run off onto the roof.	F Table 5.2a to 5.2d

 Note: Placing the outlet terminal at the ridge of the roof is the preferred option as it is not prone to wind gusts and/or certain wind directions.

Ceiling and roof junctions

Where a type-1 separating wall is used it should be continuous to the underside of the roof.	E2.55 E2.91 E2.133
The junction between the separating wall and the roof should be filled with a flexible closer which is also suitable as a fire stop.	E2.56 E2.92 E2.134

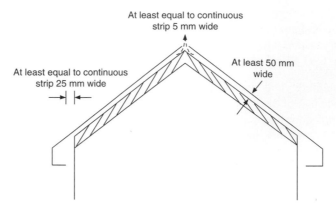

Figure 6.128 Ventilating roof void – ceiling following pitch of roof.

Wall type 1 – solid masonry

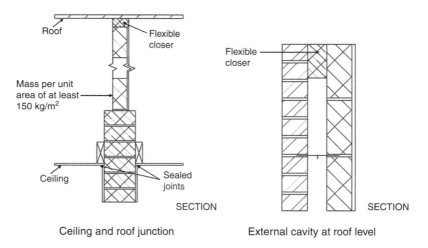

If lightweight aggregate blocks of density less than 1200 kg/ m³ are used above ceiling level, then one side should be sealed with cement, paint or plaster skim.	E2.58 E2.94 E2.138
Where the roof or loft space is not a habitable room (and there is a ceiling with a minimum mass per unit area of 10 kg/m² with sealed joints), the mass per unit area of the separating wall above the ceiling may be reduced to 150 kg/m².	E2.57
Where there is an external cavity wall, the cavity should be closed at eaves level with a suitable flexible material (e.g. mineral wool).	E2.59

A rigid connection between the inner and external wall leaves should be avoided.

Ep23

If a rigid material is used, then it should only be rigidly bonded to one leaf.

Ep23

Wall type 2 – cavity masonry

Where a type-2 separating wall is used, it should be continuous to the underside of the roof.

E2.91

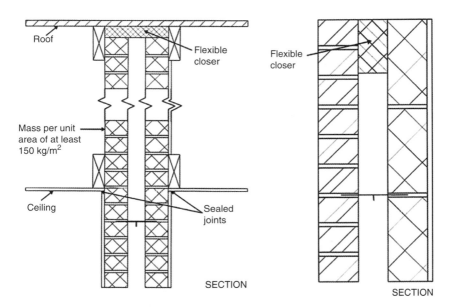

Wall type 2 – ceiling and roof junction   External cavity wall at eaves level

The junction between the separating wall and the roof should be filled with a flexible closer that is also suitable as a fire stop.

E2.92

If lightweight aggregate blocks (with a density less than $1200 \, \text{kg/m}^3$) are used above ceiling level, one side should be sealed with cement paint or plaster skim.

E2.94

The cavity of an external cavity wall should be closed at eaves level with a suitable flexible material (e.g. mineral wool).

E2.95

A rigid connection between the inner and external wall leaves should be avoided.

E2.95

If a rigid material has to be used, it should **only** be rigidly bonded to one leaf.	E2.95

Note: If the roof or loft space is not a habitable room (and there is a ceiling with a minimum mass per unit area of $10\,kg/m^2$ with sealed joints), the mass per unit area of the separating wall above the ceiling may be reduced to $150\,kg/m^2$– **but** it should still be a cavity wall.

Wall type 3 – masonry between independent panels

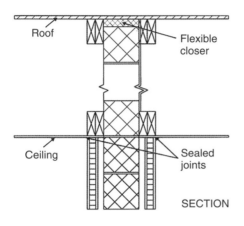

Wall types 3.1 and 3.2 – ceiling and roof junction

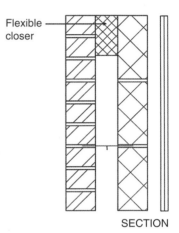

External cavity wall at eaves level

Where a type-3 separating wall is used, the masonry core should be continuous to the underside of the roof.	E2.133
The junction between the separating wall and the roof should be filled with a flexible closer that is also suitable as a fire stop.	E2.134
The junction between the ceiling and independent panels should be sealed with tape or caulked with sealant.	E2.135
If there is an external cavity wall, the cavity should be closed at eaves level with a suitable flexible material (e.g. mineral wool).	E2.136
Rigid connections between the inner and external wall leaves should be avoided where possible.	E2.136
If a rigid material is used, it should only be rigidly bonded to one leaf.	E2.136
For wall types 3.1 and 3.2 (solid masonry core):	

- if the roof or loft space is not a habitable room (and there is a ceiling with a minimum mass per unit area of 10 kg/m^2 and it has sealed joints), the independent panels may be omitted in the roof space and the mass per unit area of the separating wall above the ceiling may be a minimum of 150 kg/m^2; E2.137

- if lightweight aggregate blocks with a density less than 1200 kg/m^3 are used above ceiling level, one side should be sealed with cement paint or plaster skim. E2.138

For wall type 3.3 (cavity masonry core):

- if the roof or loft space is not a habitable room (and there is a ceiling with a minimum mass per unit area of 10 kg/m^2 and it has sealed joints), the independent panels may be omitted in the roof space, but the cavity masonry core should be maintained to the underside of the roof. E2.139

Wall type 4 – framed walls with absorbent material

Where a type-4 separating wall is used, the wall should preferably be continuous to the underside of the roof.	E2.160
The junction between the separating wall and the roof should be filled with a flexible closer.	E2.161
The junction between the ceiling and the wall linings should be sealed with tape or caulked with sealant.	E2.162
If the roof or loft space is not a habitable room (and there is a ceiling with a minimum mass per unit area of 10 kg/m^2 with sealed joints), then:	E2.162a

- either the linings on each frame may be reduced to two layers of plasterboard, each sheet with a minimum mass per unit area of 10 kg/m^2; or E2.162a

- the cavity may be closed at ceiling level without connecting the two frames rigidly together. E2.162b

External wall cavities should be closed at eaves level with suitable material.	E2.163a

Note: In which case there need only be one frame in the roof space provided there is a lining of two layers of plasterboard, each sheet of minimum mass per unit area 10 kg/m^2, on both sides of the frame.

6.10.3 Conversation of fuel and power

Note: See Annex B to this book for further information concerning the requirements for the conservation of fuel and power

Historic and traditional buildings

When undertaking work on any type of historic or traditional type of building, the aim should always be to improve energy efficiency as far as is reasonably practicable, without prejudicing the character of the host building or increasing the risk of long-term deterioration of the building fabric or fittings.

> In general, new extensions to historic or traditional buildings   L2B 3.11
> should comply with the standards of energy efficiency as
> set out in Part L:2010. The only exception would be where
> there is a particular need to match the external appearance or
> character of the extension to that of the host building.

Building inspectors normally require the use of '*sympathetic treatment*' when restoring the historic character of a building that has been subject to previous inappropriate alteration (e.g. replacing a roof or a rooflight).

   Particular issues that could warrant sympathetic treatment – and where advice from others would probably be beneficial – include:

> • Restoring the historic character of a building that has   L2B 3.11
>   been subject to previous inappropriate alteration, (e.g.
>   replacement roofs, rooflights, windows and doors, etc.);
> • rebuilding a former historic building (e.g. following a
>   fire or filling a gap site in a terrace); and
> • renovating the fabric of historic buildings to enable it
>   to '*breathe*' and to control any moisture and potential
>   long-term decay problems.

Extensions

Regulation 9 of the Building Regulations exempts some conservatory and porch extensions from the energy-efficiency requirements, **provided** that the fabric elements that separate an extension – such as a conservatory or porch – from the building are retained or, if removed, are replaced by fabrics which meet the energy-efficiency requirements and that the heating system of the building is not extended into the conservatory or porch.

Dwellings

Extensions to dwellings should either use newly constructed thermal elements or make use of the use of existing (or new) roof windows and rooflights, provided that they meet the requirements for the conservation of fuel and power.

If upgrades are proposed to the existing dwelling, such upgrades should be implemented to a standard that is no worse than that shown in column (b) of Table 6.63.

L1B 4.7

Table 6.63 Upgrading retained thermal elements

Element	(a) Threshold U-value (W/m^{2}K)	(b) Improved U-value (W/m^{2}K)
Pitched roof – insulation at ceiling level	0.35	0.16
Pitched roof – insulation between rafters	0.35	0.18
Flat roof or roof with integral insulation	0.35	0.18

Buildings other than dwellings

It is recommended that the total area of roof windows, windows and doors in extensions to existing buildings do not exceed the sum of:

- 25 per cent of the floor area of the extension; plus
- the total area of any roof windows, windows or doors which, as a result of the extension works, no longer exist – or are no longer exposed.

Fabric elements and fixed building services should satisfy minimum energy efficiency standards.

Table 6.64 sets out the worst acceptable standards for fabric properties.

L2A 4.29

Table 6.64 Limiting fabric parameters

Element	W/m^{2}K
Roof	0.25
Windows, roof windows, glazed rooflights, curtain walling and pedestrian doors	2.2
Roof ventilators (including smoke vents)	3.5
Air permeability	10.00 m^3/h m^2 at 50 Pa

> If a building extension is a conservatory or porch that is **not** exempt from the energy efficiency requirements, the effective thermal separation between the heated area in the existing building (i.e. the walls, doors and windows between the building and the extension) should be insulated and draught-proofed to at least the same extent as in the existing building.
>
> Replacement roof windows, lights and ventilators, etc., should have U-values that are no worse than those given in 6.65.

L2B 4.12

Table 6.65 Standards for controlled fittings

Fitting	Standard (W/m²K)
Windows, roof windows and glazed rooflights	1.8 for the whole unit
Plastic rooflight	1.8
Roof ventilators (including smoke extract ventilators)	3.5

 Removing and **not** replacing any of the thermal separation between the building and an existing exempt extension, or extending the building's heating system into the extension, means that the extension ceases to be exempt!

> The area of windows and rooflights in the extension should generally not exceed the percentage values given in Table 6.66.

L2B 4.4

Table 6.66 Opening areas in the extension

Building type	Windows, rooflights and personnel doors as % of exposed wall	Rooflights as % of area of roof
Residential buildings where people temporarily or permanently reside	30	20
Places of assembly, offices and shops	40	20
Industrial and storage buildings	15	20

6.11 Chimneys and fireplaces

The Clean Air Act 1993

Under the Clean Air Act 1993, local authorities may declare the whole or part of the district of the authority to be a smoke control area. It is an offence to emit smoke from a chimney of a building, from a furnace or from any fixed boiler if located in a designated smoke control area unless an authorized fuel is used. It is also an offence to acquire an *'unauthorized fuel'* for use within a smoke control area unless it is used in an *'exempt'* appliance (*'exempted'* from the controls which generally apply in the smoke control area).

Authorized fuels are fuels which are authorized by Statutory Instruments (Regulations) made under the Clean Air Act 1993. These include inherently smokeless fuels, such as gas, electricity and anthracite, together with specified brands of manufactured solid smokeless fuels. These fuels have passed tests to confirm that they are capable of burning in an open fireplace without producing smoke.

Exempt appliances are appliances (ovens, wood burners, boilers and stoves) which have been exempted by Statutory Instruments (Orders) under the Clean Air Act 1993. These have passed tests to confirm that they are capable of burning an unauthorized or inherently smoky solid fuel without emitting smoke. More information and details of authorized fuels and exempt appliances can be found at http://smokecontrol.defra.gov.uk.

6.11.1 The requirement (Building Act 1984 Section 73)

If a person erects or raises a building that is (or is going to be) taller than the chimneys and/or flues from an adjoining building that is either joined by a party wall or less than six feet away from the taller building, then the local authority may:

- if reasonably practical, require that person to build up those chimneys and flues, so that their top is of the same height as the top of the chimneys of the taller building or the top of the taller building, whichever is the higher;
- require the owner or occupier of the adjoining building to allow the person erecting or raising the building access to the adjacent building so that he can carry out such work as may be necessary to comply with the notice served on him.

The owner or occupier of the adjacent building is entitled to complete the work himself by (within 14 days) serving a 'counter-notice' that he has elected to carry out the work himself.

The building shall be constructed so that the combined dead, imposed and wind loads are sustained and transmitted by it to the ground:

- *safely;*
- *without causing such deflection or deformation of any part of the building (or such movement of the ground) as will impair the stability of any part of another building.*

(Approved Document A1)

Fire precautions (construction)

Any hidden voids in the construction shall be sealed and subdivided to inhibit the unseen spread of fire and products of combustion, in order to reduce the risk of structural failure, and the spread of fire.

(Approved Document B3)

Air supply

Combustion appliances shall be so installed that there is an adequate supply of air to them for combustion, to prevent overheating and for the efficient working of any flue.

(Approved Document J1)

Discharge of products of combustion

Combustion appliances shall have adequate provision for the discharge of products of combustion to the outside air.

(Approved Document J2)

Warning of release of carbon monoxide

Where a fixed combustion appliance is provided, appropriate provision shall be made to detect and give warning of the release of carbon monoxide.

(Approved Document J2A)

Protection of building

Combustion appliances and fluepipes shall be so installed, and fireplaces and chimneys shall be so constructed and installed, as to reduce to a reasonable level the risk of people suffering burns or the building catching fire in consequence of their use.

(Approved Document J3)

Provision of information

Where a hearth, fireplace, flue or chimney is provided or extended, a durable notice containing information on the performance capabilities of the hearth, fireplace, flue or chimney shall be affixed in a suitable place in the building for the purpose of enabling combustion appliances to be safely installed.

(Approved Document J4)

 Note: The 2010 edition of Approved Document J, *Combustion Appliances and Fuel Storage Systems* (which replaced the 2002 edition and came into force on 1 October 2010), includes:

- a new requirement (J2A) for installing a carbon monoxide alarm whenever a new or replacement fixed solid fuel appliance is installed;

- the requirement for liquid biofuel and blends of mineral oil and liquid bio-fuel to be recognized within the scope of combustion installations designed to burn oil;

and new guidance concerning:

- access for visual inspection of concealed flues when they an appliance is first commissioned and subsequently serviced;
- flue outlet clearances relative to adjacent pitched roofs;
- provision of hearths and wall clearances for solid fuel appliances;
- permanent ventilation openings for open-flued appliances in very airtight houses;
- oil tank secondary containment located in an inner protection zone as shown on the Environment Agency's groundwater sources map;

and an informative appendix (Appendix C) explaining the European designation system for certain flue and chimney products.

6.11.2 Meeting the requirement

End restraints

The ends of every wall (except single leaf walls less than 2.5 m in storey height and length) in small single-storey non-residential buildings and annexes should be bonded or otherwise securely tied throughout their full height to a buttressing wall, pier or chimney.	A1/2 (2C25)
Long walls may be provided with intermediate support, dividing the wall into distinct lengths; each distinct length is a supported wall for the purposes of this section.	A1/2 (2C25)
The buttressing wall, pier or chimney should provide support from the base to the full height of the wall.	A1/2 (2C25)
The sectional area, on plan, of chimneys (excluding openings for fireplaces and flues) should be not less than the area required for a pier in the same wall, and the overall thickness should not be less than twice the required thickness of the supported wall (Figure 6.129).	A1/2 (2C27b)
Floors and roofs should act to transfer lateral forces (see Table 6.67) from walls to buttressing walls, piers or chimneys and be secured to the supported wall as shown.	A1/2 (2C33a)

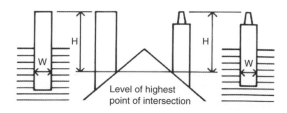

Figure 6.129 Proportions for masonry chimneys.

Floors and roofs should act to transfer lateral forces   A1/2 (2C33a)
(see Table 6.67) from walls to buttressing walls,
piers or chimneys and be secured to the supported
wall as shown.

Table 6.67 Lateral support for walls

Wall type	Wall length	Lateral support required
Solid or cavity: external compartment separating	Any length	Roof lateral support by every roof forming a junction with the supporting wall
	Greater than 3 m	Floor lateral support by every floor forming a junction with the supporting wall
Internal load-bearing wall (not being a compartment or separating wall)	Any length	Roof or floor lateral support at the top of each storey

Masonry chimneys

Where a chimney is not adequately supported by   1D1
ties or securely restrained in any way, its height H
(measured from the highest point of any chimney
pot or other flue terminal) should not exceed 4.5
times the width W (the least horizontal dimension
of the chimney measured at the same point of
intersection) – provided that the density of the
masonry is greater than 1500 kg/m^3.

The foundation of piers, buttresses and chimneys
should project as indicated in Figure 6.130 and the
projection X should never be less than P.

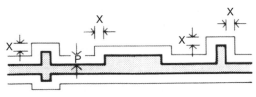

Projection X should not be less than P

Figure 6.130 Piers and chimneys.

Flues, etc.

Flue walls should have a fire resistance of at least one half of that required for the compartment wall or floor and be of non-combustible construction.	B3
If a flue (or duct containing flues and/or ventilation duct(s)) passes through a compartment wall, or is built into a compartment wall, each wall of the flue or duct should have a fire resistance of at least half that of the wall or floor (Figure 6.131).	B3 7.11 (V1) B3 10.16 (V2)

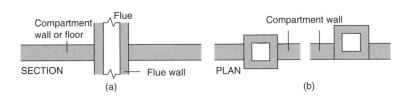

Figure 6.131 Flues penetrating compartment walls or floors. (a) Flue passing through compartment wall or floor. (b) Flue built into compartment wall.

Fire stopping

Proprietary fire-stopping and sealing systems (including those designed for service penetrations), have been shown by test to maintain the fire resistance of the wall or other element, are available and may be used. Other fire-stopping materials include:

- cement mortar;
- gypsum-based plaster;
- cement or gypsum-based vermiculite/perlite mixes;
- glass fibre, crushed rock, blast furnace slag or ceramic based products (with or without resin binders); and
- intumescent mastics.

Joints between fire-separating elements should be fire-stopped.	B3 7.12a (V1) B3 10.17a (V2)
All openings for pipes, ducts, conduits or cables to pass through any part of a fire-separating element should be:	B3 7.12b (V1) B3 10.17b (V2)
• kept as few in number as possible; • kept as small as practicable; • fire-stopped (which in the case of a pipe or duct, should allow thermal movement).	
Cables concealed in floors and walls (in certain circumstances) are required to have an earthed metal covering, be enclosed in steel conduit, or have additional mechanical protection (see BS 7671 for more information).	P App A 2d
To prevent displacement, materials used for fire-stopping should be reinforced with (or supported by) materials of limited combustibility.	B3 7.13 (V1) B3 10.18 (V2)

Chimney construction

Chimneys shall consist of a wall or walls enclosing one or more flues (Figure 6.132).	J (0.4–7)

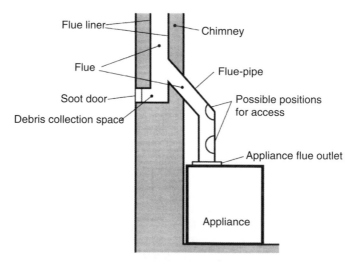

Figure 6.132 Chimneys and flues.

Note: In the gas industry, the chimney for a gas appliance is commonly called the flue.

Down-draughts that could interfere with the combustion performance of an open-flued appliance (Figure 6.133a) shall be minimized.   J (0.4–11)

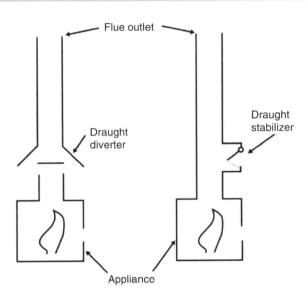

Figure 6.133 Draught diverters and draught stabilizers.

A factory-made draught stabilizer (see Figure 6.133b) is used to prevent excessive variations in the draught.   J (0.4–12)

Note: Chimneys and flues should also provide satisfactory control of water condensation.

Fireplaces

Fireplace recesses (sometimes called a 'builder's opening') shall be formed in a wall or in a chimney breast, from which a chimney leads and which has a hearth at its base.   J (0.4–17)

Simple recesses are suitable for closed appliances such as room heaters, stoves, cookers or boilers. They are **not** suitable for an open fire without a canopy.

Fireplace recesses are used for accommodating open fires and free-standing fire baskets.

Fireplace recesses are often lined with firebacks to accommodate inset open fires.

Note: Lining components and decorative treatments fitted around openings reduce the opening area. It is the finished fireplace opening area that determines the size of flue required for an open fire in such a recess.

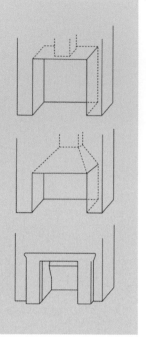

Hearths

A hearth shall safely isolate a combustion appliance from people, combustible parts of the building fabric and soft furnishings.

J (0.4–27)

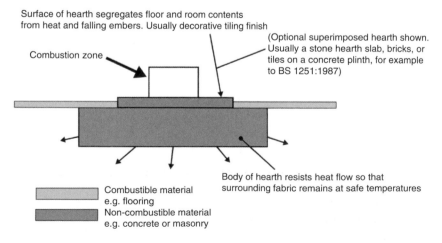

Figure 6.134 The functions of a hearth.

Flueblock chimneys

A flueblock chimney system is a factory-made set of components, made from precast concrete, clay or other masonry units, that is assembled on site to provide a complete chimney.

There are two types, one for gas-burning appliances and the other (often called a 'chimney block system') for solid-fuel-burning appliances.

Flueblock chimneys should be constructed of factory-made components suitable for the intended application and installed in accordance with manufacturer's instructions.	J (1.29)
Joints should be sealed in accordance with the flueblock manufacturer's instructions.	J (1.30)
Bends and offsets should only be formed with matching factory-made components.	J (1.30)
Flueblock chimneys should be installed with sealed joints in accordance with the flueblock manufacturer's installation instructions.	J (4.16)
Flueblocks that are not intended to be bonded into surrounding masonry should be supported and restrained in accordance with the manufacturer's installation instructions.	J (4.16)
Where a fluepipe or chimney penetrates a fire compartment wall or floor, it must not breach the fire separation requirements.	J (4.18)

Masonry chimneys (change of use)

Where a building is to be altered for a different use (e.g. it is being converted into flats) the fire resistance of walls of existing masonry chimneys may need to be improved (see Figure 6.135)	J (1.31)

Connecting flue pipes

Whenever possible, fluepipes should be manufactured from:	J (1.32)

- cast iron (BS41:1973 (1998));
- mild steel (BS1449, Part 1:1991, with a flue wall thickness of at least 3 mm);

- stainless steel (BS EN 10088–1:1995 grades 1.4401, 1.4404, 1.4432 or 1.4436 with a flue wall thickness of at least 1 mm);
- vitreous enamelled steel (BS 6999:1989 (1996)).

Fluepipes with spigot and socket joints should be fitted with the socket facing upwards to contain moisture and other condensates in the flue. J (1.33)

Joints should be made gas-tight. J (1.33)

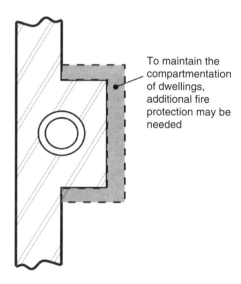

To maintain the compartmentation of dwellings, additional fire protection may be needed

Figure 6.135 Fire protection of chimneys passing through other dwellings.

Repair of flues

 If the installation and maintenance of a fireplace and/or chimney is deemed as being a material change of use, it is a **mandatory requirement** that the building **must** be brought up to the standards required by Parts J1 to J3.

 If renovation, refurbishment or repair amounts to or involves the provision of a new or replacement flue liner, it is considered 'building work' within the meaning of Regulation 3 of the Building Regulations, and must, therefore, **not** be undertaken without prior notification to the local authority. Examples of work that would need to be notified include: J (1.34–1.35)

- relining work comprising the creation of new flue walls by the insertion of new linings such as rigid or flexible prefabricated components;
- a cast in situ liner that significantly alters the flue's internal dimensions.

 Note: If you are in any doubt about this requirement, you should consult the building control department of your local authority, or an approved inspector.

Re-use of existing flues

Where it is proposed to bring a flue in an existing chimney back into use (or to re-use a flue with a different type or rating of appliance), the flue and the chimney should be checked and, if necessary, altered to ensure that they satisfy the requirements for the proposed use.   J (1.36)

Oversize flues can be unsafe. A flue may, however, be lined to reduce the flue area to suit the intended appliance.   J (1.38)

Relining

If a chimney has been previously relined using a metal lining system **and** the appliance is being replaced, then the metal liner should also be replaced unless the metal liner has been recently installed and is in good condition.   J (1.39)

 In certain circumstances, relining is considered '*building work*' within the meaning of Regulation 3 of the Building Regulations and must, therefore, **not** be undertaken without prior notification to the local authority. If you are in doubt you should consult the building control department of your local authority, or an approved inspector.

Flues should be swept to remove deposits before being relined.

 Flexible flue liners should **only** be used to reline a chimney and should **not** be used as the primary liner of a new chimney.   J (1.40)

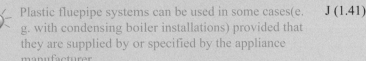 Plastic fluepipe systems can be used in some cases(e. g. with condensing boiler installations) provided that they are supplied by or specified by the appliance manufacturer.   J (1.41)

Factory-made metal chimneys

Where a factory-made metal chimney passes through a wall, sleeves should be provided to prevent damage to the flue or building through thermal expansion.　　　J (1.43)

To allow gas-tightness to be checked, joints between chimney sections should not be concealed within ceiling joist spaces or within the thicknesses of walls.　　　J (1.43)

Following installation of a factory-made metal chimney, it should be a simple measure to withdraw the appliance without having to dismantle the chimney.　　　J (1.44)

Factory-made metal chimneys should be not be installed near combustible materials.　　　J (1.45)

Where a factory-made metal chimney passes through a cupboard, storage space or a roof space, the chimney should be no closer than a distance X from combustible material, where X is defined in BS EN 1856–1:2003, as shown in Figure 6.136.　　　J (1.45)

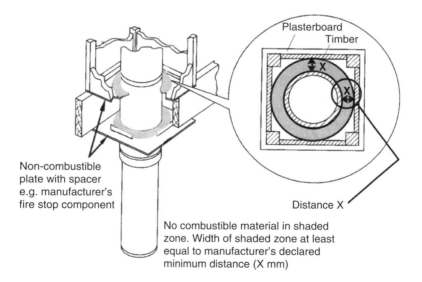

Non-combustible plate with spacer e.g. manufacturer's fire stop component

Plasterboard
Timber

Distance X

No combustible material in shaded zone. Width of shaded zone at least equal to manufacturer's declared minimum distance (X mm)

Figure 6.136 The separation of combustible material from a factory-made metal chimney meeting BS 4543: Part 1: 1990.

 Casing the chimney in non-combustible material is recommended.

Concealed flues

When a flue is routed within a void, access should be provided at strategic locations to allow the following aspects to be visually checked and confirmed whenever the appliance is first installed and subsequently when the appliance is serviced: J (1.47)

- that the flue is continuous throughout its length;
- all joints have been correctly assembled and sealed;
- the flue is adequately supported throughout its length;
- required gradients of fall back to the boiler and any other required drain points have been provided.

Access for concealed flues shall be sufficiently sized and positioned so as to permit visual inspection of the flue (particularly at any joints in the flue.)

Concealed flues should do not pass through another dwelling.

Access hatches should be at least 300 mm × 300 mm or larger where necessary to allow sufficient access to the void to look along the length of the flue.

Flue systems

Flue systems should offer least resistance to the passage of flue gases by minimizing changes in direction or horizontal length.   J (1.48)

Wherever possible flues should be built so that they are straight and vertical. except for the connections to combustion appliances with rear outlets where the horizontal section should not exceed 150 mm. Where bends are essential, they should be angled at no more than 45° to the vertical.   J (1.48)

Provisions should be made to enable flues to be swept and inspected (Figure 6.137).   J (1.49)

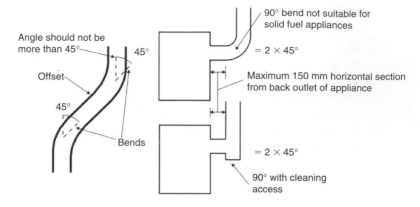

Figure 6.137 Bends in flues.

A flue should **not** have openings into more than one room or   J (1.50)
space except for the purposes of:

*   inspection and/or cleaning; or
*   fitting an explosion door, draught break, draught
    stabilizer or draught diverter.

Openings for inspection and cleaning should have an access   J (1.51)
cover that has the same level of gas-tightness as the flue
system and an equal level of thermal insulation.

After the appliance has been installed, it should be possible   J (1.50)
to sweep the **whole** flue.

Flues discharging at low level near boundaries

Flues that discharge at low level near boundaries   J (1.52)
should ensure safe flue gas dispersal (see Figures 6.138
for examples of achieving this).

Dry lining around fireplace openings

Gaps around decorative treatment, around a fireplace opening   J (1.53)
(e.g. fireplace surround, masonry cladding or dry lining)
should be sealed to prevent any leakage from the fireplace
opening into the void behind the decorative treatment.

The sealing material should be capable of remaining in   J (1.53)
place despite any relative movement between the decorative
treatment and the fireplace recess.

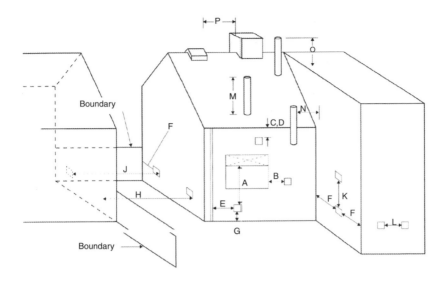

Figure 6.138 Location of outlets from flues serving oil-fired appliances (see Table 6.68 for minimum separation distances).

Table 6.68 Location of outlets from flues serving oil fired appliances

Location of outlet[1]	Minimum separation distances for terminals (mm)	
	Appliance with pressure- jet burner	Appliance with vaporizing burner
A Below an opening[2,3]	600	Should not be used
B Horizontally to an opening[2,3]	600	Should not be used
C Below a plastic/painted gutter, drainage pipe or eaves if combustible material protected[4]	75	Should not be used
D Below a balcony or a plastic/painted gutter, drainage pipe or eaves without protection to combustible material	600	Should not be used
E From vertical sanitary pipework	300	Should not be used
F From an external or internal corner or from a surface or boundary alongside the terminal	300	Should not be used
G Above ground or balcony level	300	Should not be used
H From a surface or boundary facing the terminal	600	Should not be used
J From a terminal facing the terminal	1200	Should not be used
K Vertically from a terminal on the same wall	1500	Should not be used
L Horizontally from a terminal on the same wall	750	Should not be used
M Above the highest point of an intersection with the roof	600[6]	1000[5]

Table 6.68 (Continued)

Location of outlet[1]	Minimum separation distances for terminals (mm)	
	Appliance with pressure- jet burner	Appliance with vaporizing burner
N From a vertical structure to the side of the terminal	750[6]	2300
O Above a vertical structure which is less than 750 mm (pressure jet burner) or 2300 mm (vaporizing burner) horizontally from the side of the terminal	600[6]	1000[5]
P From a ridge terminal to a vertical structure on the roof	1500	Should not be used

Notes:

1. Terminals should only be positioned on walls where appliances have been approved for such configurations when tested in accordance with BS EN 303–1: 1999 or OFTEC standards OFS A100 or OFS A101.
2. An opening means an openable element, such as an openable window, or a permanent opening such as a permanently open air vent.
3. Notwithstanding the dimensions above, a terminal should be at least 300 mm from combustible material (e.g. a window frame).
4. A way of providing protection of combustible material would be to fit a heat shield at least 750 mm wide.
5. Where a terminal is used with a vaporizing burner, the terminal should be at least 2300 mm horizontally from the roof.
6. Outlets for vertical balanced flues in locations M, N and O should be in accordance with manufacturer's instructions.

Condition of combustion installations at completion

A report should be drawn up by the person carrying out the work, to show what materials and components have been used and to confirm that flues have passed appropriate tests.   J 1.54

Note: A suggested checklist for such a report is given at Appendix A and guidance on testing is given at Appendix E to Approved Document J:2010.

On completion of work, flues should be checked to show that they are free from obstructions, satisfactorily gas-tight and constructed with materials and components of sizes which suit the intended application.   J 1.55

See Appendix A to Approved Document J:2010 for detailed checklists for checking and testing of hearths, fireplaces, flues and chimneys.

Where the building work includes the installation of a J 1.55
combustion appliance, tests should include fluepipes
and the gas-tightness of joints between fluepipes and
combustion appliance outlets.

A spillage test to check for compliance with Part J2 should J 1.55
be carried out with the appliance under fire, as part of the
process of commissioning to check for compliance with
Part L (*Conservation of Fuel and Power*), and (in relevant
cases) as required by the Gas Safety (Installation and Use)
Regulations.

Hearths should indicate the area where combustible J 1.56
materials should not encroach.

Notice plates for hearths and flues (Requirement J4)

If a hearth, fireplace (including a flue box), flue or chimney J 1.57
is installed or extended (e.g. as part of some refurbishment
work), a notice plate containing the following information
should be permanently posted in the building:

- the location of the hearth, fireplace (or flue box) or the
  location of the beginning of the flue;
- the category of the flue and generic types of appliances
  that can be safely accommodated;
- the type and size of the flue (or its liner if it has been
  relined) and the manufacturer's name;
- the installation date.

J 1.58

 Notice plates should be securely fixed either:

- next to the electricity consumer unit; or
- next to the *chimney* or *hearth* described; or
- next to the water supply stopcock.

Access to combustion appliances for maintenance

Safe access to appliances for maintenance should be J 1.60
provided.

Roof space installations of gas-fired appliances should J 1.60
comply with the requirements of BS 6798:2009.

*Additional provisions for appliances burning solid fuel
(with a rated output up to 50 kW)*

> Any room or space containing an appliance burning solid fuel   J (2.2)
> (with a rated output up to 50 kW) should have a permanent air
> vent opening of at least the size shown in Figure 6.139.

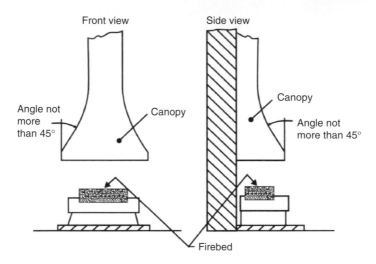

Figure 6.139 Canopy for an open solid-fuel fire.

Open fire with no throat (e.g. a fire under a canopy)

Permanently open air vent(s) should have a total free area of at least 50 per cent
of the cross-sectional area of the flue (see Figure 6.139)

Open fire with a throat and gather

Permanently open air vent(s) should have a total free area of at least 50 per cent
of the throat opening area (see Figure 6.140).

Other appliance (such as a stove, cooker or boiler)

Permanently open air vent(s).

Size of flues

> Fluepipes should have the same diameter or equivalent   J (2.4)
> cross-sectional area as the appliance's flue outlet.
>
> Flues should not be smaller than the appliance's flue outlet   J (2.5)
> or that recommended by the appliance manufacturer.

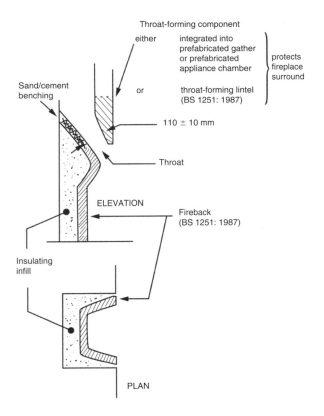

Figure 6.140 Open fireplaces – throat and fireplace components.

Flues in chimneys

Table 6.69 Size of flues in chimneys (see Figure 6.141)

Type	Dimensions
Fireplace with an opening up to 500 mm × 550 mm	200 mm diameter or rectangular/square flues having the same cross-sectional dimension not less than 175 mm
Fireplace with an opening in excess of 500 mm × 550 mm or a fireplace exposed on two or more sides	If rectangular/square flues are used, the minimum dimension should not be less than 200 mm
Closed appliance up to 20 kW rated output which: • burns smokeless or low volatile fuel; or • is an appliance that meets the requirements of the Clean Air Act when burning an appropriate bituminous coal (these appliances are known as 'exempted fireplaces')	125 mm diameter or rectangular/square flues having the same cross-sectional area and a minimum dimension not less than 100 mm for straight flues or 125 mm for flues with bends or offsets

Table 6.69 (Continued)

Other closed appliance of up to 35 kW rated output burning any fuel.	150 mm diameter or rectangular/square flues having the same cross-sectional area and a minimum dimension not less than 125 mm
Closed appliance from 30 kW and up to 50 kW rated output burning any fuel	175 mm diameter or rectangular/square flues having the same cross-sectional area and a minimum dimension not less than 150 mm
For fireplaces with openings larger than 500 mm × 550 mm or fireplaces exposed on two or more sides (e.g. a fireplace under a canopy or open on both sides of a central chimney breast) a way of showing compliance would be to provide a flue with a cross-sectional area equal to 15% of the total face area of the fireplace opening(s) using the formula:	J (2.7)

Fireplace opening area (mm^2) × Total horizontal length of fireplace opening L (mm) × Height of fireplace opening H (mm)

Examples of L and H for large and unusual fireplace openings are shown in Figure 6.141

Height of flues

Flues should be high enough (normally 4.5 m is sufficient) to ensure sufficient draught to clear the products of combustion.	J (2.8)
The outlet from a flue should be above the roof of the building so that the products of combustion can discharge freely and will not present a fire hazard, whatever the wind conditions (Figure 6.142 and Table 6.70).	J (2.10)

Flue outlet clearances – thatched or shingled roof

The clearances to flue outlets that discharge on, or are in close proximity to, roofs with surfaces which are readily ignitable (e.g. covered in thatch or shingles) should be increased to those shown in Figure 6.143.	J (2.12)

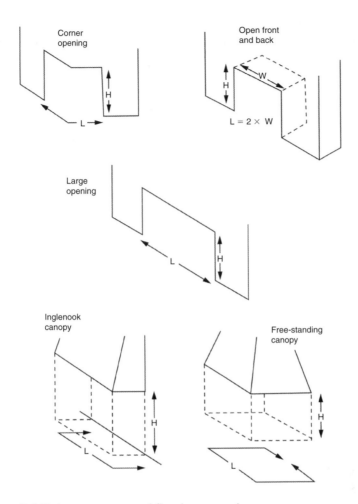

Figure 6.141 Large or unusual fireplace openings.

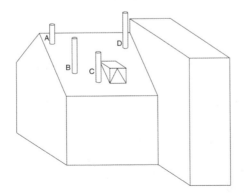

Figure 6.142 Flue outlet positions for solid fuel appliances.

Table 6.70 Flue outlet positions

Point where flue passes through weather surfaces (e.g. roof, tiles or external walls)	Clearance to flue outlet
A At or within 600 mm of the ridge	At least 600 mm above the ridge
B Elsewhere on a roof (whether pitched or flat)	At least 2300 mm horizontally from the nearest point on the weather surface and: • at least 1000 mm above the highest point of intersection of the chimney and the weather surface; or • at least as high as the ridge
C Below (on a pitched roof) or within 2300 mm horizontally to an openable rooflight, dormer window or other opening	At least 1000 mm above the top of the opening
D Within 2300 mm of an adjoining or adjacent building	At least 600 mm above the adjacent building

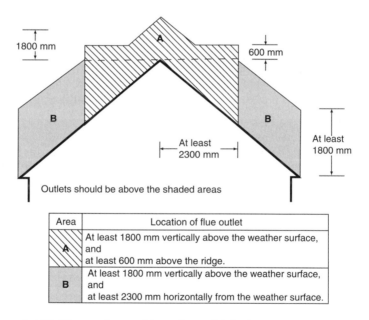

Area	Location of flue outlet
A	At least 1800 mm vertically above the weather surface, and at least 600 mm above the ridge.
B	At least 1800 mm vertically above the weather surface, and at least 2300 mm horizontally from the weather surface.

Figure 6.143 Flue outlet positions for solid fuel appliances – discharging near easily ignited roof coverings.

 Note: Thatched roofs can sometimes be vulnerable to spontaneous combustion caused by heat transferred from flues building up in thick layers of thatch in contact with the chimney. To reduce the risk it is recommended that rigid twin-walled insulated metal flue liners are used within a ventilated (top and bottom) masonry chimney void, provided they are adequately supported and not in direct contact with the masonry. Non-metallic chimneys and cast in situ flue liners can also be used, provided the heat transfer to the thatch is assessed in relation to the depth of thatch and risk of spontaneous combustion.

Spark arrestors are not generally recommended, as they can be difficult to maintain and may increase the risk of flue blockage and flue fires.

Further information and recommendations are contained in HETAS Information Paper 1/007 *Chimneys in Thatched Properties*.

Connecting flue pipes

Combustible materials in the building fabric should be protected from the heat dissipation from flues so that they are not at risk of catching fire.

Connecting fluepipes should not pass through any roof space, partition, internal wall or floor, unless they pass directly into a chimney through either a wall of the chimney or a floor supporting the chimney.	J (2.14)
Connecting fluepipes should be guarded if they are likely to be damaged or if the burn hazard they present to people is not immediately apparent.	J (2.14)
Connecting fluepipes should be located so as to avoid igniting combustible material (see Figure 6.144).	J (2.15 and 1.45)

Debris collection space

If a chimney cannot be cleaned directly through the appliance, a debris-collecting space should be provided at appropriate locations in the *chimney* for emptying and cleaning.	J (2.16) J 3.38

Masonry and flueblock chimneys

The thickness of the walls around the flues, excluding the thickness of any flue liners, should be in accordance with Figure 6.145.	J (2.17)

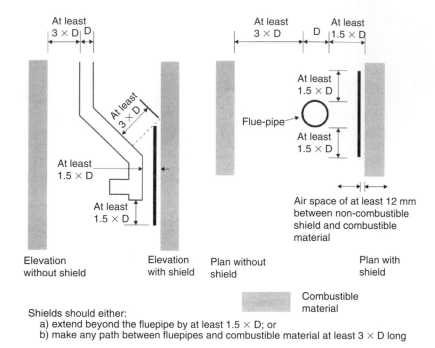

Shields should either:
   a) extend beyond the fluepipe by at least 1.5 × D; or
   b) make any path between fluepipes and combustible material at least 3 × D long

Figure 6.144 Protecting combustible material from uninsulated fluepipes for solid fuel appliances.

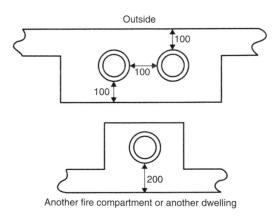

Figure 6.145 Wall thickness for masonry and flueblock chimneys. (Dimensions in mm.)

Combustible material should not be located where it could be ignited by the heat dissipating through the walls of fireplaces or flues.   J (2.18)

Figure 6.146 shows a method of meeting this requirement so that combustible material is at least:

- 200 mm from the inside surface of a flue or fireplace recess; and/or
- 40 mm from the outer surface of a masonry chimney or fireplace recess.

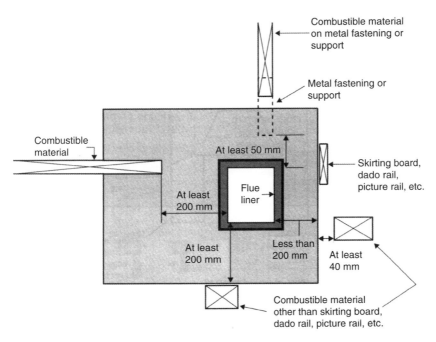

Figure 6.146 Minimum separation distances for combustible material in or near a chimney.

Construction of fireplace gathers

To minimize resistance to the proper working of flues, tapered gathers should be provided in fireplaces for open fires or corbelling of masonry, as shown in Figure 6.147.

Alternatively a suitable canopy (as shown in Figure 6.148) or a prefabricated appliance chamber incorporating a gather may be used.

This can be achieved by using prefabricated gather components built into a fireplace recess, as shown in Figure 6.149.

Lining and relining of flues in chimneys

In many circumstances, lining or relining flues **could** be considered as building work, in which case the works be brought up to the standards required by Parts J2 to J4.

J 2.20

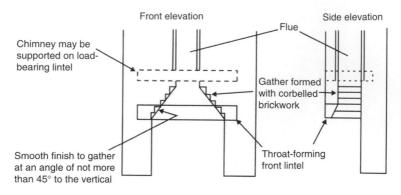

Figure 6.147 Construction of fireplace gathers – using masonry.

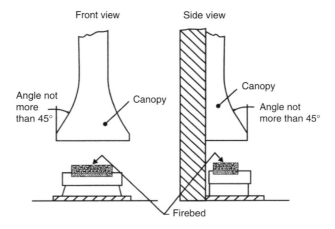

Figure 6.148 Canopy for an open fuel fire.

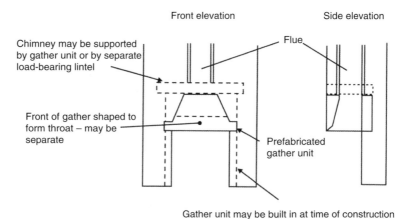

Figure 6.149 Construction of fireplace gathers – using prefabricated components.

Formation of gathers

Tapered gathers should be provided in fireplaces for open fires.	J 2.21

Construction of hearths

Hearths should be constructed so that, in normal use, they prevent combustion appliances setting fire to the building fabric and furnishings – as well as limiting the possibility of people being accidentally burnt.	J (2.22)
If the chimney is not independently supported, the. hearth should be able to accommodate the weight of the appliance and its chimney.	J (2.22)
Appliances should stand entirely above either:	J (2.23)

- hearths made of non-combustible board/sheet material; or
- tiles at least 12 mm thick or constructional hearths.

Constructional hearths should have plan dimensions as shown in Figure 6.150.	J (2.24a)

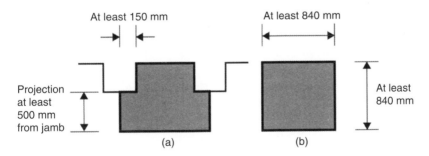

Figure 6.150 Constructional hearth suitable for solid fuel appliances (including open fires) – plan. (a) Fireplace recess. (b) Freestanding.

Constructional hearths should be made of solid, non-combustible material, such as concrete or masonry, that is at least 125 mm thick, including the thickness of any non-combustible floor and/or decorative surface.	J (2.24b)

Combustible material should not be placed beneath   J (2.25)
constructional hearths unless:

- there is an air-space of at least 50 mm between the
  underside of the hearth and the combustible material, or
- the combustible material is at least 250 mm below the
  top of the hearth (Figure 6.151).

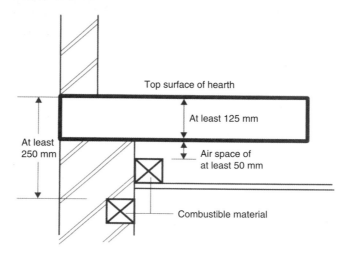

Figure 6.151 Constructional hearth suitable for solid fuel appliances (including open fires) – section.

An appliance should be located on a hearth so that it is: J (2.26)

- surrounded by a surface that is free of combustible
  material (as shown in Figure 6.152); or
- the surface of a superimposed hearth is laid wholly or
  partly upon a constructional hearth.

 Note: The edges of this surface should be marked (e.g. by a change in level) to provide a warning to the building occupants and to discourage combustible floor finishes such as carpet from being laid too close to the appliance.

Combustible material that is placed on or beside a   J 2.28
constructional *hearth* should neither extend under a
superimposed hearth by more than 25 mm or be closer than
150 mm measured horizontally to the appliance.

Note: Some ways of making these provisions are shown in Diagram 27, Approved Document J:2010.

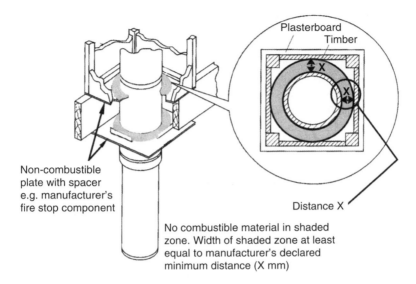

Figure 6.152 The separation of combustible material from a factory-made metal chimney meeting BS 4543: Part 1: 1990.

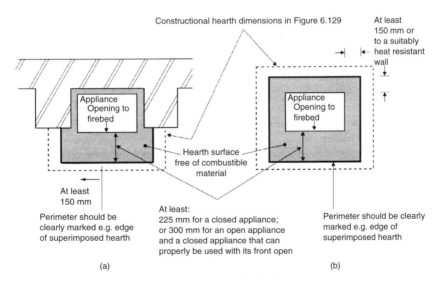

Figure 6.153 Non-combustible hearth surface surrounding a solid fuel appliance. (a) Fireplace recess. (b) Freestanding.

Fireplace recesses

Fireplaces for open fires should be constructed so that they adequately protect the building fabric from catching fire.   J (2.30)

Fireplace recesses can either: J (2.30a)

- be from masonry or concrete, as shown in Figure 6.154; or
- be a prefabricated factory-made appliance chamber   J (2.30b)
  made of insulating concrete having a density of between
  1200 kg/m^3 and 1700 kg/m^3 and with the minimum
  thickness as shown in Table 6.71.

Table 6.71 Flue outlet positions

Point where flue passes through weather surfaces (e.g. roof, tiles or external walls)	Clearance to flue outlet
A At or within 600 mm of the ridge	At least 600 mm above the ridge
B Elsewhere on a roof (whether pitched or flat)	At least 2300 mm horizontally from the nearest point on the weather surface and:
	• at least 1000 mm above the highest point of intersection of the chimney and the weather surface; or • at least as high as the ridge
C Below (on a pitched roof) or within 2300 mm horizontally to an openable rooflight, dormer window or other opening	At least 1000 mm above the top of the opening
D Within 2300 mm of an adjoining or adjacent building	At least 600 mm above the adjacent building

Fireplace lining components

Fireplace recesses containing inset open fires need to be heat protected and should be lined either with suitable firebricks or lining components as shown in Table 6.72.   J (2.31)

Table 6.72 Prefabricated appliance chambers: minimum thickness

Component	Minimum thickness (mm)
Base	50
Side section, forming wall on either side of chamber	75
Back section, forming rear of chamber	100
Top slab, lintel or gather, forming top of chamber	100

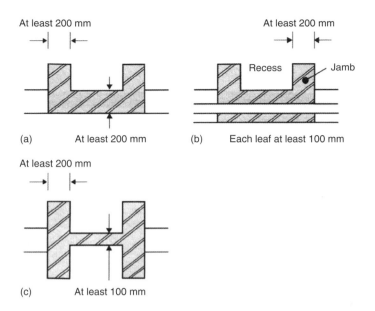

(a)   At least 200 mm   (b)   Each leaf at least 100 mm

(c)   At least 100 mm

Figure 6.154 Fireplace recesses. (a) Solid wall. (b) Cavity wall. (c) Back-to-back (within the same dwelling).

Walls adjacent to hearths

Walls that are not part of a fireplace recess or a prefabricated appliance chamber but are adjacent to hearths or appliances also need to protect the building from catching fire. A way of achieving the requirement is shown in Figure 6.156.	J (2.32)

Open fireplaces – throat and fireplace components

Carbon monoxide alarms

 This is a new 2010 **mandatory** Building Regulation requirement!

Where a new or replacement fixed solid fuel appliance is installed in a dwelling, a carbon monoxide alarm shall be provided in the room where the appliance is located.	J 2.34
Carbon monoxide alarms should comply with BS EN 50291:2001 and be powered by a battery designed to operate for the working life of the alarm.	J 2.35

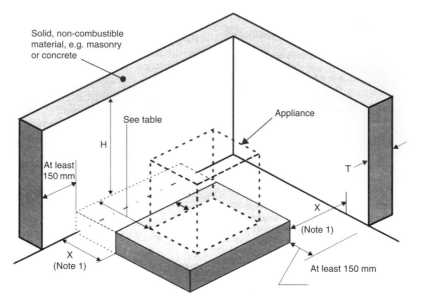

Note 1: There is no requirement for protection of the wall where X is more than 150 mm

Location of hearth or appliance	Solid, non-combustible material	
	Thickness (T)	Height (H)
Where the hearth abuts a wall and the appliance is not more than 50 mm from the wall	200 mm	At least 300 mm above the appliance and 1.2 m above the hearth
Where the hearth abuts a wall and the appliance is more than 50 mm but not more than 300 mm from the wall	75 mm	At least 300 mm above the appliance and 1.2 m above the hearth
Where the hearth does not abut a wall and is no more than 150 mm from the wall (see Note 1)	75 mm	At least 1.2 m above the hearth

Figure 6.155 Walls adjacent to hearths.

Carbon monoxide alarms should incorporate a warning device to alert users when the working life of the alarm is due to pass.   J 2.35

The carbon monoxide alarm should be located in the same room as the appliance: either   J 2.36

* on the ceiling at least 300 mm from any wall; or
* on a wall, as high up as possible (above any doors and windows) but not within 150 mm of the ceiling; and
* between 1 m and 3 m horizontally from the appliance.

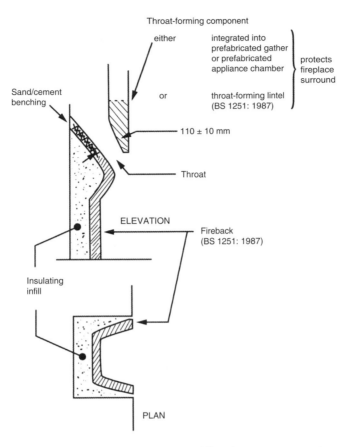

Figure 6.156 Open fireplaces – throat and fireplace components.

Additional provisions for gas-burning devices

The Gas Safety (Installation and Use) Regulations require that:

(a) gas fittings, appliances and gas storage vessels must only be installed by a person with the required competence; and

(b) any person having control to any extent of gas work must ensure that the person carrying out that work has the required competence; and

(c) any person carrying out gas installation, whether an employee or self-employed, **must** be a member of a class of persons approved by the HSE and registered with Gas Safety Register.

Important elements of the Regulations include:

Appliances installed in a room used or intended to be used as a bath or shower room must be a room-sealed type.   J (3.5a)

A gas fire, other gas space heater or gas water heater of more than 14 kW (gross) heat input (12.7 kW net heat input) must not be installed in a room used or intended to be used as sleeping accommodation:

J (3.5b)

* unless the appliance is room-sealed;

* unless it is room-sealed or equipped with a device designed to shut down the appliance before there is a build-up of a dangerous quantity of the products of combustion in the room concerned.

J (3.5c)

The restrictions in (a)–(c) above also apply in respect of any cupboard or compartment within the rooms concerned, and to any cupboard, compartment or space adjacent to and with an air vent into such a room.

Instantaneous water heaters (installed in any room) must be room-sealed or be fitted with a safety device to shut down the appliance as in (c) above.

J (3.5e)

Precautions must be taken to ensure that all installation pipework, gas fittings, appliances and flues are installed safely.

When any gas appliance is installed, checks are required for ensuring compliance with the Regulations, including the effectiveness of the flue, the supply of combustion air, the operating pressure or heat input (or where necessary both), and the operation of the appliance to ensure its safe functioning.

J (3.5f)

All flues must be installed in a safe position.

J (3.5g)

No alteration is allowed to any premises in which a gas fitting or gas storage vessel is fitted that would adversely affect the safety of that fitting or vessel, causing it no longer to comply with the Regulations.

J (3.5h)

Liquid petroleum gas (LPG) storage vessels and LPG-fired appliances fitted with automatic ignition devices or pilot lights must not be installed in cellars or basements.

J (3.5i)

Note: Outlets from flues should be situated externally so as to allow the products of combustion to dispel, and, if a balanced flue, the intake of air (Figure 6.157).

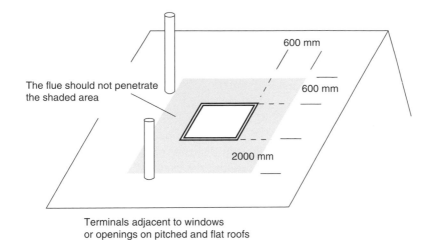

Terminals adjacent to windows
or openings on pitched and flat roofs

Figure 6.157 Location of outlets near roof windows from flues serving gas appliances.

Gas fires (other than flueless gas fires)

Gas-fired appliances should **only** be located where accidental contact is unlikely and where they can be surrounded by a non-combustible surface which provides adequate separation from combustible materials.

as fires may be installed in fireplaces which have flues designed to serve solid fuel appliances – provided it can be shown to be safe.	J (3.7)

Flueless gas appliances

A flueless instantaneous water heater should not be installed in a room or space having a volume of less than 5 m³.	J (3.8)

Flued Decorative Fuel Effect (DFE) fires

Any room or space intended to contain a DFE fire should have permanently open air vents.	J (3.11)

Air supply to flueless appliances

Rooms or spaces intended to contain flueless appliances may need: permanent ventilation and purge ventilation (e.g. openable windows) to comply with Part J; and adjustable ventilation and rapid ventilation to comply with Part F.

Height of natural draught flues for open-flued appliances

> Flues should be high enough to ensure sufficient draught   J (3.21)
> to safely clear the products of combustion.

Outlets from flues

> Outlets from flues should allow the dispersal of products   J (3.23)
> of combustion and, if a balanced flue, the intake of air.
>
> Flue outlets should be protected where flues are at   J (3.24)
> significant risk of blockage.
>
> Flues serving natural draught open-flued appliances   J (3.25)
> should be fitted with outlet terminals if the flue diameter
> is no greater than 170 mm.

In areas where nests of squirrels or jackdaws are likely, the fitting of a protective cage designed for solid fuel use and having a mesh size between 6 and 25 mm is advisable.

> Flue outlets should be protected with a guard if persons   J (3.26)
> could come into contact with it or if it could be damaged.

Bases for back boilers

> Back boilers should adequately protect the fabric of the   J (3.39)
> building from heat (see example in Figure 6.158).

Kerosene and gas-oil burning appliances

Kerosene (class C2) and gas-oil (class D) appliances have the following, additional, requirements:

> Open-fired oil appliances should not be installed in rooms   J (4.2)
> such as bedrooms and bathrooms where there is an increased
> risk of carbon monoxide poisoning.
>
> Flues should be sized to suit the intended appliance and   J (4.4)
> to ensure sufficient discharge velocity to prevent flow
> reversal problems – without imposing excessive flow
> resistances.

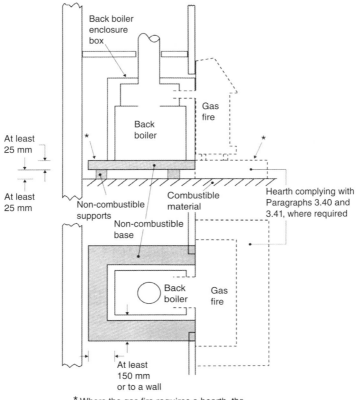

* Where the gas fire requires a hearth, the
back boiler base should be level with it

Figure 6.158 Bases for back boilers.

| The outlet from a flue should be so situated externally to ensure: | J (4.6) |

- the correct operation of a natural draft flue;
- the intake of air if a balanced flue;
- dispersal of the products of combustion.

Note: Figure 6.159 (and Table 6.73) indicates typical positioning to meet this requirement.

Relining chimney flues (for oil appliances)

In some circumstances, lining or relining flues **could** be considered as building work, in which case the works must be brought up to the standards required by Parts J2 to J4.

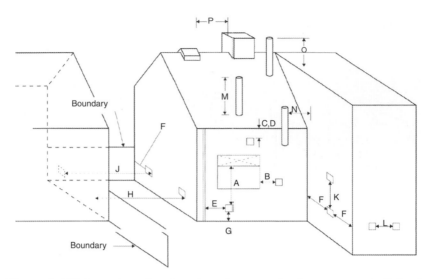

Figure 6.159 Location of outlets from flues serving kerosene and oil-fired appliances (see Table 6.73 for minimum separation distances).

Table 6.73 Location of outlets from flues serving oil fired appliances

Location of outlet[1]	Minimum separation distances for terminals (mm)	
	Appliance with pressure-jet burner	Appliance with vaporizing burner
A Below an opening[2,3]	600	Should not be used
B Horizontally to an opening[2,3]	600	Should not be used
C Below a plastic/painted gutter, drainage pipe or eaves if combustible material protected[4]	75	Should not be used
D Below a balcony or a plastic/painted gutter, drainage pipe or eaves without protection to combustible material	600	Should not be used
E From vertical sanitary pipework	300	Should not be used
F From an external or internal corner or from a surface or boundary alongside the terminal	300	Should not be used
G Above ground or balcony level	300	Should not be used
H From a surface or boundary facing the terminal	600	Should not be used
J From a terminal facing the terminal	1200	Should not be used
K Vertically from a terminal on the same wall	1500	Should not be used
L Horizontally from a terminal on the same wall	750	Should not be used
M Above the highest point of an intersection with the roof	600[6]	1000[5]
N From a vertical structure to the side of the terminal	750[6]	2300

Table 6.73 (Continued)

Location of outlet[1]	Minimum separation distances for terminals (mm)	
	Appliance with pressure-jet burner	Appliance with vaporizing burner
O Above a vertical structure which is less than 750 mm (pressure jet burner) or 2300 mm (vaporizing burner) horizontally from the side of the terminal	600[6]	1000[5]
P From a ridge terminal to a vertical structure on the roof	1500	Should not be used

Notes:

1. Terminals should only be positioned on walls where appliances have been approved for such configurations when tested in accordance with BS EN 303–1:1999 or OFTEC standards OFS A100 or OFS A101.
2. An opening means an openable element, such as an openable window, or a permanent opening such as a permanently-open air vent.
3. Notwithstanding the dimensions above, a terminal should be at least 300 mm from combustible material (e.g. a window frame).
4. A way of providing protection of combustible material would be to fit a heat shield at least 750 mm wide.
5. Where a terminal is used with a vaporizing burner, the terminal should be at least 2300 mm horizontally from the roof.
6. Outlets for vertical balanced flues in locations M, N and O should be in accordance with manufacturer's instructions.

Flexible metal flue liners should be installed in one complete length without joints within the chimney.	J.3.37 J (4.22)
Other than for sealing at the top and the bottom, the space between the chimney and the liner should be left empty (unless this is contrary to the manufacturer's instructions).	J (4.22)
Flues that may be expected to serve appliances burning class D oil (i.e. gas oil) should be made of materials that are resistant to acids.	J (4.23)

Hearths for oil appliances

Oil appliance hearths are needed to prevent the building catching fire and, while it is not a health and safety provision, it is customary to top them with a tray for collecting spilled fuel.	J (4.24)

6.12 Stairs

6.12.1 Requirements

The building shall be constructed so that the combined dead, imposed and wind loads are sustained and transmitted by it to the ground:

- *safely;*
- *without causing such deflection or deformation of any part of the building (or such movement of the ground) as will impair the stability of any part of another building.*

(Approved Document A1)

The building shall be designed and constructed so that there are appropriate provisions for the early warning of fire, and appropriate means of escape in case of fire from the building to a place of safety outside the building capable of being safely and effectively used at all material times.

(Approved Document B)

For a typical one- or two-storey dwelling, the requirement is limited to the provision of smoke alarms and to the provision of openable windows for emergency exit (see B1.i).

Airborne and impact sound

Dwellings shall be designed so that the noise from domestic activity in an adjoining dwelling (or other parts of the building) is kept to a level that:

- *does not affect the health of the occupants of the dwelling;*
- *will allow them to sleep, rest and engage in their normal activities in satisfactory conditions.*

(Approved Document E1)

Dwellings shall be designed so that any domestic noise that is generated internally does not interfere with the occupants' ability to sleep, rest and engage in their normal activities in satisfactory conditions.

(Approved Document E2)

Domestic buildings shall be designed and constructed so as to restrict the transmission of echoes.

(Approved Document E3)

Schools shall be designed and constructed so as to reduce the level of ambient noise (particularly echoing in corridors).

(Approved Document E4)

Ventilation

There shall be adequate means of ventilation provided for people in the building.

(Approved Document F)

Stairs, ladders and ramps

All stairs, steps and ladders shall provide reasonable safety between levels in a building.

(Approved Document K1)

In a public building the standard of stair, ladder or ramp may be higher than in a dwelling, to reflect the lesser familiarity and greater number of users.

This requirement only applies to stairs, ladders and ramps that form part of the building.

Pedestrian guarding should be provided for any part of a floor, gallery, balcony, roof, or any other place to which people have access and any light well, basement area or similar sunken area next to a building.

(Approved Document K2)

Requirement K2 (a) applies only to stairs and ramps that form part of the building.

Access to and use of buildings

Reasonable provisions shall be made for people to:

(a)   gain access to; and
(b)   use the building and its facilities

(Approved Document M)

In addition to the requirements of the Disability Discrimination Act 1995, precautions need to be taken to ensure that:

- new non-domestic buildings and/or dwellings (e.g. houses and flats used for student living accommodation etc.);
- extensions to existing non-domestic buildings;
- non-domestic buildings that have been subject to a material change of use (e.g. so that they become a hotel, boarding house, institution, public building or shop);

are capable of allowing people, regardless of their disability, age or gender, to:

- gain access to buildings;
- **gain access within buildings**;
- be able to use the facilities of the buildings (both as visitors and as people who live or work in them);
- use sanitary conveniences in the principal storey of any new dwelling.

Note: See Appendix A for guidance on access and facilities for disabled people.

6.12.2 Meeting the requirements

As the upper surfaces of floors and stairs are not significantly involved in a fire until it is well developed, they do not play an important part in fire spread in the early stages of a fire.

Means of escape

Except for kitchens, all habitable rooms in the upper storey(s) of a dwelling-house that are served by only one stair should be provided with:	B1 2.4 (V1) B1 2.12 (V2)
• a window (or external door); or • direct access to a protected stairway.	
If direct escape to a place of safety is impracticable, it should be possible to reach a place of relative safety such as a protected stairway within a reasonable travel distance.	B1 V (b)

Table 6.74 Limitations on distance of travel in common areas of blocks of flats

Maximum distance of travel from flat entrance door to common stair, or to stair lobby	
Escape in one direction only	Escape in more than one direction
7.5 m	30 m

Escape routes should be planned so that people do not have to pass through one stairway enclosure to reach another.	B1 2.23 (V2)
Common corridors should be protected corridors.	B1 2.24 (V2)
The wall between each flat and the corridor should be a compartment wall.	B1 2.24 (V2)
Means of ventilating common corridors/lobbies (i.e. to control smoke and so protect the common stairs) should be available.	B1 2.25 (V2)
In large buildings, the corridor or lobby adjoining the stair should be provided with a vent that is located as high as practicable, and with its top edge at least as high as the top of the door to the stair.	B1 2.26 (V2)
There should also be a vent, with a free area of at least 1.0 m² from the top storey of the stairway to the outside.	B1 2.26 (V2)

In single-stair buildings the smoke vents on the fire floor and at the head of the stair should be actuated by means of smoke detectors in the common access space providing access to the flats.

B1 2.26 (V2)

In buildings with more than one stair, the smoke vents may be actuated manually.

B1 2.26 (V2)

Note: Self-closing fire doors should be positioned so that smoke will not affect access to more than one stairway.

Stairs

Where an opening in a floor or roof for a stairway or the like adjoins a supported wall and interrupts the continuity of lateral support:

- the maximum permitted length of the opening is to be 3 m, measured parallel to the supported wall;

  A1/2 2C37a

- connections (if provided by means other than by anchors) should be throughout the length of each portion of the wall situated on each side of the opening;

  A1/2 2C37b

- connections via mild steel anchors should be spaced closer than 2 m on each side of the opening to provide the same number of anchors as if there were no opening;

  A1/2 2C37c

- there should be no other interruption of lateral support.

  A1/2 2C37d

Stairs that separate a dwelling from another dwelling (or part of the same building) shall resist:

E
E2

- the transmission of impact sound (e.g. footsteps and furniture moving);
- the flow of sound energy through walls and floors;
- the level of airborne sound;
- flanking transmission from stairs connected to the separating wall.

All new stairs constructed within a dwelling-house (flat or room used for residential purposes) – whether purpose built or formed by a material change of use – shall meet the laboratory sound insulation values set out in Table 6.75.

E0.9

Table 6.75 Dwelling-houses and flats – performance standards for separating floors and stairs that have a separating function

	Airborne sound insulation $D_{nT,w} + C_{tr}$ (dB) (minimum values)	Impact sound insulation $L9_{nT,w}$ (dB) (maximum values)
Purpose-built rooms for residential purposes	45	62
Purpose-built dwelling houses and flats	45	62
Rooms for residential purposes formed by material change of use	43	64
Dwelling-houses and flats formed by material change of use	43	64

Note:

(1)   The sound insulation values in Table 6.76 include a built-in allowance for 'measurement uncertainty', and so, if any these test values are not met, that particular test will be considered as failed.

(2)   Occasionally, a higher standard of sound insulation may be required between spaces used for normal domestic purposes and noise generated in and to an adjoining communal or non-domestic space. In these cases it would be best to seek specialist advice before committing yourself.

(3)   If the stair is not enclosed, the potential sound insulation of the internal floor will not be achieved; nevertheless, the internal floor should still satisfy Requirement E2.

(4)   In some cases it may be that an existing wall, floor or stair in a building will achieve these performance standards without the need for remedial work (e.g. if the existing construction was already compliant).

Figure 6.160 illustrates the relevant parts of the building that should be protected from airborne and impact sound in order to satisfy Requirement E2.

Sound insulation testing

The person carrying out the building work should arrange for sound insulation testing to be carried out (by a test body with appropriate third-party accreditation) in accordance with the procedure described in Annex B of Approved Document E.	E0.3 E0.4
Impact sound insulation tests should be carried out without a soft covering (e.g. carpet, foam-backed vinyl, etc.) on the stair floor.	E1.10

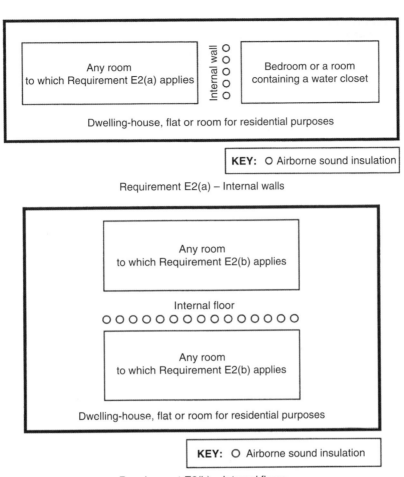

Figure 6.160 Airborne and impact sound requirements.

Testing should not be carried out between living spaces, corridors, stairwells or hallways.	E1.8
Test bodies conducting testing should preferably have UKAS accreditation (or a European equivalent) for field measurements.	E0.4

Note: Some properties, for example loft apartments, may be sold before being fitted out with internal walls and other fixtures and fittings. In these cases sound insulation measurements should be made between the available spaces.

If stairs form a separating function they are subject to the same sound insulation requirements as floors. In this case, the resistance to airborne sound depends mainly on:

- the mass of the stair;
- the mass and isolation of any independent ceiling;
- the air-tightness of any cupboard or enclosure under the stairs;
- the stair covering (which reduces impact sound at source).

Stair treatment 1

Stair treatment 1 consists of a stair covering and independent ceiling with absorbent material.

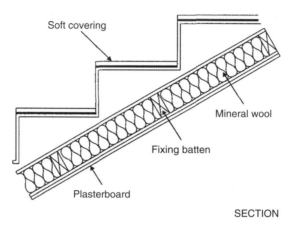

SECTION

Figure 6.161 Stair covering and independent ceiling with absorbent material.

The soft covering should be: E4.37

- at least 6 mm thick;
- laid over the stair treads;
- securely fixed (e.g. glued) so it does not become a safety hazard.

If there is a cupboard under all, or part, of the stair: E4.37

- the underside of the stair within the cupboard should be lined with plasterboard (minimum mass per unit area 10 kg/m^2) together with an absorbent layer of mineral wool (minimum density 10 kg/m^3);
- the cupboard walls should be built from two layers of plasterboard (or equivalent), each sheet with a minimum mass per unit area of 10 kg/m^2;
- a small, heavy, well-fitted door should be fitted to the cupboard.

If there is no cupboard under the stair, an independent ceiling should be constructed below the stair (see floor treatment 1).	E4.37
Where a staircase performs a separating function it shall conform to Building Regulation Part B – Fire safety.	E4.38

Reverberation

Requirement E3 requires that 'domestic buildings shall be designed and constructed so as to restrict the transmission of echoes'. The guidance notes provided in Part E cover two methods (Method A and Method B) which can be used to determine the amount of additional absorption to be used in corridors, hallways, stairwells and entrance halls that give access to flats and rooms for residential purposes. Method A is applicable to stairs and requires the following to be observed:

Cover the ceiling area with the additional absorption.	E7.10
Cover the underside of intermediate landings, the underside of the other landings, and the ceiling area on the top floor.	E7.11
The absorptive material should be equally distributed between all floor levels.	E7.12
For stairwells (or a stair enclosure), calculate the combined area of the stair treads, the upper surface of the intermediate landings, the upper surface of the landings (excluding ground floor) and the ceiling area on the top floor. Either cover an area equal to this calculated area with a class D absorber, or cover an area equal to at least 50 per cent of this calculated area with a class C absorber or better.	E7.11

 Note: Method A can generally be satisfied by the use of proprietary acoustic ceilings.

Piped services

Piped services (excluding gas pipes) and ducts that pass through separating floors should be surrounded with sound-absorbent material for their full height and enclosed in a duct above and below the floor.

Junctions with floor penetrations (excluding gas pipes)

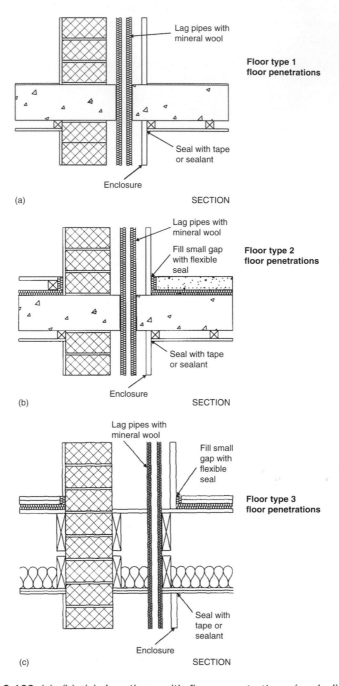

Figure 6.162 (a), (b), (c) Junctions with floor penetrations (excluding gas pipes).

Pipes and ducts that penetrate a floor separating habitable rooms in different flats should be enclosed for their full height in each flat.	E3.41 E3.79 E3.117
The enclosure should be constructed of material having a mass per unit area of at least $15\,kg/m^2$.	E3.32 E3.80 E3.118
The enclosure should either be lined or the duct (or pipe) within the enclosure wrapped with 25 mm unfaced mineral fibre.	E3.42 E3.80 E3.118
Penetrations through a separating floor by ducts and pipes should have fire protection to satisfy Building Regulation Part B – Fire safety.	E3.43 E3.82 E3.120
Fire stopping should be flexible to prevent a rigid contact between the pipe and the floor.	E3.43 E3.121
A small gap (sealed with sealant or neoprene) of about 5 mm should be left between the enclosure and the floating floor.	E3.81 E3.119
Where floating floor (a) or (b) is used the enclosure may go down to the floor base (provided that the enclosure is isolated from the floating layer).	E3.81 E3.119

Junctions with floor penetrations (including gas pipes)

Gas pipes may be contained in a separate (ventilated) duct or can remain unenclosed.	E3.43 E3.120
If a gas service is installed it shall comply with the Gas Safety (Installation and Use) Regulations 1998, SI 1998 No. 2451.	E3.43 E3.120

Note: In the Gas Safety Regulations there are requirements for ventilation of ducts at each floor where they contain gas pipes. Gas pipes may be contained in a separate ventilated duct or they can remain unducted.

Stairs, ladders and ramps

The rise of a stair shall be between 155 mm and 220 mm, with any going between 245 mm and 260 mm and a maximum pitch of 42°.	K1 (1.1–1.4)

The normal relationship between the dimensions of the rise and going is that twice the rise plus the going (2R + G) should be between 550 mm and 700 mm.

Stairs with open risers that are likely to be used by children under five years old should be constructed so that a 100 mm diameter sphere cannot pass through the open risers.

K1 (1.9)

Stairs which have more than 36 risers in consecutive flights should make at least one change of direction, between flights, of at least 30°.

K1 (1.14)

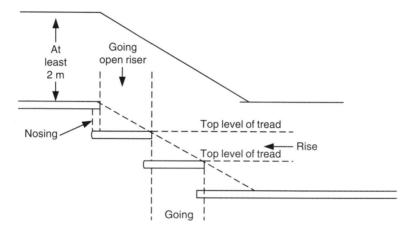

Figure 6.163 Rise and going plus headroom.

If a stair has straight and tapered treads, then the going of the tapered treads should not be less than the going of the straight tread.

K1 (1.20)

The going of tapered treads should measure at least 50 mm at the narrow end.

K1 (1.18)

The going should be uniform for consecutive tapered treads.

K1 (1.19)
K1 (1.22–1.24)

Stairs should have a handrail on both sides if they are wider than 1 m and on at least one side if they are less than 1 m wide.

K1 (1.27)

Handrail heights should be between 900 mm and 1000 mm measured to the top of the handrail from the pitch line or floor.	K1 (1.27)
Spiral and helical stairs should be designed in accordance with BS 5395.	K1 (1.21)

Steps

Steps should have level treads.	K1 (1.8)
Steps may have open risers, but treads should then overlap each other by at least 16 mm.	K1 (1.8)
Steps should be uniform with parallel nosings, the stair should have handrails on both sides and the treads should have slip-resistant surfaces.	
The headroom on the access between levels should be no less than 2 m.	K1 (1.10)
Landings should be provided at the top and bottom of every flight.	K1 (1.15)
The width and length of every landing should be the same (or greater than) the smallest width of the flight.	K1 (1.15)
Landings should be clear of any permanent obstruction.	K1 (1.16)
Landings should be level.	K1 (1.17)
Any door (entrance, cupboard or duct) that swings across a landing at the top or bottom of a flight of stairs must leave a clear space of at least 400 mm across the full width of the flight.	K1 (1.16)
Flights and landings should be guarded at the sides when there is a drop of more than 600 mm.	K1 (1.28–1.29)
For stairs that are likely to be used by children under five years old the construction of the guarding shall be such that a 100 mm sphere cannot pass through any openings in the guarding and children will not easily climb the guarding.	

| For loft conversions, a fixed ladder should have fixed handrails on both sides. | K1 (1.25) |

 While there are no recommendations for minimum stair widths, designers should bear in mind the requirements of Approved Documents B (*Means of Escape*) and M (*Access and Facilities for Disabled People*).

Ramps

All ramps shall provide reasonable safety between levels in a building (where the difference in level is more than 600 mm) and other buildings where the change of level is more than 380 mm.

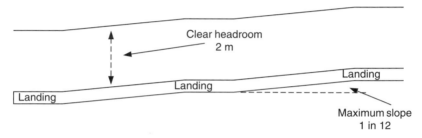

Figure 6.164 The recommended design of a ramp.

Ramps should be clear of permanent obstructions.	K1 (2.4)
The slope of a ramp shall be no more than 1:12.	K1 (2.1)
Ramps should have a handrail on both sides if they are wider than 1 m and on at least one side if they are less than 1 m wide.	K1 (2.5) M
Handrail heights should be between 900 mm and 1000 mm measured to the top of the handrail from the pitch line or floor.	K1 (2.5) M
All ramps should have landings.	K1 (2.6)
All ramps (and associated landings) should have a clear headroom throughout of at least 2 m.	K1 (2.2)
Ramps and landings should be guarded at the sides when there is a drop of more than 600 mm.	K1 (2.7)

For stairs that are likely to be used by children under five years old the construction of the guarding shall be such that a 100 mm sphere cannot pass through any openings in the guarding and children will not easily climb the guarding.

Protection from falling

All stairs, landings, ramps and edges of internal floors shall have a wall, parapet, balustrade or similar guard at least 90 mm high.	K3 (3.2)
All guarding should be capable of resisting at least the horizontal force given in BS 6399: Part 1:1996.	K3 (3.2)
If glazing is used as (or part of) the pedestrian guarding, see Approved Document N: *Glazing – Safety in Relation to Impact, Opening and Cleaning.*	N
If a building is likely to be used by children under five years old, the guarding should not have horizontal rails, should stop children from easily climbing it, and the construction should prevent a 100 mm sphere being able to pass through any opening of that guarding.	K3 (3.3)
All external balconies and edges of roofs shall have a wall, parapet, balustrade or similar guard at least 1100 mm high.	K3 (3.2)

Wall cladding

Where wall cladding is required to function as pedestrian guarding to stairs, ramps, vertical drops of 600 mm or greater or as a vehicle barrier, account should be taken of the additional imposed loading as stipulated in Part K.	A1/2 3.5
Where wall cladding is required to safely withstand lateral pressures from crowds, an appropriate design loading is given in BS 6399: Part 1 and the *Guide to Safety at Sports Grounds* (4th edition, 1997).	

Escape routes

Any storey which has more than one escape stair should be planned so that it is not necessary to pass through one stairway to reach another.	B1 3.13 (V2)
If an escape stair forms part of the only escape route from an upper storey of a large building it should not be continued down to serve any basement storey.	B1 2.44 (V2)
If there is more than one escape stair from an upper storey of a building, only one of the stairs serving the upper storeys of the building need be terminated at ground level.	B1 2.45 (V2)

Note: Other stairs may connect with the basement storey(s) if there is a protected lobby or a protected corridor between the stair(s) and accommodation at each basement level.

The basement should be served by a separate stair.

Doors on escape routes

Unless escape stairways and corridors are protected by a pressurization system complying with BS EN 12101–6:2005, every dead-end corridor exceeding 4.5 m in length should be separated by self-closing fire doors (together with any necessary associated screens) from any part of the corridor that:	B1 3.27 (V2)

- provides two directions of escape;
- continues past one storey exit to another.

A door that opens towards a corridor or a stairway should be sufficiently recessed to prevent its swing from encroaching on the effective width of the stairway or corridor.	B1 5.16 (V2)
Vision panels shall be provided where doors on escape routes subdivide corridors, or where any doors are hung to swing both ways. (See also Parts M and N.)	B1 5.17 (V2)
Revolving doors, automatic doors and turnstiles should not be placed across escape routes.	B1 5.18 (V2)

Escape stairs

An external escape stair may be used, provided that:

• there is at least one internal escape stair from every part of each storey (excluding plant areas);	B1 2.49 (V2)
• in the case of an assembly and recreation building, the route is not intended for use by members of the public;	B1 4.44 (V2)
• in the case of an institutional building, the route serves only office or residential staff accommodation;	
• all doors giving access to the stair are fire-resisting and self-closing;	B1 2.15 (a, b and c) (V1)
• any part of the external envelope of the building within 1.8 m of (and 9 m vertically below) the flights and landings of an external escape stair is of fire-resisting construction (Figure 6.165);	B1 5.25 (V2)
• there is protection by fire-resisting construction for any part of the building within 1.8 m of the escape route from the stair to a place of safety;	
• glazing is fire resistant and fixed shut.	

 Note: Glazing in any fire-resisting construction should be fire-resisting and fixed shut.

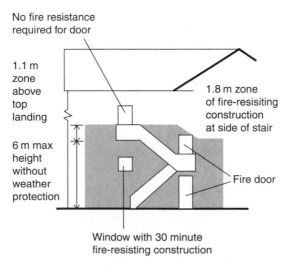

No fire resistance required for door

1.1 m zone above top landing

1.8 m zone of fire-resisiting construction at side of stair

6 m max height without weather protection

Fire door

Window with 30 minute fire-resisting construction

Figure 6.165 Fire resistance of areas adjacent to external stairs.

Escape stairs shall have a protected lobby or protected corridor at all levels (except the top storey, all basement levels and when the stair is a fire-fighting stair) if the:

B1 4.34 (V2)

- stair is the only one serving a building that has more than one storey above or below the ground storey;
- stair serves a storey that is higher than 18 m;
- building is designed for phased evacuation;

- stairway is near (or potentially next to) a place of special fire hazard.

B1 4.35 (V2)

External escape stairs greater than 6 m in vertical extent shall be protected from the effects of adverse weather conditions.

B1 2.15d

If the building (or part of the building) is served by a single access stair, that stair may be external if it:

B1 2.48 (V2)

- serves a floor not more than 6 m above the ground level; and
- meets the provisions in paragraph 5.25.

Protection of escape stairs

Escape stairs need to have a satisfactory standard of fire protection.

B1 4.31 (V2)

Internal escape stairs should be a protected stairway within a fire-resisting enclosure.

B1 4.32 (V2)

Except for bars and restaurants, an escape stair may be open provided that:

B1 4.33 (V2)

- it does not connect more than two storeys and reaches the ground storey not more than 3 m from the final exit; and
- the storey is also served by a protected stairway; or
- it is a single stair in a small premises with the floor area in any storey not exceeding 90 m².

A dwelling-house with more than one floor 4.5 m above ground level may either:

B1 2.6 (V1)

- have a protected stairway that extends to the final exit (Figure 6.166); or
- have a protected stairway that gives access to at least two escape routes at ground level, each delivering to final exits and separated from each other by fire-resisting construction and fire doors (Figure 6.166(b)); or

- have the top floor separated from the lower storeys by fire-resisting construction and have its own alternative escape route leading to its own final exit (Figure 6.166).

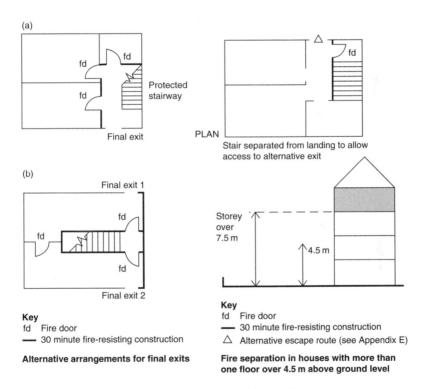

Figure 6.166 Alternative arrangements for final exits.

Construction of escape stairs

The flights and landings of every escape stair should be constructed using materials of limited combustibility, particularly if it is:

B1 5.19 (V2)

- the only stair serving the building;
- within a basement storey;
- serving any storey having a floor level more than 18 m above ground or access level;
- external;
- a fire-fighting stair.

Single steps may **only** be used on an escape route if they are prominently marked.	B1 5.21 (V2)

Helical and spiral stairs forming part of an escape route should be:	B1 5.22a (V2)

- designed in accordance with BS 5395–2:1984;
- type B (public stair) if they are intended to serve members of the public.

Fixed ladders should not be used as a means of escape for members of the public.	B1 5.22b (V2)

Note: See Part K for guidance on the design of helical and spiral stairs and fixed ladders.

If a protected stairway projects beyond, or is recessed from, or is in an internal angle adjoining an external wall of the building, the distance between any unprotected area in the external enclosures to the building and any unprotected area in the enclosure to the stairway should be at least 1800 mm (Figure 6.167).	B1 5.24 (V2)

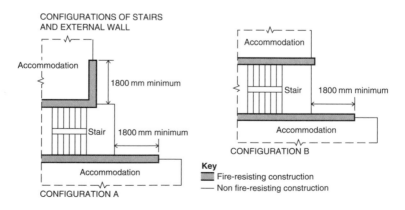

Figure 6.167 External protection to protected stairways.

The width of escape stairs should: B1 4.15 (V2)

- not be less than the width of any exit(s);
- not be less than the minimum widths given in Table 6.76;
- not exceed 1.4 m if their vertical extent is more than 30 m, **unless** it is provided with a central handrail;
- not reduce in width at any point on the way to a final exit.

In public buildings, if the width of the stair is more than 1800 mm, the stair should have a central handrail. B1 4.16 (V2)

Every escape stair should be wide enough to accommodate the number of persons needing to use it in an emergency. B1 4.18 (V2)

Note: For further guidance and worked examples see Appendix C to Part B, and Sections 4.18 (V2) to 4.25 (V2).

Table 6.76 Minimum widths of escape stairs

Stair situation	Maximum number of people served	Minimum stair width
1a In an institutional building (unless the stair is only used by staff)	150	1000 mm
1b In an assembly building and serving an area used for assembly purposes (unless the area is less than 100 m²)	220	1100 mm
1c In any other building and serving an area with an occupancy of more than 50 people	Over 2200	1000–1800 mm*
2 Any stair not described above	50	800 mm

* Depending on whether the stairs are used for simultaneous evacuation or phased evacuation (see Table 7 of Part B1 (Version 2)).

Lighting

Lighting to escape stairs should be on a separate circuit from that supplying any other part of the escape route. B1 5.36 (V2)

The installation of an escape lighting system shall be in accordance with BS 5266–1:2005 and **BS 7671** (latest edition). Full details are contained in the partner book to this publication: *Wiring Regulations in Brief.*

Number of escape stairs

The number of escape stairs required in a building (or part of a building) will be determined by: B1 4.2 (V2)

- the constraints imposed by the design of horizontal escape routes;
- whether independent stairs are required in mixed occupancy buildings;
- whether a single stair is acceptable; and
- the width for escape (and the possibility that a stair may have to be discounted because of fire or smoke).

Provided that independent escape routes are not necessary from areas in different purpose groups, single escape stairs may be used from: B1 4.6 (V2)

- small premises (other than bars or restaurants);
- office buildings comprising not more than five storeys above the ground storey;
- factories comprising not more than one storey above the ground storey if the building is of normal risk (two storeys if the building is of low risk); or
- process plant buildings with an occupant capacity of not more than 10 people.

Protected shafts

Protected shafts (i.e. spaces that connect compartments, such as stairways and service shafts) shall be protected to restrict fire spread between the compartments.   B2 8.7 (V2)

The uses of protected shafts should be restricted to stairs, lifts, escalators, chutes, ducts and pipes.   B2 8.36 (V2)

Protected shafts provide for the movement of people (e.g. stairs, lifts), or for passage of goods, air or services such as pipes or cables between different compartments. The elements enclosing the shaft (unless formed by adjacent external walls) are compartment walls and floors. Figure 6.164 shows three common examples that illustrate the principles.

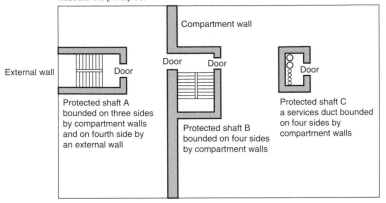

Figure 6.168 Protected shafts.

Any stairway or other shaft passing directly from one compartment to another should be enclosed in a protected shaft so as to delay or prevent the spread of fire between compartments.

B2 8.35 (V2)

An uninsulated glazed screen may be incorporated in the enclosure to a protected shaft between a stair and a lobby or corridor which is entered from the stair provided that:

B2 8.38 (V2)

* the fire resistance for the stair enclosure is not more than 60 minutes; and
* the glazed screen has at least 30 minutes fire resistance; and
* the lobby or corridor is enclosed to at least a 30 minute standard (Figure 6.169).

If a protected shaft contains a stair and/or a lift, it should **not** also contain:

B2 8.40 (V2)

* a pipe conveying oil (other than in the mechanism of a hydraulic lift); or
* a ventilating duct (other than a duct provided solely for ventilating the stairway).

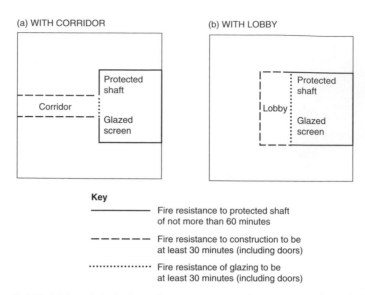

Figure 6.169 Uninsulated glazed screen separating protected shaft from lobby or corridor.

Protected stairways

Protected stairways should discharge: • directly to a final exit; or • via a protected exit passageway to a final exit.	B1 2.38 (V2)
Where two protected stairways (or exit passageways leading to different final exits) are adjacent, they should be separated by an imperforate enclosure.	B1 2.39 (V2)
A protected stairway should be relatively free of potential sources of fire.	B1 2.40 (V2)
In single-stair buildings, meters located within the stairway should be enclosed within a secure cupboard that is separated from the escape route with fire-resisting construction.	B1 2.40 (V2)
Gas service and installation pipes (together with their associated meters, etc.) should **not** be incorporated within a protected stairway unless the gas installation is in accordance with SI 1996 No. 825 and SI 1998 No. 2451.	B1 4.40 (V2) B1 2.42 (V2)
Refuse chutes and rooms provided for the storage of refuse should **not** be located within protected stairways or protected lobbies.	B1 5.55 and 5.56 (V2)

Ventilation

Separate ventilation systems should be provided for each protected stairway.	B1 5.47 (V2)
Air-circulation systems shall ensure that:	B1 2.17 (V1)
• smoke or fire is prevented from spreading into a protected stairway;	B3 7.10 (V1)
• transfer grilles are not fitted in any wall, door, floor or ceiling enclosing a protected stairway;	B1 2.18 (V2)
• any duct passing through the enclosure to a protected stairway is of rigid steel construction and all joints between the ductwork and the enclosure are fire-stopped;	B3 10.2 (V2)
• ventilation ducts supplying or extracting air directly to or from a protected stairway do not serve other areas as well;	
• any system of mechanical ventilation that recirculates air, and which serves both the stairway and other areas, is designed to shut down on the detection of smoke within the system.	

Passenger lifts

Passenger lifts that serve floors more than 4.5 m above ground level should either be located in the enclosure to the protected stairway or be contained in a fire-resisting lift shaft.	B1 2.18
Lift wells should be either:	B1 5.42 (V2)
• contained within the enclosures of a protected stairway; or	
• enclosed throughout their height with fire-resisting construction.	
In basements and enclosed (i.e. non-open-sided) car parks the lift should be approached only by a protected lobby or protected corridor (unless it is within the enclosure of a protected stairway).	B1 5.43 (V2)
Lift shafts should not be continued down to serve a basement storey if it is within the enclosure to an escape stair that is terminated at ground level.	B1 5.44 (V2)

Common stairs

All common stairs should be situated within a fire-resisting enclosure (i.e. it should be a protected stairway), to reduce the risk of smoke and heat making use of the stair hazardous.	B1 2.36 (V2)
A single common stair can be acceptable in some cases, but otherwise there should be access to more than one common stair for escape purposes.	B1 2.32 (V2)
Where a common stair forms part of the only escape route from a flat, it should **not** also serve any covered car park, boiler room, fuel storage space or other ancillary accommodation of similar fire risk.	B1 2.46 (V2)
Common stairs that do not form part of the only escape route from a flat may also serve ancillary accommodation if they are separated from the ancillary accommodation by a protected lobby or a protected corridor.	B1 2.47 (V2)
If the stair serves an enclosed (non-open-sided) car park, or place of special fire hazard, the lobby or corridor should have not less than $0.4\,m^2$ permanent ventilation or be protected by a mechanical smoke-control system.	B1 2.47 (V2)

Fire-protected stairways

Fire-protected stairways shall, as far as is reasonably possible:

• exclude all flames, smoke and gases; • be designed to provide effective 'fire-sterile' areas that lead to places of safety outside the building;	B1 1.viii
• consist of fire-resistant material and fire-resistant doors and have an appropriate form of smoke-control system;	B1 1.viii
• contain a fire-main outlet;	B5 17.10 (V2)
• have cavity barriers above the enclosures;	B1 2.14
• be free of potential sources of fire;	B1 4.38 (V2)
• discharge:	B1 4.36 (V2)

- directly to a final exit; or
- by way of a protected exit passageway to a
  final exit.

Note: Any such protected exit passageway should have the same standard of fire resistance and lobby protection as the stairway it serves.

If two protected stairways are adjacent, they (and any protected exit passageways linking them to final exits) shall be separated by an imperforate enclosure.	B1 4.37 (V2)

Fire-fighting stairs

Any stair used as a fire-fighting stair should:	B1 2.33 (V2)
• be at least 1100 mm wide (see Part B V2, Appendix C for measurement of width);	
• not have a protected lobby or protected corridor;	B1 4.34 (V2)
• be constructed using materials of limited combustibility;	B1 5.19 (V2)
• be approached from the accommodation, through a fire-fighting lobby (unless it is in blocks of flats).	B5 17.11 (V2)
All fire-fighting shafts should be equipped with fire mains having outlet connections and valves at every storey.	B5 17.12 (V2)

Refuges

Refuges are relatively safe waiting areas for short periods. They are **not** areas where disabled people should be left alone indefinitely until rescued by the fire and rescue service, or until the fire is extinguished.

A refuge (i.e. an enclosure such as a compartment, protected lobby, protected corridor, protected stairway (Figure 6.170) or an area in the open air such as a flat roof, balcony, podium or similar) should be provided for each protected stairway.	B1 4.8 (V2)

Note: The number of refuge spaces need not necessarily equal the sum of the number of wheelchair users who can be present in the building.

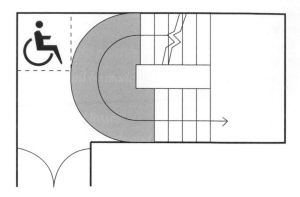

Provision where access to the wheelchair space is counter to the access flow within the stairway

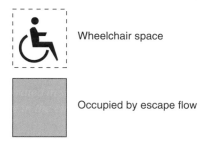

Wheelchair space

Occupied by escape flow

Figure 6.170 Refuge formed in a protected stairway.

Refuges and evacuation lifts should be clearly identified by appropriate fire-safety signs.	B1 4.10 (V2)

 Note: If a refuge is in a lobby or stairway, the sign should be accompanied by a blue mandatory sign worded 'Refuge – keep clear'.

Smoke alarms

Smoke alarms should **not** be fixed over a stair or any other opening between floors.	B1 1.16 (V1) B1 1.15 (V2)

Flats

Except for kitchens:

all habitable rooms in the upper storey(s) of a multi-storey flat that are served by only one stair should (depending on the height of the top storey) be provided with: B1 2.12 (V2)

- a window (or external door); or
- direct access to a protected stairway.

Where the vertical distance between the floor of the entrance storey and the floors above and below it does not exceed 7.5 m, multi-storey flats are required to:

- provide a protected stairway **plus** additional   B1 2.16c (V2)
  smoke alarms in all habitable rooms and a heat
  alarm in any kitchen; or

- provide a protected stairway **plus** a sprinkler sys-   B1 2.16d (V2)
  tem and smoke alarms.

An alternative exit from a flat should:

- be remote from the main entrance door to the flat;   B1 2.17 (V2)
  and
- lead to a defined exit or common stair by way of:

  - a door onto an access corridor, access lobby or
    common balcony; or
  - an internal private stair leading to an access
    corridor, access lobby or common balcony at
    another level; or
  - a door into a common stair; or
  - a door onto an external stair; or
  - a door onto an escape route over a flat roof.

Flats in mixed use buildings

The stairs of buildings which are no more than three   B1 2.50 (V2)
storeys above the ground storey may serve both flats
and other occupancies, **provided** that the stairs are
separated from each occupancy by protected lobbies at
all levels.

The stairs of buildings that are more than three storeys   B1 2.51 (V2)
above the ground storey may serve both flats and other
occupancies, **provided** that:

- the flat is ancillary to the main use of the building
  and is provided with an independent alternative
  escape route;
- the stair is separated from any other occupancies
  on the lower storeys by protected lobbies (at those
  storey levels);
- any automatic fire-detection and alarm system with
  which the main part of the building is fitted also
  covers the flat;
- any security measures should not prevent escape at
  all material times.

Galleries

Any cooking facilities within a room containing a   B1 2.12 (V1)
gallery should either:

- be enclosed with fire-resisting construction; or   B1 2.8 (V2)
- be remote from the stair to the gallery.

Loft conversions

Where the conversion of an existing roof space (e.g. a loft   B1 2.20b
conversion to a two-storey house) means that a new storey
is going to be added, the stairway will need to be protected
with fire-resisting doors and partitions.

Basements

Because of their situation, basement stairways are more likely to be filled with
smoke and heat than stairs located in ground and upper storeys. Special measures are, therefore, required in order to prevent a basement fire endangering
upper storeys.

If an escape stair forms part of the only escape route   B1 4.42 (V2)
from an upper storey of a building, it should **not** be
continued down to serve a basement storey (i.e. the
basement should be served by a separate stair).

If there is more than one escape stair from an upper   B1 4.43 (V2)
storey of a building, only **one** of the stairs serving the
upper storeys of the building need be terminated at
ground level.

 Owing to the possibility of a single stairway becoming blocked by smoke from
a fire in the basement or ground storey:

- basement storeys in a dwelling-house that contain a   B1 2.13 (V1)
  habitable room_**shall** be provided with either:

  - a protected stairway leading from the basement   B1 2.6 (V2)
    to a final exit; or
  - an external door or window suitable for egress
    from the basement.

Access and facilities for disabled people

Internal steps, stairs and ramps

Stepped access

A stepped access should have a level landing at the top and
bottom of each flight; landings should be 1200 mm long
and be unobstructed.

Doors should not swing across landings.	M (3.51a)
The surface width of flights between enclosing walls, strings or upstands should not be less than 1.2 m.	M (3.51a)
There should be no single steps.	M (3.51a)
Nosings for the tread and the riser should be 55 mm wide and of a contrasting material.	M (3.51a)
Step nosings should not project over the tread below by more than 25 mm (Figure 6.171).	M (3.51a)

The rise and going of each step should be consistent throughout a flight.	M (3.51a)
The rise of each step should be between 150 mm and 170 mm.	M (3.51a)
The going of each step should be between 280 mm and 425 mm.	M (3.51a)

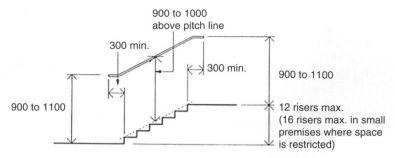

900 to 1000
above pitch line

300 min.

300 min.

900 to 1100

900 to 1100

12 risers max.
(16 risers max. in small
premises where space
is restricted)

Figure 6.171 Internal stairs – key dimensions (mm).

Rises should not be open.	M (3.51a)
There should be a continuous handrail on each side of a flight and landings.	M (3.51a)
If additional handrails are used to divide the flight into channels, they should not be less than 1 m wide or more than 1.8 m wide.	M (3.51a)
Flights between landings should contain no more than 12 risers.	M (3.51b)
The rise of each step should be between 150 mm and 170 mm.	M (3.51c)
The going of each step should be at least 250 mm.	M (3.51d)

Note: For mobility-impaired people, a going of at least 300 mm is preferred.

Materials for treads should not present a slip hazard.	M (3.50)
Areas below stairs or ramps with a soffit less than 2.1 m above ground level should be protected by guarding and low-level cane detection.	M (3.51e)
Any feature projecting more than 100 mm onto an access route should be protected by guarding that includes a kerb (or other solid barrier) that can be detected using a cane (Figure 6.172).	M (3.51e)

For school buildings, the rise should not exceed 170 mm, with a preferred going of 280 mm.

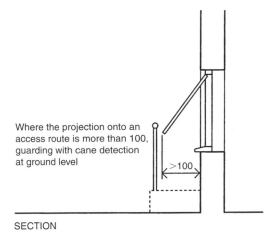

Where the projection onto an access route is more than 100, guarding with cane detection at ground level

>100

SECTION

Figure 6.172 Avoiding hazards on access routes. (Dimensions in mm.)

Internal ramps

Where an internal ramp is provided: M (3.53)

- the approach should be clearly signposted;
- the rise should be no more than 500 mm;
- if the total rise is greater than 2 m, an alternative means of access (e.g. a lift) should be provided for wheelchair users;
- the ramp surface should be slip resistant;
- the ramp surface should be of a contrasting colour with that of the landings;
- frictional characteristics of ramp and landing surfaces should be similar;
- landings at the foot and head of a ramp should be at least 1.2 m long and clear of any obstructions;
- intermediate landings should be at least 1.5 m long and clear of obstructions;
- all landings should:

  – be level;
  – have a maximum gradient of 1:60 along their length;
  – have a maximum cross-fall gradient of 1:40;

- there should be a handrail on both sides.

In addition to the guarding requirements of Part K, there should be a visually contrasting kerb on the open side of the ramp (or landing) at least 100 mm high.

Where the change in level is 300 mm or more, two or more clearly signposted steps should be provided (i.e. in addition to the ramp).

M (3.53b)

If the change in level is no greater than 300 mm, a ramp should be provided instead of a single step.

M (3.53c)

All landings should be level and have a maximum gradient of 1:60 along their entire length.

M (3.53d)

Areas below stairs or ramps with a soffit less than 2.1 m above ground level should be protected by guarding and low-level cane detection.

M (3.53e)

Any feature projecting more than 100 mm onto an access route should be protected by guarding that includes a kerb (or other solid barrier) that can be detected using a cane (Figure 6.173).

M (3.53e)

Gradients should be as shallow as practicable.

M (3.52)

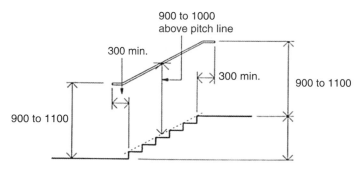

Figure 6.173 Handrails to internal steps, stairs and ramps – key dimensions (mm).

Handrails to steps, stairs and ramps

Handrails to internal stepped or ramped access should be positioned as per Figure 6.173.

M (1.37a)

Handrails to internal steps, stairs and ramps should: M (3.55)

- be continuous across flights and landings;
- extend at least 300 mm horizontally beyond the top and
  bottom of a ramped access;
- not project into an access route;
- contrast visually with the background;
- have a slip-resistant surface that is not cold to the touch;
- terminate in such a way that reduces the risk of
  clothing being caught;
- either be circular (with a diameter of between
  40 mm and 45 mm) or oval with a width of 50 mm
  (Figure 6.174).

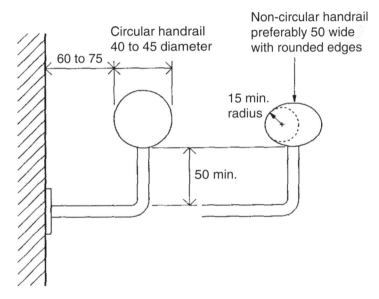

Figure 6.174 Handrail designs. (Dimensions in mm.)

Handrails to external stepped or ramped access should: M (3.55)

- not protrude more than 100 m into the surface width of
  the ramped or stepped access where this would impinge
  on the stair width requirement of Part B1;

- have a clearance of between 60 mm and 75 mm between the handrail and any adjacent wall surface;
- have a clearance of at least 50 mm between a cranked support and the underside of the handrail;
- ensure that its inner face is located no more than 50 mm beyond the surface width of the ramped or stepped access;
- be spaced away from the wall and rigidly supported in a way that avoids impeding finger grip;
- be set at heights that are convenient for all users of the building.

Common stairs in blocks of flats

The aim for all buildings containing flats should be to make reasonable provision for disabled people to visit occupants who live on any storey of the building, via a common staircase or a lift.

Common stairs

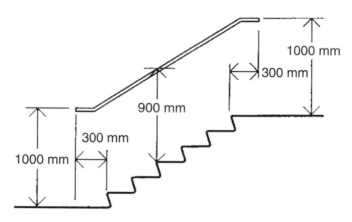

Figure 6.175 Common stairs in blocks of flats.

If there is no passenger lift to provide access between storeys, a stair (designed to suit the needs of ambulant disabled people, people with impaired sight and people with sensory impairments) should be provided.   M (9.3 and 9.4)

If a passenger lift is not installed, a common stair should be provided that has:	M (9.5a)
• step nosings with contrasting brightness;	
• top and bottom landings whose lengths are in accordance with Part K1;	M (9.5b)
• steps with suitable tread nosing profiles (Figure 6.175) with a uniform rise not more than 170 mm;	M (9.5c)
• a uniform going of each step not less than 250 mm;	M (9.5d)
• risers which are not open;	M (9.5e)
• a continuous handrail on each side of flights and landings (if the rise of the stair comprises two or more rises).	M (9.5f)
A single common stair can be acceptable in some cases, but otherwise there should be access to more than one common stair for escape purposes.	B1 2.32 (V2)
All common stairs should be situated within a fire-resisting enclosure (i.e. it should be a protected stairway), to reduce the risk of smoke and heat making use of the stair hazardous.	B1 2.36 (V2)
Where a common stair forms part of the only escape route from a flat, it should **not** also serve any covered car park, boiler room, fuel storage space or other ancillary accommodation of similar fire risk.	B1 2.46 (V2)
Common stairs which do not form part of the only escape route from a flat may also serve ancillary accommodation if they are separated from the ancillary accommodation by a protected lobby or a protected corridor.	B1 2.47 (V2)
If the stair serves an enclosed (non open-sided) car park, or place of special fire hazard, the lobby or corridor should have not less than $0.4\,m^2$ permanent ventilation or be protected by a mechanical smoke-control system.	B1 2.47 (V2)

6.13 Windows

6.13.1 Requirements

Means of escape

In an emergency, the occupants of any part of the building shall be able to escape without any external assistance.

(Approved Document B1)

Ventilation

There shall be adequate means of ventilation provided for people in the building.

(Approved Document F)

 A new Part F came into force on 1 October 2010.

Protection from falling

Pedestrian guarding should be provided for any part of a floor (including the edge below an opening window), gallery, balcony, roof (including rooflight and other openings), any other place to which people have access and any light well, basement area or similar sunken area next to a building.

(Approved Document K2)

Conservation of fuel and power

Reasonable provision shall be made for the conservation of fuel and power in buildings by:

(a) *limiting heat gains and losses –*

  (i) *through thermal elements and other parts of the building fabric; and*
  (ii) *from pipes, ducts and vessels used for space heating, space cooling and hot water services;*

(b) *providing fixed building services which –*

  (i) *are energy efficient;*
  (ii) *have effective controls; and*
  (iii) *are commissioned by testing and adjusting as necessary to ensure they use no more fuel and power than is reasonable in the circumstances; and*

(c) *providing to the owner sufficient information about the building, the fixed building services and their maintenance requirements so that the building can be operated in such a manner as to use no more fuel and power than is reasonable in the circumstances*

(Approved Document L)

 A new Part L came into force in October 2010.

Protection against impact

Glazing with which people are likely to come into contact whilst moving in or about the building, shall:

- *if broken on impact, break in a way which is unlikely to cause injury; or*
- *resist impact without breaking; or*
- *be shielded or protected from impact.*

(Approved Document N)

6.13.2 Meeting the requirement

Means of escape

Emergency egress windows

Except for kitchens, all habitable rooms on the ground floor (and the upper storey(s) of a dwelling-house that is served by only one stair) should be provided with an emergency egress window.	B1 2.3 and 2.4 (V1) B1 2.11 and 2.12 (V2)

Note: There are some other alternatives if this is not possible, such as having access to a protected route, so it is best to see the Regulations if you need confirmation of this point.

The window should be at least 450 mm high and 450 mm wide and have an unobstructed openable area of at least 0.33 m².	B1 2.8 (V1)
The bottom of the openable area should be not more than 1100 mm above the floor.	B1 2.9 (V2)
The window should enable the person escaping to reach a place free from danger of fire (e.g. a courtyard or back garden which is at least as deep as the dwelling-house is high – (Figure 6.176).	

Approved Document K (*Protection from Falling, Collision and Impact*) specifies a minimum guarding height of 800 mm, except in the case of a window in a roof where the bottom of the opening may be 600 mm above the floor.

Locks (with or without removable keys) and stays may be fitted to egress windows, **provided** that the stay is fitted with a child-resistant release catch.

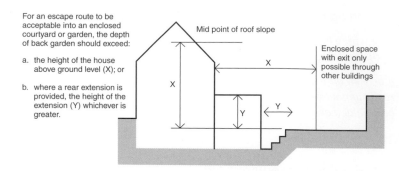

Figure 6.176 Ground or basement storey exit into an enclosed space.

Windows should be designed so that they remain in the open position without needing to be held open by the person making their escape.

Any inner room that is a kitchen, laundry or utility room, dressing room, bathroom, WC or shower room situated not more than 4.5 m above ground level and whose only escape route is through another room, shall be provided with an emergency egress window.	B1 2.9 (V1) B1 2.5 (V2)
All galleries shall be provided with an alternative exit or, where the gallery floor is not more than 4.5 m above ground level, an emergency egress window.	B1 2.12 (V1) B1 2.8 (V2)
All basement storeys in a dwelling-house that contain a habitable room shall be provided with either an external door or window suitable for egress from the basement.	B1 2.13 (V1) B1 2.6 (V2)

 Note: There are certain alternatives. See Regulations for details.

Dimensions

The height times the width of the opening part of hinged or pivot windows that are designed to open **more** than 30° and/or sliding sash windows, should be at least 1/20 of the floor area of the room.	F Tables 5.2a to 5.2d
The height times width of the opening of hinged or pivot windows designed to open **less** than 30° should be at least 1/10 of the floor area of the room.	F Tables 5.2a to 5.2d

If a room contains more than one openable window, the areas of **all** the opening parts may be added together to achieve the required floor area.

Replacement windows

Where windows are to be replaced, the replacement work should comply with the requirements of Parts L and N.	B1 2.19
If a window is currently located where, in the case of a new dwelling-house, an escape window would be necessary, the replacement window opening should be sized to provide at least the same potential for escape as the window it replaces.	B1 2.19

Note: If the original window is larger than necessary (i.e. for escape purposes) the window opening **could** be reduced down to a minimum of 450 mm high and 450 mm wide, provided that it has an unobstructed openable area of at least 0.33 m^2 and the openable area is not more than 1100 mm above the floor.

What about glazing?

Although the installation of replacement windows or glazing (e.g. by way of repair) is not considered as building work under Regulation 3 of the Building Regulations, glazing that:

- is installed in a location where there was none previously;
- is installed as part of an erection;
- is installed as part of an extension or material alteration of a building;

is subject to these requirements.

The existence of large uninterrupted areas of transparent glazing represents a significant risk of injury through collision. This risk is at its most severe between areas of a building or its surroundings that are essentially at the same level and where a person might reasonably assume direct access between locations that are separated by glazing.

The most likely places where people can sustain injuries are due to impacts with doors, door side panels (especially between waist and shoulder level) when initial impact can be followed by a fall through the glazing resulting in additional injury to the face and body. Hands, wrists and arms are particularly vulnerable

Apart from doors, walls and partitions are a low-level, high-risk area, particularly where children are concerned.

Approved Document B: (*Fire Safety*) includes guidance on fire-resisting glazing and the reaction of glass to fire.

Approved Document K: (*Protection from Falling, Collision and Impact*) covers glazing that forms part of the protection from falling from one level

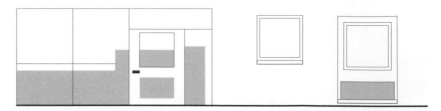

Figure 6.177 Shaded areas show critical locations in internal and external walls.

to another, and the need to ensure containment as well as limiting the risk of sustaining injury through contact.

Some glazing materials, such as annealed glass, gain strength through thickness; others, such as polycarbonates or glass blocks, are inherently strong. Some annealed glass is considered suitable for use in large areas forming fronts to shops, showrooms, offices, factories, and public buildings.

Provision of cavity barriers

Cavity barriers should be provided around window openings.	B3 6.3 (V1) B3 9.3 (V2)

Ventilation

Habitable rooms **without** openable windows may be ventilated through another habitable room (e.g. an internal room) **provided** that the other room has: • purge ventilation; • an 8000 mm² background ventilator; and • there is a permanent opening between the two rooms.	F 5.14
Habitable rooms **without** openable windows may also be ventilated through a conservatory provided that that conservatory has: • purge ventilation; • an 8000 m² background ventilator; and • there is a closable opening between the room and the conservatory that is equipped with:	F 5.15

- purge ventilation; and
- an 8000 mm² background ventilator.

Windows with night latches should **not** be used, as they are more liable to draughts as well as being a potential security risk.

F 4.19

If a fan is installed in an internal room or an office **without** an openable window, the fan should have a 15-minute overrun.

F Table 5.2c

Extensions

If the additional room is connected to an existing habitable room that now has no windows that open to the outside (or if it still has windows opening to outside, but with a total background ventilator equivalent area less than 5000 mm²), the ventilation opening (or openings) shall be greater than 8000 mm² equivalent area.

F 7.8ai

If the additional room is connected to an existing habitable room that still has windows opening to the outside (but with a total background ventilator equivalent area of at least 5000 mm² equivalent area), there should be:

F 7.8aiii

- background ventilators of at least 8000 m² equivalent area between the two rooms; and
- background ventilators of at least 8000 mm² equivalent area between the additional room and the outside.

Fabric standards

Reasonable provision would be for the proposed extension to include the following (**provided** that they meet the standards set by the Building Regulations!):

L2B 4.3

- doors, windows, roof windows, rooflights and smoke vents;
- newly constructed thermal elements; and
- existing opaque fabric which becomes a thermal element (and which have been upgraded).

Opening areas

> The area of windows and rooflights in the extension should   L2B 4.4
> generally not exceed the values given in Table 6.77.

Table 6.77 Opening areas in the extension

Building type	Windows and personnel doors as % of exposed wall	Rooflights as % of area of roof
Residential buildings where people temporarily or permanently reside	30	20
Places of assembly, offices and shops	40	20
Industrial and storage buildings	15	20
Vehicle access doors and display windows and similar glazing	As required	N/A
Smoke vents	N/A	As required

Protection from falling

> All stairs, landings, ramps and edges of internal floors shall   K3 (3.2)
> have a wall, parapet, balustrade or similar guard at least
> 900 mm high.
>
> All guarding should be capable of resisting at least the   K3 (3.2)
> horizontal force given in BS 6399: Part 1:1996.
>
> If glazing is used as (or as part of) the pedestrian guarding   N
> – then see Approved Document N: *Glazing – Safety in
> Relation to Impact, Opening and Cleaning.*
>
> If a building is likely to be used by children under   K3 (3.3)
> five years old, the guarding should **not** have horizontal
> rails, should stop children from easily climbing it, and
> the construction should prevent a 100 mm sphere being able
> to pass through any opening of that guarding.
>
> All external balconies and edges of roofs shall have a wall,   K3 (3.2)
> parapet, balustrade or similar guard at least 1100 mm high.
>
> All windows, skylights, and ventilators shall be capable   K
> of being left open without danger of people colliding with
> them, by:

> - installing windows, etc., so that projecting parts are kept away from people moving in and around the building; or
> - installing features that guide people moving in or about the building away from any open window, skylight or ventilator.
>
> Parts of windows (skylights and ventilators) that project   K either internally or externally more than about 1000 mm horizontally into spaces used by people moving in or about the building should not present a safety hazard.

Conservation of fuel and power

There have been changes in the legal requirements for the conservation of fuel and power. Regulation 9 of the Building Regulations exempts some conservatory and porch extensions from the energy-efficiency requirements. The exemption applies only for conservatories or porches:

- which are at ground level;
- where the floor area is less than $30\,m^2$;
- where the existing walls, doors and windows in the part of the dwelling which separates the conservatory are retained or, if removed, replaced by walls, windows and doors which meet the energy efficiency requirements; and
- where the heating system of the *dwelling* is not extended into the conservatory or porch.

See Appendix B to this book for further information concerning the requirements for the conservation of fuel and power.

It is recommended that the total area of windows, roof windows and doors in extensions to existing buildings do not exceed the sum of:

- 25 per cent of the floor area of the extension; plus
- the total area of any windows or doors which, as a result of the extension works, no longer exist or are no longer exposed.

Note: Responsibility for achieving compliance with the requirements of Part L rests with the person carrying out the work. That person may be, for example, a developer, a main (or sub-) contractor, or a specialist firm directly engaged by a private client.

The person responsible for achieving compliance should either provide a certificate themselves, or obtain a certificate from the subcontractor, that commissioning has been successfully carried out. The certificate should be made available to the client and the building control body.

Controlled fittings

In the context of conservation of fuel and power, the application of the term '*controlled fitting*' to a window, roof window, rooflight or door refers to a whole unit, i.e. **including** the frame. Consequently, replacing the glazing whilst retaining an existing frame is **not** providing a controlled fitting, and so such work is not notifiable and does not have to meet the Part L standards – although, where practical, it would naturally be sensible to do so! Similar arguments apply to a new door in an existing frame.

A *controlled fitting* is defined as a fitting in relation to Part G, H, J, L or P of Schedule 1 which imposes a requirement.

Domestic buildings

All windows, roof windows, rooflights and/or doors should be provided with draught-proofed units, the performance of which is no worse than as given in Table 6.78.	L1B 4.19

Table 6.78 Standards for control fittings

Fitting	Standard (W/m²K)
Windows, roof windows or rooflight	1.60
Doors with >50% of internal face glazed	1.80
Other doors	1.80

Where replacement windows are unable to meet the requirements of Table 6.78 (e.g. because of the need to maintain the external appearance of the façade and/or the character of the building) replacement windows should meet a centre pane U-value of $1.2\,\text{W/m}^2\text{k}$, or single glazing should be supplemented with low-e secondary glazing. In the latter case, the weather stripping should be on the secondary glazing to minimize condensation risk between the primary and secondary glazing.

In addition:

Insulated cavity closets should be installed where appropriate.	L1B 4.19
U-values shall be calculated (using the methods and conventions set out in BR 443) and should be based on the whole element or unit (e .g. in the case of a window, the combined performance of the glazing and the frame).	L1A 4.21

In the case of windows, the U-value can be taken as that for:

(a) the smaller of the two standard windows defined in BS EN 14351-1; or

(b) the standard configuration set out in BR 443; or

(c) the specific size and configuration of the actual window.

Note: The Table 6.79 sets out the worst acceptable standards for fabric properties. The stated value represents the area-weighted average value for **all** elements of that type. In general, the achievement of the thermal energy rating (TER) is likely to require significantly better fabric performance than is set out in Table 6.79.

Table 6.79 Limiting fabric parameters

Fitting	W/m²K
Roof	0.20
Wall	0.30
Floor	0.25
Party wall	0.20
Windows, roof windows, glazed rooflights, curtain walling and pedestrian doors	2.00
Air permeability	10.00 m³/h m² at 50 Pa

If a window is enlarged or a new one created, the area of windows, roof windows, rooflights and doors should not exceed 25 per cent of the total floor area of the *dwelling* unless compensating measures are included elsewhere in the work	L1B 4.23

Non-domestic buildings

Where windows, roof windows, rooflights or doors are to be provided, reasonable provision would be to install: • draught-proofed units whose area-weighted average performance is no worse than that given in Table 6.80; and • insulated cavity closers where appropriate.	L2B 4.24

 Where the replacement windows are unable to meet the requirements of Table 6.80 because of the need to maintain the external appearance of the façade or the character of the building, replacement windows should meet a centre pane U-value of 1.2 W/m²K, or single glazing should be supplemented with low-e secondary glazing.

Table 6.80 Standards for controlled fittings

Fitting	Standard (W/m²K)
Windows, roof windows and glazed rooflights	1.8 for the whole unit
Alternative option for windows in buildings that are essentially domestic in character	A window energy rating of band C
Plastic rooflight	1.8
Curtain walling	<1.8
Pedestrian doors where the door has more than 50% of its internal face area glazed	1.8 for the whole unit
High-usage entrance doors for people	3.5
Vehicle access and similar large doors	1.5
Other doors	1.8
Roof ventilators (including smoke extract ventilators)	3.5

> If a window, pedestrian door or rooflight is enlarged (or a new one is created), the area of windows and pedestrian doors and of rooflights expressed as a percentage of the total floor area of the building should not exceed the relevant value from Table 6.81, or should be compensated for in some other way.
>
> L2B 4.24

Table 6.81 Opening areas in the extension

Building type	Windows and personnel doors as % of exposed wall	Rooflights as % of area of roof
Residential buildings where people temporarily or permanently reside	30	20
Places of assembly, offices and shops	40	20
Industrial and storage buildings	15	20
Vehicle access doors and display windows and similar glazing	As required	N/A
Smoke vents	N/A	As required

> U-values of windows, roof-windows, rooflights and doors shall be calculated using the methods and conventions set out in BR 443 (*Conventions for U-value Calculations**) and should be based on the whole
>
> L2B 4.25

unit (i.e. in the case of a window, the combined
performance of the glazing and frame).

* http://www.bre.co.uk/filelibrary/pdf/rpts/BR_443_
(2006_Edition).pdf

 Note: In certain classes of building with high internal gains, a less demanding
U-value for glazing may be an appropriate way of reducing overall CO_2
emissions, in which case the average U-value for windows, doors and rooflights
can be relaxed from the values given in Table 6.80, but the value should not
exceed $2.7\,W/m^2\,K$.

Material change of use

Although the erection of a new dwelling is not a material change of use, the
requirements for the conservation of fuel and power still apply where a dwelling is being created in an existing building as the result of a material change of
use of all, or part of, the building.

Subsequent to a material change of use, the following will affect the building's energy status. For example, where replacement units are provided (e.g.
for an existing window, roof window, rooflight or door separating a conditioned space from an unconditioned space or external environment) which has
a U-value worse than $3.3\,W/m2\,K$.

> If the area of the openings in the newly created dwelling is   L1B 4.15
> more than 25 per cent of the total floor area, either the area
> of openings should be reduced, or the larger area should
> be compensated for in some other way using the procedure
> described in SAP 2009.

In buildings other than dwellings, a change to a building's energy status
is any change which results in a building becoming a building to which the
energy-efficiency requirements of the Building Regulations apply, where previously they did not.

 Where there is a change in the building's energy status, such work, if any,
shall be carried out to ensure that the building complies with the applicable
requirements for the conservation of fuel and power.

 Note: As well as satisfying the energy-efficiency requirements in respect of
the material change of use or change in energy status, such building work may
be one of the triggers for consequential improvements (see below).

Consequential improvements

If a building has a total useful floor area greater than $1000\,m^2$ and the proposed building work includes:

- an extension;
- the initial provision of any fixed building services; or
- an increase to the installed capacity of any fixed building services;

then, in addition to the proposed building work (i.e. the principal works), consequential improvements will have to be completed provided that they are technically, functionally and economically feasible.

For example, replacing existing windows, roof windows or rooflights (but excluding display windows) or doors (but excluding high-usage entrance doors) which have a U-value worse than $3.3\,W/m^2\,K$, will achieve a simple payback within 15 years provided that they are economically feasible and there are no unusual circumstances.

Note: *Consequential improvements* means those energy-efficiency improvements required by Regulation 17D.

Where the installed capacity per unit area of a **heating system** is increased:	L2B 6.10a
• the thermal elements within the area served which have U-values worse than those set out in column (a) of Table 6.82 should be upgraded; and	
• existing windows, roof windows or rooflights (but excluding display windows) or doors (but excluding high-usage entrance doors) within the area served and which have U-values worse than $3.3\,W/m^2\,K$ should be replaced.	L2B 6.10b
Where the installed capacity per unit area of a **cooling system** is increased:	L2B 6.11a
• thermal elements within heated areas which have U-values worse than those set out in column (a) of Table 6.82 should be upgraded; and	
• if the area of windows, roof windows (but excluding display windows) within the area served exceeds 40 per cent of the façade area (or the area of rooflights exceeds 20 per cent of the area of the roof) and, the design solar load exceeds $25\,W/m^2$, the solar control provisions should be upgraded such that at least one of the following four criteria is met:	L2B 6.11b

1. the solar gain per unit floor area averaged over the period 06:30 to 16:30 GMT is not greater than $25\,W/m^2$ when the building is subject to solar irradiances for July as given in the table of design irradiancies in *CIBSE Design Guide A*;
2. the design solar load is reduced by at least 20 per cent;
3. the effective g-value is no worse than 0.3;
4. the zone or zones satisfies the criterion 3 check in Approved Document L2A – based on calculations by an approved software tool.

Note: Any general lighting system within the area served by the relevant fixed building service which has an average lamp efficacy of less than 45 lamp-lumens per circuit-watt should be upgraded with new luminaires and/or controls following the guidance in the *Non-Domestic Building Services Compliance Guide*.

Table 6.82 Upgrading retained thermal elements

Element	(a) Threshold U-value W/m²K	(b) Improved U-value W/m²K
Wall – cavity insulation	0.70	0.55
Wall – external or cavity insulation	0.70	0.30
Floor	0.70	0.25
Pitched roof – insulation at ceiling level	0.35	0.16
Pitched roof – insulation between rafters	0.35	0.18
Flat roof or roof with integral insulation	0.35	0.18

All existing windows (less display windows), roof windows, rooflights or doors (excluding high usage entrance doors) that are within the area served by the fixed building service and which have a U-value worse than $3.3\,W/m^2K$, should be replaced.

L2B 6.10b

Where the original windows were fitted with trickle ventilators, the replacement windows should include them and they should be sized as set out Table 6.83.

F 7.3

Table 6.83 Equivalent areas for replacement windows – dwellings

Type of room	Equivalent area (mm2)
Habitable rooms	2500
Kitchen	2500
Utility room	2500
Bathroom (without WC)	2500

Where the original windows were not fitted with trickle ventilators and the room is not ventilated adequately by other installed provisions, it would be good practice to fit trickle ventilators (or an equivalent means of ventilation) to help with control of condensation and improve indoor air quality. Ventilation devices should be fitted with accessible controls.

Where there was no previous ventilation opening, or where the size of the original ventilation opening is not known, the replacement window(s) **shall** be greater than the minimum requirements shown in Tables 6.83 and 6.85.

F 7.6

Table 6.84 Equivalent areas for replacement windows – buildings other than dwellings

Type of room	Equivalent area (mm²)
Occupiable rooms with a floor areas 10 m²	2500
Occupiable rooms with a floor area 10 m²	250 per m² of floor area
Kitchens (domestic type)	2500
Bathrooms and shower rooms	2500 per bath or shower
Sanitary accommodation (and/or washing facilities)	2500 per WC

Thermal bridges

The building fabric should be constructed so that there are no reasonably avoidable thermal bridges in the insulation layers caused by gaps within the various elements, at the joints between elements, and at the edges of elements around windows and door openings.

L1A 5.9
L1B 5.3
L2A 5.3

 Provision should also be made to reduce unwanted air leakage through the new envelope parts.

Solar energy

> Reasonable provision should be made to limit solar gains   L1A 4.25

 Limiting the effects of solar gain in summer can be achieved by an appropriate combination of window size and orientation, and solar protection through shading.

Limiting the effects of solar gains in summer

The following guidance applies to all buildings, irrespective of whether or not they are air-conditioned. The intention is to limit solar gains during the summer period to either:

* reduce the need for air-conditioning; or
* reduce the installed capacity of any air-conditioning system that is installed.

The aim should be to ensure that the solar gains through the glazing aggregated over the period from April to September (inclusive) are no greater than every space defined in the National Calculation Methodology (NCM) database as being:

> **Side lit** – the reference case is an east-facing façade   L2A 4.44
> with full-width glazing to a height of 1.0 m having a
> framing factor of 10 per cent and a normal solar energy
> transmittance (g-value) of 0.68.
>
> **Top lit** (and whose average zone height is not greater than   L2A 4.44
> 6 m) – the reference case is a horizontal roof of the same
> total area that is 10 per cent glazed as viewed from the
> inside out and having rooflights that have a framing factor
> of 25 per cent and a normal solar energy transmittance (g-
> value) of 0.68.
>
> **Top lit** (and whose average zone height is greater than   L2A 4.44
> 6 m) – the reference case is a horizontal roof of the same
> total area that is 20 per cent glazed as viewed from the
> inside out and having rooflights that have a framing factor
> of 15 per cent and a normal solar energy transmittance (g-
> value) of 0.46.

 Note: For the purpose of this specific guidance, an *occupied space* means a space that is intended to be occupied by the same person for a substantial part of the day. This excludes circulation spaces, and other areas of transient occupancy (e.g. toilets), as well as spaces that are not intended for occupation (e.g. display windows).

Fabric properties

 Fabric elements and fixed building services should satisfy minimum energy-efficiency standards.

L2A 4.29

Non-domestic buildings

U-values shall be calculated (using the methods and conventions set out in BR 443) and should be based on the whole element or unit (e.g. in the case of a window, the combined performance of the glazing and the frame).

L2A 4.31

In the case of windows, the U-value can be taken as that for:

(a)   the smaller of the two standard windows defined in BS EN 14351-112; or
(b)   the standard configuration set out in BR 443; or
(c)   the specific size and configuration of the actual window.

 Note: Table 6.85 sets out the worst acceptable standards for fabric properties. The stated value represents the area-weighted average value for **all** elements of that type.

Table 6.85 Limiting fabric parameters (non domestic buildings)

Fitting	W/m²K
Roof	0.25
Wall	0.35
Floor	0.25
Windows, roof windows, glazed rooflights, curtain walling and pedestrian doors	2.20
Air permeability	10.00 m³/h m² at 50 Pa

Building extensions

In a dwelling, the total area of windows, roof windows and doors in extensions shall not exceed the sum of:

L1B 4.2

*   25 per cent of the floor area of the extension; plus
*   the total area of any windows or doors

In non-domestic buildings, the area of windows and rooflights in the extension should not exceed the values given in Table 6.86.

L2B 4.4

However, where a greater proportion of glazing is present in the part of the building to which the extension is attached, reasonable provision would be to limit the proportion of glazing in the extension so that it is no greater than the proportion that exists in the part of the building to which it is attached.

Table 6.86 Opening areas in the extension of a non domestic building

Building type	Windows and personnel doors as % of exposed wall	Rooflights as % of roof area
Residential buildings where people temporarily or permanently reside	30	20
Places of assembly, offices and shops	40	20
Industrial and storage buildings	15	20
Vehicle access doors and display Windows and similar glazing	As required	N/A
Smoke vents	N/A	As required

Conservatories and porches

 Regulation 9 of the Building Regulations exempts some conservatory and porch extensions from the energy-efficiency requirements.

This exemption applies only to conservatories or porches:

- which are at ground level;
- where the floor area is less than $30\,m^2$;
- where the existing walls, doors and windows which separate the conservatory from the building are retained or, if removed, are replaced by walls, windows and doors which meet the energy-efficiency requirements; and
- and where the heating system of the building is not extended into the conservatory or porch.

Dwellings

Some conservatory and porch extensions are exempt from the energy-efficiency requirements, such as where the existing walls, doors and windows in the part of the dwelling which separates the conservatory are retained or, if removed, replaced by walls, windows and doors which meet the energy efficiency requirements.

If a new conservatory or porch does not meet all these requirements it is **not** exempt and must comply with the relevant energy-efficiency requirements of Part L:2010,

In order to comply with Part L, conservatories or porches should provide:

Effective thermal separation between the heated area in the existing dwelling (i.e. the walls, doors, and windows between the dwelling and the extension should be insulated and draught proofed to at least the same extent as in the existing dwelling).	L1A 4.11a
Glazed elements that meet the standards set out in Table 6.87 and opaque elements that meet the standards set out in Table 6.88.	L1A 4.11b

Removing, and not replacing, any or all of the thermal separation between the dwelling and an existing exempt extension, or extending the dwelling's heating system into the extension, means the extension ceases to be exempt and should be treated as a conventional extension.

Table 6.87 Standards for control fittings in conservatories and porches (dwellings)

Fitting	Standard (W/m²K)
Window, roof window or rooflight	1.6
Doors with >50% of internal face glazed	1.8
Other doors	1.8

Table 6.88 Standards for new thermal elements in conservatories and porches (dwellings)

Element	Standard (W/m²K)
Wall	0.28
Pitched roof – insulation at ceiling level	0.16
Pitched roof – insulation at rafter level	0.18
Flat roof or roof with integral insulation	0.18
Floor	0.22
Swimming pool basin	0.25

Buildings other than dwellings

The energy-efficiency requirements given in Table 6.89 are relevant to buildings other than dwellings.

Table 6.89 Standards for controlled fittings in conservatories and porches – buildings other than dwellings

Fitting	Standard (W/m²K)
Windows, roof windows and glazed rooflights	1.8 for the whole unit
Alternative option for windows in buildings that are essentially domestic in character	A window energy rating of band C
Plastic rooflight	1.8
Curtain walling	< 1.8
Pedestrian doors where the door has more than 50% of its internal face area glazed	1.8 for the whole unit
Other doors	1.8
Roof ventilators (including smoke extract ventilators)	3.5

Historic and traditional buildings

Building inspectors normally require the use of 'sympathetic treatment' when restoring the historic character of a building that has been subject to previous inappropriate alteration (e.g. replacement doors).

Particular issues that could warrant sympathetic treatment – and where advice from others would probably be beneficial – include:

> • restoring the historic character of a building that has   L2B 3.11
>   been subject to previous inappropriate alteration (e.g.
>   replacement windows, doors and rooflights).

Renovation

Protection against impact

> Measures shall be taken to limit the risk of sustaining   N1 (0.1)
> cutting and piercing injuries.
>
> In critical locations, if glazing is damaged the breakage   N1 (0.2)
> should only result in small, relatively harmless particles.
>
> Glazing should be sufficiently robust to ensure that the risk   N1 (0.4)
> of breakage is low.

Table 6.90 U-value targets for undertaking renovation works

Proposed works	Target U-value (W/m² K)	Typical construction	Comments (reasonableness, practicability and cost-effectiveness)
Pitched roof construction			
Renewal of roof covering with living accommodation in roof space (room-in-the-roof type arrangement), with or without dormer windows	0.18	Cold structure – insulation (thickness dependent on material) placed between and below rafters Warm structure – insulation placed between and above rafters	Assess condensation risk (particularly interstitial condensation), and make appropriate provision in accordance with the requirements of Part C relating to the control of condensation Practical considerations with respect to an increase in structural thickness (particularly in terraced dwellings) may necessitate a lower performance target
Dormer window constructions			
Renewal of cladding to side walls	0.30	Insulation (thickness dependent on material) either placed between and/or fixed to outside of wall studs, or (depending on the construction) placed externally to the existing structure	Assess condensation risk and make appropriate provision in accordance with the requirements of Part C
Renewal of roof covering	–	Follow guidance on improvement to pitched or flat roofs as appropriate	Assess condensation risk and make appropriate provision in accordance with the requirements of Part C

Steps should be taken to limit the risk of contact with the   N1 (0.5)
glazing.

Glazing in critical locations should either be permanently   N1 (1.2)
protected, be in small panes or, if it breaks, break safely
(see BS 6206).

Small panes should not exceed 250 mm and an area of 0.5 m². N1 (1.6)

Transparent glazing

Transparent glazing with which people are likely to come into contact while
moving in or about the building shall incorporate features that make it apparent.

The presence of glazing should be made apparent or   N2 (0.8)
visible to people using the building.

The presence of large uninterrupted areas of transparent   N2 (0.6, 2.1,
glazing should be clearly indicated.   2.2)

In critical locations (i.e. large areas where the glazing   N2 (2.4–2.5)
forms part of internal or external walls and doors of
shops, showrooms, transoms, offices, factories, public
or other non-domestic buildings) the presence of large
uninterrupted areas of transparent glazing should be
clearly indicated by the use of broken or solid lines,
patterns or company logos at appropriate heights and
intervals.

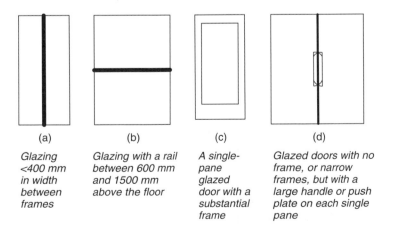

(a)	(b)	(c)	(d)
Glazing <400 mm in width between frames	Glazing with a rail between 600 mm and 1500 mm above the floor	A single-pane glazed door with a substantial frame	Glazed doors with no frame, or narrow frames, but with a large handle or push plate on each single pane

Figure 6.178 Examples of door height glazing not requiring further
identification.

Thermoplastic materials

Thermoplastic materials may be used in windows, rooflights and lighting diffusers in suspended ceilings.	B2 3.8 (V1) B2 6.10 (V2)
External windows to rooms (other than to circulation spaces) may be glazed with thermoplastic materials.	B2 3.9 (V1) B2 6.11 (V2)

Safe opening and closing of windows

Windows, skylights and ventilators that can be opened by people should be capable of being opened, closed or adjusted safely.

Where controls can be reached without leaning over an obstruction they should not be more than 1.9 m above the floor. Where there is an obstruction, the control should be lower (e.g. not more than 1.7 m where there is a 600 mm deep obstruction).	N3 (3.2)
Where controls cannot be positioned within safe reach from a permanent stable surface, a safe means of remote operation (e.g. a manual or electrical system) should be provided.	N3 (3.2)
Where there is a danger of the operator or other person falling through a window above ground floor level, suitable opening limiters should be fitted or guarding provided.	N3 (3.3)

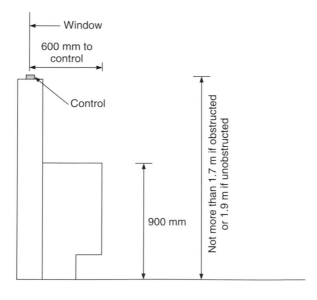

Figure 6.179 Height of controls.

Safe access for cleaning

> All windows, skylights (or any transparent or translucent walls,   N
> ceilings or roofs) of a dwelling should be safely accessible for
> cleaning.

Where glazed surfaces cannot be cleaned safely by a person standing on the ground, the requirement for a floor, or other permanent stable surface, could be satisfied by provisions such as the following:

> Safe means of access shall be provided for cleaning both   N4
> sides of glazed surfaces where there is danger of falling
> more than 2 m.
>
> Where possible, windows should be of a size and design   N4 (4.2)
> that allow the outside surface to be cleaned safely from
> inside the building.
>
> Windows that reverse for cleaning should be fitted with a   N4 (4.2)
> mechanism that holds the window in the reversed position
> (see BS 8213).
>
> For large buildings (e.g. office blocks) a firm, level surface   N4 (4.2)
> shall be provided to enable portable ladders (not more than
> 9 m long) to be used and the use of suspended cradles,
> travelling ladders, or abseiling equipment should also be
> considered.

6.14 Doors

6.14.1 Requirements

Ventilation

There shall be adequate means of ventilation provided for people in the building.

(Approved Document F)

 A new Part F came into force on 1 October 2010.

Conservation of fuel and power

Reasonable provision shall be made for the conservation of fuel and power in buildings by:

(a) *limiting heat gains and losses –*
  (i) *through thermal elements and other parts of the building fabric; and*
  (ii) *from pipes, ducts and vessels used for space heating, space cooling and hot water services;*
(b) *providing fixed building services which –*
  (i) *are energy efficient;*
  (ii) *have effective controls; and*
  (iii) *are commissioned by testing and adjusting as necessary to ensure they use no more fuel and power than is reasonable in the circumstances; and*
(c) *providing to the owner sufficient information about the building, the fixed building services and their maintenance requirements so that the building can be operated in such a manner as to use no more fuel and power than is reasonable in the circumstances.*

(Approved Document L)

 A new Part L came into force on 1 October 2010.

Fire safety

- *There shall be sufficient escape routes that are suitably located to enable persons to evacuate the building in the event of a fire.*
- *Safety routes shall be protected from the effects of fire.*
- *In an emergency, the occupants of any part of the building shall be able to escape without any external assistance.*

(Approved Document B1)

Access to and use of buildings

Reasonable provisions shall be made for people to:

(a) *gain access to; and*
(b) *use*

the building and its facilities

(Approved Document M)

In addition to the requirements of the Disability Discrimination Act 1995, precautions need to be taken to ensure that:

- new non-domestic buildings and/or dwellings (e.g. houses and flats used for student living accommodation, etc.);
- extensions to existing non-domestic buildings;
- non-domestic buildings that have been subject to a material change of use (e.g. so that they become a hotel, boarding house, institution, public building or shop);

are capable of allowing people, regardless of their disability, age or gender, to:

- gain access to buildings;
- gain access within buildings;

- be able to use the facilities of the buildings (both as visitors and as people who live or work in them);
- use sanitary conveniences in the principal storey of any new dwelling.

 Note: See Appendix A for guidance on access and facilities for disabled people.

6.14.2 Meeting the requirement

Ventilation

To ensure good transfer of air throughout the dwelling, there shall be an undercut of 7600 mm^2 (minimum) in all internal doors above the floor finish (equivalent to an undercut of 10 mm for a standard 760 mm width door).   F Table 5.2a

The height times the width of an external door (including patio doors) should be at least 1/20 of the floor area of the room.   F App B

Note: If a room contains more than one external door (or a combination of at least one external door and at least one openable window), the areas of **all** the opening parts may be added together to achieve at least 1/20 of the floor area of the room.

Conservation of fuel and power

Energy-efficiency measures shall be provided that limit the heat loss through the doors, etc., by suitable means of insulation.

Responsibility for achieving compliance with the requirements of Part L rests with main (or sub-) contractor, or a specialist firm directly engaged by a private client.

The person responsible for achieving compliance should either provide a certificate themselves, or obtain a certificate from the subcontractor, that commissioning has been successfully carried out. The certificate should be made available to the client and the building control body.

Extensions

In a dwelling, the area of doors in extensions should not exceed the sum of: L1B 15

- 25% of the floor area of the extension; **plus**
- the area of any windows or doors, which, as a result of the extension works, no longer exist or are no longer exposed.

Note: if the total floor area of the proposed extension exceeds these limits, the work should be regarded as a **new** building and the requirements of Approved Document L2A and L2B should be used.

The U-value of thermal and/or opaque doors should not exceed 3.3 W/m² K.	L2B 32
Doors between the building and an extension should be insulated and weather-stripped to at least the same extent as in the existing building.	L1B 22a L2B 32a
Glazed elements shall comply with the standards given in Table 6.91.	L1B 22c L2B 32c
The building fabric should be constructed so that any thermal bridges in the insulation layers around windows (caused by gaps and joints between the various elements) are avoided.	L1B 52 L2B 71

Table 6.91 Standards for glazed elements in conservatories

Element	Type of building	Replacement fittings (W/m² K)
Doors with more than 50% of their internal face glazed	Existing dwelling	2.2 (whole unit)
		1.2 (centre pane)
Doors with less than 50% of their internal face glazed	Existing dwelling	3.0
High-usage entrance doors	Existing building	6.0
Vehicle-access and large doors	Existing building	1.5

U-values

U-values for building fabric elements (calculated using the methods and conventions set out in BR 443) shall not exceed those shown in Table 6.92.	L1B 10 L2B 29 L2B 29c

Table 6.92 Limiting U-value standards

Element	Area-weighted dwelling average (W/m² K)	Worst individual subelement (W/m² K)
Doors	2.2	3.3
Pedestrian doors	2.2	3.0
Vehicle-access and similar large doors	1.5	4.0
High-usage entrance doors	6.0	6.0

Controlled fittings (non-domestic buildings)

Windows, roof windows, rooflights and/or doors should be provided with draught-proofed units.	L2B 75
The area-weighted average performance of draught-proofed units for new fittings in extensions and replacement fittings in an existing dwelling shall be no worse than given in Table 6.93.	L2B 75

Table 6.93 Standards for controlled fittings

Element	New fittings in an extension (W/m²K)	Replacement fittings in an existing dwelling (W/m²K)
Pedestrian doors having more than 50% of their internal face area glazed	2.2	2.2
High-usage entrance doors	6.0	6.0
Vehicle-access and large doors	1.5	1.5

Consequential improvements (non-domestic buildings)

If a building has a total useful floor area greater than $1000\,m^2$ and the proposed building work includes:

- an extension; or
- the initial provision of any fixed building services; or
- an increase to the installed capacity of any fixed building services;

then consequential improvements should be made to improve the energy efficiency of the whole building by replacing:

all existing doors (excluding high-usage entrance doors) within the area served by the fixed building service that have a U-value worse than $3.3\,W/m^2\,K$ with doors whose U-value is less than $2.2\,W/m^2\,K$.	L2B 18-7

Material changes of use (domestic buildings)

When a building is subject to a material change of use, then:

Any thermal element that is being retained should be upgraded.	L1B 27d
Any existing door which separates a conditioned space from an unconditioned space (or the external environment) and which has a U-value that is worse than 3.3 W/m² K, should be replaced by a door whose U-value is less than 2.2 W/m² K.	L1B 27e

Work on controlled services or fittings (domestic buildings)

When working on a controlled service or fitting (i.e. where the service or fitting is subject to the requirements of Part G, H, J, L or P of Schedule 1), and where windows, roof windows, rooflights and/or doors are to be provided:

Doors should be provided with draught-proofed units.	L1B 32
The area-weighted average performance of draught-proofed units for new and replacement fittings that are provided as part of the construction of an extension shall be no worse than given in Table 6.94.	L1B 32

Table 6.94 Standards for thermal elements for new fittings in an extension

Element	New fittings in an extension (W/m² K)	Replacement fittings in an existing dwelling (W/m² K)
Doors	1.8	2.0 (whole unit) 1.2 (centre pane)
Doors with more than 50% of their internal face glazed	2.2	2.2
Other doors	3.0	3.0

Dwellings should be constructed and equipped so that there are no reasonably avoidable thermal bridges in the insulation layers caused by gaps within the various elements, at the joints between elements, and at the edges of door openings.	L2A 68

Access to and use of buildings

Doors and gates on main traffic routes (and those that can be pushed open from either side) should have vision panels, unless they are low enough (e.g. 900 mm) to see over.	K5 (5.2a)
Sliding doors and gates should have a retaining rail to prevent them falling should the suspension system fail or the rollers leave the track.	K5 (5.2b)
Upward-opening doors and gates should be fitted with a device to stop them falling in a way that could cause injury.	K5 (5.2c)
Power-operated doors and gates should have safety features to prevent injury to people who are struck or trapped (e.g. a pressure-sensitive door edge that operates the power switch).	K5 (5.2d)
Power-operated doors and gates should have a readily identifiable and accessible stop switch.	K5 (5.2d)
Power-operated doors and gates should be provided with a manual or automatic opening device in the event of a power failure where and when necessary for health or safety.	K5 (5.2d)

Internal doors

The opening force for a manually operated door should not exceed 20 N.	M (3.10a)
The effective clear width through a single-leaf door (or one leaf of a double-leaf door) should be in accordance with Table 6.95 and Figure 6.180.	M (3.10b)

Table 6.95 Minimum effective clear widths of doors

Direction and width of approach	New buildings (mm)	Existing buildings (mm)
Straight-on (without a turn or oblique approach)	800	750
At right angles to an access route at least 1500 mm wide	800	750
At right angles to an access route at least 1200 mm wide	825	775
External doors to buildings used by the general public	1000	775

Effective clear width
(door stop to door leaf)

Figure 6.180 Effective clear width and visibility requirements of doors.

There should be an unobstructed space of at least 300 mm   M (3.10c)
on the pull side of the door between the leading edge of
the door and any return wall (unless the door is a powered
entrance door) (see Figure 6.180).

A space alongside the leading edge of a door should be
provided to enable a wheelchair user to reach and grip the
door handle.

Door opening furniture should:

- be easy to operate by people with limited manual   M (3.10d)
  dexterity;
- be capable of being operated with one hand using a   M (3.10d)
  closed fist (e.g. a lever handle);
- contrast visually with the surface of the door.   M (3.10e)

Door frames should contrast visually with the surrounding   M (3.10f)
wall.

The surface of the leading edge of a non-self-closing door   M (3.10g)
should contrast visually with the other door surfaces and
its surroundings.

Door leaves or side panels should be wider than 450 mm.   M (3.10h)

Vision panels towards the leading edge of the door should   M (3.10h)
include a visibility zone (or zones) between 500 mm and
1500 mm from the floor. If interrupted (e.g. to accommodate
an intermediate horizontal rail – see Figure 6.181) then this
should be 800 mm and 1150 mm above the floor.

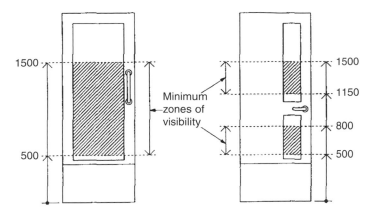

Figure 6.181 Door vision panels.

Glass entrance doors and glazed screens should be clearly marked (i.e. with a logo or sign) on the glass at two levels, 850–1000 mm and 1400–1600 mm above the floor.	M (3.10i)
It should be possible to tell between a fully glazed door and any adjacent glazed wall/partition by providing a high-contrast strip at the top and on both sides.	M (3.10j)
Fire doors (**particularly those in corridors**) should be held open with an electromagnetic device that is capable of self-closing when: • the power supply fails; • activated by smoke detectors; • activated by a hand-operated switch.	M (3.10k)
Fire doors (**particularly to individual rooms**) should be fitted with swing-free devices that close when: • activated by smoke detectors; • the building's fire alarm system is activated; • the power supply fails.	M (3.10l)
Low-energy powered door systems may be used in locations not subject to frequent use or heavy traffic.	M (3.7)
Low-energy powered swing-door systems should be capable of being operated: • in manual mode; • in powered mode; or • in power-assisted mode.	M (3.10 m)

The use of self-closing devices should be minimized as they disadvantage many people (e.g. those pushing prams or carrying heavy objects).	M (3.7)
If closing devices are needed for fire control:	M (3.7)
• they should be electrically powered hold-open devices or swing-free closing devices;	
their closing mechanism should only be activated in case of emergency.	
The presence of doors, whether open or closed, should be apparent to visually impaired people.	M (3.8)

Note: See BS 8300 for guidance on:

- electrically powered hold-open devices;
- swing-free systems;
- low-energy powered door systems.

Fire safety

Cavity barriers

Cavity barriers should be provided for all door openings.	B3 6.3 (V1) B3 9.3 (V2)
Openings in a cavity barrier should be limited to those for doors that have at least 30 minutes fire resistance.	B3 6.8 (V1) B3 9.13 (V2)

Note: Detailed guidance on door openings and fire doors is given in Appendix B of Part B.

Emergency egress doors

The door should enable the person escaping to reach a place free from danger of fire (e.g. a courtyard or back garden which is at least as deep as the dwelling-house is high – see Figure 6.182).	
In a gallery, if the floor is not provided with an alternative exit or escape window:	B1 2.12 (V1) B1 2.8 (V2)

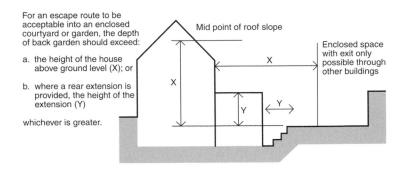

For an escape route to be acceptable into an enclosed courtyard or garden, the depth of back garden should exceed:

a. the height of the house above ground level (X); or

b. where a rear extension is provided, the height of the extension (Y)

whichever is greater.

Mid point of roof slope

Enclosed space with exit only possible through other buildings

Figure 6.182 Ground and basement storey exit into an enclosed space.

- the distance between the foot of the access stair to the gallery and the door to the room containing the gallery should not exceed 3 m.

Where an external escape stair is provided:

- all doors giving access to the stair should be fire-resisting;
- doors within 1800 mm of the escape route shall be protected by fire-resisting construction.

B1 2.15 (a, b and c)

 Note: Glazing in any fire-resisting construction should be fire-resisting and fixed shut.

Basements

Owing to the risk that a single stairway may be blocked by smoke from a fire in the basement or ground storey:

Basement storeys in a dwelling-house that contain a habitable room shall be provided with either an external door or window suitable for egress from the basement.

B1 2.13 (V1)
B1 2.6 (V2)

Fire alarm systems

When fitted, manual call points for fire alarm systems shall be adjacent to exit doors.

B1 1.29 (V2)

Fire doors

Fire doors now **only** need to be provided with self-closing devices if they are between a dwelling-house and an integral garage.

Self-closing fire doors should be positioned so that smoke will not affect access to more than one stairway.

Doors that need to be fire-resisting should meet the requirements given in Table B1 of Appendix B to Part B.	B1 5.6 (V2)
Doors on escape routes should be hung to open not less than 90°.	B1 5.15 (V2)
Doors giving access to an external escape should be fire-resisting and self-closing.	B1 5.25 (V2)
Door-closing devices for fire doors: • shall be in accordance with BS EN 1155:1997; • should take account of the needs of residents; • should have free-swing door closers in bedrooms; • should have hold-open devices in circulation spaces.	B1 3.51 (V2)
Doors on escape routes (both within and from the building) should be readily openable.	B1 5.10 (V2)
Doors on escape routes (whether or not the doors are fire doors), should either not be fitted with lock, latch or bolt fastenings, or they should only be fitted with simple fastenings that can be readily operated from the side approached by people making an escape, without the use of a key and without having to manipulate more than one mechanism.	B1 5.11 (V2)
Doors that open towards a corridor or a stairway should be sufficiently recessed to prevent swing from encroaching on the effective width of the stairway or corridor.	B1 5.16 (V2)
Fire doors should be capable of performing in accordance with Table 6.75 of Appendix B (to Volume 1 of Part B).	App B 1 (V1)
Fire doors should be fitted with a self-closing device except for: • fire doors to cupboards and to service ducts that are normally kept locked shut; and • fire doors within flats.	App B 2 (V2)

Fire doors should be marked with the appropriate fire safety sign complying with BS 5499-5:2002 according to whether the door is:	App B 8 (V2)

- to be kept closed when not in use (*Fire door – keep shut*);
- to be kept locked when not in use (*Fire door – keep locked shut*); or
- held open by an automatic-release mechanism or free-swing device (*Automatic fire door – keep clear*).

Fire doors serving an attached or integral garage should be fitted with a self-closing device.	App B 2 (V1)
Electrically powered locks should return to the unlocked position:	B1 5.11 (V2)

- on operation of the fire alarm system;
- on loss of power or system error;
- on activation of a manual door-release unit.

In assembly places, shops and commercial buildings, doors on escape routes from rooms with an occupant capacity of more than 60 should either not be fitted with lock, latch or bolt fastenings, or be fitted with panic fastenings in accordance with BS EN 1125: 1997.	B1 5.12 (V2)

 Note: See also Appendix B for guidance about door closing and 'hold-open' devices for fire doors.

No more than 25% of the length of a compartment wall should consist of door openings.	App B 5 (V2)
Revolving doors, automatic doors and turnstiles should not be placed across escape routes.	B1 5.18 (V2)
Secure doors that need to be operated by a code, combination, swipe or proximity card, biometric data or similar means should also be capable of being overridden from the side approached by people making their escape.	B1 5.11 (V2)c
Self-closing fire doors may be held open by:	App B 3 (V2)

- a fusible link; or

• an automatic-release mechanism actuated by an automatic fire detection and alarm system; or • a door-closer delay device.	
The door of any doorway or exit should be hung to open in the direction of escape.	B1 5.14 (V2)
The essential components of any hinge on which a fire door is hung should be made entirely from materials having a melting point of at least 800° C.	App B 3 (V1) App B 7 (V2)
Two fire doors may be fitted in the same opening so that the total fire resistance is the sum of their individual fire resistances, provided that each door is capable of closing the opening.	App B 4 (V2)
Vision panels shall be provided where doors on escape routes subdivide corridors, or where any doors are hung to swing both ways.	B1 5.17 (V2)
Fire doors (except those to cupboards and to service ducts) should be marked on both sides.	B1

 Note: See also Parts M and N.

Fire doors in dwellings

The minimum fire resistance for doors in buildings other than dwellings as is given in Table B1 (page 134) of Part B Volume 2.

BS 8214:1990 gives recommendations for the specification, design, construction, installation and maintenance of fire doors constructed with non-metallic door leaves.

Guidance on timber fire-resisting doorsets may be found in *Timber Fire-resisting Doorsets: Maintaining Performance under the New European Test Standard,* which is published by TRADA.

In flats where the habitable rooms do not have direct access to the entrance hall, bedrooms should be separated from the living accommodation by fire-resisting construction and fire doors (B1 2.14(V2)).

Fire-protected stairways

Fire-protected stairways shall (as far as is reasonably possible) consist of fire-resistant material and fire-resistant doors, and have an appropriate form of smoke-control system.	B1 1.viii

Fire service (access and facilities)

Doors should be provided such that there is no more than 60 m between each door and/or the end of that elevation (e.g. a 150 m elevation would need at least two doors).

B5 16.5 (V2)

Every elevation to which vehicle access is provided should have a suitable door(s), not less than 750 mm wide, giving access to the interior of the building.

B5 11.3 (V2)
B5 16.5 (V2)

Garages

Fire doors now only need to be provided with self-closing devices if they are between a dwelling-house and an integral garage.

If a door is provided between a dwelling-house and the garage, the floor of the garage should:

B3 5.5 (V1)

* be laid to allow fuel spills to flow away from the door to the outside; or
* the door opening should be positioned at least 100 mm above garage floor level.

Loft conversions

Where a loft conversion means that a new storey is going to be added, then the stairway will need to be protected with fire-resisting doors and partitions.

B1 2.20b

Means of escape

Common corridors that connect two or more storey exits should be subdivided by a self-closing fire door with, if necessary, an associated fire-resisting screen.

B1 2.28 (V2)

Every corridor more than 12 m long and which connects two or more storey exits should be subdivided by self-closing fire doors positioned approximately midway between the two storey exits.

B1 3.26 (V2)

Except for doorways, all escape routes should have a clear headroom of not less than 2 m.	B1 3.17 (V2)
Except for kitchens, all habitable rooms shall be provided with suitable means for emergency egress from each storey via doors or windows.	B1 2.1–2.4 (V1) B1 2.11–2.12 (V2)
If the only escape route from an inner room is through another room, there should be a vision panel not less 0.1 m² located in the door or walls of the inner room.	B1 3.10 (V2)
Unless escape stairways and corridors are protected by a pressurization system complying with BS EN 12101-6:2005, every dead-end corridor exceeding 4.5 m in length should be separated by self-closing fire doors.	B1 3.27 (V2)
In residential care homes, bedrooms should be enclosed in fire-resisting construction with fire-resisting doors, and every corridor serving bedrooms should be a protected corridor.	B1 3.48 (V2)
The dead-end portion of any common corridor should be separated from the rest of the corridor by a self-closing fire door.	B1 2.29 (V2)

Note: Generally, in residential care homes for the elderly, it is reasonable to assume that at least a proportion of the residents will need some assistance to evacuate.

Protected stairways

Dwelling-houses with one floor more than 4.5 m above ground level should have a protected stairway separated by fire doors.	B1 2.6 (V1)
Transfer grilles belonging to air-circulation systems shall not be fitted in any door leading to a protected stairway.	B1 2.17

Sprinkler systems

Where a sprinkler system is provided, fire doors to bedrooms need not be fitted with self-closing devices.	B1 3.52 (V2)

6.15 Vertical circulation within the building

6.15.1 The requirement

In addition to the requirements of the Disability Discrimination Act 1995 precautions need to be taken to ensure that:

- *new non-domestic buildings and/or dwellings (e.g. houses and flats used for student living accommodation etc.);*
- *extensions to existing non-domestic buildings;*
- *non-domestic buildings that have been subject to a material change of use (e.g. so that they become a hotel, boarding house, institution, public building or shop);*

are capable of allowing people, regardless of their disability, age or gender, to:

- *gain access to buildings;*
- *gain access within buildings;*
- *be able to use the facilities of the buildings (both as visitors and as people who live or work in them);*
- *use sanitary conveniences in the principal storey of any new dwelling.*

(Approved Document M)

 Note: See Appendix A for guidance on access and facilities for disabled people.

Fire safety

- *There shall be sufficient escape routes that are suitably located to enable persons to evacuate the building in the event of a fire.*
- *Safety routes shall be protected from the effects of fire.*
- *In an emergency, the occupants of any part of the building shall be able to escape without any external assistance.*

(Approved Document B1)

6.15.2 Meeting the requirement

Lifting devices

For all buildings, a passenger lift is considered the most suitable form of access for people moving from one storey to another.

Wherever possible a lifting device (e.g. a passenger lift or a lifting platform) serving all storeys should be provided in: • new developments; • existing buildings.	M (3.17, 3.24a to d)

Note: In exceptional circumstances (e.g. a listed building, or an infill site in a historic town centre) where a passenger lift cannot be accommodated, a wheelchair-platform stairlift serving an intermediate level or a single storey may be used.

The location of lifting devices that are accessible by mobility-impaired people should be clearly visible from the building entrance.	M (3.18)
Signs should be available at each landing to identify the floor reached by the lifting device.	M (3.18)
In addition to the lifting device, internal stairs (designed to suit ambulant disabled people and those with impaired sight) should always be provided.	M (3.19)

General requirements for lifting devices

There should be an unobstructed manoeuvring space of 1500 mm × 1500 mm, or a straight access route 900 mm wide, in front of each lifting device.	M (3.28a)
The landing call buttons should be located between 900 mm and 1100 mm from the floor and at least 500 mm from any return wall.	M (3.28b)
Landing call buttons and lifting-device control button symbols:	M (3.248c and d)
• should contrast visually with the surrounding face plate; • should be raised to facilitate tactile reading; • should be accessible by wheelchair users.	M (3.27)
The floor of the lifting device should not be of a dark colour.	M (3.24e)
A handrail (at 900 mm nominal) should be provided on at least one wall of the lifting device.	M (3.24f)
A suitable emergency communication system should be fitted.	M (3.24g)

Note: See also:

* Lift Regulations 1997, SI 1997/831;
* Lifting Operations and Lifting Equipment Regulations 1998, SI 1998/2307;

- Provision and Use of Work Equipment Regulations 1998, SI 1998/2306;
- Management of Health and Safety at Work Regulations 1999, SI 1999/3242;
- BS 8300.

Lifts

Lifts should **not** be used when there is a fire in the building – unless it is a fire-fighting lift.

Lifts (except suitably designed and installed evacuation lifts), are **not** considered acceptable as a means of escape.	B1 vi
Lift entrances should be separated from the floor area on every storey by a protected lobby.	B1 5.42 (V2)
Lift shafts should not be continued down to serve any basement storey if it is: • in a building served by only one escape; • within the enclosure to an escape stair that is terminated at ground level.	B1 5.44 (V2)
If a protected shaft contains a stair and/or a lift, it should **not** also contain: • a pipe conveying oil (other than in the mechanism of a hydraulic lift); or • a ventilating duct (other than a duct provided solely for ventilating the stairway).	B2 8.40 (V2)
Lift wells should be either: • contained within the enclosures of a protected stairway; or • enclosed throughout their height with fire-resisting construction.	B1 5.42 (V2)
A lift well connecting different compartments should form a protected shaft.	B1 5.42 (V2)
Lift machine rooms should be sited over the lift well whenever possible.	B1 5.45 (V2)
In basements and enclosed (i.e. non-open-sided) car parks the lift should be approached only by a protected lobby or protected corridor (unless it is within the enclosure of a protected stairway).	B1 5.43 (V2)

Evacuation lifts and refuges should be clearly   B1 4.10 (V2)
identified by appropriate fire safety signs.

Note: If a refuge is in a lobby or stairway the sign should be
accompanied by a blue mandatory sign worded *Refuge –
keep clear*.

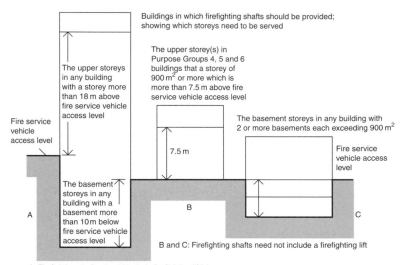

Figure 6.183 Provision of fire-fighting shafts.

Fire-fighting lifts

Buildings with a floor at more than 18 m above   B5 17.2 (V2)
(or with a basement at more than 10 m below) fire   B5 17.8 (V2)
and rescue service vehicle access level should
be provided with at least two fire-fighting shafts
containing fire-fighting lifts.

Other than in blocks of flats, all fire-fighting lifts   B5 17.11 (V2)
should be approached from the accommodation
through a fire-fighting lobby.

Passenger lifts

Lift sizes should be chosen to suit the anticipated density of use of the building
and the needs of disabled people.

Passenger lifts should conform to the requirements of:

* the Lift Regulations 1997, SI 1997/831;
* the relevant British Standards, EN 81 series of   M (3.34a)
standards;
* BS EN 81-70:2003 (*Safety Rules for the Construction
and Installation of Lifts*).

Passenger lifts should:

* be accessible from the remainder of the storey;   M (3.34b)
* have power-operated horizontal sliding doors which   M (3.34e)
provide an effective clear width of at least 800 mm
(nominal);
* have doors fitted with timing devices (and re-opening   M (3.34f)
activators) to allow enough time for people and any
assistance dogs to enter or leave;
* have the car controls located between 900 mm and   M (3.34g)
1200 mm (preferably 1100 mm) from the car floor and
at least 400 mm from any return wall;
* have all landing call buttons located between 900 mm   M (3.34h)
and 1100 mm from the floor of the landing and at least
500 mm from any return wall;
* be fitted with lift landing and car doors that are   M (3.34i)
visually distinguishable from adjoining walls;
* include (in the lift car and the lift lobby) audible and   M (3.31
visual information to tell passengers that a lift has   and 3.34j)
arrived, which floor it has reached and where in a
bank of lifts it is located;
* allow all glass areas to be easily identified by people   M (3.34 k)
with impaired vision;
* conform to BS 5588-8 if the lift is to be used to   M (3.34l)
evacuate disabled people in an emergency.

If the lift is not large enough to allow a wheelchair user   M (3.34d)
to turn around within the lift car, then a mirror should be
provided that enables the wheelchair user to see behind
the wheelchair.

The minimum dimensions of the lift cars should be   M (3.34c)
1100 mm wide and 1400 mm deep (Figure 6.184);

Visually and acoustically reflective wall surfaces should   M (3.32)
not be used.

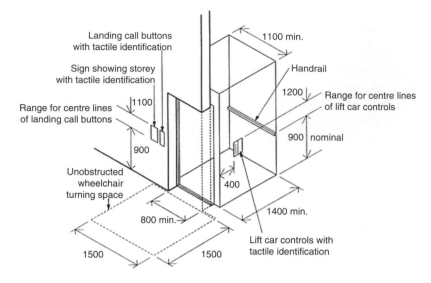

Figure 6.184 Key dimensions associated with passenger lifts (mm).

Where possible, lift cars (used for access between two levels only) may be provided with opposing doors to allow a wheelchair user to leave without reversing out.	M (3.33)
Passenger lifts in dwelling-houses (which serve any floor more than 4.5 m above ground level) should either be located in the enclosure to a protected stairway or a fire-resisting lift shaft.	B1 2.18

Lifting platforms

A lifting platform should only be provided to transfer wheelchair users, people with impaired mobility and their companions vertically between levels or storeys (M3.35).

Lifting platforms should: • conform to the requirements of the Supply of Machinery (Safety) Regulations 1992, SI 1992/3073, the relevant British Standards and the EN81 series of standards;	M (3.43a)

- restrict the vertical travel distance to: M (3.43b)

  - not more than 2 m if there is no liftway enclosure
    and/or floor penetration;
  - more than 2 m if there is a liftway enclosure;
- restrict the rated speed of the platform so that it does   M (3.43c)
  not exceed 0.15 m/s;
- have controls located between 80 mm and 1100 mm   M (3.43d)
  from the floor of the lifting platform and at least
  400 mm from any return wall;
- have all landing call buttons located between 900 mm   M (3.43f)
  and 1100 mm from the floor of the landing and at least
  500 mm from any return wall;
- have continuous pressure controls (e.g. push buttons);   M (3.43e)
- have doors with an effective clear width of at least: M (3.43h)

  - 900 mm for an 1100 mm wide and 1400 mm deep
    lifting platform;
  - 800 mm in other cases;
- be fitted with clear instructions for use;   M (3.43i)
- have their entrances accessible from the remainder of   M (3.43j)
  the storey;
- have doors visually distinguishable from adjoining   M (3.43 k)
  walls;
- have an audible and visual announcement of platform   M (3.43l)
  arrival and level reached;
- have areas of glass that are identifiable by people with   M (3.43m)
  impaired vision.

The minimum clear dimensions of the platform should be: M (3.43g)

- 800 mm wide by 1250 mm deep (where the lifting
  platform is not enclosed and provision has been made
  for an unaccompanied wheelchair user);
- 900 mm wide by 1400 mm deep (where the lifting
  platform is enclosed and provision has been made for
  an unaccompanied wheelchair user);
- 1100 mm wide by 1400 mm deep (where two doors
  are located at 90° relative to each other, the lifting
  platform is enclosed or where provision is being made
  for an accompanied wheelchair user).

All users including wheelchair users should be able to   M (3.36)
reach and use the controls that summon and direct the
lifting platform.

Where possible, lifting platforms (used for access between two levels only) may be provided with opposing doors to allow a wheelchair user to leave without reversing out.	M (3.41)
Visually and acoustically reflective wall surfaces should not be used.	M (3.42)

Wheelchair platform stairlifts

Wheelchair platform stairlifts are only intended for the transportation of wheelchair users and should only be considered for conversions and alterations where it is not practicable to install a conventional passenger lift or a lifting platform.

Wheelchair platform stairlifts should conform to the requirements of the Supply of Machinery (Safety) Regulations 1992, SI 1992/3073, the relevant British Standards and EN81 series of standards.	M (3.49a)
Buildings with single stairways shall maintain the required clear width.	M (3.49b)
The speed of the platform should not exceed 0.15 m/s.	M (3.49c)
Continuous pressure controls (e.g. joystick) should be provided.	M (3.47 and 3.49d)
The platform should have minimum clear dimensions of 800 mm wide and 1250 mm deep.	M (3.49e)
Wheelchair platform stairlifts should:	
• be fitted with clear instructions for use;	M (3.49f)
• provide an effective clear width of at least 800 mm;	M (3.49g)
• not be installed where their operation restricts the safe use of the stair by other people.	M (3.44)

Passenger lifts in blocks of flats

If a passenger lift is installed, it should:	
• be suitable for an unaccompanied wheelchair user;	M (9.4)
• have a minimum load capacity of 400 kg;	M (9.6)

- have a clear landing at least 1500 mm wide and 1500 mm long in front of its entrance;   M (9.7a)
- have a door (or doors) with a clear opening width of at least 800 mm;   M (9.7b)
- have a car at least 900 mm wide and 1250 mm long;   M (9.7c)
- have landing and car controls between 900 mm and 1200 mm above the landing and the car floor and at least 400 mm from the front wall;   M (9.7d)
- have suitable tactile indication (on the landing and adjacent to the lift call button) to identify the storey;   M (9.7e)
- have suitable tactile indication (on or adjacent to lift buttons within the car) to confirm the floor selected;   M (9.7f)
- incorporate a signalling system that provides visual notification that the lift is answering a landing call;   M (9.7g)
- have a dwell time of five seconds before its doors begin to close after they are fully open;   M (9.7g)
- provide a visual and audible indication of the floor reached (when the lift serves more than three storeys).   M (9.7h)

Vertical circulation within the entrance storey of a dwelling

A stair providing vertical circulation within the entrance/principal storey of a dwelling should have: M (7.7a to c)

- flights with clear widths of at least 900 mm;
- a suitable continuous handrail on each side of the flight (and any intermediate landings where the rise of the flight comprises three or more rises);
- the rise and going in accordance with the guidance in Approved Document K for private stairs in dwelling.

6.16 Corridors and passageways

6.16.1 The requirement

Fire safety

There shall be sufficient escape routes that are suitably located to enable persons to evacuate the building in the event of a fire.

Safety routes shall be protected from the effects of fire.

In an emergency, the occupants of any part of the building shall be able to escape without any external assistance.

(Approved Document B1)

6.16.2 Meeting the requirement

General requirements for corridors

In large buildings, the corridor adjoining the stair should be provided with a vent that is located as high as is practicable and with its top edge at least as high as the top of the door to the stair.	B1 2.26 (V2)
In buildings containing flats, the wall between each flat and the corridor should be a compartment wall.	B1 2.24 (V2)
Stores and other ancillary accommodation should **not** form part of the only common escape route from a flat on the same storey as that ancillary accommodation.	B1 2.30 (V2)
Every corridor more than 12 m long that connects two or more storey exits, should be subdivided by self-closing fire doors positioned approximately midway between the two storey exits.	B1 3.26 (V2)
Unless escape stairways and corridors are protected by a pressurization system complying with BS EN 12101-6:2005, every dead-end corridor exceeding 4.5 m in length should be separated by self-closing fire doors (together with any necessary associated screens) from any part of the corridor that: • provides two directions of escape; • continues past one storey exit to another.	B1 3.27 (V2)
Corridors associated with a common stair serving an enclosed (non-open-sided) car park or place of special fire hazard shall have not less than 0.4 m^2 permanent ventilation or be protected by a mechanical smoke-control system.	B1 2.47 (V2)

Protected corridors

Protected corridors shall be installed for: • corridors serving bedrooms; • dead-end corridors (excluding recesses not exceeding 2 m deep); • any corridor that is common to more than one different occupancy.	B1 3.24 (V2)

Escape stairs shall have a protected lobby or protected corridor at all levels (except the top storey, all basement levels and when the stair is a fire-fighting stair) if: B1 4.34 (V2)

- the stair is the only one serving a building that has more than one storey above or below the ground storey;
- the stair serves any storey at a height greater than 18 m; or
- the building is designed for phased evacuation.

Corridors serving bedrooms in residential care homes should be protected corridors.   B1 3.48 (V2)

Fire safety

Common corridors should be protected corridors.   B1 2.24 (V2)

Means of ventilating common corridors should be available.   B1 2.25 (V2)

Common corridors that connect two or more storey exits should be subdivided by a self-closing fire door with, if necessary, an associated fire-resisting screen (see Figure 6.185).   B1 2.28 (V2)

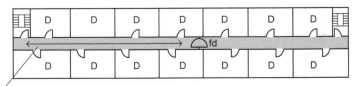

30 m max.

(a) Corridor access without dead ends

Note:
The arrangements shown also apply to the top storey.

Key
D   Dwelling
fd   Fire door
▢   Shaded area indicates zone where ventilation should be provided in accordance with paragraph 2:26
(An external wall vent or smoke shaft located anywhere in the shaded area)

Figure 6.185 Flats served by more than one common stair – corridor access **without** dead ends.

The dead-end portion of any common corridor should   B1 2.29 (V2)
be separated from the rest of the corridor by a self-
closing fire door (see Figure 6.186).

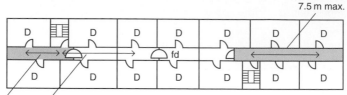

(b) Corridor access with dead ends
The central door may be omitted if maximum travel distance is not
more than 15 m

Note:
The arrangements shown also apply to the top storey.

Key

D   Dwelling
fd   Fire door
     Shaded area indicates zone where ventilation
     should be provided in accordance with
     paragraph 2:26
     (An external wall vent or smoke shaft located
     anywhere in the shaded area)

Figure 6.186 Flats served by more than one common stair – corridor access **with** dead ends.

 Note: Self-closing fire doors should be positioned so that smoke will not affect access to more than one stairway.

Disabled access

Corridors and passageways should:

- be wide enough to allow wheelchair users, people   M (3.11)
  with buggies, people carrying cases and/or people on
  crutches to pass others on the access route;
- **not** have projecting elements such as columns,   M (3.14a)
  radiators and fire hoses;
- have an unobstructed width of at least 1200 mm;   M (3.14b)

 Note: for school buildings, the preferred corridor width is 2700 mm where there are lockers within the corridor.

- have passing places at least 1800 mm long and at least   M (3.14c)
  1800 mm wide at corridor junctions;
- have a floor level no steeper than 1:60;   M (3.14d)
- have an internal ramp in accordance with Table 6.96   M (3.14d)
  and Figure 6.187 for floors with a gradient of 1:20 or
  steeper.

Table 6.96 Limits for ramp gradients

Going of a flight (m)	Maximum gradient	Maximum rise (mm)
10	1:20	500
5	1:15	333
2	1:12	16

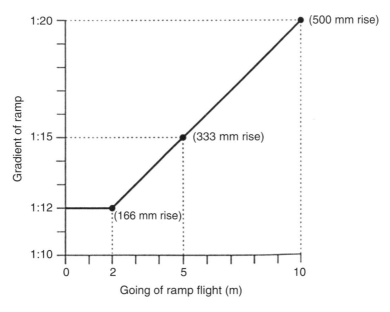

Figure 6.187 Relationship of ramp gradients.

Corridors should be at least 1800 mm wide where doors   M (3.14h)
from a unisex wheelchair-accessible toilet project open
into that corridor.

If a section of the floor has a gradient steeper than 1:60 (but less than 1:20) it should rise no more than 500 mm without a level rest area at least 1500 mm long.	M (3.14e)
Sloping sections should extend the full width of the corridor or have the exposed edge protected by guarding.	M (3.14f)
Doors opening towards a corridor that is a major access route or an escape route should be recessed.	M (3.14 g)
On a major access route (or an escape route) the wider leaves of double doors should be on the same side of the corridor.	M (3.14i)
The use of floor surface finishes which have patterns that could be mistaken for steps or changes of level should be avoided.	M (3.14j)
Floor finishes should be slip-resistant.	M (3.14k)
Glazed screens alongside a corridor should be clearly marked (i.e. with a logo or sign) on the glass at two levels, 850–1000 mm and 1400–1600 mm above the floor.	M (3.14l)
The acoustic design should be neither too reverberant nor too absorbent.	M (3.13)

Corridors, passageways and internal doors within the entrance storey of a dwelling

The objective is to make it easy for wheelchair users, ambulant disabled people, people of either sex with babies and small children, or people with luggage to gain access to an entrance and/or principal storey of a dwelling, into habitable rooms and/or a room containing a WC on that level.

Corridors and passageways in the entrance storey should be sufficiently wide for a wheelchair user to circumnavigate.	M (7.2)
Internal doors should be wide enough for wheelchairs to go through with ease.	M (7.4)

Permanent obstructions in a corridor (e.g. a radiator) should be no longer than 2 m (provided that the width of the corridor is not less than 750 mm and the obstruction is not opposite a door to a room).

M (7.2 and 7.5b)

Corridors and/or other access routes in the entrance storey should have an unobstructed width in accordance with Table 6.97 and Figure 6.188.

M (7.5a)

Table 6.97 Minimum widths of corridors and passageways for a range of doorway widths

Doorway clear opening width (mm)	Corridor/passageway width (mm)
750 or wider	900 (when approached head on)
750	1200 (when not approached head on)
775	1050 (when not approached head on)
800	900 (when not approached head on)

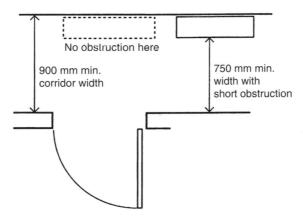

Figure 6.188 Corridors, passages and internal doors.

Doors to habitable rooms and/or rooms containing a WC should have minimum clear opening widths shown in Table 6.97 and Figure 6.187.

M (7.5c)

6.17 Facilities in buildings other than dwellings

6.17.1 The requirement

Conservation of fuel and power

Reasonable provision shall be made for the conservation of fuel and power in buildings by:

(a) limiting heat gains and losses –

  (i)   through thermal elements and other parts of the building fabric; and
  (ii) from pipes, ducts and vessels used for space heating, space cooling and hot water services;

(b) providing fixed building services which –

  (i)   are energy efficient;
  (ii) have effective controls; and
  (iii) are commissioned by testing and adjusting as necessary to ensure they use no more fuel and power than is reasonable in the circumstances; and

(c) providing to the owner sufficient information about the building, the fixed building services and their maintenance requirements so that the building can be operated in such a manner as to use no more fuel and power than is reasonable in the circumstances

(Approved Document L)

A new Part L came into force on 1st October 2010.

Fire safety

* *There shall be sufficient escape routes that are suitably located to enable persons to evacuate the building in the event of a fire.*
* *Safety routes shall be protected from the effects of fire.*
* *In an emergency, the occupants of any part of the building shall be able to escape without any external assistance.*
* *There shall be an early-warning fire alarm system for persons in the building.*

(Approved Document B1)

Access to and use of buildings

Reasonable provisions shall be made for people to:

(a) gain access to; and
(b) use;
the building and its facilities

(Approved Document M)

In addition to the requirements of the Disability Discrimination Act 1995, precautions need to be taken to ensure that:

- new non-domestic buildings and/or dwellings (e.g. houses and flats used for student living accommodation, etc.);
- extensions to existing non-domestic buildings;
- non-domestic buildings that have been subject to a material change of use (e.g. so that they become a hotel, boarding house, institution, public building or shop);

are capable of allowing people, regardless of their disability, age or gender, to:

- gain access to buildings;
- **gain access within buildings**;
- be able to use the facilities of the buildings (both as visitors and as people who live or work in them);
- use sanitary conveniences in the principal storey of any new dwelling.

 Note: See Appendix A for guidance on access and facilities for disabled people.

Sanitary conveniences

*All dwellings (and buildings containing one or more rooms for residential purposes) will be provided with a bathroom which contains either a fixed bath or a shower **and** a washbasin.*

(Approved Document G5)

Ventilation

There shall be adequate means of ventilation provided for people in the building.

(Approved Document F)

 A new Part F came into force on 1 October 2010.

6.17.2 Meeting the requirement

Conservation of energy and power

If a door is enlarged or a new one created in a building other than a dwelling, the area of the pedestrian door, expressed as a percentage of the total floor area of the building, should not exceed the relevant value from Table 6.98, or should be compensated for in some other way.	L2B 4.24

Table 6.98 Opening areas in the extension

Building type	Personnel doors as % of exposed wall	Rooflights as % of area of roof
Places of assembly, offices and shops	40	20
Industrial and storage buildings	15	20
Vehicle access doors	As required	N/A

> The area of windows and rooflights in the extension should generally not exceed the values given in Table 6.99.   L2B 4.4

Table 6.99 Opening areas in the extensions to buildings other than dwellings

Building type	Windows and personnel doors as % of exposed wall	Rooflights as % of area of roof
Places of assembly, offices and shops	40	20
Industrial and storage buildings	15	20
Vehicle access doors and display windows and similar glazing	As required	N/A
Smoke vents	N/A	As required

Ventilation

> Fresh air supplies should be protected from contaminants that would be injurious to health.   F 6.3

Offices

> All office sanitary accommodation, washrooms and food- and beverage-preparation areas shall be provided with intermittent air-extract ventilation capable of meeting the requirements of Table 100.   F 6.101
>
> Extract fans that are located in an internal room does not have an openable window, should have a 15 minute overrun.   F Tables 5.2a to 5.2d
>
> Extract ventilators should be located as high as practicable and preferably not less than 400 mm below the ceiling level.   F Tables 5.2a to 5.2d

Passive stack ventilators (PSVs) should be located in the ceiling of the room.	F Tables 5.2a to 5.2d
Purge ventilation shall be provided in each office.	F 6.12
Purged air should be taken directly outside and should not be recirculated to any other part of the building.	F 6.12
PSV can be used as an alternative to a mechanical extract fan for office sanitary and washrooms and food-preparation areas.	F Tables 5.2a to 5.2d
PSV controls can be either manual or automatic.	F Tables 5.2a to 5.2d
The controls for extract fans can be either manual or automatic.	F Tables 5.2a to 5.2d
Printers and photocopiers that are being used in large numbers and which are in almost constant use (i.e. greater than 30 minutes per hour) shall: • be located in a separate room; • have extract facilities capable of providing an extract rate greater than 20 l/s per machine during use (Table 6.100).	F 6.12

Note: The air flow rates given in Table 6.100 can mainly be provided by natural ventilation.

Table 6.100 Extract ventilation rate

Room	Air extract rate
Rooms containing printers and photocopiers in substantial use (greater than 30 minutes per hour)	20 l/s per machine during use
Office sanitary accommodation and washrooms	15 l/s per shower/bath 6 l/s per WC/urinal
Food- and beverage-preparation areas (not commercial kitchens)	15 l/s with microwave and beverages only 30 l/s adjacent to the hob with cooker(s) 60 l/s elsewhere with cooker(s)
Specialist buildings and spaces (e.g. commercial kitchens, fitness rooms)	See Table 2.3 of Approved Document F

The outdoor air supply rates shown in Table 6.100 for offices are based on controlling body odours with low levels of other pollutants.

The whole-building ventilation rate for the supply of air to the offices should be greater than 10 l/s per person (Table 6.101).

F 6.12

Table 6.101 Whole-building ventilation rate for air supply to offices

Offices	Air supply rate
Total outdoor air supply rate for offices (no smoking and no significant pollutant sources)	10 l/s per person

Other types of building

The ventilation requirements for other buildings (e.g. assembly halls, broadcasting studios, computer rooms, factories, hospitals, hotels, museums, schools, sports centres and warehouses, etc.) are listed in Table 2.3 of Approved Document F, which also provides a link to the relevant controlling Acts of Parliament, Statutory Instruments, British Standards, CIBSE and HSE standards, practices and recommendations.

Car parks

Where a car park is well ventilated there is a low probability of fire spread from one storey to another. Ventilation, however, is an important factor and, as heat and smoke cannot be dissipated so readily from a car park that is not open-sided, fewer concessions are made. For more guidance see Section 11 of Part B3 (Volume 2).

Non-combustible materials should be used in the construction of any 'open sided' car park.

Underground car parks, enclosed car parks and multi-storey car parks should be designed to limit the concentration of carbon monoxide to not more than 30 parts per million, averaged over an eight-hour period and peak concentrations.

F 6.18

Naturally ventilated car parks shall have openings at each car-parking level:

F 6.20a

• at least 1/20 of the floor area at that level;
• with a minimum of 25 per cent on each of two opposing walls.

Ramps and exits shall not go above 90 parts per million for periods not exceeding 15 minutes.	F 6.18
Car parks should have a separate and independent extraction system and the extracted air should not be recirculated.	B1 5.50 (V2)
Mechanically ventilated car parks can either have natural ventilation openings that are not less than 1/40 of the floor area or a mechanical ventilation system capable of at least three air changes per hour (ach).	F 6.20b
Mechanically ventilated basement car parks shall be capable of at least six air changes per hour (ach).	F 6.20b
Mechanically ventilated exits and ramps (i.e. where cars queue inside the building with engines running) shall be capable of at least 10 air changes per hour (ach).	F 6.20b
Where a common stair forms part of the only escape route from a flat, it should **not** also serve any covered car park.	B1 2.46 (V2)
In enclosed (i.e. non-open-sided) car parks, lifts should be approached by a protected lobby or protected corridor (unless it is within the enclosure of a protected stairway).	B1 5.43 (V2)
Car parks are not normally expected to be fitted with sprinklers.	B5 18.13 (V2)

Fire safety

Hospitals

 HTM 05 'Firecode' should be the design of hospitals and similar health-care premises.

Offices

In small premises: • floor areas should be generally undivided (except for ancillary offices and stores) to ensure that exits are clearly visible from all parts of the floor areas;	B1 3.34 (V2)

- clear-glazed areas should be provided in any par-   B1 3.36 (V2)
  titioning separating an office from the open floor
  area to enable any person within the office to obtain
  early visual warning of an outbreak of fire.

Disabled facilities

The overall aim should be that **all** people can have access to (and be able to use) **all** of the facilities provided within a building. Everyone (no matter their disability) should be able to fully participate in lecture/conference facilities as well as be able to enjoy entertainment, leisure and social venues – not just as spectators, but also as participants and/or staff. To achieve these aims:

All floor areas (even when located at different levels)   M (4.3)
should be accessible.

In hotels, motels and student accommodation: M (4.4)

- a proportion of the sleeping accommodation should
  be designed for wheelchair users;
- the remainder should include facilities suitable
  for people with sensory, dexterity or learning
  difficulties.

If there is a reception point: M (3.2 to 3.5)

- it should be easily accessible and convenient to
  use;
- information about the building should be clearly
  available from notice boards and signs;
- the floor surface should be slip-resistant.

Disabled people should be able to have: M (4.2)

- a choice of seating location at spectator events;
- a clear view of the activity taking place (whilst not
  obstructing the view of others).

Bars and counters in refreshment areas should be at a   M (4.3)
suitable level for wheelchair users.

Audience and spectator facilities

Audience and spectator facilities fall primarily into three categories:

- lecture/conference facilities;
- entertainment facilities (e.g. theatres/cinemas);
- sports facilities (e.g. stadia).

Wheelchair users (as well as those with mobility and/or sensory problems) may need to see or listen from a particular side, or sit at the front to lip read or read sign interpreters.

For this reason, they should be provided with a selection of spaces into which they can manoeuvre easily and that offer them a clear view of an event – taking particular care that these do not become segregated into 'special areas'.

The route to wheelchair spaces should be accessible to users.	M (12a)
Stepped access routes to audience seating should be provided with fixed handrails.	M (12b)
Handrails to external stepped or ramped access should be positioned as shown in Figure 6.189.	M (4.12a)

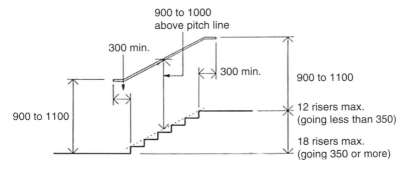

Figure 6.189 Handrails to external stepped and ramped access – key dimensions (mm).

Handrails to external stepped or ramped access should: M (4.12b)

- be continuous across flights and landings;
- extend at least 300 mm horizontally beyond the top and bottom of a ramped access;
- not project into an access route;
- contrast visually with the background;
- have a slip-resistant surface which is not cold to the touch;
- terminate in such a way that reduces the risk of clothing being caught;
- either be circular (with a diameter of between 40 mm and 45 mm) or oval (with a width of 50 mm) (Figure 6.190).

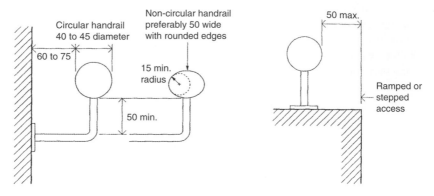

Figure 6.190 Handrail – key dimensions (mm).

Handrails to external stepped or ramped access should: M (4.12b)

- not protrude more than 100 mm into the surface width of the ramped or stepped access where this would impinge on the stair width requirement of Part B1;
- have a clearance of between 60 mm and 75 mm between the handrail and any adjacent wall surface;
- have a clearance of at least 50 mm between a cranked support and the underside of the handrail;
- have their inner face located no more than 50 mm beyond the surface width of the ramped or stepped access;
- be spaced away from the wall and rigidly supported in a way that avoids impeding finger grip;
- be set at heights that are convenient for all users of the building.

The minimum number of permanent and removable   M (4.12c)
spaces provided for wheelchair users is as given in Table 6.102.

Table 6.102 Provision of wheelchair spaces for audience seating

Seating capacity	Minimum provision of spaces for wheelchairs	
	Permanent	Removable
Up to 600	1% of total seating capacity (rounded up)	Remainder to make a total of 6
Over 600 but less than 10,000	1% of total seating capacity (rounded up)	Additional provision if desired

Some wheelchair spaces should be provided in pairs, with standard seating on at least one side (Figure 6.191).

M (4.12d)

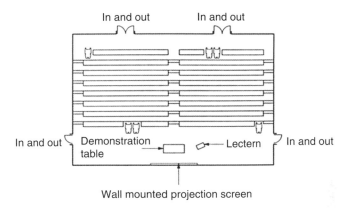

Figure 6.191 An example of wheelchair spaces in a lecture theatre.

If more than two wheelchair spaces are provided, they should be located so as to give a range of views of the event.

M (4.12e)

The minimum clear space for:
- access to wheelchair spaces should be 900 mm;
- wheelchair spaces in a parked position should be 900 mm wide by 1400 mm deep.

M (4.12f)

M (4.12g)

The floor of each wheelchair space should be horizontal.

M (4.12h)

Seats at the ends of rows **and** next to wheelchair spaces should have detachable (or lift-up) arms.

M (4.12j)

For seating on a stepped terraced floor:

Wheelchair spaces at the back of a stepped terraced floor should be in accordance with Figure 6.192.

M (4.12k)

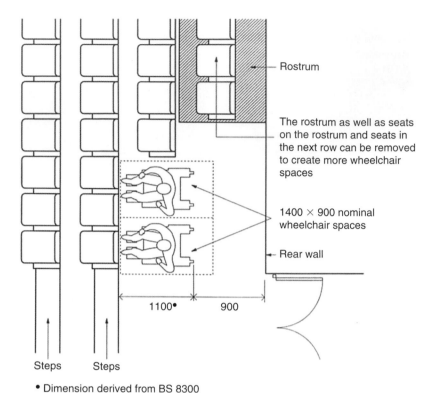

Figure 6.192 Possible location of wheelchair spaces in front of a rear a isle. (Dimensions in mm.)

Lecture/conference facilities

All people should be able to use presentation facilities.	M (4.9)
People with hearing impairments should be able to participate fully in conferences, committee meetings and study groups.	M (4.9)
The design of the acoustic environment should ensure that audible information can be heard clearly.	M (4.9)
Artificial lighting should be designed to give good colour rendering of all surfaces.	M (4.9)
Glare and reflections from shiny surfaces, and large repeating patterns, should be avoided in spaces where visual accuracy is critical.	M (4.9)

Uplighters mounted at low or floor level should be avoided as they can disorientate some visually impaired people.   M (4.9)

Where a podium or stage is provided, wheelchair users should have access to it by means of a ramp or lifting platform.   M (4.12.1)

> A clearly audible public address system should be supplemented by visual information.

Hearing enhancement systems should be installed: M (4.12.1)

- in rooms and spaces designed for meetings, lectures;
- classes, performances, spectator sports or films;
- at service or reception counters (especially when situated in noisy areas or behind glazed screens).

The availability of an induction loop or infrared hearing enhancement system should be indicated by the standard symbol.   M (4.12.1)

Telephones suitable for hearing aid users should: M (4.12.1)

- be clearly indicated by the standard ear symbol;
- incorporate an inductive coupler and volume control.

Text telephones for deaf and hard of hearing people should be clearly indicated by the standard symbol.   M (4.12.1)

Artificial lighting should be designed to be compatible with other electronic and radio frequency installations.   M (4.12.1)

Entertainment, leisure and social facilities

Toilets are available that have been adapted for people with mobility impairments

Changing rooms are available for people with mobility impairments

Assistance dogs are welcome on the premises

Figure 6.193 Examples of facility signs.

In theatres and cinemas (where seating is normally closely packed together) special care should be given to the design and location of wheelchair spaces.　　　　M (4.10)

Sports facilities

 See *Guide to Safety at Sports Grounds*.

Refreshment facilities

Restaurants and bars should be designed so that they can be reached and used by all people independently or with companions.

All people should have access to:

- all parts of the facility;　　　　M (4.13)
- staff areas;
- public areas (e.g. lavatory accommodation, public　　　M (4.14)
  telephones and external terraces);
- self-service facilities (when provided).

 Changes of floor level are permitted, provided that　　M (4.16b)
all the different levels are accessible and raised
thresholds are avoided.

Part of the working surface of a bar or serving counter should:

- be permanently accessible to wheelchair users;
- be at a level of not more than 850 mm above the floor.

 Note: If this is unavoidable, the total height should not be more than 15 mm, with a minimum number of upstands and slopes and with any upstands higher than 5 mm chamfered or rounded.

In addition:　　　　M (4.16c)

- the worktop of a shared refreshment facility (e.g.
  for tea making) should be 850 mm above the floor
  with a clear space beneath at least 700 mm above
  the floor (see Figure 6.194);
- basin taps should either be controlled automatically
  or be capable of being operated using a closed fist
  (e.g. by lever action);

- all terminal fittings should comply with Guidance Note G18.5 of the Guidance Document relating to Schedule 2: Requirements for Water Fittings, of the Water Supply (Water Fittings) Regulations 1999, SI 1999/1148.

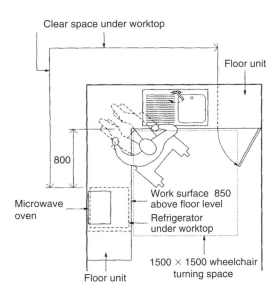

Figure 6.194 An example of a typical shared refreshment facility. (Dimensions in mm.)

Sleeping accommodation

Sleeping accommodation in hotels, motels and student accommodation should be convenient for all types of people.	M (4.17)
A proportion of rooms should be available for wheelchair users.	M (4.17)
Wheelchair users should be able to:	
• reach all the facilities available within the building;	M (4.18)
• manoeuvre around and use the facilities in the room, and operate switches and controls.	M (4.19)
En-suite sanitary facilities are the preferred option for wheelchair-accessible bedrooms.	M (4.19)

There should be at least as many en-suite shower rooms as en-suite bathrooms – some mobility-impaired people may find it easier to use a shower than a bath).   M (4.19)

In all bedrooms, built-in wardrobes and shelving should be accessible and convenient to use.   M (4.20)

Bedrooms not designed for independent use by a person in a wheelchair should have an outer door wide enough to be accessible to a wheelchair user.   M (4.21)

For all bedrooms:

- built-in wardrobe swing doors should open through 180°;   M (4.24b)
- handles on hinged and sliding doors: M (4.24c)

  - should be easy to grip and operate;
  - should contrast visually with the surface of the door;
- windows and window controls should be: M (4.24d)

  - located between 800 and 1000 mm above the floor;
  - easy to operate without using both hands simultaneously;
- a visual fire alarm signal should be provided in addition to the requirements of Part B;   M (4.24e)
- room numbers should be embossed characters;   M (4.24f)
- the effective clear width of the door from the access corridor should comply with Table 6.103.   M (4.24a)

At least one wheelchair-accessible bedroom should be provided for every 20 bedrooms

Table 6.103 Minimum effective clear widths of doors

Direction and width of approach	New buildings (mm)	Existing buildings (mm)
Straight on (without a turn or oblique approach)	800	750
At right angles to an access route at least 1500 mm wide	800	750
At right angles to an access route at least 1200 mm wide	825	775
External doors to buildings used by the general public	1000	775

Wheelchair-accessible bedrooms should:

- be located on accessible routes;   M (4.24 h)
- be designed to provide a choice of location;   M (4.24i)
- have a standard of amenity equivalent to that of other   M (4.24i)
  bedrooms;
- (for en-suite bathroom and shower room doors) have   M (4.24k)
  an effective clear width complying with Table 6.103;
- be large enough to enable a wheelchair user to   M (4.24l)
  manoeuvre with ease (Figure 6.195).

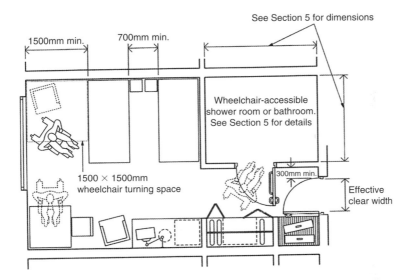

Figure 6.195 Example of a wheelchair-accessible hotel bedroom. (Dimensions in mm.)

In wheelchair-accessible bedrooms:

- if wide-angle viewers are provided in the entrance   M (4.24n)
  door, they should be located at 1050 mm and 1500 mm
  above floor level;

- if a balcony is provided, it should: M (4.24o)

  - comply with Table 6.103;
  - have a level threshold;
  - have no horizontal transoms between 900 mm and
    1200 mm above the floor;

- there should be no permanent obstructions in a zone   M (4.24p)
  1500 mm back from any balcony doors;

- emergency assistance alarms should be provided;   M (4.24q)

- the door from the access corridor to a wheelchair-   M (4.24j)
  accessible bedroom should:
- not require more than 20 N opening force;
- have an effective clear width through a single-leaf
  door (or one leaf of a double-leaf door) in accordance
  with Table 6.103 and Figure 6.196;
- have an unobstructed space of at least 300 mm on the
  pull side of the door Figure 6.196).

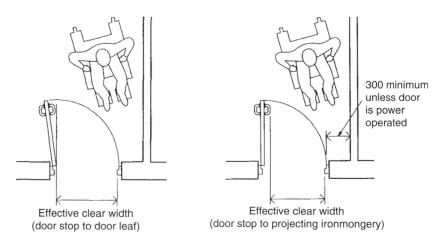

Figure 6.196 Effective clear width of doors. (Dimensions in mm.)

Sanitary facilities, en suite to a wheelchair-accessible   M (4.24 m)
bedroom, should comply with the provisions of M5.15
to M5.21 for *Wheelchair-accessible Bathrooms* or
Wheelchair-accessible Shower Facilities.

Smoke alarms

Smoke alarms should not be fixed in bathrooms, showers, cooking areas or garages.	B1 1.17 (V1) B1 1.16 (V2)

Switches, outlets and controls

The aim should be to ensure that all switches, outlets and controls are easy to operate, visible and free from obstruction.

Light switches should: • have large push pads; • align horizontally with door handles; • be within 900 mm and 1100 mm from the entrance door opening.	M (4.30h and I)
Switches and controls should be located between 750 mm and 1200 mm above the floor.	M (4.30c and d)
The operation of all switches, outlets and controls should not require the simultaneous use of both hands (unless necessary for safety reasons).	M (4.30j)
Switched socket outlets should indicate whether they are 'ON'.	M (4.30k)
Mains and circuit isolator switches should clearly indicate whether they are 'ON' or 'OFF'.	M (4.30l)
Individual switches on panels and on multiple socket outlets should be well separated.	M (4.29)
All socket outlets should be wall mounted.	M (4.30a and b)
All telephone points and TV sockets should be located between 400 mm and 1000 mm above the floor (or 400 mm and 1200 mm above the floor for permanently wired appliances).	M (4.30a and b)
Socket outlets should be located no nearer than 350 mm from room corners.	M (4.30g)
Controls that need close vision (e.g. thermostats) should be located between 1200 mm and 1400 mm above the floor.	M (4.30f)

Emergency-alarm pull-cords should: M (4.30e)

- be coloured red;
- be located as close to a wall as possible;
- have two red 50 mm diameter bangles.

Front plates should contrast visually with their   M (4.30m)
backgrounds.

The colours red and green should **not** be used in   M (4.28)
combination as indicators of 'ON' and 'OFF' for
switches and controls.

Aids to communication

A hearing system is available   Mini-com and/or text   Staff have received disability
in certain locations   phone facility available   awareness training

Figure 6.197 Examples of facility signs.

The design of the acoustic environment should ensure that   M (4.33)
audible information can be heard clearly.

A clearly audible public address system should be
supplemented by visual information.

 Note: Assisting people with impaired hearing to fully participate in public discussions (meetings and performances) may require an advanced sound level system (e.g. induction loop, infrared or radio system) to be installed.

Hearing enhancement systems should be installed: M (4.36b)

- in all rooms and spaces designed for meetings, lectures, classes, performances, spectator sport or films;
- at service or reception counters (especially when they are situated in noisy areas or they are behind glazed screens.

The availability of an induction loop or infrared hearing enhancement system should be indicated by the standard symbol.	M (4.36c)
Telephones suitable for hearing aid users should: • be clearly indicated by the standard ear and 'T' symbol; • incorporate an inductive coupler and volume control.	M (4.36d)
Text telephones for deaf and hard of hearing people should be clearly indicated by the standard symbol.	M (4.36e)
Artificial lighting should be designed to be compatible with other electronic and radio-frequency installations.	M (4.36f)
Artificial lighting should be designed to give good colour rendering of all surfaces.	M (4.34)
Glare and reflections from shiny surfaces, and large repeating patterns, should be avoided in spaces where visual acuity is critical.	M (4.32)
Uplighters mounted at low or floor level should be avoided as they can disorientate some visually impaired people.	M (4.34)

Detailed guidance on surface finishes, visual, audible and tactile signs, as well as the characteristics and appropriate choice and use of hearing-enhancement systems, is available in BS 8300.

6.18 Water (and earth) closets, bathrooms and showers

6.18.1 The requirement

Sanitary conveniences (provided in separate rooms or in bathrooms) shall be:

• *separated from places where food is prepared;*
• *provided with washbasins plumbed with hot and cold water;*
• *designed and installed so as to allow effective cleaning.*

(Building Act 1984 Section 26)

All plans for buildings must include at least one (or more) water or earth closets **unless** the local authority is satisfied in the case of a particular building that one is not required (e.g. in a large garage separated from the house).

If you propose using an earth closet, the local authority cannot reject the plans unless they consider that there is insufficient water supply to that earth closet.

Sanitary conveniences

(1)　Adequate and suitable sanitary conveniences must be provided in rooms provided to accommodate them or in bathrooms.

(2)　Adequate stand washing facilities must be provided in:

(a)　rooms containing sanitary conveniences; or

(b)　rooms or spaces adjacent to rooms containing sanitary conveniences.

(3)　Any room containing a sanitary convenience, a bidet, or any facility for washing hands provided in accordance with paragraph (2)(b) must be separated from any kitchen or any area where food is prepared.

(Approved Document G4)

There must be a suitable installation for the provision of water of suitable quality to any sanitary convenience fitted with a flushing device.

(Approved Document G1)

*All dwellings (and buildings containing one or more rooms for residential purposes) will be provided with a bathroom which contains either a fixed bath or a shower **and** a washbasin.*

(Approved Document G5)

Ventilation

There shall be adequate means of ventilation provided for people in the building.

(Approved Document F)

Access to and use of buildings

Reasonable provisions shall be made for people to:

(c)　gain access to; and

(d)　use;

the building and its facilities

(Approved Document M)

In addition to the requirements of the Disability Discrimination Act 1995, precautions need to be taken to ensure that:

- new non-domestic buildings and/or dwellings (e.g. houses and flats used for student living accommodation, etc.);
- extensions to existing non-domestic buildings;
- non-domestic buildings that have been subject to a material change of use (e.g. so that they become a hotel, boarding house, institution, public building or shop);

are capable of allowing people, regardless of their disability, age or gender, to:

- gain access to buildings;
- **gain access within buildings**;
- be able to use the facilities of the buildings (both as visitors and as people who live or work in them);
- use sanitary conveniences in the principal storey of any new dwelling.

Note: See Appendix A for guidance on access and facilities for disabled people.

If the proposed building is going to be used as a workplace or a factory in which persons of both sexes are going to be employed, separate closet accommodation **must** be provided unless the local authority approves otherwise.

The Building Act 1984 Sections 64–68

Under existing regulations, all buildings (except factories and buildings used as workplaces) shall be provided with sufficient closet accommodation (or privy) according to the intended use of that building and the amount of people using that building. The only exceptions are if the building (in the view of the local authority) has an insufficient water supply and a sewer is not available.

If a building already has a sufficient water supply and sewer available, the local authority has the authority to insist that the owner of the property replaces any other closet (e.g. an earth closet) with a water closet. In these cases the owner is entitled to claim 50 per cent of the expense of doing this from the local authority.

If the local authority completes the work, they are entitled to claim 50 per cent back from the owner.

The owner of the property has **no** right of appeal in these cases.

In the Greater London area, a 'water closet' can **also** be taken to mean a 'urinal'.

Business premises

There may be some additional requirements regarding the number, types and siting of appliances in business premises. If this applies to you, then you will need to look at:

- the Offices, Shops and Railway Premises Act 1963;
- the Factories Act 1961; or
- the Food Hygiene (General) Regulations 1970.

6.18.2 Meeting the requirement

Although the above are the minimum requirements for meeting sanitary convenience and washing facility regulations, other local authority regulations may apply, and it is worth seeking the advice of the local planning officer before proceeding.

Sanitary conveniences

Sanitary conveniences of an appropriate type for the sex and age of the persons using the building shall be provided in sufficient numbers, and hand-washing facilities shall be provided in, or adjacent to, rooms containing sanitary conveniences.

 Where hot and cold taps are provided on a sanitary appliance, the hot tap should be on the **left**!

Water intended for sanitary conveniences shall:	G1

- be reliable;
- be either wholesome, softened wholesome or of suitable quality;
- possess a pressure and flow rate sufficient for the operation of the sanitary appliances;
- convey water to sanitary appliances and locations without waste, misuse, undue consumption or contamination of wholesome water.

It may be provided from the following alternative sources: G1-6

- water abstracted from wells, springs, boreholes or water courses;
- harvested rainwater;
- reclaimed greywater; or
- reclaimed industrial-process water.

It should incorporate measures to minimize the impact on   G1-6
water quality from:

- failure of any components;
- failure due to lack of maintenance;
- power failure; and
- any other measures identified in a risk assessment.

Any system/unit used to supply dwellings with water from   G1-14
alternative sources should be subject to a risk assessment by
the system designer and the manufacturer.

 A record of all sanitary appliances used in the water consumption calculation and installed in the dwelling shall be maintained.

Scale of provision and layout in dwellings

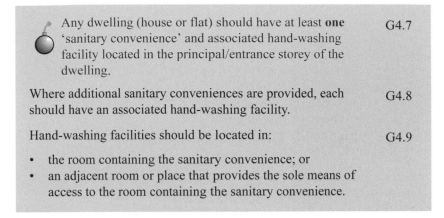

Any dwelling (house or flat) should have at least **one** 'sanitary convenience' and associated hand-washing facility located in the principal/entrance storey of the dwelling.

G4.7

Where additional sanitary conveniences are provided, each should have an associated hand-washing facility.

G4.8

Hand-washing facilities should be located in:

G4.9

- the room containing the sanitary convenience; or
- an adjacent room or place that provides the sole means of access to the room containing the sanitary convenience.

Provided that it is **not** used for the preparation of food:

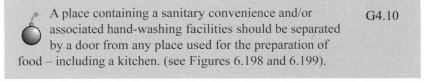

A place containing a sanitary convenience and/or associated hand-washing facilities should be separated by a door from any place used for the preparation of food – including a kitchen. (see Figures 6.198 and 6.199).

G4.10

Scale of provision and layout in buildings other than dwellings

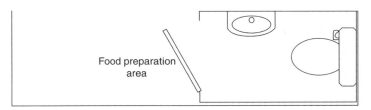

Figure 6.198 Separation between hand washbasin/WC and food preparation area – single room.

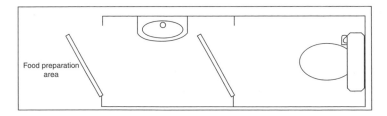

Figure 6.199 Separation between hand washbasin/WC and food preparation area – two rooms.

The Workplace (Health, Safety and Welfare) Regulations 1992 require that a minimum number of sanitary conveniences **must** be provided in workplaces.

A sanitary convenience may be provided in: G4.16

- a self-contained room which also contains hand-washing facilities;
- in a cubicle with shared hand-washing facilities located in a room containing a number of cubicles; or
- in a self-contained room with hand-washing facilities provided in an adjacent room.

A place containing a sanitary convenience and/or associated   G4.17
hand-washing facilities should be separated by a door from any place used for the preparation of food (including a kitchen).

Urinals, WC cubicles and hand-washing facilities may be in   G4.16
the same room.

Chemical and composting toilets

Chemical toilets or composting toilets may be used where:

- suitable arrangements can be made for the disposal of the   G4.19
  waste either on or off the site; and
- the waste can be removed from the premises without carrying it through any living space or food-preparation areas (including a kitchen); and
- no part of the installation is installed in any places where it might be rendered ineffective by the entry of flood water.

Composting toilets should **not** be connected to an   G4.21
energy source other than for purposes of ventilation or sustaining the composting process.

Discharges to drains

Any WC fitted with flushing apparatus should discharge to   G4.22
an adequate system of drainage.

A urinal fitted with flushing apparatus should discharge through a grating, a trap or mechanical seal and a branch pipe to a discharge stack or a drain.   G4.23

A WC fitted with a macerator and pump may be connected to a small-bore drainage system discharging to a discharge stack if: G4.24

- there is also access to a WC discharging directly to a gravity system; and
- the macerator and pump meets the requirements of BS EN 12050-1:2001 or BS EN 12050-3:2001.

Sanitary appliances used for personal washing: G5.9

- should discharge through a grating, a trap and a branch discharge pipe to an adequate system of drainage;
- that are fitted with are a macerator and pump may be connected to a small-bore drainage system discharging to a discharge stack if there is also access to washing facilities discharging directly to a gravity system.

Bathrooms

All dwellings (and buildings containing one or more rooms for residential purposes) will be provided with a bathroom which contains either a fixed bath or a shower **and** a washbasin.

Any dwelling (house or flat) must have at least one bathroom with a fixed bath or shower, and a washbasin.   G5.6

Prevention of scalding

Where hot and cold taps are provided on a sanitary appliance, **the hot tap should be on the left**.   G4.6

Note: In some of the other Approved Documents it also states that the hot tap must be on the left. I think that the reasoning is that always having the hot tap on the left-hand side will stop people inadvertently scalding themselves – particularly children and poorly sighted people. To be on the safe side, therefore, it is recommended that you always ensure the hot water taps are located on the left-hand side of all of your basins and baths.

The hot water supply temperature to a bath should be limited to a maximum of 48° C by use of an in-line blending valve or other appropriate temperature-control device, with a maximum temperature stop and a suitable arrangement of pipework.	G3.65
In-line blending valves and composite thermostatic mixing valves should be compatible with the hot and cold water sources that serve them.	G3.67
The length of supply pipes between in-line blending valves and outlets should be kept to a minimum in order to prevent the colonization of waterborne pathogens.	G3.68

Extract ventilation

Extract ventilation concerns the removal of air directly from a space or spaces to outside. Extract ventilation may be by natural means such as Passive Stack Ventilation (PSV) or by mechanical means (e.g. by an extract fan or central system).

The main requirements are that:

- All sanitary accommodation shall be provided with extract ventilation to the outside that is capable of operating either intermittently or continuously with a minimum extract rate of 6 l/s.
- Ventilation devices designed to work continuously shall not have automatic controls such as a humidity control when used for sanitary accommodation.
- As odour is the main pollutant, humidity controls should not be used for intermittent extract in sanitary accommodation.
- Common outlet terminals and/or branched ducts shall not be used for wet rooms such as WCs.
- PSV can be used as an alternative to a mechanical extract fan for office sanitary and washrooms.

Extract to outside is required in all office sanitary accommodation and washrooms

Passive stack ventilation (PSV) devices shall have a minimum internal duct diameter of 100 mm and a minimum cross-sectional area of 8000 mm².	F Tables 5.2a to 5.2d
Where there was no previous ventilation opening, or where the size of the original ventilation opening is not known, replacement window(s) shall have an equivalent area greater than 2500 mm² per WC.	F 7.6

In buildings other than dwellings, fresh air supplies should be protected from contaminants that could be injurious to health.	F 6.3
All office sanitary accommodation and washrooms shall be provided with intermittent air extract ventilation.	F 6.10
All bathrooms shall be provided with extract ventilation to the outside which is capable of operating either: • intermittently at a minimum extract rate of 15 l/s; or • continuously with a minimum extract rate of 8 l/s.	F.6
In bathrooms (without a WC) where there was no previous ventilation opening, or where the size of the original ventilation opening is not known, replacement window(s) shall have an equivalent area greater than 2500 mm².	F 7.6
Common outlet terminals and/or branched ducts shall **not** be used for wet rooms such as a bathroom.	F Tables 5.2a to 5.2d
Wall/ceiling-mounted centrifugal fans (which are fitted with a 100 mm diameter flexible duct or rectangular duct and which are designed to achieve 15 l/s for kitchens) should not be ducted further than 6 m and should have no more than one 90° bend.	F Tables 5.2a to 5.2d

Appliances fitted in bathrooms

Any appliance installed in a room used or intended to be used as a bathroom must be of the room-sealed type.

Open-flued oil-fired appliances should **not** be installed in bathrooms where there is an increased risk of carbon monoxide poisoning.	J 4.2

If locating combustion appliances in such rooms cannot be avoided, a way of meeting the requirements is to provide room-sealed appliances.

Fire safety

Bathrooms, WCs or showers that are situated less than 4.5 m above ground level and whose only escape route is through another room shall be provided with an emergency egress window.	B1 2.9 (V1) B1 2.5 (V2)
Smoke alarms should **not** be fixed in bathrooms.	B1 1.17 (V1) B1 1.16 (V2)

Workplace conveniences

If the building is a workplace used by both sexes, sufficient and satisfactory accommodation is required for persons of each sex.

This requirement does not apply to premises to which the Offices, Shops and Railway Premises Act 1963 applies.

Loan of temporary sanitary conveniences

If the local authority is maintaining, improving or repairing drainage systems and this requires the disconnection of existing buildings from these sanitary conveniences, then, on request from the occupier of the building, the local authority is required to supply (on temporary loan and at no charge) sanitary conveniences:

* if the disconnection is caused by a defect in a public sewer;
* if the local authority has ordered the replacement of earth closets (see above);
* for the first seven days of any disconnection.

Erection of public conveniences

You are not allowed to erect a public sanitary convenience in (or on) any location that is accessible from a street, without the consent of the local authority. Any person who contravenes this requirement is liable to a fine and can be made (at his own expense) to remove or permanently close the convenience.

This requirement does not apply to sanitary conveniences erected by a railway company within their railway station, yard or approaches, or erected by dock undertakers on land belonging to them.

Protection of openings for pipes

Pipes that pass through a compartment wall or compartment floor (unless the pipe is in a protected shaft), or through a cavity barrier, should conform to one of the following alternatives:

Proprietary seals (any pipe diameter) that maintain the fire resistance of the wall, floor or cavity barrier.	B3 (11.5–11.6)
Pipes with a restricted diameter where fire-stopping is used around the pipe, keeping the opening as small as possible.	B3 (11.5 and 11.7)
Sleeving – a pipe of lead, aluminium, aluminium alloy, fibre-cement or UPVC, with a maximum nominal internal diameter of 160 mm, may be used with a sleeving of non-combustible pipe as shown in Figure 6.200.	B3 (11.5 and 11.8)

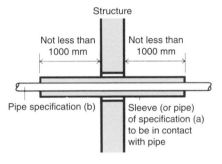

Figure 6.200 Pipes penetrating a structure.

 Make the opening in the structure as small as possible and provide fire-stopping between pipe and structure.

Joints between fire-separating elements should be fire-stopped and all openings for pipes that pass through any part of a fire-separating element should be:	B3 7.12 (V1) B3 10.17 (V2)
• kept as few in number as possible; and • kept as small as practicable; and • fire-stopped (which in the case of a pipe or duct should allow thermal movement).	

Disabled access

The aim of the amended Approved Document M is that suitable sanitary accommodation should be available to *everybody*, including sanitary accommodation specifically designed for wheelchair users, ambulant disabled people, people of either sex with babies and small children, and/or people with luggage.

Provision of toilet accommodation

Where sanitary facilities are provided in a building, at least one wheelchair-accessible unisex toilet should be available.	M (5.7b)
If there is only space for one toilet in a building, then it should be a wheelchair-accessible unisex type.	M (5.7a)
In separate-sex toilet accommodation, at least one WC cubicle should be provided for ambulant disabled people.	M (5.7c)
In separate-sex toilet accommodation having four or more WC cubicles, at least one should be an enlarged cubicle for use by people who need extra space.	M (5.7d)
Wheelchair-accessible unisex toilets should always be provided in addition to any wheelchair-accessible accommodation in separate-sex toilet washrooms.	M (5.5)

If there is only space for one toilet in a building:

• it should be a wheelchair-accessible unisex type;	M (5.7a)
• its width should be increased from 1.5 m to 2 m;	M (5.7e)
• it should include a standing-height wash basin, in addition to the finger-rinse basin associated with the WC.	M (5.7e)

For specific guidance on the provision of sanitary accommodation in sports buildings, refer to Appendix A to this book (Access for Disabled People).

Sanitary accommodation generally

Sanitary accommodation and washrooms may be included in protected shafts.

Doors

Doors to WC cubicles and wheelchair-accessible unisex toilets should:	M (5.3)

•   (ideally) open outwards;
•   be operable by people with limited strength or manual dexterity;
•   be capable of being opened if a person has collapsed against them while inside the cubicle.

Doors to wheelchair-accessible unisex toilets, changing rooms or shower rooms should:

- be fitted with light-action privacy bolts; M (5.4d)
- be capable of being opened using a force no greater than 20 N; M (5.4d)
- have an emergency release mechanism so that they can be opened outwards (from the outside) in case of emergency. M (5.4e)

Door-opening furniture should: M (5.4c)

- be easy to operate by people with limited manual dexterity;
- be easy to operate with one hand using a closed fist (e.g. a lever handle);
- contrast visually with the surface of the door.

Doors, when open, should not obstruct emergency escape routes. M (5.4f)

Sanitary fittings

The surface finish of sanitary fittings and grab bars should contrast visually with background wall and floor finishes. M (5.4 k)

Taps should be operable by people with limited strength and/or manual dexterity. M (5.3)

Bath and wash basin taps should either be controlled automatically or be capable of being operated using a closed fist (e.g. by lever action). M (5.4a)

All terminal fittings should comply with Guidance Note G18.5 of the Guidance Document relating to Schedule 2: Requirements for Water Fittings, of the Water Supply (Water Fittings) Regulations 1999, SI 1999/1148. M (5.4b)

Outlets, controls and switches

The aim is to ensure that all controls and switches should be easy to operate, visible and free from obstruction: M (5.4i)

- they should be located between 750 mm and 1200 mm above the floor;

- they should not require the simultaneous use of both hands (unless necessary for safety reasons) to operate;
- light switches should:
  - have large push pads;
  - align horizontally with door handles;
  - be within 900 mm to 1100 mm from the entrance door opening;
- switched socket outlets should indicate whether they are 'ON';
- mains and circuit isolator switches should clearly indicate whether they are 'ON' or 'OFF';
- individual switches on panels and on multiple socket outlets should be well separated;
- controls that need close vision (e.g. thermostats) should be located between 1200 mm and 1400 mm above the floor;
- **emergency-alarm pull-cords** should be:
  - coloured **red**;
  - located as close to a wall as possible;
  - have two **red** 50 mm diameter bangles;
- front plates should contrast visually with their backgrounds.

Heat emitters should either be screened or have their exposed surfaces kept at a temperature below 43° C.	M (5.4j)
Where possible, light switches with large push pads should be used in preference to pull cords.	M (5.3)
The colours **red** and green should not be used in combination as indicators of 'ON' and 'OFF' for switches and controls.	M (5.3)

Smoke alarms

Smoke alarms should be positioned in the circulation spaces between sleeping spaces and places where fires are most likely to start (e.g. in kitchens and living rooms).	B1 1.11

Wheelchair-accessible unisex toilets

Where sanitary facilities are provided in a building, at least one wheelchair-accessible unisex toilet should be available.

Wheelchair-accessible unisex toilets should **not** be used for baby changing.   M (5.5)

Wheelchair-accessible unisex toilets should:

- be located as close as possible to the entrance and/or waiting area of the building;   M (5.10a)

- not be located in a way that compromises the privacy of users;   M (5.10b)

- be located in a similar position on each floor of a multi-storey building;   M (5.10c)

- allow for right- and left-hand transfer on alternate floors;   M (5.10c and d)

- be located on accessible routes that are direct and obstruction free;   M (5.10f)

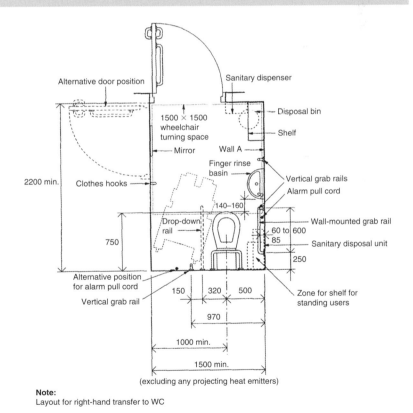

Note:
Layout for right-hand transfer to WC

Figure 6.201 Example of a unisex wheelchair-accessible toilet with a corner WC. (Dimensions in mm.)

• always be provided in addition to any wheel-chair-accessible accommodation in separate-sex toilet washrooms;	M (5.5)
• not be used for baby changing.	M (5.5)
The minimum overall dimensions and arrangement of fittings within a wheelchair-accessible unisex toilet should comply with Figure 6.201.	M (5.10i)

Accessibility

The approach to a unisex toilet should be separate from other sanitary accommodation.	M (5.9)
Wheelchair users should:	
• not have to travel more than 40 m on the same floor to reach a unisex toilet;	M (5.9 and 5.10 h)
• not have to travel more than the combined horizontal distance where the unisex toilet accommodation is on another floor of the building (accessible by passenger lift);	M (5.9.5.10h)
• be able to approach, transfer to, and use the sanitary facilities provided within a building.	M (5.8)

Heights and arrangements

The heights and arrangement of fittings in a wheelchair-accessible unisex toilet should comply with Figure 6.201 and (as appropriate) Figure 6.202.	M (5.10)

The space provided for manoeuvring should enable wheelchair users to adopt various transfer techniques that allow independent or assisted use.	M (5.8)
The transfer space alongside the WC should be kept clear to the back wall.	M (5.8)
The relationship of the WC to the finger-rinse basin and other accessories should allow a person to wash and dry hands while seated on the WC.	M (5.8)
Heat emitters (if located) should not restrict:	

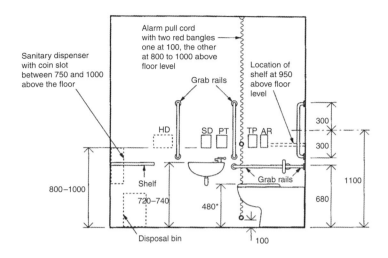

* Height subject to manufacturing tolerance of WC pan

HD: Possible position for automatic hand dryer
SD: Soap dispenser
PT: Paper-towel dispenser
AR: Alarm reset button
TP: Toilet-paper dispenser

* Height of drop-down rails to be the same as the other horizontal grab rails

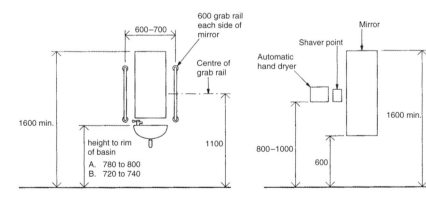

Height of independent wash basin
and location of associated fittings, for
wheelchair users and standing people

A. For people standing
B. For use from WC

Mirror located away from wash basin
suitable for seated and standing people
(Mirror and associated fittings used
within a WC compartment or serving a
range of compartments)

Figure 6.202 Typical heights of arrangement of fittings in a unisex wheelchair-accessible toilet. (Dimensions in mm.)

> • the minimum clear wheelchair manoeuvring space;   M (5.10p)
> • the space beside the WC used for transfer from the
>   wheelchair to the WC.

Doors

> Doors should:
>
> • preferably open outwards;   M (5.10 g)
> • be fitted with a horizontal closing bar fixed to the
>   inside face.

Support rails

The rail on the open side can be a drop-down rail, but on   M (5.8)
the wall side it can either be a wall-mounted grab rail or,
alternatively, a second drop-down rail in addition to the
wall-mounted grab rail.

If the horizontal support rail (on the wall adjacent to the   M (5.10j)
WC) is set at the minimum spacing from the wall, an
additional drop-down rail should be provided on the wall
side, 320 mm from the centre line of the WC.

If the horizontal support rail (on the wall adjacent to   M (5.10k)
the WC) is set so that its centre line is 400 mm from the
centre line of the WC, there is no additional drop-down
rail.

Emergency assistance

> An emergency assistance alarm system should be
> provided that has:
> • an outside emergency-assistance call signal that   M (5.10n)
>   can be easily seen and heard by those able to give
>   assistance;
> • visual and audible indicators to confirm that an emer-   M (5.10m)
>   gency call has been received;
> • a reset control reachable from a wheelchair, WC, or   M (5.10m)
>   from a shower/changing seat;

- a signal that is distinguishable visually and audibly   M (5.10m)
  from the fire alarms provided.

Emergency-assistance pull cords should: M (5.10o)

- be easily identifiable;
- be reachable from the WC and from the floor, close to
  the WC;
- be coloured;
- be located as close to a wall as possible;
- have two **red** 50 mm diameter bangles.

Fire safety

There shall be an early warning fire alarm system for persons in the building.

Alarms

Fire alarms should emit an audio and visual signal to warn   M (5.4g)
occupants with hearing or visual impairments.

Emergency assistance alarm systems should have: M (5.4h)

- visual and audible indicators to confirm that an emergency call has been received;
- a reset control reachable from a wheelchair, WC, or
  from a shower/changing seat;
- a signal that is distinguishable visually and audibly
  from the fire alarm.

WC pans

WC pans should:

- conform to BS 5503-3 or BS 5504-4;   M (5.10q)
- be able to accept a variable-height toilet seat riser;   M (5.10q)
- have a flushing mechanism positioned on the open or   M (5.10r)
  transfer side of the space – irrespective of handing.

See BS 8300 for more detailed guidance on the various techniques used to
transfer from a wheelchair to a WC, as well as appropriate sanitary and other
fittings.

Toilets in separate-sex washrooms

There should be at least the same number of WCs (for women) as urinals (for men).

Ambulant disabled people should have the opportunity to use a WC compartment within any separate-sex toilet washroom.	M (5.11)
A wheelchair-accessible compartment (where provided) shall have the same layout and fittings as the unisex toilet.	M (5.14f)
Where a separate-sex toilet washroom can be accessed by wheelchair users, it should be possible for them to use both a urinal (where appropriate) and a washbasin at a lower height than is provided for other users.	M (5.13)
Consideration should be given to providing a low-level urinal for children in male washrooms.	M (5.13)
Separate-sex toilet washrooms above a certain size should include an enlarged WC cubicle for use by people who need extra space (e.g. parents with children and babies, people carrying luggage, and ambulant disabled people).	M (5.12)
The minimum dimensions of compartments for ambulant disabled people should comply with Figure 6.203.	M (5.14b)

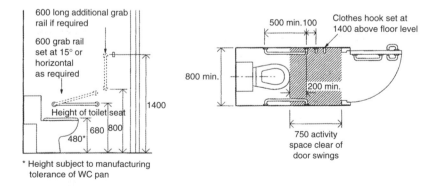

Figure 6.203 Example of a WC cubicle for an ambulant disabled person. (Dimensions in mm.)

Accessibility

The approach to a unisex toilet should be separate to other sanitary accommodation.   M (5.14e)

Wheelchair users should: M (5.14e)

- not have to travel more than 40 m on the same floor to reach a unisex toilet;
- not have to travel more than combined horizontal distance where the unisex toilet accommodation is on another floor of the building (accessible by passenger lift);
- be able to approach, transfer to, and use the sanitary facilities provided within a building.

Height and arrangement

Compartments used for ambulant disabled people should: M (5.14d)

- be 1200 mm wide;
- include a horizontal grab bar adjacent to the WC;
- include a vertical grab bar on the rear wall;
- include space for a shelf and fold-down changing table.

A wheelchair-accessible washroom (where provided) shall have: M (5.14g)

- at least one washbasin with its rim set at 720 mm to 740 mm above the floor;
- for men, at least one urinal with its rim set at 380 mm above the floor, with two 600 mm long vertical grab bars with their centre lines at 1100 mm above the floor, positioned either side of the urinal.

The compartment should: M (5.11)

- be fitted with support rails;
- include a minimum activity space to accommodate people who use crutches, or otherwise have impaired leg movements.

Doors

Doors to compartments for ambulant disabled people should:	M (5.14d)
• preferably open outwards; • be fitted with a horizontal closing bar fixed to the inside face.	
The swing of any inward-opening doors to standard WC compartments should enable a 450 mm diameter manoeuvring space to be maintained between the swing of the door, the WC pan and the side wall of the compartment.	M (5.14a)

WC pans

WC pans should:	
• conform to BS 5503-3 or BS 5504-4;	M (5.14e)
• accommodate the use of a variable-height toilet seat riser.	M (5.14e)

Note: More detailed guidance on appropriate sanitary and other fittings is given in BS 8300.

Wheelchair-accessible changing and shower facilities

A choice of shower layout, together with correctly located shower controls and fittings, will enable disabled people to independently make use of the facilities – or be assisted by others where necessary.

In large building complexes (e.g. retail parks and large sports centres) there should be one wheelchair-accessible unisex toilet capable of including an adult changing table.	M (5.17)
The dimensions of the self-contained compartment should allow sufficient space for a helper.	M (5.16)
A combined facility should be divided into distinct 'wet' and 'dry' areas.	M (5.16)

For changing and shower facilities

A choice of layouts suitable for left-hand and right-hand transfer should be provided when more than one individual changing compartment or shower compartment is available.   M (5.18a)

Wall-mounted drop-down support rails and wall-mounted slip-resistant tip-up seats (not spring-loaded) should be provided.   M (5.18b)

Subdivisions (with the same configuration of space and equipment as for self-contained facilities) should be provided for communal shower facilities and changing facilities.   M (5.18c)

In addition to communal separate-sex facilities, individual self-contained shower and changing facilities should be available in sports amenities.   M (5.18d)

An emergency-assistance alarm system should be provided and should have: M (5.18f)

- visible and audible indicators to confirm that an emergency call has been received;
- a reset control reachable from a wheelchair, WC, or from a shower/changing seat;
- a signal that is distinguishable visually and audibly from the fire alarm.

An **emergency-assistance pull cord** should be provided that should: M (5.18e)

- be easily identifiable and reachable from the wall-mounted tip-up seat (or from the floor);
- be located as close to a wall as possible;
- be coloured;
- have two **red** 50 mm diameter bangles.

Facilities for limb storage should be included for the benefit of amputees.   M (5.18)

For changing facilities

The floor of a changing area should be level and slip-resistant when dry or when wet – particularly when associated with shower facilities.   M (5.18i)

There should be a manoeuvring space 1500 mm deep in front of lockers in self-contained and/or communal changing areas.   M (5.18j)

The minimum overall dimensions of (and the arrangement of equipment and controls within) individual self-contained changing facilities should comply with Figure 6.204.   M (5.18h)

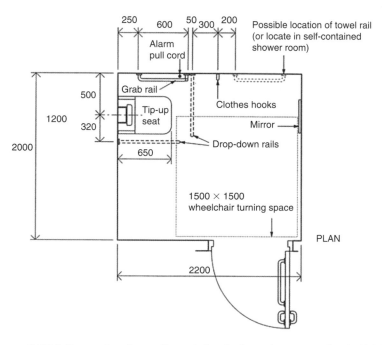

Figure 6.204 Example of a self-contained changing room for individual use. (Dimensions in mm.)

For shower facilities

A shower curtain (enclosing the seat and rails and which can be operated from the shower seat) should be provided.   M (5.18m)

A shelf (that can be reached from the shower seat and/or from the wheelchair) should be provided for toiletries.   M (5.18n)

The floor of the shower and shower area should be slip-resistant and self-draining.   M (5.18o)

The shower controls should be positioned between 750 mm and 1000 mm above the floor in all communal area wheelchair-accessible shower facilities.   M (5.18q)

If showers are provided in commercial developments for the benefit of staff, at least one wheelchair-accessible shower compartment (complying with Figure 6.205) should be made available.   M (5.18l)

Individual self-contained shower facilities should comply with Figure 6.205.   M (5.18k)

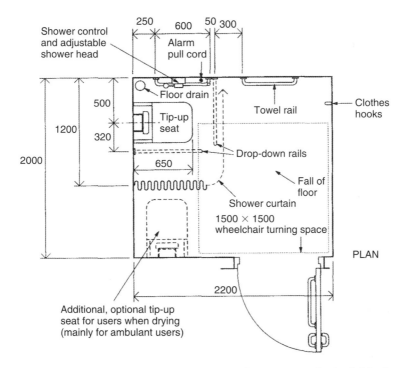

Figure 6.205 Example of a self-contained shower room for individual use. (Dimensions in mm.)

For shower facilities incorporating a WC

A choice of left-hand and right-hand transfer layouts should be available when more than one shower area includes a corner WC.   M (5.18s)

The minimum overall dimensions of (and the arrangement   M (5.18r)
of fittings within) an individual self-contained shower
area incorporating a corner WC (e.g. in a sports building)
should comply with Figure 6.206.

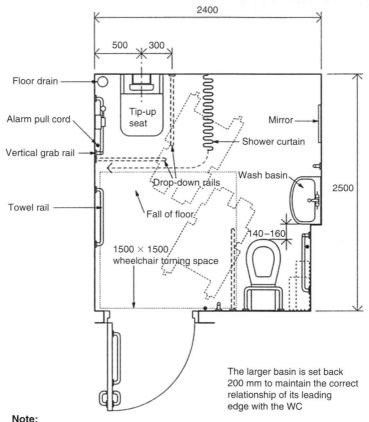

Note:
Layout shown for right-hand transfer to shower seat and WC

Figure 6.206 Example of a shower room incorporating a corner WC for individual use. (Dimensions in mm.)

 Note: More detailed guidance on appropriate sanitary and other fittings is given in BS 8300.

Wheelchair-accessible bathrooms

Wheelchair users and ambulant disabled people (in hotels, motels, relatives' accommodation in hospitals, and student accommodation and sports facilities) should be able to wash or bathe either independently or with assistance from others.

The minimum overall dimensions of (and the arrangement   M (5.21a)
of fittings within) a bathroom for individual use
incorporating a corner WC should comply with Figure
6.207.

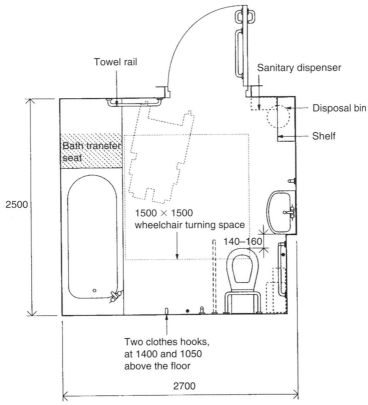

Towel rail

Sanitary dispenser

Disposal bin

Shelf

Bath transfer seat

2500

1500 × 1500
wheelchair turning space

140–160

Two clothes hooks,
at 1400 and 1050
above the floor

2700

Note:
Layout shown for right-hand transfer to bath and WC.

Figure 6.207 Example of a bathroom containing a WC. (Dimensions in mm.)

A choice of layouts suitable for left-hand and right-hand   M (5.21b)
transfer should be provided when more than one bathroom
is available.

The floor of a bathroom should be slip-resistant when dry or when wet.

M (5.21c)

The bath should be provided with a transfer seat that is 400 mm deep and equal to the width of the bath.

M (5.21d)

Outward opening doors, fitted with a horizontal closing bar fixed to the inside face, should be provided.

M (5.21e)

An **emergency assistance pull cord** should be provided which should:

M (5.21f)

- be easily identifiable and reachable from the wall-mounted tip-up seat (or from the floor);
- be located as close to a wall as possible;
- be coloured;
- have two **red** 50 mm diameter bangles.

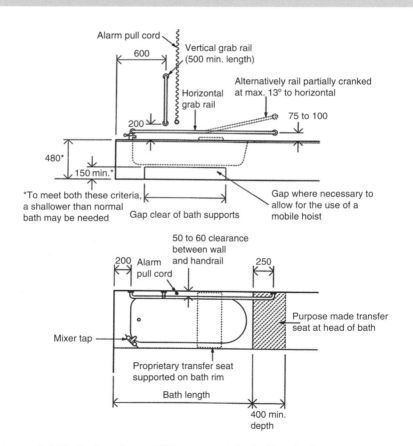

Figure 6.208 Grab rails and fitting associated with a bath.

Note:

(1) More detailed guidance on appropriate sanitary and other fittings, including facilities for the use of mobile and fixed hoists, is given in BS 8300.

(2) Guidance on the slip resistance of floor surfaces is given in Annex C of BS 8300:2001.

WC provision in the entrance storey of the dwelling

Whenever possible, a WC should be provided in the entrance storey of the dwelling so that there is no need to negotiate a stair to reach it from the habitable rooms in that storey.

If there is a bathroom in the principal storey, then a WC may be collocated with it.	M (10.2)
The door to the WC compartment should:	M (10.3b)
• open outwards; • be positioned so as to allow wheelchair users access to it; • have a clear opening width in accordance with Table 6.104.	
The WC compartment should:	M (10.3c)
• provide a clear space for wheelchair users to access the WC; • position the washbasin so that it does not impede access.	

Table 6.104 Minimum widths of corridors and passageways for a range of doorway widths

Doorway clear opening width (mm)	Corridor/passageway width (mm)
750 or wider	900 (when approached head on)
750	1200 (when not approached head on)
775	1050 (when not approached head on)
800	900 (when not approached head on)

For further information see the website of the Equality and Human Rights Commission (http://www.equalityhumanrights.com).

Appliances fitted in bathrooms and shower rooms

Any appliance installed in a room used, or intended to be used, as a bath or shower room must be of the room-sealed type.

Open-flued oil-fired appliances should **not** be installed in bathrooms (or bedrooms for that matter) where there is an increased risk of carbon monoxide poisoning.	J 4.2

If locating combustion appliances in such rooms cannot be avoided, a way of meeting the requirements is to provide room-sealed appliances.

Ventilation ducts and flues, etc.

If a flue passes through a compartment wall or compartment floor, or is built into a compartment wall, each wall of the flue should have a fire resistance of at least half that of the wall or floor (see Figure 6.209).	B1 7.11 (V1) B3 10.16 (V2)

(a) FLUE PASSING THROUGH COMPARTMENT WALL OR FLOOR

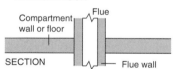

Flue walls should have a fire resistance of at least one half of that required for the compartment wall or floor, and be of non-combustible construction.

(b) FLUE BUILT INTO COMPARTMENT WALL

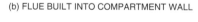
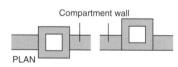

In each case flue walls should have a fire resistance at least one half of that required for the compartment wall and be of non-combustible construction.

Figure 6.209 Flues penetrating compartment walls or floors.

6.19 Electrical safety

6.19.1 The requirement

Although:

- Part B (Fire Safety)
- Part E (Resistance to the Passage of Sound)
- Part J (Combustion Appliances and Fuel Storage Systems)
- Part K (Protection from Falling, Collision and Impact)
- Part G (Sanitation, Hot Water Safety and Water Efficiency)

all have a number of requirements concerning electrical safety and electrical installations (see below for details), the main requirements are contained in:

- Part P (Electrical Safety) together with:
- Part L (Conservation of Fuel and Power) and, where necessary,
- Part M (Access and use of buildings).

Electrical work

Reasonable provision shall be made in the design and installation of electrical installations in order to protect persons operating, maintaining or altering the installations from fire or injury.

(Approved Document P1)

Note: Although Part P makes requirements for the safety of fixed electrical installations, this does not cover system functionality (such as electrically powered fire alarm systems, fans and pumps) which are covered in other parts of the Building Regulations and other legislation.

Conservation of fuel and power

Reasonable provision shall be made for the conservation of fuel and power in buildings by:

(a) limiting heat gains and losses –
> *(i) through thermal elements and other parts of the building fabric; and*
> *(ii) from pipes, ducts and vessels used for space heating, space cooling and hot water services;*

(b) providing fixed building services which –
> *(i) are energy efficient;*
> *(ii) have effective controls; and*
> *(iii) are commissioned by testing and adjusting as necessary to ensure they use no more fuel and power than is reasonable in the circumstances; and*

(c) providing to the owner sufficient information about the building, the fixed building services and their maintenance requirements so that the building can be operated in such a manner as to use no more fuel and power than is reasonable in the circumstances

(Approved Document L)

A new Part L came into force on 1 October 2010.

Responsibility for achieving compliance with the requirements of Part L rests with the person carrying out the work. That 'person' may be, for example, a developer, a main (or sub-) contractor, or a specialist firm directly engaged by a private client.

Note: The person responsible for achieving compliance should either themselves provide a certificate, or obtain a certificate from the subcontractor, that commissioning has been successfully carried out. The certificate should be made available to the client and the building control body.

Access to and use of buildings

Reasonable provisions shall be made for people to:

(a)   gain access to; and
(b)   use;

the building and its facilities

(Approved Document M)

In addition to the requirements of the Disability Discrimination Act 1995 precautions need to be taken to ensure that:

- new non-domestic buildings and/or dwellings (e.g. houses and flats used for student living accommodation etc.);
- extensions to existing non-domestic buildings;
- non-domestic buildings that have been subject to a material change of use (e.g. so that they become a hotel, boarding house, institution, public building or shop);

are capable of allowing people, regardless of their disability, age or gender, to:

- gain access to buildings;
- **gain access within buildings;**
- be able to use the facilities of the buildings (both as visitors and as people who live or work in them);
- use sanitary conveniences in the principal storey of any new dwelling.

Note: See Appendix A for guidance on access and facilities for disabled people.

6.19.2 Meeting the requirement

Water safety devices

Water safety devices normally consist of:

- non-self-resetting energy cut-outs;
- temperature and pressure relief devices.

Non-self-resetting energy cut-outs

Non-self-resetting energy cut-outs may only be used where they would have the effect of instantly disconnecting the supply of energy to the storage vessel.	G3.28
Where an electrical device, such as a relay or motorized valve, is connected to the energy cut-out, the device should operate to interrupt the supply of energy if the electrical power supply is disconnected.	G3.31
Where there is more than one energy cut-out, each non-self-resetting energy cut-out should be independent and have a separate motorized valve and a separate temperature sensor.	G3.32
Where an energy cut-out is fitted, each heat source should have a separate non-self-resetting energy cut-out.	G3.33

Temperature and pressure relief devices

Where relevant, appropriate pressure-, temperature-, or temperature- and pressure-activated safety devices should be fitted in addition to a safety device such as energy cut-out.	G3.34
Temperature relief valves and combined temperature and pressure relief valves should not be used in systems which do not automatically replenish the stored water (e.g. unvented primary thermal storage vessels).	G3.35 G3.36

 In such cases there should be a second non-self-resetting energy cut-out independent of the one provided.

Temperature relief valve(s) or combined temperature and pressure relief valve(s) should:	G3.37 G3.38
• give a discharge rating at least equal to the total power input to the hot water storage system; • be located directly on the storage vessel, to ensure the stored water does not exceed 100° C.	
In hot water storage system units and packages, the temperature relief valve(s):	G3.39
• should be factory fitted; • should not be disconnected other than for replacement; and • should not be relocated in any other device or fitting installed.	

Fixed building services, including controls, should be commissioned by testing and adjusting as necessary to ensure that they use no more fuel and power than is reasonable in the circumstances.

Electric water heating

Electric fixed immersion heaters should comply with BS EN 60335-2-73:2003.	G3.43
Electric instantaneous water heaters should comply with BS EN 60335-2-35:2002.	G3.44
Electric storage water heaters should comply with BS EN 60335-2-21:2003.	G3.45

Solar water heating

Factory-made solar water heating systems should comply with BS EN 12976-1:2006.	G3.46
Other solar water heating systems should comply with prEN/ TS 12977-1:2008 or BS 5918:1989.	G3.47
Where solar water heating systems are used, an additional heat source should be available in order to maintain the water temperature and thus to restrict microbial growth.	G3.48
As some solar hot water systems operate at high temperatures and pressures, all components should be rated to the appropriate temperatures and pressures.	

Electrical installations

Electrical installations must be inspected and tested during and at the end of installation, and before they are taken into service to verify that they:	P 1.7
• comply with Part P (and any other relevant parts) of the Building Regulations;	
• are safe to use, maintain and alter;	
• meet the relevant equipment and installation standards;	P 0.1b
• meet the requirements of the Building Regulations.	P 3.1
Any proposal for a new mains supply installation (or where significant alterations are going to be made to an existing mains supply) **must** be agreed with the electricity distributor.	P 1.2

Design

Where it is critical for electrical circuits to be able to continue to function during a fire, protected circuits (meeting the requirements of BS EN 50200:2006) shall be installed.	B1 5.38 (V2)
Electrically powered locks should return to the unlocked position:	
• on operation of the fire alarm system;	B1 5.11 (V2)
• on loss of power or system error;	
• on activation of a manual door-release unit.	

 Electrical installations must be designed and installed (suitably enclosed and appropriately separated) so that they:

• comply with Part P (and any other relevant parts) of the Building Regulations;	P 1.7 and 3.1
• comply with the relevant equipment and installation standards;	P 0.1
• do not present an electric shock or fire hazard to people;	P 0.1a
• provide mechanical and thermal protection;	P 1.3
• provide adequate protection against mechanical and thermal damage;	P 0.1a
• provide adequate protection for persons against the risks of electric shock, burn or fire injuries;	P1.3
• are safe to use, maintain and alter.	P 1.7
When downlighters, loudspeakers and other electrical accessories are installed, additional protection may be required to maintain the integrity of a wall or floor.	B1 2.17
Transfer grills of air-circulation systems should **not** be fitted in any wall, door, floor or ceiling enclosing a protected stairway.	
Electrical installations must be designed and installed (suitably enclosed and appropriately separated) so that they comply with the requirements of BS 7671, as amended.	P 1.4

 Note: See Appendix A of Part P to the Building Regulations for details of some of the types of electrical services normally found in dwellings, some of the ways they can be connected, and the complexity of wiring and protective systems that can be used to supply them.

Extensions, material alterations and material changes of use

In accordance with Regulation 4(2) the **whole** of an existing installation **does not** have to be upgraded to current standards, but only to the extent necessary for the new work to meet current standards, except where upgrading is required by the energy-efficiency requirements of the Building Regulations.

Where any electrical installation work is classified as an extension, a material alteration or a material change of use, the work must consider and include:

• confirmation that the mains supply equipment is suitable and can carry the additional loads envisaged;	P 2.1b–2.2
• the earthing and bonding systems are satisfactory and meet the requirements;	P 2.1a–2.2c
• the necessary additions and alterations to the circuits which feed them;	P 2.1a
• the protective measures required to meet the requirements;	P 2.1a–2.2b
• the rating and the condition of existing equipment (belonging to both the consumer and the electricity distributor) is sufficient.	P 2.2a

See Figure 6.210 for details of some of the types of electrical service normally found in dwellings, some of the ways they can be connected, and the complexity of wiring and protective systems that can be used to supply them.

Note: Appendix C to Part P of the Building Regulations offers guidance on some of the older types of installation that might be encountered during alteration work, and Appendix D provides guidance on the application of the now harmonized European cable identification system.

New habitable rooms that are the result of a material alteration and which are above ground floor level (or at ground floor level where no final exit has been provided) shall be equipped with:	B1 1.8 (V1) B1 1.6 (V2)

• a fire detection and fire alarm system;
• smoke alarms (in accordance with Paragraphs 1.10 to 1.18) in the circulation spaces.

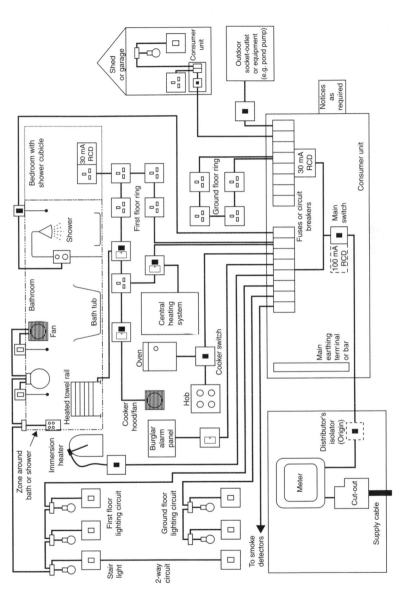

Figure 6.210 Typical fixed installations that might be encountered in new (or upgraded) existing dwellings.

Electricity distributor's responsibilities

> The electricity distributor is responsible for: P 1.2
>
> • evaluating and agreeing proposals for new installations or
>   significant alterations to existing ones;
> • installing the cut-out and meter in a safe location;   P 1.5
> • ensuring that it is mechanically protected and can be safely
>   maintained;
> • taking into consideration the possible risk of flooding.   P 1.5

 Note: See the DCLG publication *Preparing for Flooding*, which can be downloaded from http://www.dglc.gov.uk.

> Distributors are also required to: P 3.8
>
> • maintain the supply within defined tolerance limits;
> • provide an earthing facility for new connections;
> • provide certain technical and safety information to con-
>   sumers to enable them to design their installations.
>
> Distributors (and meter operators) must ensure that their   P 3.9
> equipment on consumers' premises:
>
> • is suitable for its purpose;
> • is safe in its particular environment;
> • clearly shows the polarity of the conductors.
>
> Distributors are prevented by the Regulations from   P 3.12
> connecting installations to their networks which do not
> comply with BS 7671.
>
> Distributors may disconnect consumers' installations which   P 3.12
> are a source of danger or cause interference with their
> networks or other installations.

Electrical installation work

> Electrical installation work:
>
> • is to be carried out professionally;   P 1.1
> • is to comply with the Electricity at Work Regulations 1989
>   as amended;
> • may only be carried out by persons that are competent   P 3.4a
>   to prevent danger and injury while doing it, or who are
>   appropriately supervised.

Note: Persons installing domestic combined heat and power equipment **must** advise the local distributor of their intentions before (or at the time of) commissioning the source.

Consumer units

Accessible consumer units should be fitted with a childproof cover or installed in a lockable cupboard.	P 1.6

Earthing

All electrical installations shall be properly earthed.	P App C

Note: The most usual type is an electricity distributor's earthing terminal, provided for this purpose near the electricity meter.

All lighting circuits shall include a circuit protective conductor.	P App C
All socket outlets which have a rating of 32 A or less, and which may be used to supply portable equipment for use outdoors, shall be protected by a Residual Current Device (RCD).	P App C
Distributors are required to provide an earthing facility for new connections.	P 3.8
It is not permitted to use a gas, water or other metal service pipe as a means of earthing for an electrical installation (this does not rule out, however, equipotential bonding conductors being connected to these pipes).	P App C
New or replacement, non-metallic light fittings, switches and/or other components must not require earthing (e.g. non-metallic varieties) unless new circuit protective (earthing) conductors are provided.	P App C
Socket outlets that will accept unearthed (two-pin) plugs must **not** use supply equipment that needs to be earthed.	P App C
Where electrical installation work is classified as an extension, a material alteration or a material change of use, the work must consider and include earthing and bonding systems that are satisfactory and meet the requirements.	P 2.1a–2.2c

See Figure 6.211 for details of some earth and bonding conductors that might be part of an electrical installation.

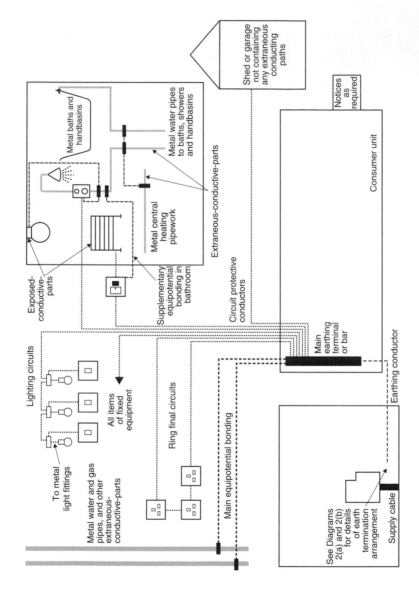

Figure 6.211 Typical earth and bonding conductors that might be part of an electrical installation.

Equipotential bonding conductors

Main equipotential bonding conductors are required to all water service pipes, gas installation pipes, oil supply pipes plus certain other 'earthy' metalwork that may be present on the premises.	P App C
The installation of supplementary equipotential bonding conductors is required for installations and locations where there is an increased risk of electric shock (e.g. such as bathrooms and shower rooms).	P App C
The minimum size of supplementary equipotential bonding conductors (without mechanical protection) is $4\,\text{mm}^2$.	P App C

Types of wiring or wiring system

Cables concealed in floors (and walls in certain circumstances) are required to have an earthed metal covering, be enclosed in steel conduit, or have additional mechanical protection (see BS 7671 for more information).	P App C
Cables to an outside building (e.g. garage or shed) if run underground, should be routed and positioned so as to give protection against electric shock and fire as a result of mechanical damage to a cable.	P App C
Heat-resisting flexible cables are required for the final connections to certain equipment (see maker's instructions).	P App C

PVC insulated and sheathed cables are likely to be suitable for much of the wiring in a typical dwelling.

Electrical components

All electrical work should be inspected (during installation as well as on completion) to verify that the components have:	
• been made in compliance with appropriate British Standards or harmonized European Standards;	P 1.11a i
• been selected and installed in accordance with BS 7671;	P 1.11a ii

- been evaluated against external influences (such as   P 1.11a ii
  the presence of moisture);
- been tested as per sections 71–74 of BS 7671:2001;   P 1.9
- been tested using appropriate and accurate   P 1.12
  instruments;
- been tested to check satisfactory performance with   P 1.11b
  respect to continuity of conductors, insulation resistance, separation of circuits, polarity, earthing and
  bonding arrangements, earth fault loop impedance
  and functionality of all protective devices, including
  Residual Current Devices (RCDs);
- not been visibly damaged (or are defective) so as to   P 1.11a iii
  be unsafe;   P 1.12
- had their test results recorded using the models in
  Appendix 6 of BS 7671 (see Annex D, Appendices
  1–4 for examples).

Note: Inspections and testing of DIY work should **also** meet these
requirements.

Socket outlets

Where necessary, socket outlets should comply with the requirements of Part M.	P 1.6
Socket outlets should be wall mounted.	M 4.30a and b
Socket outlets that will accept unearthed (two-pin) plugs must not be used to supply equipment that needs to be earthed.	P App C
Socket outlets which have a rating of 32 A or less and which may be used to supply portable equipment for use outdoors shall be protected by a residual current device (RCD).	P App C
Socket outlets should be located no nearer than 350 mm from room corners.	M 4.30
Switched socket outlets should indicate whether they are 'on'.	M 4.30k
Front plates should contrast visually with their backgrounds.	M 4.30m
Mains and circuit isolator switches should clearly indicate whether they are 'ON' or 'OFF'.	M 4.30l

Older types of socket outlet designed non-fused plugs must not be connected to a ring circuit.   P App C

The colours **red** and green should **not** be used in combination as indicators of 'ON' and 'OFF' for switches and controls.   M 4.28

Wall sockets

Wall sockets shall meet the requirements listed in Table 6.105.

Table 6.105 Building Regulations requirements for wall sockets

Type of wall	Requirement	Section
Timber framed	Power points may be set in the linings provided there is a similar thickness of cladding behind the socket box	E p14
	Power points should not be placed back to back across the wall	E p14
Solid masonry	Deep sockets and chases should **not** be used in separating walls	E 2.32
	The position of sockets on opposite sides of the separating wall should be staggered	E 2.32f
Cavity masonry	The position of sockets on opposite sides of the separating wall should be staggered	E 2.65e
	Deep sockets and chases should **not** be used in a separating wall	E 2.65d2
	Deep sockets and chases in a separating wall should **not** be placed back to back	E 2.65d2
Framed walls with absorbent material	Sockets should:	
	• be positioned on opposite sides of a separating wall;	E 2.146b
	• not be connected back to back;	E 2.146b2
	• be staggered a minimum of 150 mm edge to edge	E 2.146b2

Wall-mounted switches and socket outlets

Where necessary, wall-mounted switches and socket outlets should comply with the requirements of Part M.   P 1.6

A cable or stud detector shall be used when attempting to drill into walls, floors or ceilings.   P App C

Individual switches on panels and on multiple socket outlets should be well separated.   M 4.29

Switches and socket outlets for lighting and other equip-   M 8.2
ment should be located so that they are easily reachable.

Switches and socket outlets for lighting and other equipment   M 8.3
in habitable rooms should be located between 450 mm and
1200 mm from finished floor level (see Figure 6.212).

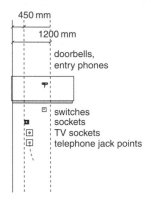

Figure 6.212 Heights of switches and sockets, etc.

The aim is to help people with limited reach (e.g. seated in a wheelchair)
access a dwelling's wall-mounted switches and socket outlets.

Outlets, controls and switches

The aim should be to ensure that all controls and switches   M 5.4i
are easy to operate, visible, free from obstruction and:

- are located between 750 mm and 1200 mm above the
  floor;
- do not require the simultaneous use of both hands (unless
  necessary for safety reasons) to operate;
- switched socket outlets indicate whether they are 'ON';
- switched socket outlets indicate whether they are 'ON';
- mains and circuit isolator switches clearly indicate
  whether they are 'ON' or 'OFF';
- individual switches on panels and on multiple socket
  outlets are well separated;
- controls that need close vision (e.g. thermostats) are
  located between 1200 mm and 1400 mm above the floor;
- front plates contrast visually with their backgrounds.

The operation of all switches, outlets and controls should not   M 4.30j
require the simultaneous use of both hands (unless necessary
for safety reasons).

Where possible, light switches with large push pads should   M 5.3
be used in preference to pull cords.

The colours **red** and green should **not** be used in   M 5.3
combination as indicators of 'on' and 'off' for switches and
controls.

Light switches

Light switches should: M 4.30 h and l

- have large push pads;
- align horizontally with door handles;
- be within 900 mm and 1100 mm from the entrance
  door opening.

Switches and controls should be located between   M 4.30c and d
750 mm and 1200 mm above the floor.

Where possible, light switches with large push pads   M 5.3
should be used in preference to pull cords.

The colours **red** and green should not be used   M 5.3
in combination as indicators of 'on' and 'off' for
switches and controls.

Heat emitters

In toilets and bathrooms designed for disabled people,   M 5.10
heat emitters (if located) should not restrict:

- the minimum clear wheelchair manoeuvring space;
- the space beside a WC used to transfer from the
  wheelchair to the WC.

Heat emitters should either be screened or have their   M 5.4j
exposed surfaces kept at a temperature below 43°C.

Telephone points and TV sockets

All telephone points and TV sockets should be   M 4.30a and b
located between 400 mm and 1000 mm above the
floor (or 400 mm and 1200 mm above the floor for
permanently wired appliances).

Thermostats

> Controls that need close vision (e.g. thermostats) should be   M 4.30f
> located between 1200 mm and 1400 mm above the floor.

Fixed internal lighting (domestic buildings)

In order that dwelling occupiers may benefit from the installation of efficient electric lighting, whenever

- a dwelling is extended, or
- a new dwelling is created from a material change of use, or
- an existing lighting system has been replaced as part of rewiring works,

The rewiring works **must** comply with Part P.

Lighting fittings (including lamp, control gear, housing, reflector, shade, diffuser or other device for controlling the output light) should only take lamps with a luminous efficiency greater than 40 lumens per circuit-watt.

Note: Light fittings in less-frequented areas such as cupboards and other storage areas do not count.

> New or replacement, non-metallic light fittings, switches   App C
> or other components must not require earthing unless new
> circuit protective (earthing) conductors are provided.

Fixed external lighting (i.e. lighting that is fixed to an external surface of the dwelling and which is powered from the dwelling's electrical system) should either:

- have a lamp capacity not exceeding 150 W per light fitting that automatically switches off:
  − when there is enough daylight, and
  − when it is not required at night; or
- include sockets that can only be used with lamps which have an efficiency greater than 40 lumens per circuit-watt.

> All lighting circuits shall include a circuit protective   P App C
> conductor.

Fixed energy-efficient light fittings (one per 25 m^2 dwelling floor area (excluding garages) and one per four fixed light fittings) should be installed in the most frequented locations in the dwelling.

See GIL 20, *Low Energy Domestic Lighting*, BRECSU, 1995 for further guidance.

Office, industrial and storage areas (non-domestic buildings)

All areas that involve predominantly desk-based tasks (i.e. classrooms, seminar and conference rooms – **including** those in schools) shall have an average efficiency of not less than 45 luminaire-lumens/circuit-watt (averaged over the whole area).

 Note: In other spaces (i.e. other than office or storage spaces) less efficient lamps may be used **provided** that the installed lighting has an average initial (100 hour) lamp plus ballast efficacy of not less than 50 lamp lumens per circuit-watt.

Lighting controls (non-domestic buildings)

The distance between the local switch and the luminaire it controls should generally be not more than 6 m, or twice the height of the luminaire above the floor if this is greater.

Local switches should be:

- located in easily accessible positions within each working area (or at boundaries between working areas and general circulation routes);
- operated by the deliberate action of the occupants (referred to as occupant control), either manually or remotely;
- located within 6 m (or twice the height of the light fitting above the floor if this is greater) of any luminaire it controls.

Local (manual) switching can be supplemented by automatic controls which:

- switch the lighting off when they sense the absence of occupants; or
- dim (or switch off) the lighting when there is sufficient daylight.

L2A 58

Occupant control of local switching may be supplemented by automatic systems which:

- switch the lighting off when they sense the absence of occupants; or
- dim (or switch) the lighting off when there is sufficient daylight.

In addition:

- automatically switched lighting systems should be subject to a risk assessment;
- Lighting controls should be provided to switch off the lighting during daylight hours and when the area is unoccupied;
- if the space is daylit space served by side windows, the perimeter row of luminaires should be separately switched;
- manually operated local switches should be in easily accessible positions within each working area, at boundaries between working areas, and at general circulation routes.

Display lighting in all types of space (non-domestic buildings)

Display lighting should have an average initial (100 hour) efficiency of not less than 15 lamp-lumens per circuit-watt.

Note: In spaces where it would be reasonable to expect cleaning and restocking outside public access hours, general lighting should also be provided.

Where possible, display lighting should be connected in dedicated circuits that can be switched off at times when people will not be inspecting exhibits or merchandise or attending entertainment events.

Note: In a retail store, for example, this could include timers that switch off the display lighting outside store opening hours.

Emergency escape lighting (non-domestic buildings)

Emergency escape lighting, specialist process lighting and vertical transportation systems are not subject to the requirements of Part L.

General lighting efficiency in all other types of space (non-domestic buildings)

Lighting (over the whole of these areas) should have an average initial efficacy of not less than 45 luminaire-lumens/circuit-watt.

Lighting systems serving other types of space may use lower powered and less efficient lamps.

Limiting the effects of solar gains in summer – buildings other than dwellings

In occupied spaces that are not served by a comfort cooling system, the combined solar and internal casual gains (people, lighting and equipment) per unit floor area averaged over the period of daily occupancy should not be greater than 35 W/m² calculated over a perimeter area not more than 6 m from the window wall and averaged during the period 06:30–16:30 hours GMT.

Consequential improvements (non-domestic buildings)

If a building has a total useful floor area greater than 1000 m² and the proposed building work includes:

- an extension; or
- the initial provision of any fixed building services; or
- an increase to the installed capacity of any fixed building services, then:

any general lighting system serving an area greater than 100 m² which has an average lamp efficacy of less than 40 lamp-lumens per circuit-watt, should be upgraded with new luminaires or improved controls.

Inspection and commissioning of the building services systems (non-domestic buildings)

When building services systems are commissioned, the systems and their controls shall be left in their intended working order and are capable of operating efficiently regarding the conservation of fuel and power.

Systems should be provided with meters to efficiently manage energy use and to ensure that at least 90 per cent of the estimated annual energy consumption of each fuel is assigned to the various end-use categories (heating, lighting, etc.).

Whenever a cooling plant (i.e. a chiller) is being replaced, cooling loads should (if practicable and cost-effective) be improved through solar control and/or more efficient lighting.

Air-handling systems should be capable of achieving a specific fan power at 25 per cent of the design flow rate.

Commissioning (non-domestic buildings)

Building services systems should be commissioned so that, at completion, the system(s) and their controls are left in working order and operate efficiently (i.e. for the purposes of the conservation of fuel and power).

Controlled services (non-domestic buildings)

Where the work involves the provision of a controlled service, new lighting systems should be provided with controls that achieve reasonable standards of energy efficiency.

External lighting fixed to the building

External lighting (including lighting in porches, but not lighting in garages and carports) should:

- automatically extinguish when there is enough daylight, and when not required at night;
- have sockets that can only be used with lamps having an efficacy greater than 40 lumens per circuit-watt (such as fluorescent or compact fluorescent lamp types, and **not** GLS tungsten lamps with bayonet cap or Edison screw bases).

Emergency alarms

Emergency assistance alarm systems should have:	M 5.4h
• visual and audible indicators to confirm that an emergency call has been received; • a reset control reachable from a wheelchair, WC, or from a shower/changing seat; • a signal that is distinguishable visually and audibly from the fire alarm.	
Emergency alarm pull cords should be:	M 4.30e
• coloured **red**; • located as close to a wall as possible; • have two **red** 50 mm diameter bangles.	
Front plates should contrast visually with their backgrounds.	M 4.30m
The colours **red** and green should **not** be used in combination as indicators of 'ON' and 'OFF' for switches and controls.	M 4.28

Smoke alarms

With the introduction of Part B 2007:

- smoke alarms now need to be installed in accordance with BS 5839-6:2004;
- all smoke alarms should have a standby power supply;
- the provision of smoke alarms shall be based on an assessment of the risk to the occupants in the event of fire.

 Note: If a dwelling-house is extended, smoke alarms should be provided in all circulation spaces.

Smoke alarms should:	B1 1.4
• be mains-operated and conform to BS 5446-1:2000; • have a standby power supply such as a rechargeable (or non-rechargeable) battery;	

• be designed and installed in accordance with BS 5839-6:2004;	B1 1.10 (V1) B1 1.9 (V2)
• be positioned in the circulation spaces between sleeping spaces and places where fires are most like to start (e.g. in kitchens and living rooms);	B1 1.11 (V1) B1 1.10 (V2)
• be installed on every storey of a dwelling-house.	B1 1.12 (V1)

Where the kitchen area is not separated from the stairway or circulation space by a door, there should be a compatible interlinked heat detector or heat alarm in the kitchen, in addition to whatever smoke alarms are needed in the circulation space(s).	B1 1.11 (V2) B1 1.13 (V1) B1 1.12 (V2)

If more than one alarm is installed, they should be linked so that the detection of smoke or heat by one unit operates the alarm signal in all of them.	B1 1.14 (V1) B1 1.13 (V2)
Smoke alarms/detectors should be sited so that:	B1 1.15a (V1)
• there is a smoke alarm in the circulation space within 7.5 m of the door to every habitable room;	B1 1.14a (V2)
• they are ceiling-mounted and at least 300 mm from walls and light fittings;	B1 1.15b (V1) B1 1.14b (V2)
• the sensor in ceiling-mounted devices is between 25 mm and 600 mm below the ceiling (25–150 mm in the case of heat detectors or heat alarms).	B1 1.15c (V1) B1 1.14c (V2)
Smoke alarms should **not** be fixed:	B1 1.16 (V1)
• over a stair or any other opening between floors;	B1 1.15 (V2)
• next to or directly above heaters or air-conditioning outlets;	B1 1.17 (V1) B1 1.16 (V2)
• in bathrooms, showers, cooking areas or garages;	B1 1.17 (V1) B1 1.16 (V2)
• in any place where steam, condensation or fumes could give false alarms;	B1 1.17 (V1) B1 1.16 (V2)
• in places that get very hot (such as a boiler room);	B1 1.18 (V1) B1 1.17 (V2)
• in places that get very cold (such as an unheated porch);	B1 1.18 (V1) B1 1.17 (V2)
• to surfaces which are normally much warmer or colder than the rest of the space.	B1 1.18 (V1) B1 1.17 (V2)

Smoke detectors – power supplies

The power supply for a smoke alarm system should:	B1 1.19
• be derived from the dwelling-house's mains electricity supply; • comprise a single independent circuit at the dwelling-house's main distribution board (consumer unit or a single regularly used local lighting circuit).	
It should be possible to isolate the power to the smoke alarms without isolating the lighting.	B1 1.19
The electrical installation should comply with Approved Document P (Electrical Safety).	B1 1.20
Any cable suitable for domestic wiring may be used for the power supply and interconnection to smoke alarm systems (except in large buildings where the cable needs to be fire-resistant (see BS 5839-6:2004)).	B1 1.21
Conductors used to interconnect alarms (e.g. signalling) should be colour coded so as to distinguish them from those supplying mains power.	B1 1.21
Mains powered smoke alarms may be interconnected using radio-links, provided that this does not reduce the lifetime or duration of any standby power supply below 72 hours.	B1 1.21

Where the vertical distance between the floor of the entrance storey and the floors above and below it does not exceed 7.5 m, multi-storey flats are required to:

• provide a protected stairway plus additional smoke alarms in all habitable rooms and a heat alarm in any kitchen; or	B1 216c (V2)
• provide a protected stairway plus a sprinkler system and smoke alarms.	B1 2.16d (V2)

Smoke alarms in thatched roofs

In thatched roofs:	B4 10.9 (V1)
• the rafters should be overdrawn with construction having not less than 30 minutes fire-resistance; • a smoke alarm should be installed in the roof space.	B4 14.9 (V2)

Fire alarms

A visual and audible fire alarm signal should be provided in buildings where it is anticipated that one or more persons with impaired hearing may be in relative isolation (e.g. hotel bedrooms and sanitary accommodation).	B1 1.34 (V2)
Fire alarms should emit an audio and visual signal to warn occupants with hearing or visual impairments.	M 5.4 g

 All fire-detection and fire-warning systems shall be properly designed, installed and maintained.

All buildings should have arrangements for detecting fire.	B1 1.27 (V2)
All buildings should have the means of raising an alarm in case of fire (e.g. rotary gongs, handbells or shouting 'fire') or be fitted with a suitable electrically operated fire warning system (in compliance with BS 5839).	B1 1.28 (V2)
The fire warning signal should be distinct from other signals that may be in general use.	B1 1.32 (V2)
In premises used by the general public (e.g. large shops and places of assembly), a staff alarm system (complying with BS 5839) may be used.	B1 1.33 (V2)
In small buildings, raising the alarm may be a comparatively simple matter, but when this does not apply, the building should be provided with an electrically operated fire warning system with manual call points adjacent to exit doors which shall comply with BS 5839-1:2002.	B1 1.29 and 1.30 (V2)
Call points for electrical alarm systems should be installed in accordance with BS 5839-1 and comply with either BS 5839-2:1983, or type A of BS EN 54-11:2001.	B1 1.31 (V2)
Type B call points should **only** used with the approval of the Building Control Body.	
• Where it is critical for electrical circuits to be able to continue to function during a fire, protected circuits (meeting the requirements of BS EN 50200:2006) shall be installed.	B1 5.38 (V2)

Electrically powered locks

Electrically powered locks should:	B1 5.11 (V2)
• return to the unlocked position;	
• on operation of the fire alarm system;	
• on loss of power or system error;	
• on activation of a manual door release unit;	
• comply with Part P (and any other relevant parts) of the Building Regulations;	P 1.7 and 3.1
• comply with the relevant equipment and installation standards;	P 0.1b
• not present an electric shock or fire hazard to people;	P 0.1a
• provide mechanical and thermal protection;	P 1.3
• provide adequate protection against mechanical and thermal damage;	P 0.1a
• provide adequate protection for persons against the risks of electric shock, burn or fire injuries;	P1.3
• be safe to use, maintain and alter.	P 1.7

Power-operated doors

Power-operated doors and gates should:

• be provided with a manual or automatic opening device in the event of a power failure where and when necessary for health or safety;	K5 5.2d
• have safety features to prevent injury to people who are struck or trapped (e.g. a pressure-sensitive door edge which operates the power switch);	
• have a readily identifiable and accessible stop switch.	

Power-operated entrance doors

Doors to accessible entrances shall be provided with a power-operated door opening and closing system if a force greater than 20 N is required to open or shut a door.	M 2.13a

Once open, all doors to accessible entrances should be wide enough to allow unrestricted passage for a variety of users, including wheelchair users, people carrying luggage, people with assistance dogs, and those with pushchairs and small children.

The effective clear width through a single leaf door (or one leaf of a double-leaf door) should be in accordance with Table 6.106.	M 2.13b

Table 6.106 Minimum effective clear widths of doors

Direction and width of approach	New buildings (mm)	Existing buildings (mm)
Straight-on (without a turn or oblique approach)	800	750
At right angles to an access route at least 1500 mm wide	800	750
At right angles to an access route at least 1200 mm wide	825	775
External doors to buildings used by the general public	1000	775

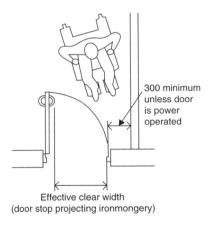

300 minimum unless door is power operated

Effective clear width
(door stop projecting ironmongery)

Figure 6.213 Effective clear width and visibility requirements of doors.

Power-operated entrance doors should have a sliding, swinging or folding action controlled manually (by a push pad, card swipe, coded entry, or remote control) or automatically controlled by a motion sensor or proximity sensor such as a contact mat.

Power-operated entrance doors should:

• open towards people approaching the doors;	M 2.21a
• provide visual and audible warnings that they are operating (or about to operate);	M 2.21c
• incorporate automatic sensors to ensure that they open early enough (and stay open long enough) to permit safe entry and exit;	M 2.21c
• incorporate a safety stop that is activated if the doors begin to close when a person is passing through;	M 2.21b
• revert to manual control (or fail safe) in the open position in the event of a power failure;	M 2.21d
• when open, should **not** project into any adjacent access route;	M 2.21e
• ensure that its manual controls:	M 2.21f
– are located between 750 mm and 1000 mm above floor level;	
– are operable with a closed fist;	
• be set back 1400 mm from the leading edge of the door when fully open;	M 2.21 g
• be clearly distinguishable against the background;	M 2.21 g
• contrast visually with the background.	M 2.19 and 2.21 g

 Note: Revolving doors are **not** considered 'accessible' as they create particular difficulties for (and possible injury to) people who are visually impaired, people with assistance dogs or mobility problems, and people with children and/or pushchairs.

Cellars or basements

LPG storage vessels and LPG-fired appliances fitted with automatic ignition devices or pilot lights must not be installed in cellars or basements.	J 3.5i

Lecture/conference facilities

Artificial lighting should be designed to:

• give good colour rendering of all surfaces;	M 4.9 and 4.34
• be compatible with other electronic and radio-frequency installations.	M 4.12.1

Swimming pools and saunas

Swimming pools and saunas are subject to special requirements specified in Part 6 of BS 7671:2001.   P App A

Inspection and testing

Electrical installations must be inspected and tested:

- during, at the end of installation and before they are taken into service to verify that they are reasonably safe **and** that they comply with BS 7671:2001;   P 1.6
- to verify that they meet the relevant equipment and installation standards.   P 0.1b

All electrical work should be inspected (during installation as well as on completion) to verify that the components have:

- been selected and installed in accordance with BS 7671;   P 1.8a (ii)
- been made in compliance with appropriate British Standards or harmonized European Standards;   P 1.8a (i)
- been evaluated against external influences (e.g. the presence of moisture);   P 1.8a (ii)
- not been visibly damaged (or are defective) so as to be unsafe;   P 1.8a (iii)
- been tested to check satisfactory performance with respect to continuity of conductors, insulation resistance, separation of circuits, polarity, earthing and bonding arrangements, earth fault loop impedance and functionality of all protective devices including residual current devices (RCDs);   P 1.8b
- been inspected for conformance with section 712 of BS 7671:2001;   P 1.9
- been tested as per section 713 of BS 7671:2001;   P 1.10
- been tested using appropriate and accurate instruments;   P 1.10
- had their test results recorded using the model in Appendix 6 of BS 7671;   P 1.10
- had their test results compared with the relevant performance criteria to confirm compliance.   P 1.10

Note: Inspections and testing of DIY work should **also** meet the above requirements.

Inspection and testing of non-notifiable work

Although it is not necessary for non-notifiable electrical installation work to be checked by a building control body, it nevertheless **must** be carried out in accordance with the requirements of BS 7671:2001.	P 1.30
Installers who are qualified to complete BS 7671 installation certificates and who carry out non-notifiable work should issue the appropriate electrical installation certificate for all but the simplest of like-for-like replacements.	P 1.32

Certification

BS 7671 Installation certificates

Compliance with Part P can be demonstrated by the issue of the appropriate BS 7671 electrical installation certificate.	P 1.8
An electrical installation certificate may **only** be issued by the installer responsible for the installation work.	P 1.28
Inspection and testing should be carried out in compliance with BS 7671:2001.	P 1.9
Section 712 of BS 7671:2001 provides a list of all the inspections that may be necessary, while Section 713 provides a list of tests.	
Tests should be carried out using appropriate and accurate instruments under the conditions given in BS 7671, and the results compared with the relevant performance criteria to confirm compliance.	P 1.15
A copy of the installation certificate should be supplied to the person ordering the work.	P 1.9
Certificates may only be made out and signed by someone with the appropriate qualifications, knowledge and experience to carry out the inspection and test procedures.	P 1.9
Certificates should show that the electrical installation work has been:	P 1.11a

- inspected during erection as well as on completion, and   P 1.11b
  that components have been:
  - made in compliance with appropriate British Standards
    or harmonized European Standards;
  - selected and installed in accordance with BS 7671
    (including consideration of external influences such as
    the presence of moisture);
  - not visibly damaged or defective so as to be unsafe;
- tested for continuity of conductors, insulation resistance,
  separation of circuits, polarity, earthing and bonding
  arrangements, earth fault loop impedance and functionality of all protective devices (including Residual Current
  Devices (RCDs)).

A full electrical installation certificate should be used for the   P 1.13
replacement of consumer units.

Minor works certificate

Appropriate tests (according to the nature of the work) should   P 1.16
be carried out.

A minor works certificate should be issued whenever   P 1.13
inspection and testing has been carried out, irrespective of the
extent of the work undertaken.

A minor works certificate shall **not** be used for the   P 1.13
replacement of consumer units or similar items.

Building Regulations compliance certificates

The following are additional certificates that are issued on   P 1.17
completion of notifiable works as evidence of compliance
with the Building Regulations:

- A Building Regulations Compliance Certificate (issued
  by Part P competent person scheme installers) – see
  Appendix 4 to Annex D for an example);
- completion certificates (issued by local authorities);
- final notices (issued by approved inspectors).

These documents are **different** documents to the BS 7671 installation
certificate and are used to attest compliance with **all** relevant requirements of
the Building Regulations – not just to Part P.

Certification of notifiable work

Only installers registered with a Part P competent person   P 1.18
self-certification scheme are qualified to complete BS 7671
installation certificates and should do so in respect of every
job they undertake.

A copy of the certificate should always be given to the person   P 1.18
ordering the electrical installation work.

A Building Regulations Compliance Certificate must be   P 1.19
issued to the occupant (and the building control body) either
by the installer or the installer's registration body within 30
days of the work being completed.

If notifiable electrical installation work is going to be carried   P 1.21
out by a person **not** registered with a Part P competent person
self-certification, the work should be notified to a building
control body (the local authority or an approved inspector)
before work starts.

Note: The building control body then becomes responsible for making sure
the work is safe and complies with all relevant requirements of the Building
Regulations.

On satisfactory completion of all work, the building   P 1.23
control body will issue a Building Regulation Completion
Certificate (if they are the local authority) or a Final
Certificate (if they are an approved inspector).

If notifiable electrical installation work is going to be carried   P 1.24
out by installers who are not qualified to issue BS 7671
completion certificates (e.g. subcontractors or DIYers),
the building control body must be notified before the work
starts.

Note: The building control body then becomes responsible for making sure that
the work is safe and complies with all relevant requirements of the Building
Regulations – **but** not at the householder's expense!

Third-party certification

Unregistered installers should not themselves arrange for a   P 1.28
third party to carry out final inspection and testing.

A third party may only sign a BS 7671 Periodic Inspection   P 1.29
Report (or similar) to indicate that electrical safety tests
had been carried out on the installation which met BS 7671
criteria.

Third parties are not entitled to verify that the installation
complies, fully, with BS 7671 requirements (e.g. with regard
to routing of hidden cables).

Provision of information

Sufficient information should be left with the occupant to   P 1.33
ensure that persons wishing to operate, maintain or alter
an electrical installation can do so with reasonable
safety.

This information should include: P 1.34

- all items called for by BS 7671;
- an Electrical Installation Certificate (describing the installation and giving details of work carried out);
- permanent labels (e.g. on earth connections and bonds and
  on items of electrical equipment such as consumer units
  and Residual Current Devices (RCDs));
- operating instructions and log books;
- detailed plans (but only for unusually large or complex
  installations).

Where can I get more information?

Further guidance concerning the requirements for electrical safety is available
from:

- the IET (Institution of Engineering and Technology) at http://www.theiet.
  org;
- the NICEIC (National Inspection Council for Electrical Installation Contracting) at http://www.niceic.org.uk;
- the ECA (Electrical Contractors' Association) at http://www.niceic.org.uk
  or http://www.eca.co.uk.

To download pdf copies of Part P, go to http://www.planningportal.
gov.uk;

For details of fixed wire colour changes, go to http://www.niceic.org.uk.

6.20 Combustion appliances

6.20.1 The requirements

Air supply

Combustion appliances shall be so installed that there is an adequate supply of air to them for combustion, to prevent overheating and for the efficient working of any flue.

(Approved Document J1)

 A new Part J came into force on 1 October 2010.

Discharge of products of combustion

Combustion appliances shall have adequate provision for the discharge of products of combustion to the outside air.

(Approved Document J2)

Warning of release of carbon monoxide

J2A. Where a fixed combustion appliance is provided, appropriate provision shall be made to detect and give warning of the release of carbon monoxide.

(Approved Document J2)

 A new Part J came into force on 1 October 2010.

Protection of building

Combustion appliances and fluepipes shall be so installed, and fireplaces and chimneys shall be so constructed and installed, as to reduce to a reasonable level the risk of people suffering burns or the building catching fire in consequence of their use.

(Approved Document J3)

 A new Part J came into force on 1 October 2010.

Provision of information

Where a hearth, fireplace, flue or chimney is provided or extended, a durable notice containing information on the performance capabilities of the hearth, fireplace, flue or chimney shall be affixed in a suitable place in the building for the purpose of enabling combustion appliances to be safely installed.

(Approved Document J4)

 A new Part J came into force on 1 October 2010.

Protection of liquid fuel storage systems

Liquid fuel storage systems and the pipes connecting them to combustion appliances shall be so constructed, and separated from buildings and the boundary of the premises as to reduce to a reasonable level the risk of the fuel igniting in the event of fire in adjacent buildings or premises.

(Approved Document J5)

 A new Part J came into force on 1 October 2010.

Protection against pollution

Oil storage tanks and the pipes connecting them to combustion appliances shall:

- *be so constructed and protected as to reduce to a reasonable level the risk of the oil escaping and causing pollution; and*
- *have affixed in a prominent position a durable notice containing information on how to respond to an oil escape so as to reduce to a reasonable level the risk of pollution.*

(Approved Document J6)

 A new Part J came into force on 1 October 2010.

6.20.2 Meeting the requirement

Combustion appliances require ventilation to supply them with air for combustion. Ventilation is also required to ensure the proper operation of flues or, in the case of flueless appliances, to ensure that the products of combustion are safely dispersed to the outside air.

Air supplies for combustion installations

> A room containing an open-flued appliance may need permanently open air vents (Figure 6.214(a) and (c)). J (1.4)

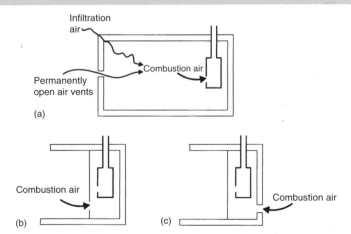

Figure 6.214 Air for combustion and operation of the flue (open flued). (a) Appliance in room. (b) Appliance in appliance compartment with internal vent. (c) Appliance in appliance compartment with external vent.

Appliance compartments that enclose open-flued combustion appliances should be provided with vents large enough to admit all the air required by the appliance for combustion and proper flue operation, whether the compartment draws its air from a room directly from outside or not (Figure 6.214(b) and (c)). J (1.5)

Where appliances require cooling air, appliance compartments should be large enough to enable air to circulate, and high- and low-level vents should be provided (Figure 6.215). J (1.6)

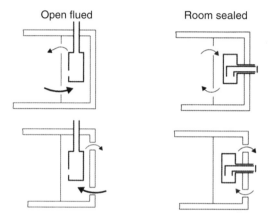

Figure 6.215 Combustion and operation requiring air cooling.

 Where appliances are to be installed within balanced compartments, special provisions will be necessary.

 Note: In a flueless situation, air for combustion (and to carry away its products) can be achieved as shown in Figure 6.216.

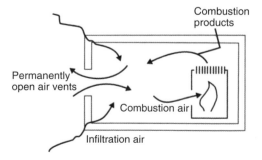

Figure 6.216 Air for combustion and operation of the flue (flueless).

If an appliance is room-sealed but takes its combustion air   J (1.8)
from another space in the building (e.g. the roof void), or if a
flue has a permanent opening to another space in the building
(e.g. where it feeds a secondary flue in the roof void), that
space should have ventilation openings directly to outside.

Where flued appliances are supplied with combustion air   J (1.9)
through air vents which open into adjoining rooms or spaces,
the adjoining rooms or spaces should have air vent openings
of at least the same size direct to the outside.

 Note: Air vents for flueless appliances, however, should open directly to the outside air.

Any hidden voids in the construction shall be sealed   B3
and subdivided to inhibit the unseen spread of fire and
products of combustion.

Air vents

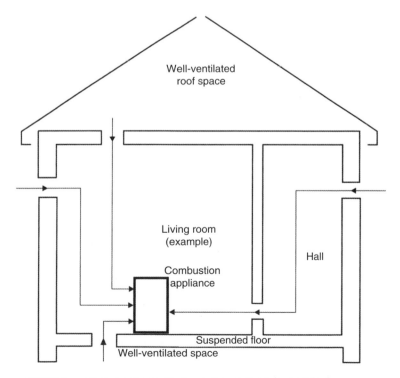

Figure 6.217 Locating permanent air vent openings (examples).

Permanently open air vents should be non-adjustable, sized to admit sufficient air for the purpose intended and positioned where they are unlikely to become blocked.   J (1.10)

Air vents should be sufficient for the appliances to be installed (taking account where necessary of obstructions such as grilles and anti-vermin mesh).   J (1.11)

Air vents should be sited outside fireplace recesses and beyond the hearths of open fires so that dust or ash will not be disturbed by draughts.   J (1.11a)

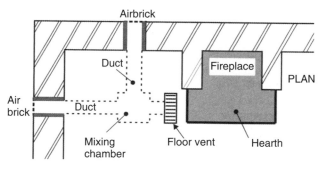

Figure 6.218 Air vent openings in a solid floor.

Air vents should be sited in a location unlikely to cause discomfort from cold draughts.   J (1.11b)

Grilles or meshes protecting air vents from the entry of animals or birds should have aperture dimensions no smaller than 5 mm.   J (1.15)

 In noisy areas, it may be necessary to install proprietary noise attenuated ventilators to limit the entry of noise into the building

In buildings where it is intended to install open-flued combustion appliances and extract fans, the combustion appliances should be able to operate safely whether or not the fans are running.   J (1.20)

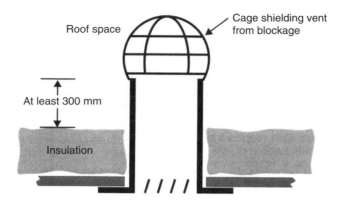

Figure 6.219 Ventilator used in a roof space (e.g. a loft).

For gas appliances where a kitchen contains an open-flued appliance, the extract rate of the kitchen extract fan should not exceed 20 l/s (72 m³/h).	J (1.20a)
When installing ventilation for solid fuel appliances, avoid installing extract fans in the same room.	J (1.20c)

Note: Discomfort from cold draughts can be avoided by placing vents close to appliances (e.g. by using floor vents), by drawing air from intermediate spaces (e.g. hallways) or by ensuring good mixing of incoming cold air by placing air vents close to ceilings.

Flues

Appliances other than flueless appliances should incorporate or be connected to suitable flues which discharge to the outside air.	J (1.24)
Chimneys and flues should provide satisfactory control of water condensation.	J (1.26)
New chimneys should be constructed with flue liners (clay, concrete or pre-manufactured) and masonry (bricks, medium weight concrete blocks or stone) suitable for the intended application.	J (1.27)
Liners should be selected to form the flue without cutting and joints should be kept to a minimum.	J (1.28)

Liners need to be placed with the sockets or rebate ends uppermost to contain moisture and other condensates in the flue.	J (1.28)
Joints should be sealed with fire cement, refractory mortar or installed in accordance with their manufacturer's instructions.	J (1.28)
Spaces between the lining and the surrounding masonry should not be filled with ordinary mortar.	J (1.28)

 Connecting fluepipes and factory-made chimneys should always be guarded if there is a possibility of them being damaged or if they could present a burn hazard (that is not immediately apparent) for people.

Ventilation ducts and flues, etc.

If a flue passes through a compartment wall or compartment floor, or is built into a compartment wall, each wall of the flue should have a fire resistance of at least half that of the wall or floor (see Figure 6.220).	B1 7.11 (V1) B3 10.16 (V2)

(a) FLUE PASSING THROUGH COMPARTMENT WALL OR FLOOR

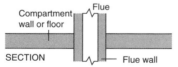

Flue walls should have a fire resistance of at least one half of that required for the compartment wall or floor, and be of non-combustible construction.

(b) FLUE BUILT INTO COMPARTMENT WALL

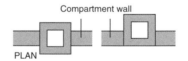

In each case flue walls should have a fire resistance at least one half of that required for the compartment wall and be of non-combustible construction.

Figure 6.220 Flues penetrating compartment walls or floors.

6.21 Hot water storage

6.21.1 The requirement

Hot water supplies and systems

(1) *There must be a suitable installation for the provision of heated whole-some water or heated softened wholesome water to:*

(a) *any washbasin or bidet provided in or adjacent to a room containing a sanitary convenience;*
(b) *any washbasin, bidet, fixed bath and shower in a bathroom; and*
(c) *any sink provided in any area where food is prepared.*

(2) *A hot water system, including any cistern or other vessel that supplies water to or receives expansion water from a hot water system, shall be designed, constructed and installed so as to resist the effects of temperature and pressure that may occur either in normal use or has the event of such malfunctions as may reasonably be anticipated, and must be adequately supported.*

(3) *A hot water system that has a hot water storage vessel shall incorporate precautions to:*

(a) *prevent the temperature of the water stored in the vessel at any time exceeding 100° C; and*
(b) *ensure that any discharge from safety devices is safely conveyed to where it is visible but will not cause a danger to persons in or about the building.*

(Approved Document G3)

Note: Requirement G3(3) does not apply to a system which heats or stores water for the purposes only of an industrial process.

(4) *The hot water supply to any fixed bath must be so designed and installed as to incorporate measures to ensure that the temperature of the water that can be delivered to that bath does not exceed 48° C.*

(Approved Document G3)

Note: Requirement G3(4) applies only when a dwelling is:

(a) erected;
(b) formed by a material change of use.

Hot water storage

The hot water system shall:

- *be installed by a competent person;*
- *not exceed 100° C;*

- *discharge safely;*
- *not cause danger to persons in or about the building.*

(Approved Document G3)

Conservation of fuel and power

Reasonable provision shall be made for the conservation of fuel and power in buildings by:

(a) *limiting heat gains and losses –*

   (i)   *through thermal elements and other parts of the building fabric; and*
   (ii) *from pipes, ducts and vessels used for space heating, space cooling and hot water services;*

(b) *providing fixed building services which –*

   (i)   *are energy efficient;*
   (ii) *have effective controls; and*
   (iii) *are commissioned by testing and adjusting as necessary to ensure they use no more fuel and power than is reasonable in the circumstances; and*

(c) *providing to the owner sufficient information about the building, the fixed building services and their maintenance requirements so that the building can be operated in such a manner as to use no more fuel and power than is reasonable in the circumstances*

(Approved Document L)

 A new Part L came into force on 1 October 2010.

Cold water supply

(1) *There must be a suitable Installation for the provision of;*

   (a) *wholesome water to any place where drinking eater is drawn off;*
   (b) *wholesome water or softened wholesome water to any washbasin or bidet provided in or adjacent to a room containing a sanitary convenience;*
   (c) *wholesome water or softened wholesome water to any washbasin, bidet, fixed bath or shower In a bathroom; and*
   (d) *wholesome water to any sink provided in any area where food is prepared.*

(2) *There must be a suitable installation for the provision of water of suitable quality to any sanitary convenience fitted with a flushing device.*

(Approved Document G1)

6.21.2 Meeting the requirement

Hot water supply and systems

All electrical work associated with hot water systems should be carried out in accordance with BS 7671:2008 (*Requirements for Electrical Installations*, commonly referred to as the *IEE Wiring Regulations 17th Edition*).

All fixed building services, including controls, should be commissioned by testing and adjusting as necessary to ensure that they use no more fuel and power than is reasonable in the circumstances.

Hot water (heated wholesome water or heated softened water) shall be supplied to the sanitary appliances and locations specified in the requirement without waste, misuse or undue consumption of water.	G3
The hot water outlet temperature device used to limit the maximum temperature that can be supplied at the outlet cannot be easily altered by building users.	G3(4)

All components of the hot water system (including any cistern that supplies water to, or receives expansion water from, the hot water system) shall continue to safely contain the hot water:

• during normal operation of the hot water system; • following failure of any thermostat used to control temperature; and • during operation of any of the safety devices.	G3(2)

A hot water storage system that has a vented storage vessel:

• shall have a suitable vent pipe connecting the top of the vessel to a point open to the atmosphere above the level of the water in the cold water storage cistern;	G3(3)
• in addition to any thermostat, either the heat source, or the storage vessel itself shall be with a device that will prevent the temperature of the stored water at any time exceeding 100° C;	G3(3)
• shall have pipework that allows hot water from the safety devices to discharge into a place open to the atmosphere where it will cause no danger to persons in or about the building.	G3(3)

A hot water storage system that has an unvented storage vessel:

• shall have at least two independent safety devices that release pressure and, in doing so, prevent the temperature of the stored water at any time exceeding 100° C;	G3(3)
• shall have pipework that is capable of discharging hot water from safety devices so that it is visible at some point and safely conveys it to an appropriate place open to the atmosphere where it will cause no danger to persons in or about the building.	G3(3)

Hot water storage systems

Pipework should be designed and installed in such a way as to minimize the transfer time between the hot water storage system and hot water outlets.

Hot water storage systems should be designed and installed in accordance with BS 6700:2006 or BS EN 12897:2006.	G3.10
Hot water storage vessels should conform to BS 853–1:1, BS 1566–1:2002 or BS 3198:1981, or other relevant national standards as appropriate.	G3.11
The temperature relief valve(s) in hot water storage system units and packages should: • be factory fitted; • not be disconnected other than for replacement; and • not be relocated in any other device or fitting installed.	G3.39

Hot water system

All components of the hot water system (including any cistern that supplies water to, or receives expansion water from, the hot water system) shall continue to safely contain the hot water: • during normal operation of the hot water system; • following failure of any thermostat used to control temperature; and • during operation of any of the safety devices.	G3(2)

Vented storage system

A hot water storage system with a vented storage vessel shall:	G3(3)

- have a suitable vent pipe connecting the top of the vessel to a point open to the atmosphere above the level of the water in the cold water storage cistern;
- in addition to any thermostat, the heat source or the storage vessel shall be fitted with a device to prevent the temperature of the stored water exceeding 100° C;
- have pipework that will enable hot water from the safety devices to discharge into a place open to the atmosphere where it will cause no danger to persons in or about the building.

Vented hot water storage systems should incorporate a vent pipe, not less than 19 mm internal diameter, connecting the top of the hot water storage.	G3.12

Vented hot water storage systems should incorporate either:	G3.13

- for all direct heat sources, a non-self-resetting energy cut-out in the event of the storage system overheating; and
- an overheat cut-out in the event of the stored water overheating so that the temperature of the stored water does not exceed 100° C; or
- a safety device (e.g. a temperature relief valve or a combined temperature and pressure relief valve) should safely discharge the water in the event of overheating either directly or by way of a manifold via a short length of metal pipe to a tun dish.

Vent pipes should discharge over a cold water storage cistern conforming to BS 417–2:1987 or BS 4213:2004, as appropriate.	G3.14

Unvented storage system

The installation of an unvented system is notifiable building work which **must** be reported to the building control body before work commences, unless the installer is registered with a competent person scheme, in which case the installer may self-certify that the work complies with all relevant requirements in the Building Regulations and provide the building owner/occupier with a Building Regulations Certificate of Compliance.

A hot water system that has an unvented storage vessel shall: G3(3)

- have at least two independent safety devices that release pressure and, in doing so, prevent the temperature of the stored water at any time exceeding 100° C; and
- have pipework that is capable of visibly discharging hot water from safety devices and safely conveying it to an appropriate place open to the atmosphere where it will cause no danger to persons in or about the building.

In addition to any thermostat provided to control the desired temperature of the stored water, unvented hot water storage systems should incorporate a minimum of two independent safety devices, for example: G3.17 and G3.18

- a non-self-resetting energy cut-out;
- a temperature relief valve or a combined temperature and pressure relief valve.

Any unvented hot water storage system unit or package should be indelibly marked with the following information: G3.23

- the manufacturer's name and contact details;
- the model reference;
- the rated storage capacity;
- the operating pressure of the system;
- the operating pressure of the expansion valve;
- relevant operating data on each of the safety devices fitted; and
- the maximum primary circuit pressure and flow temperature of indirect hot water storage system units or packages.

The information shown in Figure 6.221 should be indelibly marked on the hot water storage system unit or package.   G3.24

Unvented hot water storage systems – systems over 500 litres capacity or over 45 kW power input

Systems over 500 litres capacity will generally be bespoke designs for specific projects, and as such are inappropriate for approval by a third-party accredited product conformity certification scheme.

WARNING TO USER

a.   Do not remove or adjust any component part of this unvented water heater–
     contact the Installer.

b.   If this unvented water heater develops a fault, such as a flow of hot water from
     the discharge pipe, switch the heater off and contact the Installer.

WARNING TO INSTALLER

a.   This Installation is subject to the Building Regulations.

b.   Use only appropriate components for Installation or maintenance. Installed by:

Name...

Address..

Tel. No ....

Completion date

Figure 6.221 Hot water warning sign.

Where this is the case, the unvented hot water storage system   G3.26
should incorporate a minimum of two independent safety
devices, such as:

- a non-self-resetting energy cut-out;
- a temperature relief valve or a combined temperature and
  pressure relief valve.

Note: Any unvented hot water storage system having a power input of more
than 45 kW, but a capacity of 500 litres or less should be in the form of a proprietary hot water storage system unit or package.

Any hot water outlet temperature device being used to limit   G3(4)
the maximum temperature supplied at the outlet shall not be
capable of being easily altered by building users.

The delivered hot water can be considered as heated wholesome water or
heated softened wholesome water where:

- the cold water supply to the hot water system is wholesome or softened
  wholesome; and
- the installation complies with the requirements of the Water Supply (Water
  Fittings) Regulations 1999 (SI 1999/1148 as amended).

The installation shall convey hot water to the sanitary appliances and locations specified in the requirement without waste, misuse or undue consumption of water; and the water supplied is heated wholesome water or heated softened water.	G3
Pipework should be designed and installed in such a way as to minimize the transfer time between the hot water storage system and hot water outlets.	G3.7

Commissioning heating and hot water systems

 Any heating system more than 15 years old should either be replaced or be equipped with improved controls

Extensions

Where the installed capacity per unit area of a heating system is increased, existing doors (but excluding high-usage entrance doors) within the area served and which have U-values worse than 3.3 W/m²K should be replaced.	L2B 6.10 L2B 6.10

 Replacing these sorts of doors will normally achieve a simple payback within 15 years.

When commissioning heating and hot water systems, the person carrying out the commission should ensure that:

- the performance of the building fabric and the heating and hot water systems are no worse than the design limits;
- the system(s) and their controls have been left in working order and are capable of operating efficiently for the purposes of the conservation of fuel and power;
- independent temperature and ON/OFF controls to all heating appliances have been provided;
- the heating system uses heat raising appliances that have an efficiency not less than that recommended in the *Heating Compliance Guide ICCM*;
- if both heating and cooling are provided, they are capable of being controlled so as to not operate simultaneously;
- energy meters have been included so as to allow building occupants to assign at least 90 per cent of the estimated annual energy consumption of each fuel used for heating and lighting, etc.;
- meters have been provided to enable installed low- or zero-carbon (LZC) systems to be separately monitored;
- automatic meter reading and data collection have been provided in all buildings with a total useful floor area that is greater than 1000 m².

The person carrying out the work shall provide the local authority with a notice confirming that all fixed building services have been properly commissioned in accordance with a procedure approved by the Secretary of State.

In non-domestic buildings:

- new HVAC systems should be provided with controls that are capable of achieving a reasonable standard of energy efficiency;
- separate control zones should be capable of independent switching and control set-point;
- if both heating and cooling are provided, then they should not operate simultaneously;
- the central plant serving zone-based systems should:

  - only operate as and when required;
  - have a default condition that is off

Conservatories and porches

Regulation 9 of the Building Regulations exempts some conservatory and porch extensions from the energy-efficiency requirements, if doors and windows which separate the conservatory from the building are retained or, if removed, are replaced by walls, windows and doors which meet the energy-efficiency requirements; and where the heating system of the building is not extended into the conservatory or porch

(1) Under regulation 17D of the Building Regulations, the construction of an extension may trigger the requirement for consequential improvements.

(2) If a new conservatory or porch does not meet all these requirements, it is not exempt and **must** comply with the relevant energy-efficiency requirements of Part L: 2010, and particular attention should be paid to independent temperature and ON/OFF controls to any heating system installed within the extension.

(3) Removing and not replacing any of the thermal separation between the building and an existing exempt extension, or extending the building's heating system into the extension, means that the extension ceases to be exempt!

Consequential improvements (non-domestic buildings)

If work being undertaken on an existing buildings which has a total useful floor area of over $1000\,m^2$, then, in addition to the principal works (which **must** still comply with the energy-efficiency requirements detailed in Part L:2010 in the normal way), consequential improvements (where technically, functionally and economically feasible) will have to be completed provided that they are:

Where the installed capacity per unit area of a heating system is increased, existing doors (but excluding high-usage entrance doors) within the area served and which have U-values worse than 3.3 W/m² K should be replaced.	L2B 6.10 L2B 6.10

Replacing these sorts of doors will normally achieve a simple payback within 15 years.

Insulation of pipes, ducts and vessels

For buildings other than dwellings:

- insulation should not be less than those shown in the *Non-Domestic Heating, Cooling and Ventilation Compliance Guide*;
- hot and chilled water pipework, storage vessels, refrigerant pipework and ventilation ductwork should be insulated so as to conserve energy and to maintain the temperature of the heating or cooling service.

Cooling plant

In buildings other than dwellings, cooling systems should have a suitably efficient cooling plant and an effective control system.

Air-handling plant

For buildings other than dwellings:

- the air-handling plant should be an efficient and effective control system;
- the system should be capable of achieving a specific fan power at 25 per cent of design flow rate;
- ventilation system fans rated at more than 1100 W should be equipped with variable-speed drives;
- ventilation ductwork should be made and assembled so as to be reasonably airtight;
- replacement and new air-handling plants should be economical and energy efficient;
- fans that are rated at more than 1100 W and which form part of the environmental control system should be equipped with variable-speed drives.

A room thermostat for a ducted warm air heating system should be mounted in the living room, at a height between 1370 mm and 1830 mm, and its maximum setting should not exceed 27° C.	B1 2.17 (V1) B1 2.18 (V2) B3 7.10 (V1) B3 10.2 (V2)

6.22 Liquid fuel

6.22.1 The requirement

Protection of liquid fuel storage systems

Liquid fuel storage systems and the pipes connecting them to combustion appliances shall be so constructed, and separated from buildings and the boundary of the premises as to reduce to a reasonable level the risk of the fuel igniting in the event of fire in adjacent buildings or premises.

(Approved Document J5)

6.22.2 Meeting the requirement

Storage and supply

Oil and LPG fuel storage installations (including the pipework connecting them to the combustion appliances in the buildings they serve) shall:

• be located and constructed so that they are reasonably protected from fires that may occur in buildings or beyond boundaries;	J (5.1a)
• be reasonably resistant to physical damage and corrosion;	J (5.1a)
• be designed and installed so as to minimize the risk of oil escaping during the filling or maintenance of the tank;	J (5.1bi)
• incorporate secondary containment when there is a significant risk of pollution;	J (5.1bii)
• contain labelled information on how to respond to a leak.	J (5.1biii)

Oil pollution

The Control of Pollution (Oil Storage) (England) Regulations 2001 (SI 2001/2954) came into force on 1 March 2002. They apply to a wide range of oil storage installations in England, but they do not apply to the storage of oil on any premises used wholly or mainly as one or more private dwellings, if the capacity of the tank is 3500 litres or less.

Provisions where there is a risk of oil pollution

The main problems with regard to leakage concerns inland freshwater streams, rivers, reservoirs and lakes, as well as ditches and ground drainage (e.g. perforated drainage pipes) that feed into them.

When secondary containment is considered necessary, a way   J 5.10
of meeting the requirement would be to:

• provide an integrally bunded prefabricated tank; or
• construct a bund from masonry or concrete.

Bunds (whether part of a prefabricated tank system or   J 5.11
constructed on site) should have a capacity of at least 110 per
cent of the largest tank they contain.

An oil storage installation should carry a label in a prominent   J 5.12
position giving advice on what to do if an oil spill occurs,
and the telephone number of the Environment Agency's
Emergency Hotline (see Appendix F to Part J:2010 for further
details).

Table 107 Limits on design flexibility for mechanical ventilation

System type	Performance
Specific fan power (SFP) for continuous supply only and continuous extract only	0.8 l/s W
SFP for balanced systems	2.0 l/s W
Heat-recovery efficiency	66%

LPG storage

Liquid Petroleum Gas (LPG) installations are controlled by legislation enforced by the HSE which includes the following requirements applicable to dwellings:

 Note: Oil storage below ground is not recommended if other options are available, as underground tanks are difficult to inspect and leaks may not be immediately obvious.

The LPG tank should be installed outdoors and not within   J (5.15)
an open pit.

The tank should be adequately separated from buildings,   J (5.15)
the boundary and any fixed sources of ignition to enable
safe dispersal in the event of venting or leaks and in the
event of fire, to reduce the risk of fire spreading (see
Figure 6.222).

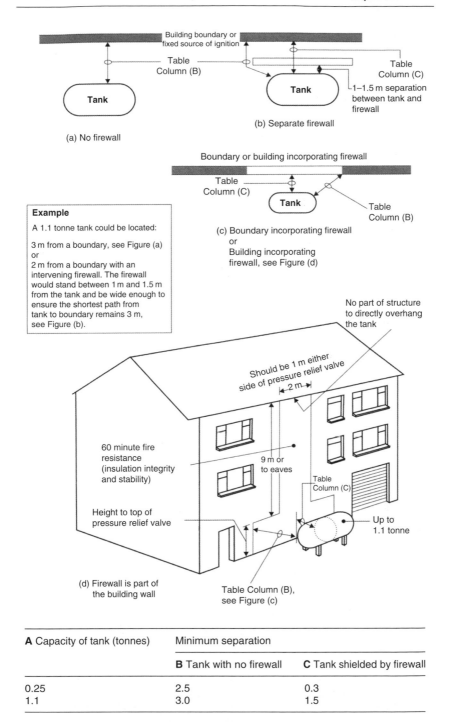

Example

A 1.1 tonne tank could be located:

3 m from a boundary, see Figure (a)
or
2 m from a boundary with an intervening firewall. The firewall would stand between 1 m and 1.5 m from the tank and be wide enough to ensure the shortest path from tank to boundary remains 3 m, see Figure (b).

A Capacity of tank (tonnes)	Minimum separation	
	B Tank with no firewall	C Tank shielded by firewall
0.25	2.5	0.3
1.1	3.0	1.5

Figure 6.222 Separation or shielding of LPG tanks from the building, boundaries and fixed sources of ignition.

Firewalls may be free-standing, built between the tank and the building, boundary and fixed source of ignition (see Figure 6.222b) or a part of the building, or a fire resistance (insulation, integrity and stability) boundary wall belonging to the property.

J (5.16)

Where a firewall is part of the building or a boundary wall, it should be located in accordance with Figure 6.222c.

J (5.16)

If the firewall is part of the building, it should be constructed as shown in Figure 6.222d.

J (5.16)

Firewalls should be imperforate and of solid masonry, concrete or similar construction.

J (5.17)

Firewalls should have a fire resistance (insulation, integrity and stability) of at least 30 minutes.

J (5.17)

If firewalls are part of the building as shown in Figure 6.222d, they should have a fire resistance (insulation, integrity and stability) of at least 60 minutes.

J (5.17)

To ensure good ventilation, firewalls should not normally be built on more than one side of a tank.

J (5.17)

A firewall should be at least as high as the pressure relief valve.

J (5.18)

Any pipe carrying natural gas or LPG should be:

B2 8.40 (V2)

- of screwed steel or of all welded steel construction;
- installed in accordance with SI 1996 No. 825 and SI 1998 No. 2451.

A protected shaft conveying piped flammable gas should be adequately ventilated direct to the outside air by ventilation openings at high and low level in the shaft.

B2 8.41 (V2)

If a door is provided between a dwelling-house and the garage, the floor of the garage should:

B3 5.5 (V1)

- be laid so as to allow fuel spills to flow away from the door to the outside; or
- the door opening should be positioned at least 100 mm above garage floor level.

Where an LPG storage installation consists of a set of cylinders, a way of meeting the requirements is as shown in Figure 6.223.

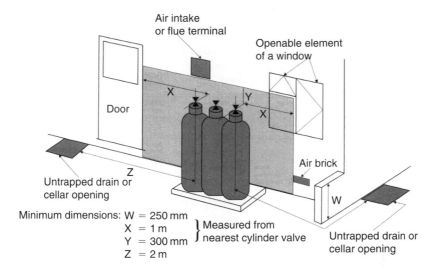

Figure 6.223 Location of LPG cylinders.

Cylinders should stand upright and be secured by straps (or chains) against a wall outside the building, in a well-ventilated position at ground level.	J (5.20)
Cylinders should be provided with a firm, level base, such as concrete at least 50 mm thick or paving slabs bedded on mortar.	J (5.20)

6.23 Cavities and concealed spaces

6.23.1 The requirement

Internal fire spread (structure)

* *The building should be sub-divided by elements of fire-resisting construction into compartments.*
* *Any hidden voids in the construction shall be sealed and sub-divided to inhibit the unseen spread of fire and products of combustion, in order to reduce the risk of structural failure, and the spread of fire.*

(Approved Document B3)

6.23.2 Meeting the requirement

Concealed spaces or cavities in walls, floors, ceilings and roofs will provide an easy route for smoke and flame spread which, because it is obscured, will present a greater danger than would be more obvious from a weakness in the fabric of the building. To overcome this danger, buildings shall be designed and constructed so that the unseen spread of fire and smoke within concealed spaces in its structure and fabric is inhibited.

With the introduction of Part B:2007, window and door frames are now only suitable for use as cavity barriers if they are constructed of steel or timber of an appropriate thickness.

Provision of cavity barriers

Cavity barriers should be provided:

• at the edges of cavities;	B3 6.3 (V1)
• around window and door openings;	B3 9.3 (V2)
• at the junction between an external cavity wall and a compartment wall;	
• at the junction between an external cavity wall and a compartment floor;	
• at the top of such an external cavity wall;	
• at the junction between an internal cavity wall and any assembly which forms a fire-resisting barrier;	
• above the enclosures to a protected stairway in a dwelling-house with a floor more than 4.5 m above ground level.	B1 2.14

Cavity barriers need not be provided between double-skinned corrugated or profiled insulated roof sheeting, if the sheeting is a material of limited combustibility. | B3 6.4 (V1) B3 9.5 (V2) |

 Note: Separate rules exist for bedrooms in institutional and other residential buildings (see B3 9.7 (V2)).

Construction and fixings for cavity barriers

Every cavity barrier should be constructed to provide at least 30 minutes fire resistance.	B3 6.5 (V1) B3 9.13 (V2)
A cavity barrier should, wherever possible, be tightly fitted to a rigid construction and mechanically fixed in position.	B3 6.6 (V1) B3 9.14 (V2)

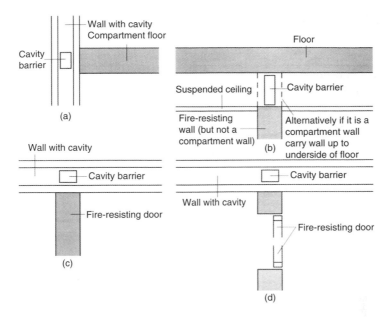

Figure 6.224 Interrupting concealed spaces and cavities. (a, b) Sections. (c, d) Plans.

Cavity barriers should be fixed so that their performance is unlikely to be made ineffective by:	B3 6.7 (V1) B3 9.15 (V2)

- movement of the building due to subsidence, shrinkage or temperature change;
- movement of the external envelope due to wind;
- collapse in a fire of any services penetrating them;
- failure in a fire of their fixings;
- failure in a fire of any material or construction which they abut.

Cavity barriers in a stud wall or partition (or provided around openings) may be made of:

- steel at least 0.5 mm thick; or
- timber at least 38 mm thick; or
- polythene-sleeved mineral wool, or mineral-wool slab (in either case under compression);
- calcium silicate, cement-based or gypsum-based boards at least 12 mm thick.

Extensive cavities in floor voids should be subdivided with cavity barriers.	B3 (V2)
The need for cavity barriers in some concealed floor or roof spaces can be reduced by using a fire-resisting ceiling below the cavity.	B2 3.6 (V1) B2 6.6 (V2)

Openings in cavity barriers

Openings in a cavity barrier should be limited to those for:	B3 6.8 (V1)
• doors which have at least 30 minutes fire resistance;	B3 9.13 (V2)
• the passage of pipes which meet the provisions in Part P Section 7;	
• the passage of cables or conduits containing one or more cables;	
• openings fitted with a suitably mounted automatic fire damper;	
• ducts which are fire-resisting or are fitted with a suitably mounted automatic fire damper where they pass through the cavity barrier.	

6.24 Kitchens and utility rooms

6.24.1 The requirement

Sanitary conveniences

(1) *Adequate and suitable sanitary conveniences must be provided in rooms provided to accommodate them or in bathrooms.*

(2) *Adequate stand washing facilities must be provided in:*

    (a) *rooms containing sanitary conveniences; or*

    (b) *rooms or spaces adjacent to rooms containing sanitary conveniences.*

(3) *Any room containing a sanitary convenience, a bidet, or any facility for washing hands provided in accordance with paragraph (2)(b) must be separated from any kitchen or any area where food is prepared.*

(Approved Document G4)

Food preparation

A suitable sink must be provided in any area where food is prepared.

(Approved Document G6)

Fire safety

- *There shall be an early warning fire alarm system for persons in the building.*
- *There shall be sufficient escape routes that are suitably located to enable persons to evacuate the building in the event of a fire.*
- *Safety routes shall be protected from the effects of fire.*
- *In an emergency, the occupants of any part of the building shall be able to escape without any external assistance.*

(Approved Document B1)

- *The spread of flame over the internal linings of the building shall be restricted.*
- *The heat released from the internal linings shall be restricted.*

(Approved Document B2)

Ventilation

There shall be adequate means of ventilation provided for people in the building.

(Approved Document F)

6.24.2 Meeting the requirement

Kitchens

A suitable sink must be provided in any area where food is prepared.

A sink should be provided in any kitchen or place used for the preparation of food.	G6.1
Where a dishwasher is provided in a separate room that is not the principal place for the preparation of food, an additional sink need not be provided in that room.	G6.2
Any sink should discharge through a grating, a trap and a branch discharge pipe to an adequate system of drainage.	G6.5
In buildings where the Food Hygiene (England) Regulations 2006 (SI 2006114) and the Food Hygiene (Wales) Regulations 2006 (SI 2006131 W5) apply, separate hand-washing facilities may be needed.	G6.4

Note: This is in addition to any hand-washing facilities associated with WCs in accordance with Requirement G4.

Fire safety

In small premises:

- store rooms should be enclosed with fire-resisting construction;   B1 3.35 (V2)

- clear glazed areas should be provided in any partitioning separating a kitchen from the open floor area to enable any person within the kitchen to obtain early visual warning of an outbreak of fire.   B1 3.36 (V2)

Smoke alarms

Smoke alarms should be positioned in the circulation spaces **between** sleeping spaces and places where fires are most likely to start (e.g. in kitchens and living rooms).   B1 1.11 (V1)
B1 1.10 (V2)

Where the kitchen area is not separated from the stairway or circulation space by a door, there should be a compatible interlinked heat detector or heat alarm in the kitchen, in addition to whatever smoke alarms are needed in the circulation space(s).   B1 1.13 (V1)
B1 1.12 (V2)

Inner rooms

Any inner room that is a kitchen or utility room that is situated not more than 4.5 m above ground level, whose only escape route is through another room, shall be provided with an emergency egress window.   B1 2.9 (V1)
B1 2.5 (V2)

Ancillary accommodation

Ancillary accommodation such as: B1 3.50 (V2)

- kitchens;
- staff changing and locker rooms; and
- store rooms;

should be enclosed by fire-resisting construction.

Non-domestic kitchens:

• should have separate and independent extraction systems; and • extracted air should not be recirculated.	B1 5.50 (V2)

Ventilation

Extract ventilation concerns the removal of air directly from a space or spaces to outside. Extract ventilation may be by natural means (e.g. by passive stack ventilation) or by mechanical means (e.g. by an extract fan or central system).

In kitchens, any automatic control must provide sufficient flow during cooking with fossil fuel (e.g. gas) to avoid build-up of combustion products.	F Table 5.2a
Manual boost controls should be provided in kitchens to guard against the possibility of a single centrally located switch being left in an incorrect mode of operation.	F Table 5.2d

 The 2010 version of Approved Document F has now provided further guidance for ventilation when a kitchen or bathroom in an existing dwelling is refurbished and now makes it mandatory that:

If you carry out any 'building work', and there is an existing extract fan or Passive Stack Ventilator (PSV) (or cooker hood extracting to outside in the kitchen), you should retain or replace it.	F 7.23
All kitchens and utility rooms shall be provided with extract ventilation to the outside which is capable of operating either: • intermittently at a minimum extract rate of 30 l/s (adjacent to hob) and 60 l/s elsewhere; or • continuously with a minimum extract rate of 13 l/s.	F 5.5

Common outlet terminals and/or branched ducts shall **not** be used for wet rooms such as a kitchen or a utility room.	F Tables 5.2a–5.2d

PSV devices shall have a minimum: F Tables 5.2a–5.2d

*   internal duct diameter of 125 mm for kitchens, 100 mm for utility rooms; and
*   a cross-sectional area of 12,000 mm^2 for kitchens, 8000 mm^2 for utility rooms.

Where there was no previous ventilation opening, or where the size of the original ventilation opening is not known, replacement window(s) shall have an equivalent area greater than 2500 mm^2.	F 7.6
In buildings other than dwellings, fresh air supplies should be protected from contaminants that would be injurious to health.	F 6.3
Wall/ceiling-mounted centrifugal fans (which are fitted with a 100 mm diameter flexible duct or rectangular duct and which are designed to achieve 60 l/s for kitchens) should not be ducted further than 3 m and should have no more than one 90° bend.	F Tables 5.2a–5.2d
Automatic controls for ventilators that are designed to work continuously in kitchens must be capable of providing sufficient flow during cooking with fossil fuels (e.g. gas) to avoid the build-up of combustion products.	F Tables 5.2a–5.2d
All office sanitary accommodation, washrooms and food- and beverage-preparation areas shall be provided with intermittent air extract ventilation.	F 6.10

If any of the work being carried out in the kitchen or bathroom of an existing building is 'building work', as defined in Regulation 3 of the Building Regulations, the Regulations require that you comply with the appropriate requirements of the Regulations, and in doing so you do not make compliance with other requirements of the Regulations, including ventilation, worse than before, the Regulations also require that, before you start work, the work is notified to a building control body, except in certain circumstances.

6.25 Storage of food

6.25.1 The requirement
(Building Act 1984 Sections 28 and 70)

All houses or buildings that have been converted into houses **must** provide sufficient and suitable accommodation for storing food or 'sufficient and suitable space for the provision of such accommodation by the occupier'.

This could prove to be a problem when submitting plans for approval, and you would be wise to consider the possibilities.

Table 6.108 Ventilation of rooms containing openable windows (i.e. located on an external wall)

Room	Rapid ventilation (e.g. opening windows)	Background ventilation (mm^2)	Extract ventilation fan rates or Passive Stack Ventilation (PSV)
Habitable room	1/20 of floor area	8000	
Kitchen	Opening window (no minimum size)	4000	30 l/s adjacent to a hob or 60 l/s elsewhere or PSV
Utility room	Opening window (no minimum size)	4000	30 l/s or PSV
Bathroom (with or without WC)	Opening window (no minimum size)	4000	15 l/s or PSV
Sanitary accommodation (separate from bathroom)	1/20 of floor area or mechanical extract at 6 l/s	4000	

6.26 Refuse facilities

6.26.1 The requirement (Building Act 1984 Section 23)

Probably due to new EU agreements, local authorities have become far stricter in seeing that the requirements contained in the Building Act for storage and collection of refuse are applied. This means that you have to ensure that the building is equipped with a satisfactory method for storing refuse (with a house this would normally be a simple dustbin; with a block of flats or a factory, however, the system for storage would have to be more sophisticated).

The local council will also need to be able to collect and remove this refuse easily, and this should be borne in mind when siting the refuse collection point.

Under the Building Act 1984 it is **unlawful** for any person (except with the consent of the local authority) to close or obstruct the means of access by which refuse or faecal matter is removed from a building.

Fire safety

- *Buildings shall be sub-divided by elements of fire-resisting construction into compartments.*
- *Openings in fire-separating elements shall be suitably protected in order to maintain the integrity of the element (i.e. the continuity of the fire separation).*

(Approved Document B3)

Airborne and impact of sound

Dwellings shall be designed so that the noise from domestic activity in an adjoining dwelling (or other parts of the building) is kept to a level that:

- *does not affect the health of the occupants of the dwelling;*
- *will allow them to sleep, rest and engage in their normal activities in satisfactory conditions.*

(Approved Document E1)

Dwellings shall be designed so that any domestic noise that is generated internally does not interfere with the occupants' ability to sleep, rest and engage in their normal activities in satisfactory conditions.

(Approved Document E2)

Domestic buildings shall be designed and constructed so as to restrict the transmission of echoes.

(Approved Document E3)

Schools shall be designed and constructed so as to reduce the level of ambient noise (particularly echoing in corridors).

(Approved Document E4)

6.26.2 Meeting the requirement

Refuse chutes and storage

Refuse storage chambers, refuse chutes and refuse hoppers should be sited and constructed in accordance with BS 5906.	B1 5.54 (V2)
In buildings containing flats, walls that enclose a refuse storage chamber should be constructed as a compartment wall.	B2 8.13 (V2)
Refuse chutes and rooms provided for the storage of refuse should: • be approached either directly from the open air or by way of a protected lobby; • be separated from other parts of the building by fire-resisting construction;	B1 5.55 and 5.56 (V2)

- **not** be located within protected stairways or protected lobbies.

The access to refuse storage chambers should not be sited adjacent to escape routes/final exits, or near to windows of flats.	B1 5.57 (V2)
A wall separating a habitable room or kitchen and a refuse chute should have mass (including any finishes) of at least 1320 kg/m².	E (2.28)
A wall separating a non-habitable room, which is in a dwelling, from a refuse chute should have a mass (including any finishes) of at least 220 kg/m².	E (2.28)

6.27 Loft conversions

6.27.1 The requirement

Fire safety

- *There shall be an early warning fire alarm system for persons in the building.*
- *There shall be sufficient escape routes that are suitably located to enable persons to evacuate the building in the event of a fire.*
- *Safety routes shall be protected from the effects of fire.*
- *In an emergency, the occupants of any part of the building shall be able to escape without any external assistance.*

(Approved Document B1)

- *The spread of flame over the internal linings of the building shall be restricted.*
- *The heat released from the internal linings shall be restricted.*

(Approved Document B2)

Ventilation

There shall be adequate means of ventilation provided for people in the building.

(Approved Document F)

 A new Part F came into force in October 2010.

Stairs, ladders and ramps

All stairs, steps and ladders shall provide reasonable safety between levels in a building.

(Approved Document K1)

Protection from falling

Pedestrian guarding should be provided for any part of a floor (including the edge below an opening window) gallery, balcony, roof (including rooflight and other openings), any other place to which people have access and any light well, basement area or similar sunken area next to a building.

(Approved Document K2)

Requirement K2(a) applies only to stairs and ramps that form part of the building.

6.27.2 Meeting the requirement

Fire safety

Where the conversion of an existing roof space (e.g. a loft conversion to a two-storey house) means that a new storey is going to be added, then the stairway will need to be protected with fire-resisting doors and partitions.

The floor(s), both old and new, shall have the full 30 minute standard of fire resistance shown in Part B Appendix A, Table A1, unless:	B3 4.7 (V1)

* only one storey is being added;
* the new storey contains no more than two habitable rooms; and
* the total area of the new storey is less than 50 m².

In those places where the floor only separates rooms (and not circulation spaces), a modified 30 minute standard of fire resistance may be applied.	B3 4.7 (V1)

New habitable rooms that are the result of a material alteration and which are above ground floor level (or at ground floor level where no final exit has been provided) shall be equipped with:	B1 2.20a (V1) B1 1.8 (V1)

* a fire detection and fire alarm system;
* smoke alarms in accordance with BS 5839–6.

Loft conversions

The following guidance (extracted from the 2006 version of Approved Document F) has been included here for information purposes

Fans and/or ducting placed in or passing through an unheated void or loft space should be insulated to reduce the possibility of condensation forming.	F(2006) App E

The inner radius of any bend should be greater than or equal to the diameter of the ducting being used (see Figure 6.225).	F(2006) App E
Vertical duct rises may need to be fitted with a condensation trap in order to prevent the backflow of any moisture.	F(2006) App E
The circular profile of a flexible duct should be maintained throughout the full length of the duct run (see Figure 6.225).	F(2006) App E

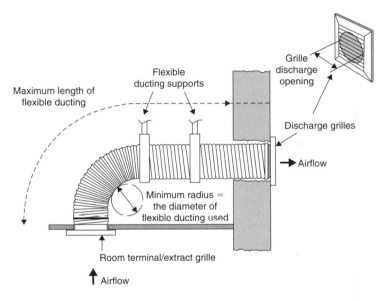

Figure 6.225 Correct installation of ducting.

If a back-draught device is used it may be incorporated into the fan itself.	F(2006) App E
Flexible ducting should be installed without any peaks or troughs (see Figure 6.226).	F(2006) App E

Stairs, ladders and ramps

The rise of a stair shall be between 155 mm and 220 mm with any going between 245 mm and 260 mm and a maximum pitch of 42°.	K1 (1.1–1.4)

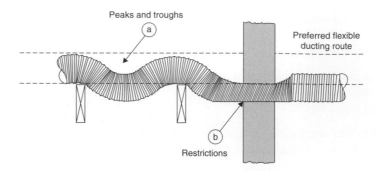

Figure 6.226 Incorrect installation of ducting.

The normal relationship between the dimensions of the rise and going is that twice the rise plus the going (2R + 1G) should be between 550 mm and 700 mm.

Stairs with open risers that are likely to be used by children under five years old should be constructed so that a 100 mm diameter sphere cannot pass through the open risers.   K1 (1.9)

Stairs that have more than 36 risers in consecutive flights should make at least one change of direction, between flights, of at least 30°.   K1 (1.14)

If a stair has straight and tapered treads, then the going of the tapered treads should not be less than the going of the straight tread.   K1 (1.20)

The going of tapered treads should measure at least 50 mm at the narrow end.   K1 (1.18)

The going should be uniform for consecutive tapered treads.   K1 (1.19)
                                                                K1 (1.22–1.24)

Stairs should have a handrail on both sides if they are wider than 1 m and on at least one side if they are less than 1 m wide.   K1 (1.27)

Handrail heights should be between 900 mm and 1000 mm measured to the top of the handrail from the pitch line or floor.   K1 (1.27)

Spiral and helical stairs should be designed in accordance with BS 5395.   K1 (1.21)

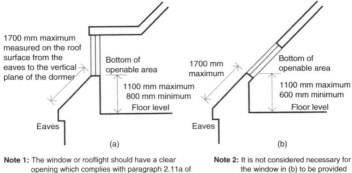

Note 1: The window or rooflight should have a clear opening which complies with paragraph 2.11a of Approved Document B and quoted on page 523.

Note 2: It is not considered necessary for the window in (b) to be provided with safety glazing.

Figure 6.227 Position of dormer window or rooflight that is suitable for emergency purposes from a loft conversion of a two-storey dwelling-house. (a) Dormer window (the window may be in the end wall of the house, instead of the roof as shown). (b) Rooflight or roof window.

Steps

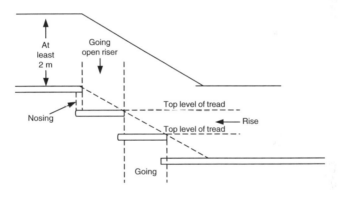

Figure 6.228 Rise and going plus headroom.

Steps should have level treads.	K1 (1.8)
Steps may have open risers, but treads should then overlap each other by at least 16 mm.	K1 (1.8)
Steps should be uniform with parallel nosings, the stair should have handrails on both sides and the treads should have slip-resistant surfaces.	
The headroom on the access between levels should be no less than 2 m.	K1 (1.10)

Landings should be provided at the top and bottom of every flight.	K1 (1.15)
The width and length of every landing should be the same (or greater than) the smallest width of the flight.	K1 (1.15)
Landings should be clear of any permanent obstruction.	K1 (1.16)
Landings should be level.	K1 (1.17)
Any door (entrance, cupboard or duct) that swings across a landing at the top or bottom of a flight of stairs must leave a clear space of at least 400 mm across the full width of the flight.	K1 (1.16)
Flights and landings should be guarded at the sides when there is a drop of more than 600 mm.	K1 (1.28–1.29)
For stairs that are likely to be used by children under five years old the construction of the guarding shall be such that a 100 mm sphere cannot pass through any openings in the guarding and children will not easily climb the guarding.	
For loft conversions, a fixed ladder should have fixed handrails on both sides.	K1 (1.25)

While there are no recommendations for minimum stair widths, designers should bear in mind the requirements of Approved Documents B (*Means of Escape*) and M (*Access for Disabled People*).

Protection from falling

All stairs, landings, ramps and edges of internal floors shall have a wall, parapet, balustrade or similar guard at least 900 mm high.	K3 (3.2)
All guarding should be capable of resisting at least the horizontal force given in BS 6399: Part 1:1996.	K3 (3.2)
If glazing is used as (or part of) the pedestrian guarding, see Approved Document N: Glazing – Safety in Relation to Impact, Opening and Cleaning.	N
If a building is likely to be used by children under five years old, the guarding should not have horizontal rails, should stop children from easily climbing it, and the construction should prevent a 100 mm sphere being able to pass through any opening of that guarding.	K3 (3.3)
All external balconies and edges of roofs shall have a wall, parapet, balustrade or similar guard at least 1100 mm high.	K3 (3.2)

6.28 Extensions and additions to buildings

Under Regulation 17D of the Building Regulations, the construction of an extension may trigger the requirement for consequential improvements.

For example, if a building has a total useful floor area greater than $1000 \, m^3$ and the proposed building work includes an extension, or the initial provision of any fixed building service, or an increase to the installed capacity of any fixed building services, consequential improvements should be made to improve the energy efficiency of the whole building and:

- thermal units with high U-values should be upgraded;
- existing windows (but not display windows), roof windows, rooflights and doors (excluding high-usage entrance doors) within the area served by the fixed building service with an increased capacity should be replaced;
- heating systems, cooling systems and air-handling systems that are more than 15 years old should either be replaced or be equipped with improved controls;
- any general lighting system serving an area greater than $100 \, m^2$ which has an average lamp efficacy of less than 40 lamp-lumens per circuit-watt, should be upgraded with new luminaires or improved controls;
- energy metering should be installed if less than 10 per cent of the building's energy demand is provided by a low- or zero-carbon (LZC) energy system; and
- the building should be upgraded with an additional LZC energy system, **provided** that the system would achieve a simple payback within seven years or less.

6.28.1 The requirement

Ventilation

There shall be adequate means of ventilation provided for people in the building.
(Approved Document F)

A new Part F came into force on 1 October 2010.

Conservation of fuel and power

Reasonable provision shall be made for the conservation of fuel and power in buildings by:

(a) *limiting heat gains and losses –*

　　(i) *through thermal elements and other parts of the building fabric; and*

　　(ii) *from pipes, ducts and vessels used for space heating, space cooling and hot water services;*

(b)   *providing fixed building services which –*

   (i)   *are energy efficient;*
   (ii)   *have effective controls; and*
   (iii) *are commissioned by testing and adjusting as necessary to ensure they use no more fuel and power than is reasonable in the circumstances; and*

(c)   *providing to the owner sufficient information about the building, the fixed building services and their maintenance requirements so that the building can be operated in such a manner as to use no more fuel and power than is reasonable in the circumstances*

(Approved Document L)

 A new Part L came into force on 1 October 2010.

Responsibility for achieving compliance with the requirements of Part L rests with the person carrying out the work. That person may be, for example, a developer, a main (or sub-) contractor, or a specialist firm directly engaged by a private client.

The person responsible for achieving compliance should either themselves provide a certificate, or obtain a certificate from the subcontractor, that commissioning has been successfully carried out. The certificate should be made available to the client and the building control body.

Fire risk analysis

Part B:2007 now includes a requirement for the responsible person (i.e. the person carrying out work to a building) to make available to the owner (other than houses occupied as single private dwellings) 'fire safety information' concerning the design and construction of the building or extension, plus details of the services, fittings and equipment that have been provided in order that they (when required under the new Regulatory Reform (Fire Safety) Order 2005 – Statutory Instrument 2005 No. 1541) may complete a fire risk analysis.

The sort of information required must include basic advice on the proper use and maintenance of systems provided in the building such as:

* emergency egress windows;
* fire doors;
* smoke alarms;
* sprinklers, etc.

6.28.2 Meeting the requirement

Changes in the legal requirements

The Building Regulations exempts some extensions from the energy-efficiency requirements. This exemption from the requirements of Part L:2010 only applies where the existing doors are retained (or replaced) and where the extension is:

- at ground level;
- has a floor area is less than $30\,m^2$;
- has either retained the existing walls, doors and windows of the dwelling or, if removed, have been replaced by walls, windows and doors which meet the energy-efficiency requirements of Part L:2010; and
- where the heating system of the dwelling has not been extended into the extension.

 If a new extension does not meet all these requirements then it is not exempt and **must** comply with the relevant energy-efficiency requirements of Part L:2010, and reasonable provision should be provided to ensure that:

effective thermal separation between the heated area in   L1A 4.11
the existing dwelling (i.e. the wails, doors, and windows
between the dwelling and the extension) have been
insulated and draught-proofed to at least the same extent
as in the existing dwelling.

 Doors with >50% of their internal face glazed
should not exceed $1.8\,W/m^2\,K$

If a building extension is **not** exempt from the energy-efficiency requirements, then:

- the effective thermal separation between the heated   L2B 4.12
  area in the existing building (i.e. the walls, doors and
  windows between the building and the extension)
  should be insulated and draught-proofed to at least the
  same extent as in the existing building;
- particular attention should be paid to independent
  temperature and ON/OFF controls to any heating
  system installed within the extension; and
- glazed elements should meet the standards set out in
  Table 6.109 and opaque elements should meet the
  standards set out in Table 6.110.

 Removing and not replacing any of the thermal separation between the building and an existing exempt extension, or extending the building's heating system into the extension, means that the extension ceases to be exempt!

Table 6.109 Standards for controlled fittings

Fitting	Standard (W/m²K)
Windows, roof windows and glazed rooflights	1.8 for the whole unit
Alternative option for windows in buildings that are essentially domestic in character	Window energy rating of B and C
Plastic rooflight	1.8
Curtain walling	<1.8
Pedestrian doors where the door has more than 50% of its internal face area glazed	1.8 for the whole unit
High-usage entrance doors for people	3.5
Vehicle access and similar large doors	1.5
Other doors	1.8
Roof ventilators (including smoke extract ventilators)	3.5

Ventilation

If the additional room is connected to an existing habitable room:

- which now has no windows opening to outside, then the ventilation opening (or openings) shall be greater than 8000 mm² equivalent area;   F 3.8a(i)

- which still has windows opening to outside, but with a total background ventilator equivalent area of at least 5000 mm² equivalent area, then the ventilation opening (or openings) shall be greater than 8000 mm² equivalent area;   F 3.8a(ii)

- which still has windows opening to outside, but with a total background ventilator equivalent area of at least 5000 mm² equivalent area, then there should be background ventilators of at least 8000 mm² equivalent area: F 3.8a(iii)

    - between the two rooms; and
    - between the additional room and outside.

If the extension is the addition of a habitable room to an existing building, then:

Table 6.110 Standards for new thermal elements

Element	Standard (W/m² K)
Wall	0.262
Pitched roof – insulation at ceiling level	0.16

If a building extension is a conservatory or porch that is *not* exempt from the energy efficiency requirements, then:

- the effective thermal separation between the heated area in the existing building (i.e. the walls, doors and windows between the building and the extension) should be insulated and draught proofed to at least the same extent as in the existing building;
- particular attention should be paid to independent temperature and ON/OFF controls to any heating system installed within the extension; and
- glazed elements should meet the standards set out in Table RLT-L11 and opaque elements should meet the standards set out in Table RLT-L2.

L2B 4.12

Fitting	Standard (W/m² K)
Windows, roof windows and glazed rooflights	1.8 for the whole unit

Alternative option for windows in buildings that are essentially domestic in character
A window energy rating of Band C

Plastic rooflight	1.8
Curtain walling	< 1.8
Pedestrian doors where the door has more than 50% of its internal face area glazed	1.8 for the whole unit
High-usage entrance doors for people	3.5
Vehicle access and similar large doors	1.5
Other doors	1.8
Roof ventilators (including smoke extract ventilators)	3.5

Table 6.110 (Continued)

Element	Standard (W/m² K)

Table RLT-L11 Standards for controlled fittings

Removing and not replacing any of the thermal separation between the building and an existing exempt extension, or extending the building's heating system into the extension, means that the extension ceases to be exempt!

Element	Standard (W/m² K) (W/m2N
Wall	0.262
Pitched roof – insulation at ceiling level	0.16
Pitched roof – insulation at rafter level	0.18
Flat roof or roof with integral insulation	0.18
Floors	0.224
Swimming pool basin	0.25

Table RLT-L2 Standards for new thermal elements

Element	Standard (W/m² K)
Pitched roof – insulation at rafter level	0.18
Flat roof or roof with integral insulation	0.18
Floors	0.224
Swimming pool basin	0.25

Internal doors between the wet room and the existing building   F 3.13
should have an undercut of at least minimum area 7600 mm²
(equivalent to an undercut of 10 mm above the floor finish for
a standard 760 mm width door).

Whole-building and extract ventilation can be provided by: F 3.12

- an intermittent extract and a background ventilator of at
  least 2500 mm² equivalent area; or
- a single room heat recovery ventilator; or
- a passive stack ventilator; or
- a continuous extract fan.

Historic and traditional buildings

When undertaking work on any historic or traditional building, the aim should
always be to improve energy efficiency as far as is reasonably practicable without prejudicing the character of the host building or increasing the risk of long-
term deterioration of the building fabric or fittings.

In general, new extensions to historic or traditional   L2B 3.11
buildings should comply with the standards of energy
efficiency as set out in Part L:2010. The **only** exception
would be where there is a particular need to match the
external appearance or character of the extension to that of
the host building.

Building inspectors normally require the use of 'sympathetic treatment' when
restoring the historic character of a building that has been subject to previous
inappropriate alteration (e.g. replacement doors).

Particular issues that could warrant sympathetic treatment – and where
advice from others would probably be beneficial – include:

- restoring the historic character of a building that has   L2B 3.11
  been subject to previous inappropriate alteration (e.g.
  replacement windows, doors and rooflights);
- rebuilding a former historic building (e.g. following a
  fire or filling a gap site in a terrace); and
- renovating the fabric of historic buildings to 'breathe' to
  control moisture and potential long-term decay problems.

Thermal elements

Extensions to dwellings should either use newly constructed thermal elements (that meet the requirements for the conservation of fuel and power) or use existing (or new) doors, windows, roof windows and rooflights that meet these standards (see Table 6.111.

Table 6.111 Standards for control fittings

Fitting	Standard (W/m²K)
Window, roof window or rooflight	1.6
Doors with >50% of internal face glazed	1.8
Other doors	1.8

In most circumstances the most reasonable provision would be to limit the total area of windows, roof windows and doors in extensions so that it does not exceed the sum of:

- 25 per cent of the floor area of the extension; plus
- the total area of any windows or doors which, as a result of the extension works, no longer exist or are no longer exposed.

Where the proposed extension has a total useful floor area that is both:

- greater than $1000\,\text{m}^2$; and
- greater than 25 per cent of the total useful floor area of the existing building;

then the work should be regarded as a **new building**.

One way of complying would be to show that the area-weighted U-value of all the elements in the extension is no greater than that of an extension of the same size and shape.

If upgrades are proposed to the existing dwelling, such upgrades should be implemented to a standard that is no worse than that shown in column (b) of Table 6.112.	L1B 4.7

Opening areas of rooflights and windows

The area of windows and rooflights in the extension should generally not exceed the values given in Table 6.113.	L2B 4.4

Table 6.112 Upgrading retained thermal elements

Element	(a) Threshold U-value (W/m²K)	(b) Improved U-value (W/m²K)
Wall – cavity insulation	0.70	0.55
Wall – external or cavity insulation	0.70	0.30
Floor	0.70	0.25
Pitched roof – insulation at ceiling level	0.35	0.16
Pitched roof – insulation between rafters	0.35	0.18
Flat roof or roof with integral insulation	0.35	0.18

Table 6.113 Opening areas in the extension

Building type	Windows and personnel doors as % of exposed wall	Rooflights as % of area of roof
Residential buildings where people temporarily or permanently reside	30	20
Places of assembly, offices and shops	40	20
Industrial and storage buildings	15	20
Vehicle access doors and display windows and similar glazing	As required	N/A
Smoke vents	N/A	As required

6.29 External balconies

6.29.1 The requirement

Protection from falling

Pedestrian guarding should be provided for any part of a floor (including the edge below an opening window) gallery, balcony, roof (including rooflight and other openings), any other place to which people have access and any light well, basement area or similar sunken area next to a building.

(Approved Document K2)

Requirement K2(a) applies only to stairs and ramps that form part of the building.

6.29.2 Meeting the requirement

All external balconies and edges of roofs shall have a wall, parapet, balustrade or similar guard at least 1100 mm high.	K3 (3.2)
If a balcony or flat roof is provided for escape purposes, guarding may be required.	B1 2.11(V1)

6.30 Garages

6.30.1 The requirement

Electrical work

Reasonable provision shall be made in the design and installation, operation, maintenance or alteration of electrical installations in order to protect persons from fire or injury.

(Approved Document P1)

Fire precautions

As a fire precaution, all materials used for internal linings of a building should have a low rate of surface flame spread and (in some cases) a low rate of heat release.

(Approved Document B2)

6.30.2 Meeting the requirement

Cables to an outside building (e.g. garage or shed), if run underground, should be routed and positioned so as to give protection against electric shock and fire as a result of mechanical damage to a cable.	P App A 2d
Cables concealed in floors and walls (in certain circumstances) are required to have an earthed metal covering, be enclosed in steel conduit, or have additional mechanical protection (see BS 7671 for more information).	P App A 2d

Wall and ceiling linings

In general terms (but see paragraphs 3.2 to 3.14 of Part B V1 and 6.2 to 6.14 of Part B V2 for more details), the surface linings of walls and ceilings should meet the classifications given in Table 6.114.

Table 6.114 Classification of linings

Location	National class	European class
Domestic garages less than 40 m²	3	D-s3, d2
Other rooms (including garages)	1	C-s3, d2

For the purpose of this requirement for surface lining, walls and ceiling are as defined in Table 6.115.

Table 6.115 Surface lining requirements – wall and ceiling

Element	Includes	Does not include
Walls	The surface of glazing except glazing in doors Any part of a ceiling which slopes at an angle of more than 70° to the horizontal	Doors and door frames Window frames and frames in which glazing is fitted Architraves, cover moulds, picture rails, skirtings and similar narrow members Fireplace surrounds, mantle shelves and fitted furniture
Ceilings	The surface of glazing Any part of a wall which slopes at an angle of 70° or less to the horizontal The underside of a gallery The underside of a roof exposed to the room below	Trapdoors and their frames The frames of windows or rooflights and frames in which glazing is fitted Architraves, cover moulds, picture rails Exposed beams and similar narrow members

Compartmentation

> Compartment walls and compartment floors should be provided if a domestic garage is attached to (or forms an integral part of) a dwelling-house, and the garage should be separated from the rest of the dwelling-house, as shown in Figure 6.229.
>
> B3 5.4 (V1)

Note: The wall and any floor between the garage and the house shall have a 30 minute standard of fire resistance. Any opening in the wall should be at least 100 mm above the garage floor level with an FD30 door.

> If a door is provided between a dwelling-house and the garage, the floor of the garage should:
>
> B3 5.5 (V1)
>
> • be laid so as to allow fuel spills to flow away from the door to the outside; or

- the door opening should be positioned at least
  100 mm above garage floor level.

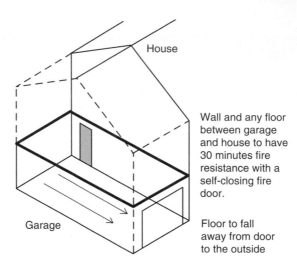

Figure 6.229 Separation between garage and dwelling-house.

Self-closing devices

Fire doors now **only** need to be provided with self-closing devices **if** they are between a dwelling-house and an integral garage.

6.31 Conservatories

6.31.1 The requirement

Ventilation

There shall be adequate means of ventilation provided for people in the building.
(Approved Document F)

 A new Part F came into force on 1 October 2010.

Conservation of fuel and power

Reasonable provision shall be made for the conservation of fuel and power in buildings by:

(a)   *limiting heat gains and losses –*

    (iii) *through thermal elements and other parts of the building fabric; and*

    (iv) *from pipes, ducts and vessels used for space heating, space cooling and hot water services;*

(b)   *providing fixed building services which –*

    (iv) *are energy efficient;*

    (v) *have effective controls; and*

    (vi) *are commissioned by testing and adjusting as necessary to ensure they use no more fuel and power than is reasonable in the circumstances; and*

(c)   *providing to the owner sufficient information about the building, the fixed building services and their maintenance requirements so that the building can be operated in such a manner as to use no more fuel and power than is reasonable in the circumstances.*

(Approved Document L)

A new Part L came into force on 1 October 2010.

6.31.2 Meeting the requirement

Ventilation

Habitable rooms without an openable window may be ventilated through a conservatory (see Figure 6.230) provided that that conservatory has:	F 1.14

- purge ventilation; and
- an 8000 mm^2 background ventilator; and
- there is a closable opening between the room and the conservatory that is equipped with:

    – purge ventilation; and
    – an 8000 mm^2 background ventilator.

The general ventilation rate for conservatories with a floor area greater than 30 m^2 conservatory (and adjoining rooms) can be achieved by the use of background ventilators.	F 3.18

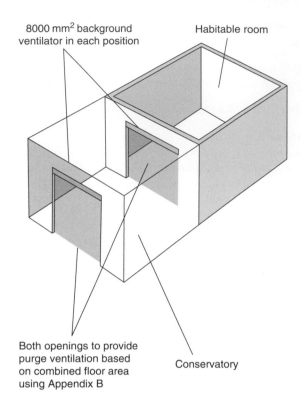

8000 mm² background
ventilator in each position

Habitable room

Both openings to provide
purge ventilation based
on combined floor area
using Appendix B

Conservatory

Figure 6.230 A habitable room ventilated through a conservatory.

Conservation of fuel and power

Under Regulation 17D of the Building Regulations, the construction of a conservatory or porch may trigger the requirement for consequential improvements.

Changes in the legal requirements

Regulation 9 of the Building Regulations exempts some conservatory and porch extensions from the energy-efficiency requirements. This exemption from the requirements of Part L:2010 are only applicable where the existing doors are retained (or replaced) and where the conservatory or porch is:

- at ground level;
- has a floor area is less than 30 m²;
- has either retained the existing walls, doors and windows of the dwelling or, if removed, these have been replaced by walls, windows and doors which meet the energy-efficiency requirements of Part L:2010, **and**

- where the heating system of the dwelling has **not** been extended into the conservatory or porch.

If a new conservatory or porch does not meet all these requirements, it is not exempt and **must** comply with the relevant energy-efficiency requirements of Part L:2010, and reasonable provision should be provided to ensure that:

• the effective thermal separation between the heated area in the existing building (i.e. the walls, doors and windows between the building and the extension) should be insulated and draught-proofed to at least the same extent as in the existing building;	L1A 4.11 L2B 4.12
• particular attention has been paid to independent temperature and ON/OFF controls to any heating system installed within the extension; and	
• glazed elements meet the standards set out in Table 6.116 and opaque elements meet the standards set out in Table 6.117.	

Table 6.116 Standards for controlled fittings

Fitting	Standard (W/m² K)
Windows, roof windows and glazed rooflights	1.8 for the whole unit
Alternative option for windows in buildings that are essentially domestic in character	Window energy rating of B and C
Plastic rooflight	1.8
Curtain walling	<1.8
Pedestrian doors where the door has more than 50% of its internal face area glazed	1.8 for the whole unit
Other doors	1.8
Roof ventilators	3.5

Removing and not replacing any of the thermal separation between the building and an existing exempt extension, or extending the building's heating system into the extension, means that the extension ceases to be exempt!

Conservatories and porches which are classified as an extension to a dwelling should either use newly constructed thermal elements (that meet the

Table 6.117 Standards for new thermal elements

Element	Standard (W/m²K)
Wall	0.262

0.16

If a building extension is a conservatory or porch that is *not* exempt from the energy efficiency requirements, then:

- the effective thermal separation between the heated area in the existing building (i.e. the walls, doors and windows between the building and the extension) should be insulated and draught proofed to at least the same extent as in the existing building;
- particular attention should be paid to independent temperature and ON/OFF controls to any heating system installed within the extension; and
- glazed elements should meet the standards set out in Table RLT-L11 and opaque elements should meet the standards set out in Table RLT-L2.

L2B 4.12

Pitched roof – insulation at ceiling level		
Fitting		Standard (W/m².K)
Windows, roof windows and glazed rooflights		1.8 for the whole unit
Alternative option for windows in buildings that are essentially domestic in character		
A window energy rating of Band C		
Plastic rooflight		1.8
Curtain walling		< 1.8
Pedestrian doors where the door has more than 50% of its internal face area glazed		1.8 for the whole unit
High-usage entrance doors for people		3.5
Vehicle access and similar large doors		1.5
Other doors		1.8
Roof ventilators (including smoke extract ventilators)		3.5

Table RLT-L11 Standards for controlled fittings

Removing and not replacing any of the thermal separation between the building and an existing exempt extension, or extending the building's heating system into the extension, means that the extension ceases to be exempt!

Table 6.117 (Continued)

Element	Standard (W/m²K)
Pitched roof – insulation at rafter level	0.18
Flat roof or roof with integral insulation	0.18
Floors	0.224

requirements for the conservation of fuel and power) or make use of existing (or new) doors, windows, roof windows and rooflights that meet the standards shown in Table 6.118.

Table 6.118 Standards for control fittings

Fitting	Standard (W/m²K)
Window, roof window or rooflight	1.6
Doors with >50% of internal face glazed	1.8
Other doors	1.8

In most circumstances the most reasonable provision would be to limit the total area of windows, roof windows and doors in extensions so that it does not exceed the sum of:

- 25 per cent of the floor area of the extension; plus
- the total area of any windows or doors which, as a result of the extension works, no longer exist or are no longer exposed.

 One way of complying would be to show that the area-weighted U-value of all the elements in the extension is no greater than that of an extension of the same size and shape.

If upgrades are proposed to the existing dwelling, such upgrades should be implemented to a standard that is no worse than that shown in column (b) of Table 6.119.   L1B 4.7

Table 6.119 Upgrading retained thermal elements

Element	(a) Threshold U-value (W/m²K)	(b) Improved U-value (W/m²K)
Wall – cavity insulation	0.70	0.55
Wall – external or cavity insulation	0.70	0.30
Floor	0.70	0.25
Pitched roof – insulation at ceiling level	0.35	0.16
Pitched roof – insulation between rafters	0.35	0.18
Flat roof or roof with integral insulation	0.35	0.18

6.32 Rooms for residential purposes

6.32.1 The requirement

Rooms for residential purposes are defined in Regulation 2 of the Building Regulations 2010 (2010: SI 2214). Such rooms need to conform with the applicable Approved Documents and meet the requirements for airborne and impact sound insulation shown in Table 120.

Table 6.120 Dwelling houses and flats – performance standards for separating walls, separating floors and stairs that have a separating function

	Airborne sound insulation $D_{nT,w}$ 1 C_{tr} (dB) (minimum values)	Impact sound insulation $L9_{nT,w}$ (dB) (maximum values)
Purpose-built rooms for residential purposes		
Walls	43	–
Floors and stairs	45	62

6.32.2 Meeting the requirement

Separating walls in new buildings containing rooms for residential purposes

Separating wall types 1 and 3 are considered to be the most suitable for use in new buildings containing rooms for residential purposes.

Note: Wall types 2 and 4 can be used **provided** that care is taken to maintain isolation between the leaves. Specialist advice may be needed.

Wall type 1 (solid masonry)

When using a solid masonry wall, the resistance to airborne sound depends mainly on the mass per unit area of the wall. As shown in Table 121, there are three different categories of solid masonry walls:

Note: Plasterboard may be used as an alternative wall finish, provided a sheet of minimum mass per unit area of $10 \, \text{kg/m}^2$ is used on each room face.

Table 6.121 Wall type 1 – categories

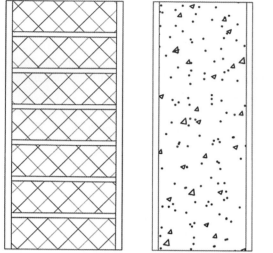
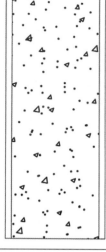
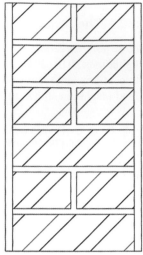

Wall type 1.1	**Wall type 1.2**	**Wall type 1.3**
Solid masonry (dense aggregate concrete block, plaster on both room faces)	**Dense aggregate concrete** (dense aggregate concrete cast in situ, plaster on both room faces)	**Brick** (plaster on both room faces)

Wall type 3 (masonry between independent panels)

Wall types 3.1 and 3.2 provide a high resistance to the transmission of both air-borne sound and impact sound on the wall. Their resistance to sound depends partly on the type (and mass) of the core and partly on the isolation and mass of the panels.

Corridor walls and doors in new buildings containing rooms for residential purposes

Separating walls as described in Table 6.122 should be used between rooms for residential purposes and corridors in order to control flanking transmission and to provide the required sound insulation between the dwelling and the corridor. Sound insulation will be reduced by the presence of a door.

All corridor doors shall have a good perimeter sealing (including the threshold where practical).	E6.6
All corridor doors shall have a minimum mass per unit area of 25 kg/m².	E6.6

Table 6.122 Wall Type 3 - categories

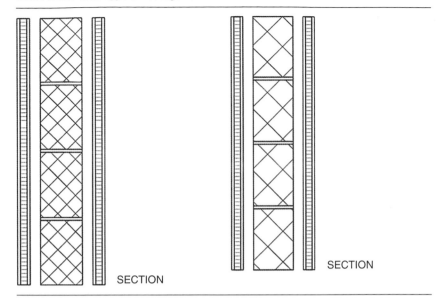

| Solid masonry core (dense aggregate concrete block), independent panels on both room faces | Solid masonry core (lightweight concrete block), independent panels on both room faces |

All corridor doors shall have a minimum sound reduction index   E6.6
of 29 dB Rw (measured according to BS EN ISO 140–3:1995
and rated according to BS EN ISO 717–1:1997).

All corridor doors shall meet the requirements for fire safety as   E6.6
described in Building Regulations Part B – Fire safety.

Noisy parts of the building should preferably have a lobby,   E6.7
double door or high-performance doorset to contain the noise.

Separating floors in new buildings containing rooms for residential purposes

Although only one of the separating floor types described in Section 3 of Part E is considered most suitable for use in new buildings containing rooms for residential purposes, provided that floating floors and ceilings are not continuous between rooms, floor types 2 and 3 can also be used.

Table 6.123 Floor type 1 – categories

Floor type 1.1C **(with ceiling treatment C)** Solid concrete slab (cast in situ or with permanent shuttering), soft floor covering	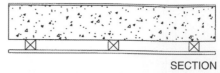 SECTION
Floor type 1.2 **(with ceiling treatment B)** Concrete planks (solid or hollow), soft floor covering	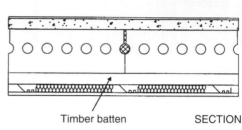 Timber batten SECTION

 Specialist advice may be required!

6.33 Rooms for residential purposes resulting from a material change of use

The erection of a new dwelling is not a material change of use. However, the requirements for the conservation of fuel and power still apply where a dwelling is being created in an existing building as the result of a material change of use of all, or part of, the building.

A material change of use in an existing dwelling is where, after the change, where replacement units are provided for an existing door separating a conditioned space from an unconditioned space/external environment has a U-value worse than 3.3 W/m² K

6.33.1 The requirement

Rooms for residential purposes formed by material change of use need to conform with the applicable Approved Document and meet the requirements for airborne and impact sound insulation shown in Table 6.124.

Table 6.124 Dwelling houses and flats – performance standards for separating walls, separating floors and stairs that have a separating function

	Airborne sound insulation $D_{nT,w}1C_{tr}$ (dB) (minimum values)	Impact sound insulation $L9_{nT,w}$ (dB) (maximum values)
Rooms for residential purposes formed by material change of use		
Walls	43	–
Floors and stairs	43	64

6.33.2 Meeting the requirement

In some cases it may be that an existing wall, floor or stair in a building will achieve these performance standards without the need for remedial work (e.g. if the existing construction was already compliant). If this is not the case, the building work should be in compliance with the Regulations concerning walls and floors as described previously in this book.

Rooms for residential purposes

Junction details

If there is a junction between a solid masonry separating wall type 1 and the ceiling void and roof space, the solid wall need not be continuous to the underside of the structural floor or roof provided that:	E6.14a
• there is a ceiling consisting of two or more layers of plasterboard, of minimum total mass per unit area 20 kg/m²;	
• there is a layer of mineral wool (minimum thickness 200 mm, minimum density 10 kg/m³) in the roof void;	E6.14b
• the ceiling is not perforated.	E6.14c

As shown in Figure 6.231, the ceiling joists and plasterboard sheets should not be continuous between rooms for residential purposes.

Note: The ceiling void and roof space detail can only be used where the requirements of Building Regulation Part B – Fire safety can also be satisfied. The requirements of Building Regulation Part L – *Conservation of Fuel and Power* should also be satisfied.

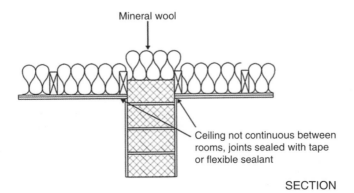

Mineral wool

Ceiling not continuous between rooms, joints sealed with tape or flexible sealant

SECTION

Figure 6.231 Ceiling void and roof space (only applicable to rooms for residential purposes).

Room layout and building services – design considerations

As internal noise levels are affected by room layout, building services and sound insulation, the actual layout of rooms should be considered at the design stage, particularly to avoid placing noise-sensitive rooms next to rooms in which noise is generated.

Note: See also:
- BS 8233:1999 *Sound Insulation and Noise*;
- *Reduction for Buildings – Code of Practice and Sound Control for Homes.*

6.34 Reverberation in the common internal parts of buildings containing flats or rooms for residential purposes

6.34.1 The requirement

Requirement E3 requires that 'Domestic buildings shall be designed and constructed so as to restrict the transmission of echoes.'

6.34.2 Meeting the requirement

The guidance notes provided in Part E cover two methods. These methods assist in determining the amount of additional absorption to be used in corridors, hallways, stairwells and entrance halls that give access to flats and rooms for residential purposes.

Table 6.125 Methods assist in determining the amount of additional absorption

Method	Entrance halls	Corridors	Hallways	Stairwells
A	Yes	Yes	Yes	Yes
B	Yes	Yes	Yes	No

Method A

Cover the ceiling area with the additional absorption.	E7.10
Cover the underside of intermediate landings, the underside of the other landings, and the ceiling area on the top floor.	E7.11
The absorptive material should be equally distributed between all floor levels.	E7.12

For stairwells (or a stair enclosure), calculate the combined   E7.11
area of the stair treads, the upper surface of the intermediate
landings, the upper surface of the landings (excluding the
ground floor) and the ceiling area on the top floor. Either
cover an area equal to this calculated area with a class D
absorber, or cover an area equal to at least 50 per cent of this
calculated area with a class C absorber or better.

Note: Method A can generally be satisfied by the use of proprietary acoustic ceilings.

Method B

In comparison to method A, method B takes account of the existing absorption provided by all the surfaces. Section 7 of Part E provides details of how to calculate the total absorption area based on the material's absorption coefficient (α).

This can become a fairly specialist area, and it would probably be advisable to seek professional advice and/or peruse BS EN 20354:1993 *Acoustics – Measurement of Sound Absorption in a Reverberation Room* or BS EN ISO 11654:1997 *Acoustics – Sound Absorbers for use in Buildings – Rating of Sound Absorption* for detailed information.

Absorption areas should be calculated for each octave band (in square metres) using the formula:

$$A_T = \alpha_1 S_1 + \alpha_2 S_2 + \ldots \alpha_n S_n$$

where:

A_T = total absorption area in sq. m
$\alpha_1 S_1$ = absorption coefficient for material 1
$\alpha_2 S_2$ = absorption coefficient for material 2
$\alpha_n S_n$ = absorption coefficient for the last type of material n.

Requirement E3 will be satisfied when:

For entrance halls, provide a minimum of $0.20\,\text{m}^2$ total   E7.17
absorption area per cubic metre of the volume.

For corridors or hallways, provide a minimum of $0.25\,\text{m}^2$ total   E7.18
absorption area per cubic metre of the volume.

Table 6.126 Absorption coefficient data for common materials in buildings

Material	Sound absorption coefficient (a) in octave frequency bands (Hz)				
	250	500	1000	2000	4000
Fair-faced concrete or plastered masonry	0.01	0.01	0.02	0.02	0.03
Fair-faced brick	0.02	0.03	0.04	0.05	0.07
Painted concrete brick	0.05	0.06	0.07	0.09	0.08
Windows glass façade	0.08	0.05	0.04	0.03	0.02
Doors (timber)	0.10	0.08	0.08	0.08	0.08
Glazed tile/marble	0.01	0.01	0.01	0.02	0.02
Hard floor coverings (e.g. lino, parquet) on concrete floor	0.03	0.04	0.05	0.05	0.06
Soft floor coverings (e.g. carpet) on concrete floor	0.03	0.06	0.15	0.30	0.40
Suspended plaster or plasterboard ceiling (with large airspace behind)	0.15	0.10	0.05	0.05	0.05

6.35 Internal walls and floors (new buildings)

6.35.1 The requirement

Dwellings shall be designed so that the noise from domestic activity in an adjoining dwelling (or other parts of the building) is kept to a level that:

- *does not affect the health of the occupants of the dwelling;*
- *will allow them to sleep, rest and engage in their normal activities in satisfactory conditions.*

(Approved Document E1)

Dwellings shall be designed so that any domestic noise that is generated internally does not interfere with the occupants' ability to sleep, rest and engage in their normal activities in satisfactory conditions.

(Approved Document E2)

Domestic buildings shall be designed and constructed so as to restrict the transmission of echoes.

(Approved Document E3)

6.35.2 Meeting the requirement

Note: To avoid air paths between rooms, all gaps around internal walls and floors should be filled.

There are four main types of internal wall and three types of internal floor as detailed below.

Internal wall type A

Timber or metal frames with plasterboard linings on each side of frame

PLAN

- Each lining to be two or more layers of plasterboard
- Each sheet of minimum mass per unit area $10\,kg/m^2$
- Linings fixed to timber frame with a minimum distance between linings of 75 mm (or metal frame with a minimum distance between linings of 45 mm)
- All joints are to be sealed

Internal wall type B

Timber or metal frames with plasterboard linings on each side of frame and absorbent material

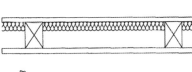

PLAN

- Single layer of plasterboard of minimum mass per unit area $10\,kg/m^2$
- Linings fixed to timber frame with a minimum distance between linings of 75 mm (or metal frame with a minimum distance between linings of 45 mm)
- An absorbent layer of unfaced mineral wool batts or quilt (minimum thickness 25 mm, minimum density $10\,kg/m^3$) suspended in the cavity (may be wire reinforced)
- All joints well sealed

Internal wall type C
Concrete block wall, plaster or plasterboard finish on both sides

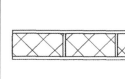

SECTION

- Minimum mass per unit area, excluding finish 120 kg/m^2
- All joints well sealed
- Plaster or plasterboard finish on both sides

Internal wall type D
Aircrete block wall, plaster or plasterboard finish on both sides

 Note: Internal wall type D should only be used with the separating walls described in section Ei (i.e. where there is no minimum mass requirement on the internal masonry walls) and it should not be used as a load-bearing wall or be rigidly connected to the separating floors.

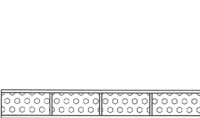

SECTION

- For plaster finish, minimum mass per unit area, including finish 90 kg/m^2
- For plasterboard finish, minimum mass per unit area, including finish 75 kg/m^2
- All joints well sealed

Internal floor type A
Concrete planks

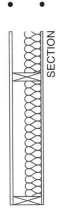

SECTION

 Note: Insulation against impact sounds can be improved by adding a soft covering (e.g. carpet).

- Minimum mass per unit area $180 \, \text{kg/m}^2$
- Regulating screed optional
- Ceiling finish optional

Internal floor type B
Concrete beams with infilling blocks, bonded screed and ceiling

SECTION

Note: Insulation against impact sounds can be improved by adding a soft covering (e.g. carpet).

- Minimum mass per unit area of beams and blocks $220 \, \text{kg/m}^2$
- Bonded sand cement screeds with a minimum thickness of 40 mm
- Ceiling finish required

Internal floor type C
Timber or metal joist, with wood-based board and plasterboard ceiling, and absorbent material

SECTION

Note: Insulation against impact sounds can be improved by adding a soft covering (e.g. carpet).

- Floor of timber or wood-based board, minimum mass per unit area $15 \, \text{kg/m}^2$
- Ceiling treatment of single layer of plasterboard, minimum mass per unit area $10 \, \text{kg/m}^2$, fixed using any normal fixing method
- An absorbent layer of mineral wool (minimum thickness 100 mm, minimum density $10 \, \text{kg/m}^2$) laid in the cavity

The resistance to airborne sound depends on the type of wall or internal floor as shown in Table 6.127:

Table 6.127 Resistance to airborne sound

Type	Resistance
Internal wall type A or B *Timber or metal frame*	The resistance to airborne sound depends on the mass per unit area of the leaves, the cavity width, frame material and the absorption in the cavity between the leaves.
Internal wall type C or D *Concrete or aircrete block*	The resistance to airborne sound depends mainly on the mass per unit area of the wall.
Internal floor type A or B *Concrete planks or concrete beams with infilling blocks*	The resistance to airborne sound depends on the mass per unit area of the concrete base or concrete beams and infilling blocks. A soft covering will reduce impact sound at source.
Internal floor type C *Timber or metal joist*	The resistance to airborne sound depends on the structural floor base, the ceiling and the absorbent material. A soft covering will reduce impact sound at source.

Other considerations

Doors

Lightweight doors with poor perimeter sealing provide a lower standard of sound insulation than walls.

This will reduce the effective sound insulation of the internal wall. Ways of improving sound insulation include ensuring that there is good perimeter sealing or using a doorset.

Stairs

If the stair is not enclosed, the potential sound insulation of the internal floor will not be achieved; nevertheless, the internal floor should still satisfy Requirement E2.

Noise reduction

It is good practice to consider the layout of rooms at the design stage to avoid placing noise-sensitive rooms next to rooms in which noise is generated. Guidance on layout is provided in BS 8233:1999 *Sound Insulation and Noise Reduction for Buildings – Code of Practice.*

Electrical cables

Electrical cables give off heat when in use, and special precautions may be required when they are covered by thermally insulating materials. See BRE BR 262 *Thermal Insulation: Avoiding Risks*, Section 2.3.

6.36 Regulation 7 – materials and workmanship

6.36.1 The requirement

Building work shall be carried out –

(a)   with adequate and proper materials which –

> *(i)   are appropriate for the circumstances in which they are used;*
> *(ii)   are adequately mixed or prepared; and*
> *(iii) are applied, used or fixed so as adequately to perform the functions for which they are designed; and*

(b)   in a workmanlike manner.

(Regulation 7)

Parts A to K and N of Schedule 1 to these regulations do not require anything to be done except for the purpose of securing reasonable standards of health and safety for persons in or about buildings (and any others who may be affected by buildings or matters connected with buildings).

6.36.2 Meeting the requirements

Materials (including products, components, fittings, naturally occurring materials (e.g. stone, timber and thatch), items of equipment, and backfilling for excavations in connection with building work) shall be:

* of a suitable nature and quality in relation to the purposes and conditions of their use;
* adequately mixed or prepared (where relevant); and
* applied, used or fixed so as to perform adequately the functions for which they are intended.

Environmental impact of building work

The environmental impact of building work can be minimized by careful choice of materials and the use of recycled and recyclable materials.	Reg. 7 (0.2)
The use of such materials must not have any adverse implications for the health and safety standards of the building work.	Reg. 7 (0.2)

Limitations

Reasonable standards of health or safety for persons in or about the building shall be assured.	Reg. 7 (0.3)
Materials and workmanship shall ensure that fuel and power is conserved.	Reg. 7 (0.3)
Access and facilities for disabled people shall be provided.	Reg. 7 (0.3)

Note: There are no provisions under the Building Regulations for continuing control over the use of materials following the completion of building work.

Fitness of materials

The suitability of material used for a specific purpose shall be assessed using:	Reg. 7 (1.2 to 1.7)

- British Standards;
- European Standards (ENs);
- other national (and international) technical specifications (from other European Community member states);
- technical approvals (covered by a national or European certificate issued by a European technical approvals issuing body);
- CE marking (which provides a presumption of conformity with the stated minimum legal requirements);
- independent (UK) product certification schemes (such as those accredited by UKAS);
- tests and calculations (e.g. those in conformance with UKAS's Accreditation Scheme for Testing Laboratories);
- past experience (that the material can be shown, by experience, to be capable of performing the function for which it is intended);
- sampling (of materials by local authorities).

Short-lived materials and materials susceptible to changes in their properties should only be used in works where these changes do not adversely affect their performance.

Resistance to moisture

> Any material that is likely to be adversely affected   Reg. 7 (1.8)
> by condensation, moisture from the ground, rain or
> snow shall either resist the passage of moisture to the
> material or be treated or otherwise protected from
> moisture.

Resistance to substances in the subsoil

> Any material in contact with the ground (or in the   Reg. 7 (1.9)
> foundations) shall be capable of resisting attacks by
> harmful materials in the subsoil such as sulphates.

Establishing the adequacy of workmanship

The adequacy (and competence) of workmanship will normally be established by using:

- the British Standard code of practice;
- workmanship specified and covered by a national or European certificate issued by a European technical approvals issuing body;
- the recommendations of an integrated management system (e.g. ISO 9001: latest edition);
- past experience (of workmanship that is capable of performing the function for which it is intended);
- tests (e.g. the local authority has the power to test sewers and drains in or in connection with buildings).

Workmanship on building sites

In the main, local authorities will use the codes of practice as contained and detailed in BS 8000 *Workmanship on Building Sites* to monitor and inspect all building work. These codes of practice consist of:

- Part 1: 1989 *Code of Practice for Excavation and Filling*
- Part 2: *Code of Practice for Concrete Work*, Section 2.1:1990 *Mixing and Transporting Concrete*; Section 2.2:1990 *Sitework with In Situ and Precast Concrete*
- Part 3:1989 *Code of Practice for Masonry*
- Part 4:1989 *Code of Practice for Waterproofing*
- Part 5:1990 *Code of Practice for Carpentry, Joinery and General Fixings*
- Part 6:1990 *Code of Practice for Slating and Tiling of Roofs and Claddings*
- Part 7:1990 *Code of Practice for Glazing*
- Part 8:1994 *Code of Practice for Plasterboard Partitions and Dry Linings*

- Part 9:1989 *Code of Practice for Cement/Sand Floor Screeds and Concrete Floor Toppings*
- Part 10:1995 *Code of Practice for Plastering and Rendering*
- Part 11: *Code of Practice for Wall and Floor Tiling*, Section 11.1:1989 *Ceramic Tiles, Terrazzo Tiles and Mosaics* (confirmed 1995), Section 11.2:1990 *Natural Stone Tiles*
- Part 12:1989 *Code of Practice for Decorative Wall Coverings and Painting*
- Part 13:1989 *Code of Practice for Above Ground Drainage and Sanitary Appliances*
- Part 14:1989 *Code of Practice for Below Ground Drainage*
- Part 15:1990 *Code of Practice for Hot and Cold Water Services (Domestic Scale)*
- Part 16:1997 *Code of Practice for Sealing Joints in Buildings Using Sealants*.

6.37 Work on existing constructions

If an existing building is going to be converted into dwelling-houses (and/or flats) via a *material change of use*, a certain amount of remedial work to the existing construction will first have to be completed. In the paragraphs below are listed the most important areas that should be addressed, but, if you are contemplating carrying out this sort of work, it would be best to have a chat with your local authority as it might have specific requirements and rulings for your own particular area.

The erection of a new dwelling is not a material change of use. However, the requirements for the conservation of fuel and power still apply where a dwelling is being created in an existing building as the result of a material change of use of all, or part of, the building.

A material change of use in an existing dwelling is where, after the change, where replacement units are provided for existing doors separating a conditioned space from an unconditioned space/external environment having a U-value worse than $3.3\,\text{W/m}^2\text{K}$.

If work is being undertaken on an existing building which has a total useful floor area of over $1000\,\text{m}^2$, then, in addition to the principal works (which must still comply with the energy-efficiency requirements detailed in Part L:2010 in the normal way), consequential improvements (where technically, functionally and economically feasible) will have to be completed.

Where the installed capacity per unit area of a heating system is increased, existing doors (but excluding high-usage entrance doors) within the area served and which have U-values worse than $3.3\,\text{W/m}^2\text{K}$ should be replaced.	L2B 6.10 L2B 6.10

Replacing these sorts of doors will normally achieve a simple payback within 15 years.

> If upgrades are proposed to the existing dwelling, such upgrades should be implemented to a standard that is no worse than that shown in column (b) of Table 6.128.   L1B 4.7

Table 6.128 Upgrading retained thermal elements

Element	(a) Threshold U-value (W/m²K)	(b) Improved U-value (W/m²K)
Wall – cavity insulation	0.70	0.55
Wall – external or cavity insulation	0.70	0.30
Floor	0.70	0.25
Pitched roof – insulation at ceiling level	0.35	0.16
Pitched roof – insulation between rafters	0.35	0.18
Flat roof or roof with integral insulation	0.35	0.18

Large extensions

Where the proposed extension has a total useful floor area that is both:

- greater than 1000 m²; and
- greater than 25% of the total useful floor area of the existing building;

then the work should be regarded as a new building.

6.37.1 Walls

In any type of building, the amount of sound resistance provided by a wall depends on:

- its construction;
- the type of independent panel(s) it uses (if any);
- the isolation of these panel(s);
- the type of absorbent material that has been used.

If the existing wall is masonry (and has a thickness of at least 100 mm and is plastered on both faces), the following wall treatment (commonly referred to as *Wall treatment 1: Independent panel(s) with absorbent material*) is recommended.

In particular:

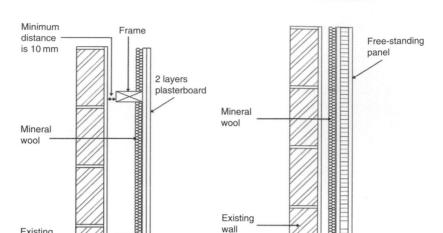

Figure 6.232 Wall treatment 1: independent panel(s) with absorbent material.

The independent panel and its supporting frame should not be in contact with the existing wall.	E4.25a
The perimeter of the independent ceiling should be sealed with tape or sealant.	E4.25b
The absorbent material should not be tightly compressed as this may bridge the cavity.	E4.25

In addition:

- the minimum mass per unit area of panel (excluding any supporting framework) should be $20\,kg/m^2$;
- each panel should consist of at least two layers of plasterboard with staggered joints;
- if the panels are free-standing, they should be at least 25 mm from masonry core;
- if the panels are supported on a frame, there should be a gap of at least 10 mm between the frame and the face of the existing wall;

- mineral wool (minimum density $10 \, kg/m^3$ and minimum thickness 35 mm) should be used in the cavity between the panel and the existing wall.
- if a wall is common to two or more buildings, it must be designed and constructed so that it adequately resists the spread of fire between those buildings;
- external walls must be constructed so as to have a low rate of heat release and thereby be capable of reducing the risk of ignition from an external source and the spread of fire over their surfaces;
- the amount of unprotected area in the sides of the building shall be restricted so as to limit the amount of thermal radiation that can pass through the wall.

(Extracted from Approved Document B and B4)

 Note: Wall linings may be required to reduce flanking transmission.

Table 6.129 Dwelling houses and flats – performance standards for separating walls, separating floors and stairs that have a separating function

	Airborne sound insulation $D_{nT,w}$ 1 C_{tr} (dB) (minimum values)	Impact sound insulation $L9_{nT,w}$ (dB) (maximum values)
Purpose-built rooms for residential purposes		
Walls	43	–
Floors and stairs	45	62

6.37.2 Floors

In buildings, the amount of resistance to airborne and impact sound will depend on:

- the combined mass of the existing floor and the independent ceiling;
- the amount of absorbent material;
- the isolation of the independent ceiling;
- the airtightness of the whole construction.

Two types of floor treatment are recommended dependent on the type of construction and material that has been used for the existing floor. The following requirements are common to both treatments:

- if the existing floor is timber, then gaps in the floor boarding should be sealed by overlaying with hardboard (alternatively, the gap should be filled with sealant);

- if floor boards are going to be replaced, boarding (minimum thickness 12 mm) and mineral wool (minimum thickness 100 mm, minimum density 10 kg/m^3) should be laid between the joists in the floor cavity;
- if the existing floor is concrete (and the mass per unit area of the concrete floor is less than 300 kg/m^2), the mass of the floor should be increased to at least 300 kg/m^2;
- any air gaps through the concrete floor should be sealed.

Floor treatment 1: independent ceiling with absorbent material

The resistance to airborne and impact sound from floor treatment 1 depends on:

- the combined mass of the existing floor;
- the independent ceiling;
- the absorbent material;
- the isolation of the independent ceiling;
- the airtightness of the whole construction.

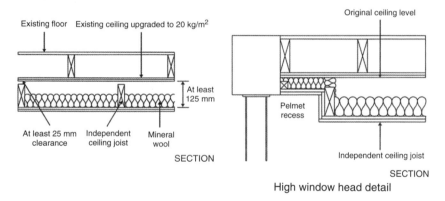

Figure 6.233 Floor treatment 1: independent ceiling with absorbent material.

Specifically

The ceiling should have: E4.27

- at least two layers of plasterboard with staggered joints;
- a minimum total mass per unit area 20 kg/m^2;
- an absorbent layer of mineral wool laid on the ceiling (minimum thickness 100 mm, minimum density 10 kg/m^3).

The ceiling should be supported by either: E2.47

- independent joists fixed only to the surrounding walls;
- independent joists fixed to the surrounding walls with
  additional support provided by resilient hangers attached
  directly to the existing floor base.

> **Note:** A clearance of at least 25 mm should be left
> between the top of the independent ceiling joists and
> the underside of the existing floor construction.

Where a window head is near to the existing ceiling, the new   E4.29
independent ceiling may be raised to form a pelmet recess.

The perimeter of the independent ceiling should be sealed   E4.30
with tape or sealant.

A rigid or direct connection should not be created between the   E4.30
independent ceiling and the floor base.

The absorbent material should not be tightly compressed as   E4.30
this may bridge the cavity.

Floor treatment 2: platform floor with absorbent material

With floor treatment 2, the resistance to airborne and impact sound depends
on:

- the total mass of the floor;
- the effectiveness of the resilient layer;
- the absorbent material.

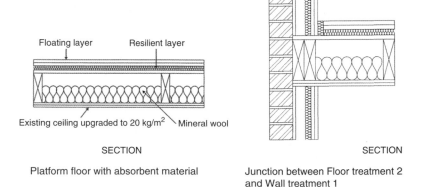

Floating layer   Resilient layer

Existing ceiling upgraded to 20 kg/m² \ Mineral wool

SECTION

Platform floor with absorbent material

SECTION

Junction between Floor treatment 2
and Wall treatment 1

Figure 6.234 Floor treatment 2: platform floor with absorbent material.

Specifically:

Where this treatment is used to improve an existing timber floor, a layer of mineral wool (100 mm, minimum density 10 kg/m³) should be laid between the joists in the floor cavity.	E4.32
The floating layer should be a: • minimum of two layers of board material; • minimum total mass per unit area 25 kg/m².	E4.32
Each layer should be: • a minimum thickness of 8 mm; • fixed together (e.g. spot bonded or glued/screwed) with joints staggered; • laid loose on a resilient layer.	E4.32
The resilient layer should be: • mineral wool (minimum thickness 25 mm, density 60–100 kg/m³); • paper faced on the underside.	E4.32
The correct density of resilient layer (to carry the anticipated load) should be ensured.	E4.32
The probable movement of materials after laying (e.g. expansion of chipboard) should be taken into account.	E4.32
The resilient layer should be carried up at all room edges to isolate the floating layer from the wall surface.	E4.32
A 5 mm gap should be left between the skirting and floating layer and filled with a flexible sealant.	E4.32
Resilient materials should be laid in sheets with joints tightly butted and taped.	E4.32
The perimeter of any new ceiling should be sealed with tape or sealant.	E4.32
The floating layer and the base or surrounding walls should not be bridged with services or fixings that penetrate the resilient layer.	E4.32

Load-bearing elements of structure

When altering an existing two-storey, single-family dwelling-house to provide additional storeys, the floor(s), both old and new, shall have the full 30 minute standard of fire resistance.	B3 4.7 (V1)
All load-bearing elements of a structure shall have a minimum standard of fire resistance.	B3 4.1 (V1) B3 7.1 (V2)
Structural frames, beams, floor structures and gallery structures should have at least the fire resistance given in Appendix A of Approved Document B.	B3 4.2 (V1) B3 7.2 (V2)

Fire resistance – compartmentation

The division of the building into discrete fire zones offers perhaps the most effective means of limiting fire damage. Designed to contain the fire to within the zone of origin, this approach provides at least some protection for the rest of the building and its occupants, even if first-aid fire-fighting measures are used and fail. It also delays the spread of fire prior to the arrival of the fire brigade.

To prevent the spread of fire within a building, whenever possible the building should be subdivided into compartments separated from one another by walls and/or floors of fire-resisting construction.	B3 5.1 (V1) B3 8.1 (V2)
Parts of a building that are occupied mainly for different purposes should be separated from one another by compartment walls and/or compartment floors.	B3 5.3 (V1) B3 8.11 (V2)
The wall and any floor between the garage and the house shall have a 30 minute fire resistance. Any opening in the wall to be at least 100 mm above the garage floor level with an FD30 door.	B3
In buildings containing flats or maisonettes, compartment walls or compartment floors shall be constructed between: • every floor (unless it is within a maisonette); • one storey and another within one dwelling;	B3 8.13 (V2)

- every wall separating a flat or maisonette from any other part of the building;
- every wall enclosing a refuse storage chamber.

Every compartment floor should: B3 5.6 (V1)

- form a complete barrier to fire between the com-   B3 8.20 (V2)
  partments they separate; and
- have the appropriate fire resistance as indicated in
  Appendix A of Approved Document B, Tables A1
  and A2.

Where a compartment wall or compartment floor   B3 5.9 (V1)
meets another compartment wall, or an external wall,   B3 8.25 (V2)
the junction should maintain the fire resistance of the
compartmentation.

Junctions between a compartment floor and an   B2 8.26 (V2)
external wall that has no fire resistance (e.g. a curtain
wall) should be restrained at floor level to reduce
the movement of the wall away from the floor when
exposed to fire.

Compartment walls should be able to accommodate   B2 8.27 (V2)
the predicted deflection of the floor above by either:

- having a suitable head detail between the wall and
  the floor, that can deform but maintain integrity
  when exposed to a fire; or
- the wall may be designed to resist the additional
  vertical load from the floor above as it sags under
  fire conditions and thus maintain integrity.

Underfloor voids

Extensive cavities in floor voids should be subdivided with cavity   B3
barriers.

6.37.3 Ceilings

If there is an existing lath and plaster ceiling it should be retained as long as it satisfies Building Regulation Part B – Fire safety.

If the existing ceiling is not lath and plaster, it should be upgraded to provide at least two layers of plasterboard, with joints staggered and a total mass per unit area of $20\,kg/m^2$.

Note: Care should be taken at the design stage to ensure that adequate ceiling height is available in all rooms to be treated.

Ceiling linings

To inhibit the spread of fire within the building, ceiling internal linings shall:

*   adequately resist the spread of flame over their surfaces; and
*   have, if ignited, a rate of heat release or a rate of fire growth which is reasonable in the circumstances.

Note: Flame spread over wall or ceiling surfaces is controlled by ensuring that the lining materials or products meet given performance levels, which are measured in terms of performance with reference to Tables A1 and A3 of Part B.

6.37.4 Corridor walls and doors

A separating wall should be used between the new dwellings (and/or flats) and the adjacent corridor in order to control flanking transmission and provide the required amount of sound insulation.

Note: It is likely that the sound insulation will be reduced by the presence of a door.

In particular, measures should be taken to ensure that any door:

• has good perimeter sealing (including the threshold where practical);	E4.20
• has a minimum mass per unit area of 25 kg/m^2 or a minimum sound reduction index of 29 dB R_W (measured according to BS EN ISO 1403:1995 and rated according to BS EN ISO 7171:1997);	
• satisfies the requirements of Building Regulation Part B – Fire safety.	E4.20

Noisy parts of the building should preferably have a lobby, double door or high-performance doorset to contain the noise.

Note: These facilities should also comply with Building Regulation Part M – *Access to and Use of Buildings*.

6.37.5 Stairs

If stairs form a separating function, they are subject to the same sound insulation requirements as floors. In all cases, the resistance to airborne sound depends mainly on:

- the mass of the stair;
- the mass and isolation of any independent ceiling;
- the airtightness of any cupboard or enclosure under the stairs;
- the staircovering (which reduces impact sound at source).

The following wall treatment (commonly referred to as *Stair treatment 1: stair covering and independent ceiling with absorbent material)* is recommended.
It should be noted that:

The soft covering should be:	E4.37

- at least 6 mm thick;
- laid over the stair treads;
- be securely fixed (e.g. glued) so it does not become a safety hazard.

If there is a cupboard under all, or part, of the stair: E4.37

- the underside of the stair within the cupboard should be lined with plasterboard (minimum mass per unit area 10 kg/m^2) together with an absorbent layer of mineral wool (minimum density 10 kg/m^3);
- the cupboard walls should be built from two layers of plasterboard (or equivalent), with each sheet having a minimum mass per unit area of 10 kg/m^2;
- a small, heavy, well-fitted door should be fitted to the cupboard.

If there is no cupboard under the stair, an independent ceiling   E4.37
should be constructed below the stair (see floor treatment 1).

Where a staircase performs a separating function, it shall   E4.38
conform to Building Regulation Part B – Fire safety.

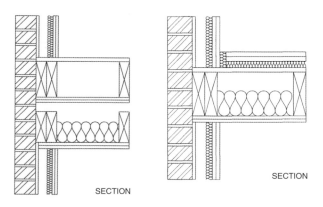

Figure 6.235 Stair treatment 1: stair covering and independent ceiling with absorbent material.

6.37.6 Junction requirements for material change of use

There are three recommended types of junctions that can be made:

Junctions with abutting construction

The perimeter of any new ceiling should be sealed with tape or caulked with sealant.	E4.41
For floating floors:	E4.39
• the resilient layer shall be carried up at all room edges to isolate the floating layer from the wall surface;	
• a 5 mm gap, filled with flexible sealant, should be left between the skirting and floating layer.	E4.40

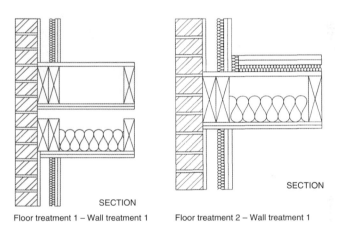

Floor treatment 1 – Wall treatment 1   Floor treatment 2 – Wall treatment 1

Figure 6.236 Junctions with abutting construction.

Junctions with external or load-bearing walls

If the adjoining masonry wall has a mass per unit area less than $375\,kg/m^2$, the walls should be lined with either:	
• an independent layer of plasterboard; or	E4.43a
• a laminate of plasterboard and mineral wool.	E4.43b

Note: Specialist advice may be needed on the diagnosis and control of flanking transmission.

Junctions with floor penetrations

Piped services (excluding gas pipes) and ducts which pass through separating floors in conversions should be surrounded with sound-absorbent material for their full height and enclosed in a duct above and below the floor.	E4.45
The joint between casings and ceiling should be sealed with tape, or a nominal 5 mm gap, sealed with sealant, should be left between the casing and any floating layer.	E4.46 E4.46
Pipes and ducts that penetrate a floor separating habitable rooms in different flats should be enclosed for their full height in each flat.	E4.46
The enclosure should be constructed of material having a mass per unit area of at least 15 kg/m².	E4.47
Either the enclosure should be lined or the duct or pipe within the enclosure should be wrapped with 25 mm unfaced mineral wool.	E4.48
The enclosure may go down to the floor base if floor treatment 2 is used, but ensure isolation from the floating layer.	E4.49
Penetrations through a separating floor by ducts and pipes should have fire protection to satisfy Building Regulation Part B – Fire safety.	E4.50
Fire-stopping should be flexible and be capable of preventing a rigid contact between the pipe and floor.	E4.50
Gas pipes may be contained in a separate (ventilated) duct or can remain unenclosed.	E3.120
If a gas service is installed it shall comply with the Gas Safety (Installation and Use) Regulations 1998 (SI 1998 No. 2451).	E3.120

 Note: All **these** facilities should also comply with Building Regulation Part M – *Access to and Use of Buildings*.

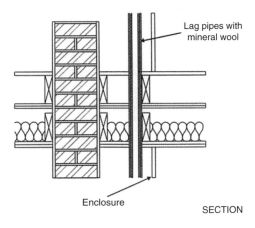

Figure 6.237 Floor penetrations.

Appendix A_____

Access and facilities for disabled people

(A copy of this Appendix is available on www.routledge.com/books/details/ 9780415809696/)

A.1 Background

During 2002/3, Approved Document M was thoroughly overhauled and restructured in order to meet the changed requirements of the Disability Discrimination Act 1995.

This major rewrite of Part M became effective on 1 May 2004 and was as a result of amendments made to Section 6 of the Disability Discrimination Act 1995 (DDA) which previously stated that 'reasonable adjustments to physical features of premises . . . shall be made . . . in certain circumstances' and 'shall apply to all employers with 15 or more employees'. As such, an employer with only a few employees (as well as occupations such as police, firefighters and prison officers) was not required to alter the physical characteristics of a building that met existing requirements at the time the building works were carried out and **still** continued to meet those particular requirements.

From 1 October 2004, however, under the Disability Discrimination Act 1995 (Amendment) Regulations 2003 (SI 2003/1673), this exemption ceases.

According to the Department of Works and Pensions, it is estimated that the new requirements will affect almost one and a half million private service and over 100,000 public service providers.

If this situation affects you, then it would be best to seek advice from either the Social Services or the Local Planning Officer.

In a similar manner, up to 30 September 2004, there is no requirement for service providers to make any adjustments to the physical features of their premises.

From 1 October 2004, however, 'all those who provide services to the public, irrespective of their size' are required 'to take reasonable steps to remove, alter or provide a reasonable means of avoiding a physical feature of their premises, which makes it unreasonably difficult or impossible for disabled people to make use of their services'.

In rewriting Part M, the drafting committee also took into consideration BS 8300:2001 (which provides guidance on the design of domestic and non-domestic buildings and their approaches), recent ergonomic research conducted in support of this standard and the need for all people (no matter their disability) to have access and be able to use buildings and their facilities.

 Note: BS 8300:2001 (*Design of buildings and their approaches to meet the needs of disabled people – Code of Practice*) supersedes BS 5619:1978 and BS 5810:1979.

These amendments have naturally resulted in Part M being completely over-hauled and it now covers:

- the conversion of a building for use as a shop being redefined as a 'material change of use';
- amendments to omit specific references to (and a definition of) disabled people;
- expansion of the terms to include parents with children, elderly people and people with all types of disabilities (e.g. mobility, sight and hearing etc.);
- the use of a building to disabled people as residents, visitors, spectators, customers or employees, or participants in sports events, performances and conferences (which resulted in amendments being made to M1 (accessibility), M2 (sanitary accommodation) and M3 (audience and/or spectator seating).

The current edition, therefore, no longer primarily concentrates on wheelchair users, but includes people using walking aids, people with impaired sight (and other mobility and sensory problems), mothers with prams as well as people with luggage etc.

A.1.1 Benefits resulting from these changes

In producing these changes, the government considers that the following benefits to society will result:

Benefits to disabled people:

- improved legal rights of access to goods, facilities and services;
- better opportunity to play as full a role as possible in the economy and in society;
- reduced financial cost of injuries resulting from negotiating inadequately accessible premises;
- reduced travelling costs as more services closer to home/work become accessible;
- wider range of facilities disabled people can enjoy with carers, friends and family.

Benefits to other people with difficulties such as:

- people with young children;
- elderly people;

- people encumbered with luggage, shopping bags etc.;
- people with temporary impairments (e.g. people with broken limbs).

Benefits to business/service providers:

- better public image;
- easier movement of goods;
- reduced need for home support services;
- reduction in accidents where there are lifts, safer stairs, handrails, better lighting, fewer obstructions and even floor surfaces;
- increased tourism where there is sufficient accommodation for wheelchair users.

A.2 The requirements

M1 – Access and use

Reasonable provision shall be made for people to:

(a)   *gain access to; and*

(b)   *use*

the buildings and its facilities.

(Approved Document M1)

M2 – Access to extensions of buildings other than dwellings

Suitable independent access shall be provided in any building that is to be extended. Reasonable provision shall be made within the extension for sanitary convenience.

(Approved Document M2)

M3 – Sanitary conveniences in extensions to buildings other than dwellings

If sanitary conveniences are provided in any building that is to be extended, reasonable provision shall be made within the extension for sanitary conveniences.

(Approved Document M3)

M4 – Sanitary conveniences in dwellings

Reasonable provision shall be made in the entrance storey for sanitary conveniences, or where the entrance contains no habitable rooms, reasonable provision for sanitary convenience shall be made in either the entrance storey or principal storey.

(Approved Document M4)

A.2.1 Requirement in a nutshell

In addition to the requirements of the *Disability Discrimination Act 1995 (Amendment) Regulations 2003*, precautions need to be taken to ensure that:

- new non-domestic buildings and/or dwellings (e.g. houses and flats used for student living accommodation etc.);
- extensions to existing non-domestic buildings;
- non-domestic buildings that have been subject to a material change of use (e.g. so that they become a hotel, boarding house, institution, public building or shop);

are capable of allowing people, regardless of their disability, age or gender, to:

- gain access to buildings;
- gain access within buildings;
- be able to use the facilities of the buildings (both as visitors and as people who live or work in them).

Note: The requirements of this part do **not** apply to:

(a)   an extension of, or material alteration of, a (domestic) dwelling; or
(b)   any part of a building which is used solely to enable the building or service (or fitting in the building) to be inspected, repaired or maintained.

The requirements are also limited if:

- a building is listed as being of architectural or historic interest;
- the cost of providing a fully accessible route is disproportionate to the cost of the intended facility;
- the physical restraints imposed by the building make full compliance impossible or impracticable.

A.2.2 Changes to Approved Document M

Extensions to non-domestic buildings

With the new revision of Part M, an extension to a non-domestic building should now be treated in the same manner as a new building for compliance with Approved Document M which means that:

- there must be 'suitable independent access to the extension where reasonably practicable';
- if a building is to be extended, 'reasonable provision must be made within the extension for sanitary conveniences'.

Note: This requirement does not apply if it is possible for people using the extension to gain access to and be able to use sanitary conveniences in the existing building.

A.2.3 Material alterations of non-domestic buildings

Under Regulation 4, where an alteration of a non-domestic building is a material alteration:

- the work itself must comply, where relevant, with Requirement M1;
- reasonable provision must be made for people to gain access to and use new or altered sanitary conveniences.

A.2.4 Material changes of use

Where there is a material change of use of a whole building to a hotel, boarding house, institution, public building or a shop (restaurant, bar or public house) the building must be upgraded, if necessary, so as to comply with M1 (Access and use).

If an existing building does undergo a change of use so that part of it can be used as a hotel, boarding house, institution, public building or a shop, the work being carried out must ensure that:

- people can gain access from the site boundary and any on-site car parking space;
- sanitary conveniences are provided in that part of the building or it is possible for people (no matter their disability) to use sanitary conveniences elsewhere in the building.

A.2.5 What requirements apply

Buildings other than dwellings

1. Regardless of disability, age or gender it should be possible for people:

    - to reach the principal entrance to the building from the site boundary and from car parking within the site and from other buildings on the same site (such as a university campus, a school or a hospital);
    - to have access into and within any storey of the building;
    - to have access into and use of the building's facilities.

2. The structure and amenities of a building should not constitute a hazard to users (especially people with impaired sight).
3. Suitable accommodation should be made available for people in wheelchairs (or people with other disabilities), in audience or spectator seating.
4. People with a hearing or sight impairment should be provided with some form of aid to communication in auditoria, meeting rooms, reception areas, ticket offices and at information points.
5. Sanitary accommodation should be available for **all** users of the building.

Dwellings

People, including disabled people, should be able:

- to reach the principal, or suitable alternative, entrance to the dwelling from the point of access;
- to gain access into and within the principal storey of the dwelling;
- to gain access to sanitary conveniences at no higher storey than the principal storey.

A.2.6 Educational establishments

From 1 April 2001, maintained schools ceased to have exemption from the Building Regulations and school-specific standards have now been incorporated into the latest editions of Approved Documents.

Purpose-built student living accommodation (including flats) should thus be treated as hotel/motel accommodation in respect of space requirements and internal facilities.

A.2.7 Access statements

To assist building control bodies it is recommended that an 'Access Statement' is also provided when plans are deposited, a building notice is given, or details of a project are provided to an approved inspector.

 Note: A building control file should also be prepared for all new buildings, changes of use and where extensive alterations are being made to existing buildings.

In its simplest form, an Access Statement should show where an applicant wishes to deviate from the guidance in Approved Document M, either to:

- make use of new technologies (e.g. infrared activated controls);
- provide a more convenient solution; or
- address the constraints of an existing building.

The Access Statement should include:

- the reasons for departing from the guidance;
- the rationale for the design approach adopted;
- constraints imposed by the existing structure and its immediate environment (why it is not practicable to adjust the existing entrance or provide a suitable new entrance);
- convincing arguments that an alternative solution will achieve the same, a better, or a more convenient outcome (e.g. why a fully compliant independent access is considered impracticable);
- evidence (e.g. current validated research) to support the design approach;

- the identification of buildings (or particular parts of buildings) where access needs to be restricted (e.g. processes that are carried out which might create hazards for children, disabled people or frail, elderly people).

 Note: Further guidance on Access Statements is available on the Disability Rights Commission's website at www.drc-gb.org.

A.3 Meeting the requirements

A.3.1 Access

Access (i.e. approach, entry or exit) to a building is frequently a problem for wheelchair users, people who need to use walking aids, people with impaired sight and mothers with prams etc. In designing the approach to a building (and routes between buildings within a complex) the following should, therefore, always be taken into consideration:

- changes in level between the entrance storey and the site entry point should be minimized;
- access routes should be wide enough to let people pass each other;
- potential hazards (e.g. windows from adjacent buildings opening onto access routes) should be avoided.

For all new buildings, therefore, the primary aim should be to make it reasonably possible for a disabled person to approach and gain access into the dwelling from the point of alighting from a vehicle (which may be within or outside the plot). In particular the following should be observed.

Access to a building

Approach

- the principal entrance should be accessible to disabled people;
- access from the boundary of the site to the principal entrance should be level;
- access from car parking designated for disabled people to the principal entrance should be level;
- risks to people when entering the building should be minimal;
- access routes should be wide enough to let people pass each other;
- the route to the principal entrance should be clearly identified and well lit;
- uncontrolled vehicular crossing points should be identified.

Approach gradients

- should ideally be no steeper than 1:60 along its whole length;
- cross-fall gradients should be no steeper than 1:40.

Building perimeters

- the perimeter of the building should be well lit.

Surface of access routes

The surface of access routes should:

- allow people to travel along them easily, without excessive effort and without the risk of tripping or falling;
- be at least 1.5 m wide;
- be firm, durable and slip resistant;
- have undulations not exceeding 3 mm;
- be made of the same material and similar frictional characteristics.

Joints

Joints should:

- be filled flush or (if recessed) no deeper than 5 mm;
- be no wider than 10 mm or (if unfilled) no wider than 5 mm;
- have a difference in level (at joints between paving units) no greater than 5 mm.

Passing places

Passing places should be:

- free of obstructions to a height of 2.1 m;
- at least 1.8 m wide and at least 2 m long.

On-site car parking and setting down – parking bays

At least one parking bay designated for disabled people should be provided as close as possible to the principal entrance of the building and these should:

- be clearly signposted;
- have the setting down point located on firm, level ground;
- have the surface of the parking bay firm, durable and slip resistant;
- have ticket machines located nearby and positioned so that a person in a wheelchair (or a person of short stature) is able to reach the controls;
- not allow the plinth of the ticket machine to project in front of the face of the machine.

Ramped access

- the approach should be clearly signposted;
- the going should be no greater than 10 m;
- the rise should be no more than 500 mm;
- the surface width should be at least 1.5 m;
- gradients should be as shallow as practicable;

- the ramp surface should be slip resistant and of a contrasting colour to that of the landings;
- frictional characteristics of ramp and landing surfaces should be similar;
- landings at the foot and head of a ramp should be at least 1.2 m long and clear of any obstructions;
- intermediate landings should be at least 1.5 m long and clear of obstructions;
- all landings should be:

  - level;
  - have a maximum gradient of 1:60 along their length;
  - have a maximum cross fall gradient of 1:40;
  - have a handrail on both sides.

Stepped access

A stepped access should:

- have a level landing at the top and bottom of each flight which is at least 1200 mm long and unobstructed;
- have a corduroy hazard warning surface at top and bottom landings of a series of flights;
- ensure that any side accesses (onto intermediate landings) include a deep corduroy hazard warning surface;
- have no doors that swing across landings;
- have the rise of a flight between landings contain no more than 18 risers;
- have nosings (for the tread and the riser) of a contrasting material;
- not allow step nosings to project over the tread below by more than 25 mm;
- ensure that the rise and going of each step is consistent throughout a flight;
- include a continuous handrail on each side of a flight and landings;
- ensure that tread materials do not present a slip hazard.

Handrails to external stepped and ramped access

Handrails to external stepped or ramped access should:

- be continuous across flights and landings;
- extend at least 300 mm horizontally beyond the top and bottom of a ramped access;
- not project into an access route;
- contrast visually with the background;
- have a slip resistant surface which is **not** cold to the touch;
- terminate in such a way that reduces the risk of clothing being caught;
- either be circular (with a diameter of between 40 and 45 mm) or oval with a width of 50 mm;
- not protrude more than 100 mm into the surface width of the ramped or stepped access;
- have a clearance of between 60 and 75 mm between the handrail and any adjacent wall surface;

- have a clearance of at least 50 mm between a cranked support and the underside of the handrail;
- ensure that its inner face is located no more than 50 mm beyond the surface width of the ramped or stepped access;
- be spaced away from the wall and rigidly supported in a way that avoids impeding finger grip;
- be set at heights that are convenient for all users of the building.

Hazards on access routes

- no 'feature' of a building (e.g. windows and doors) should obstruct an access route;
- areas below stairs or ramps with a soffit less than 2.1 m above ground level should be protected by guarding and low-level cane detection;
- any feature projecting more than 100 mm onto an access route should be protected by guarding that includes a kerb (or other solid barrier) that can be detected using a cane.

Accessible entrances

Accessible entrances should:

- be clearly signposted;
- be easily identifiable;
- not present a hazard for visually impaired people;
- have a level landing at least 1500 × 1500 mm, clear of any door swings, immediately in front of the entrance;
- avoid raised thresholds;
- ensure that all door entry systems are accessible to deaf and hard of hearing people, plus people who cannot speak;
- not have internal floor surface material (e.g. coir matting) adjacent to the threshold;
- not have changes in floor materials that could create a potential trip hazard;
- have the route from the exterior across the threshold weather protected.

Doors to accessible entrances

Doors to the principal (or alternative accessible) entrance should:

- be easily reached (particularly for wheelchair users and people with limited physical dexterity);
- be capable of being held closed when not in use;
- not require a great deal of force to open or shut a door;
- be wide enough to allow unrestricted passage for a variety of users, including wheelchair users, people carrying luggage, people with assistance dogs, and parents with pushchairs and small children;
- enable people to see other people approaching from the opposite direction;
- incorporate vision panels if the door leaves, and side panels are wider than 450 mm.

Entrance lobbies

Entrance lobbies should:

- be large enough to allow a wheelchair user or a person pushing a pram to move clear of one door before opening the second door;
- be capable of accommodating a companion helping a wheelchair user to open doors and guide the wheelchair through.

Within the lobby:

- glazing should not create distracting reflections;
- floor surface materials should not impede the movement of wheelchairs etc.;
- changes in floor materials should not create a potential trip hazard;
- the floor surface should assist in removing rainwater from shoes and wheelchairs;
- any columns and ducting etc. that projects into the lobby by more than 100 mm should be protected by a visually contrasting guard rail.

Manually operated non-powered entrance doors

- the opening force at the leading edge of the door should be no greater than 20 N;
- a space alongside the leading edge of a door should be provided to enable a wheelchair user to reach and grip the door handle;
- door opening furniture should:

  - be easy to operate by people with limited manual dexterity;
  - be capable of being operated with one hand using a closed fist (e.g. a lever handle);
  - contrast visually with the surface of the door;
  - not be cold to the touch;

- there should be an unobstructed space of at least 300 mm on the pull side of the door between the leading edge of the door and any return wall.

Power-operated entrance doors

Power-operated entrance doors should:

- have a sliding, swinging or folding action that can be operated manually, remotely or automatically;
- open towards people approaching the doors;
- provide visual and audible warnings that they are operating (or about to operate);
- incorporate automatic sensors to ensure that they open early enough (and stay open long enough) to permit safe entry and exit;
- incorporate a safety stop that is activated if the doors begin to close when a person is passing through;

- revert to manual control (or fail-safe) in the open position in the event of a power failure;
- when open, not project into any adjacent access route;
- ensure that its manual controls:

  - are located between 750 mm and 1000 mm above floor level;
  - are operable with a closed fist;

- be set back 1400 mm from the leading edge of the door when fully open;
- be clearly distinguishable against the background;
- contrast visually with the background.

Glass entrance doors and glazed screens

- the presence of the door should be apparent not only when it is shut but also when it is open;
- glass entrance doors and glazed screens should:

  - be clearly marked (i.e. with a logo or sign) on the glass at two levels;
  - be provided with a high contrast strip at the top and side when adjacent to (or forming part of) a glazed screen;
  - be capable of being held open;
  - be protected by guarding to prevent the leading edge from becoming a possible hazard.

Access to a dwelling

Approach

For dwellings:

- an external door providing access for disabled people should have a minimum clear opening width of 775 mm;
- the door opening width should be sufficient to enable a wheelchair user to manoeuvre into the dwelling;
- the approach to a principal entrance should be reasonably level;
- wheelchair users (having approached the entrance) should be able to gain access into the dwelling-house and/or entrance level of flats;
- a suitable approach should be provided from the point of access to the entrance of the dwelling;
- the whole, or part, of the approach may be a driveway;
- the approach should:

  - not have crossfalls greater than 1 in 40;
  - be safe and convenient for disabled people as is reasonably possible;
  - ideally be level or ramped;
  - be firm enough to support the weight of the user and their wheelchair;
  - be smooth enough to permit easy manoeuvre;
  - not be made up of loose-laid materials (such as gravel and shingle);
  - take account of the needs of stick and crutch users.

Level approach

A 'level' approach should:

- be no steeper than 1 in 20;
- have a firm and even surface;
- have a width not less than 900 mm.

Ramped approach

If a plot gradient exceeds 1 in 20 a ramped approach may be provided, in which case:

- the surface should be firm and even;
- the flights should have unobstructed widths of at least 900 mm;
- individual flights should be no longer than 10 m (for gradients not steeper than 1 in 15) or 5 m for gradients not steeper than 1 in 12;
- it should have top and bottom landings (and intermediate landings if necessary) not less than 1.2 m in length – exclusive of the swing of any door or gate which opens onto it.

Stepped approach

A stepped approach should:

- be used if the plot gradient is greater than 1 in 15;
- have flights with an unobstructed width of at least 900 mm;
- have a flight rise of not more than 1.8 m;
- have a top and bottom (and if necessary intermediate landings) not less than 900 mm in length;
- have steps with suitable tread nosing profiles and a uniform rise between 75 mm and 150 mm;
- ensure that the going of each step is not less than 280 mm.

Access into the dwelling

- an accessible threshold should be provided into the entrance;
- the point of access should be reasonably level;
- where the approach to the entrance consists of a level or ramped approach, an accessible threshold at the entrance should be provided.

Vertical circulation within the building

For all buildings, a passenger lift is considered the most suitable form of access for people moving from one storey of the building to another. The actual size of the lift should be chosen to suit the anticipated density of use of the building and the needs of disabled users, but in all cases should **not** be less than 1100 mm wide and 1400 mm deep. In some circumstances a lifting platform may be provided but only to transfer wheelchair users and people with impaired mobility (plus their companions) vertically between levels or storeys.

Lifting devices

- wherever possible a passenger lift (or a lifting platform) serving all storeys should be provided;
- the location of lifting devices should be clearly visible from the building entrance;
- signs should be available at each landing to identify the floor reached by the lifting device;
- in addition to the lifting device, internal stairs (designed to suit ambulant disabled people and those with impaired sight) should always be provided.

General requirements for lifting devices

- an unobstructed manoeuvring space should be available in front of each lifting device;
- landing call buttons and lifting device control button symbols:

  - should contrast visually with the surrounding face plate;
  - should be raised to facilitate tactile reading;
  - should be accessible for wheelchair users;

- the floor of the lifting device should be of a light colour;
- a handrail should be provided on at least one wall;
- an emergency communication system should be fitted.

Passenger lifts

Passenger lifts should:

- be accessible from the remainder of the storey;
- have power-operated horizontal sliding doors with an effective clear width of at least 800 mm;
- have doors fitted with timing devices (and reopening activators);
- locate the car controls and landing call buttons within easy reach of wheelchair users;
- be fitted with lift landing and car doors that are visually distinguishable from the adjoining walls;
- include (in the lift car **and** the lift lobby) audible and visual information that a lift has arrived, which floor it has reached and where in a bank of lifts it is located;
- ensure that all glass areas can be easily identified by people with impaired vision;
- conform to BS 5588-8 if the lift is to be used to evacuate disabled people in an emergency;
- not have visually and acoustically reflective wall surfaces.

Lifting platforms

Lifting platforms should:

- restrict the vertical travel distance to no more than 2 m if there is no lift-way enclosure and/or floor penetration;
- restrict the rated speed of the platform so that it does not exceed 0.15 m/s;
- locate their controls between 800 mm and 1100 mm from the floor of the lifting platform and at least 400 mm from any return wall;
- locate all landing call buttons within easy reach of wheelchair users;
- have continuous pressure controls (e.g. push buttons);
- have doors with an effective clear width of at least 800 mm;
- be fitted with clear instructions for use;
- have their entrances accessible from the remainder of the storey;
- have doors that are visually distinguishable from adjoining walls;
- have an audible and visual announcement of platform arrival and level reached;
- ensure that all areas of glass can be identified by people with impaired vision.

Wheelchair platform stairlifts

- wheelchair platform stairlifts are only intended for the transportation of wheelchair users and should only be considered for conversions and alterations where it is not practicable to install a conventional passenger lift or a lifting platform;
- the speed of the platform should not exceed 0.15 m/s;
- continuous pressure controls (e.g. joystick) should be provided;
- the platform should have minimum clear dimensions of 800 mm wide and 1250 mm deep;
- wheelchair platform stairlifts should:

  - be fitted with clear instructions for use;
  - provide a clear width of at least 800 mm;
  - not be installed where their operation restricts the safe use of the stair by other people.

Internal steps, stairs and ramps

Stepped access

- a stepped access should have a level landing (1200 mm long) at the top and bottom of each flight;
- doors should not be allowed to swing across landings;
- the surface width of flights between enclosing walls should not be less than 1.2 m;
- there should be no single steps;
- tread and riser nosings should be 55 mm wide and of a contrasting material;
- step nosings should not project over the tread below by more than 25 mm;

- the rise and going of each step should be consistent (e.g. between 150 mm and 170 mm) throughout a flight;
- the rise of each step should be between 150 mm and 170 mm;
- the going of each step should be at least 250 mm;
- rises should not be open;
- there should be a continuous handrail on each side of a flight and landings;
- flights between landings should contain no more than 12 risers;
- areas below stairs or ramps with a soffit less than 2.1 m above ground level should be protected by guarding and low-level cane detection;
- features projecting more than 100 mm onto an access route should be protected by guarding that includes a kerb.

Internal ramps

- if the change in level is no greater than 300 mm, a ramp should be provided instead of a single step;
- if the change in level is greater than 300 mm, at least two clearly signposted steps should be provided;
- the approach should be clearly signposted;
- the going should be no greater than 10 m;
- the rise should be no more than 500 mm;
- the ramp surface should be slip resistant and of a contrasting colour to that of the landings;
- the frictional characteristics of ramp and landing surfaces should be similar;
- landings at the foot and head of a ramp should be at least 1.2 m long and clear of any obstructions;
- intermediate landings should be at least 1.5 m long and clear of obstructions;
- all landings should be level;
- there should be a handrail on both sides;
- there should be a visually contrasting kerb on the open side of the ramp or landing;
- areas below stairs should be protected by guarding and/or low-level cane detection;
- no feature should project more than 100 mm onto an access route unless protected by a guard;
- gradients should be as shallow as practicable.

Handrails to internal steps, stairs and ramps

Handrails to external stepped or ramped access should:

- be continuous across flights and landings;
- extend at least 300 mm horizontally beyond the top and bottom of a ramped access;
- not project into an access route;
- contrast visually with the background;
- have a slip resistant surface which is not cold to the touch;

- terminate in such a way that reduces the risk of clothing being caught;
- either be circular or oval;
- not protrude more than 100 mm into the surface width of the ramped or stepped access;
- have a clearance of between 60 and 75 mm between the handrail and any adjacent wall surface;
- have a clearance of at least 50 mm between a cranked support and the underside of the handrail;
- ensure that its inner face is located no more than 50 mm beyond the surface width of the ramped or stepped access;
- be spaced away from the wall and rigidly supported in a way that avoids impeding finger grip;
- be set at heights that are convenient for all users of the building.

Entrance hall and reception area

- if there is a reception point it should be:
  - easily accessible and convenient to use;
  - located away from the principal entrance;
  - easily identifiable from the entrance doors or lobby;
  - designed to accommodate both standing and seated visitors;
  - provided with a hearing enhancement system;
- relevant information about the building should be clearly available from noticeboards and signs;
- the floor surface should be slip resistant;
- the approach should:
  - be direct and free from obstructions;
  - allow space for wheelchair users to gain access;
- there should be a clear manoeuvring space in front of the reception desk (at least 1200 mm deep and 1800 mm wide);
- if there is a knee recess it should be at least 500 mm.

Provision of toilet accommodation

General

- wheelchair-accessible unisex toilets should always be provided in addition to any wheelchair-accessible accommodation in separate-sex toilet washrooms;
- where sanitary facilities are provided in a building, at least one wheelchair-accessible unisex toilet should be available;
- at least one WC cubicle should be in separate-sex toilet accommodation;
- if there is only space for **one** toilet in a building:
  - then it should be a wheelchair-accessible unisex type;
  - its width should be increased from 1.5 m to 2 m;
  - it should include a standing height washbasin, in addition to the finger rinse basin associated with the WC.

Sanitary accommodation generally

Doors

- doors to WC cubicles and wheelchair-accessible unisex toilets should:
  - (ideally) open outwards;
  - be operable by people with limited strength or manual dexterity;
  - be capable of being opened if a person has collapsed against them while inside the cubicle;
- doors to wheelchair-accessible unisex toilets/changing rooms or shower rooms should:
  - be fitted with light action privacy bolts;
  - be capable of being opened using a force no greater than 20 N;
  - have an emergency release mechanism so that they are capable of being opened outwards (from the outside) in case of emergency;
- door opening furniture should:
  - be easy to operate by people with limited manual dexterity;
  - be easy to operate with one hand using a closed fist (e.g. a lever handle);
  - contrast visually with the surface of the door;
- doors when open should not obstruct emergency escape routes.

Sanitary fittings

- surface finish of sanitary fixtures/fittings should contrast visually with wall and floor finishes;
- taps should be operable by people with limited strength and/or manual dexterity;
- bath and washbasin taps should either be controlled automatically or capable of being operated using a closed fist, e.g. by lever action.

Alarms

- fire alarms should emit an audio and visual signal to warn occupants with hearing or visual impairments;
- emergency assistance alarm systems should have:
  - visual and audible indicators to confirm that an emergency call has been received;
  - a reset control reachable from a wheelchair, WC, or from a shower/changing seat.

Outlets, controls and switches

- all controls and switches should:
  - be easy to operate, visible and free from obstruction;
  - be located between 750 mm and 1200 mm above the floor;
  - not require the simultaneous use of both hands (unless necessary for safety reasons) to operate;

- light switches should:
    - have large push pads;
    - align horizontally with door handles;
    - be within the 900 to 1100 mm from the entrance door opening;
- mains/circuit isolator switches and switched socket outlets should clearly indicate whether they are 'ON';
- individual switches on panels and on multiple socket outlets should be well separated;
- controls that need close vision (e.g. thermostats) should be located between 1200 mm and 1400 mm above the floor;
- **emergency alarm pull cords** should:
    - be coloured red;
    - be located as close to a wall as possible;
    - have two red 50 mm diameter bangles;
- front plates should contrast visually with their backgrounds;
- heat emitters should either be screened or have their exposed surfaces kept at a temperature below 43° C;
- where possible, light switches with large push pads should be used in preference to pull cords;
- the colours red and green should not be used in combination as indicators of 'ON' and 'OFF' for switches and controls.

Wheelchair-accessible unisex toilets

General

- where sanitary facilities are provided in a building, at least one wheel-chair-accessible unisex toilet should be available;
- wheelchair-accessible unisex toilets should:
    - be located as close as possible to the entrance and/or waiting area of the building;
    - not be located in a way that compromises the privacy of users;
    - be located in a similar position on each floor of a multi-storey building;
    - allow for right- and left-hand transfer on alternate floors;
    - be located on accessible routes that are direct and obstruction-free;
    - always be provided in addition to any wheelchair-accessible accommodation in separate-sex toilet washrooms;
    - not be used for baby changing.

Accessibility

- the approach to a unisex toilet should be separate to other sanitary accommodation;
- wheelchair users should:
    - not have to travel more than 40 m on the same floor to reach a unisex toilet;

- not have to travel more than combined horizontal distance where the unisex toilet accommodation is on another floor of the building (accessible by passenger lift);
- be able to approach, transfer to, and use the sanitary facilities provided within a building.

Doors

Doors should:

- preferably open outwards;
- be fitted with a horizontal closing bar fixed to the inside face.

Support rails

- a drop-down and/or wall-mounted grab rail should be provided.

Heights and arrangements

- the space provided for manoeuvring should enable wheelchair users to adopt various transfer techniques that allow independent or assisted use;
- the transfer space alongside the WC should be kept clear to the back wall;
- the relationship of the WC to the finger rinse basin and other accessories should allow a person to wash and dry hands while seated on the WC;
- heat emitters (if located) should not restrict:

  - the minimum clear wheelchair manoeuvring space;
  - the transfer space beside the wheelchair.

Emergency assistance

- emergency assistance alarm systems should have:

  - an outside emergency assistance call signal;
  - visual and audible indicators to confirm that an emergency call has been received;
  - a reset control reachable from a wheelchair, WC, or from a shower/changing seat;
  - a signal that is distinguishable visually and audibly from the fire alarms provided;

- **emergency assistance pull cords** should:

  - be easily identifiable;
  - be reachable from the WC and from the floor close to the WC;
  - be coloured red;
  - be located as close to a wall as possible;
  - have two red 50 mm diameter bangles.

Toilets in separate-sex washrooms

General

- there should be at least the same number of WCs (for women) as urinals (for men) and vice versa;
- ambulant disabled people should have the opportunity to use a WC compartment within any separate-sex toilet washroom;
- wheelchair-accessible compartments shall have the same layout and fittings as the unisex toilet;
- where a separate-sex toilet washroom can be accessed by wheelchair users, it should be possible for them to use both a urinal and a washbasin at a lower height than is provided for other users;
- where possible, a low-level urinal for children should be provided in male washrooms;
- separate-sex toilet washrooms above a certain size should include an enlarged WC cubicle for use by people who need extra space, e.g. parents with children and babies, people carrying luggage and also ambulant disabled people.

Doors

- doors to compartments for ambulant disabled people should:

  - preferably open outwards;
  - be fitted with a horizontal closing bar fixed to the inside face.

Accessibility

- the approach to a unisex toilet should be separate to other sanitary accommodation;
- wheelchair users should:

  - not have to travel more than 40 m on the same floor to reach a unisex toilet;
  - not have to travel more than combined horizontal distance where the unisex toilet accommodation is on another floor of the building (accessible by passenger lift);
  - be able to approach, transfer to, and use the sanitary facilities provided within a building.

Heights and arrangements

- compartments used for ambulant disabled people should:

  - be at least 1200 mm wide;
  - include a horizontal grab bar adjacent to the WC;
  - include a vertical grab bar on the rear wall;
  - include space for a shelf and fold-down changing table;

- a wheelchair-accessible washroom (where provided) shall have:

  - at least one washbasin;
  - and, for men, at least one urinal;

- the compartment should:

  - be fitted with support rails;
  - include a minimum activity space to accommodate people who use crutches, or otherwise have impaired leg movements.

Wheelchair-accessible changing and shower facilities

General

- in large building complexes (e.g. retail parks and large sports centres) there should be at least one wheelchair-accessible unisex toilet;
- a combined facility should be divided into distinct 'wet' and 'dry' areas.

For changing and shower facilities

A choice of layouts suitable for left-hand and right-hand transfer should be provided when more than one individual changing compartment or shower compartment is available:

- wall-mounted drop-down support rails and slip-resistant tip-up seats should be provided;
- subdivisions should be provided for communal shower and changing facilities;
- individual self-contained shower and changing facilities should be available in sports amenities;
- emergency assistance alarm systems should be provided which should have:

  - visual and audible indicators to confirm that an emergency call has been received;
  - a reset control reachable from a wheelchair, WC, or from a shower/ changing seat;
  - a signal that is distinguishable visually and audibly from the fire alarm;

- an **emergency assistance pull cord** should be provided which should:

  - be easily identifiable and reachable from the wall-mounted tip-up seat (or from the floor);
  - be located as close to a wall as possible;
  - be coloured **red**;
  - have two red 50 mm diameter bangles;

- facilities for limb storage should be included for the benefit of amputees.

For changing facilities

- the floor should be level and slip resistant when dry or when wet;
- there should be a manoeuvring space 1500 mm deep in front of lockers.

For shower facilities

- if showers are provided in commercial developments for the benefit of staff, at least one wheelchair-accessible shower compartment should be made available;
- a shower curtain should be provided;
- a shelf should be provided for toiletries;
- the floor of the shower and shower area should be slip resistant and self-draining;
- shower controls should be positioned between 750 and 1000 mm above the floor.

For shower facilities incorporating a WC

General

- a choice of left-hand and right-hand transfer layouts should be available when more than one shower area includes a corner WC.

Wheelchair-accessible bathrooms

General

- a choice of layouts suitable for left-hand and right-hand transfer should be provided when more than one bathroom is available;
- the floor should be slip resistant when dry or when wet;
- the bath should be provided with a transfer seat;
- outward opening doors, fitted with a horizontal closing bar fixed to the inside face, should be provided;
- an **emergency assistance pull cord** should be provided which should:

  - be easily identifiable and reachable from the wall-mounted tip-up seat (or from the floor);
  - be located as close to a wall as possible;
  - be coloured red;
  - have two red 50 mm diameter bangles.

WC provision in the entrance storey of the dwelling

- if there is a bathroom in the principal storey, then a WC may be collocated with it;
- the door to the WC compartment should:

  - open outwards;
  - be positioned so as to allow wheelchair users access to it;
  - have a clear opening width;

- the WC compartment should:

  - provide a clear space for wheelchair users to access the WC;
  - position the washbasin so that it does not impede access.

Accessible switches and socket outlets in the dwelling

- switches and socket outlets for lighting and other equipment should be:

  - located so that they are easily reachable;
  - located between 450 mm and 1200 mm from finished floor level.

Facilities in buildings other than dwellings

General

- all floor areas should be accessible;
- in hotels, motels and student accommodation:

  - a proportion of the sleeping accommodation should be designed for wheelchair users;
  - the remainder should include facilities suitable for people with sensory, dexterity or learning difficulties;

- if there is a reception point:

  - it should be easily accessible and convenient to use;
  - information about the building should be clearly available from notice-boards and signs;
  - the floor surface should be slip resistant;

- disabled people should be able to have:

  - a choice of seating location at spectator events;
  - a clear view of the activity taking place (whilst not obstructing the view of others);

- bars and counters in refreshment areas should be at a suitable level for wheelchair users.

Audience and spectator facilities

General

- disabled people should be provided with a selection of spaces into which they can manoeuvre easily and which offer them a clear view of an event – taking particular care that these do not become segregated into 'special areas'.

For audience seating generally

- the route to wheelchair spaces should be accessible to users;
- stepped access routes to audience seating should be provided with fixed handrails.

Handrails to external stepped or ramped access should:

- be continuous across flights and landings;
- extend at least 300 mm horizontally beyond the top and bottom of a ramped access;
- not project into an access route;
- contrast visually with the background;
- have a slip-resistant surface which is not cold to the touch;
- terminate in such a way that reduces the risk of clothing being caught;
- either be circular or oval with a width of 50 mm;
- not protrude more than 100 mm into the surface width;
- have a clearance of between 60 and 75 mm between:

  - the handrail and any adjacent wall surface;
  - a cranked support and the underside of the handrail;

- ensure that its inner face is located no more than 50 mm beyond the surface width of the ramped or stepped access;
- should be spaced away from the wall and rigidly supported;
- should be set at heights that are convenient for all users of the building.

Seating

- some wheelchair spaces should be provided in pairs, with standard seating on at least one side;
- if more than two wheelchair spaces are provided, they should be located so as to give a range of views of the event;
- the minimum clear space for:

  - access to wheelchair spaces should be 900 mm;
  - wheelchair spaces in a parked position should be 900 mm wide by 1400 mm deep;

- the floor of each wheelchair space should be horizontal;
- seats at the ends of rows and next to wheelchair spaces should have detachable (or lift-up) arms.

Lecture/conference facilities

- all people should be able to use presentation facilities;
- people with hearing impairments should be able to participate fully in conferences, committee meetings and study groups;
- acoustic environment should ensure that audible information can be heard clearly;
- artificial lighting should give good colour rendering of all surfaces;
- glare and reflections from shiny surfaces should be avoided;
- uplighters mounted at low or floor level should be avoided;
- wheelchair users should have access to the podium or stage;
- an audible public address system should be supplemented by visual information;

- the availability of an induction loop or infrared hearing enhancement system should be indicated by the standard symbol;
- hearing enhancement systems should be installed:

  - in rooms and spaces designed for meetings, lectures, classes, performances, spectator sports or films;
  - at service or reception counters (especially when situated in noisy areas or behind glazed screens);

- telephones suitable for hearing aid users should be:

  - clearly indicated by the standard ear symbol;
  - incorporate an inductive coupler and volume control;

- text telephones should be clearly indicated by the standard symbol;
- artificial lighting should be designed to be compatible with other electronic and radio frequency installations.

Entertainment, leisure and social facilities

Theatres and cinemas

- in theatres and cinemas special care should be given to the design and location of wheelchair spaces.

Refreshment facilities

- restaurants and bars should be designed so that they can be reached and used by all people independently or with companions;
- all people should have access to:

  - all parts of the facility;
  - staff areas;
  - public areas (e.g. lavatory accommodation, public telephones and external terraces);
  - self-service facilities (when provided);

- all (changes to) floor levels should be accessible;
- raised thresholds should be avoided;
- part of the working surface of a bar or serving counter should:

  - be permanently accessible to wheelchair users;
  - be at a level of not more than 850 mm above the floor;
  - have worktops of shared refreshment facilities (e.g. tea making) 850 mm above the floor with a clear space beneath at least 700 mm above the floor;
  - have basin taps that can either be controlled automatically or be capable of being operated using a closed fist (e.g. by lever action).

Sleeping accommodation

General

- sleeping accommodation in hotels, motels and student accommodation should be convenient for all types of people;
- a proportion of rooms should be available for wheelchair users;
- wheelchair users should be able to:

  - reach all the facilities available within the building;
  - manoeuvre around and use the facilities in the room;
  - operate switches and controls;

- en-suite sanitary facilities are the preferred option for wheelchair-accessible bedrooms;
- there should be at least as many en-suite shower rooms as en-suite bathrooms;
- built-in wardrobes and shelving should be accessible and convenient to use;
- bedrooms not designed for independent use by a person in a wheelchair should have an outer door wide enough to be accessible to a wheelchair user.

For all bedrooms

- built-in wardrobe swing doors should open through 180°;
- handles on hinged and sliding doors should:

  - be easy to grip and operate;
  - contrast visually with the surface of the door;

- windows and window controls should be:

  - located between 800 and 1000 mm above the floor;
  - easy to operate without using both hands simultaneously;

- a visual fire alarm signal should be provided;
- room numbers should be embossed characters.

Wheelchair-accessible bedrooms should:

- be located on accessible routes;
- be designed to provide a choice of location;
- have a standard of amenity equivalent to that of other bedrooms;
- be large enough to enable a wheelchair user to manoeuvre with ease;
- if a balcony is provided:

  - have a level threshold;
  - have no horizontal transoms between 900 mm and 1200 mm above the floor;

- have no permanent obstructions in a zone 1500 mm back from any balcony doors;
- be provided with emergency assistance alarms;

- have the door from the access corridor to a wheelchair-accessible bedroom that should:

  - not require more than 20 N opening force;
  - have an effective clear width through a single leaf door;
  - have an unobstructed space of at least 300 mm on the pull side of the door.

Switches, outlets and controls

- light switches should:

  - have large push pads;
  - align horizontally with door handles;
  - be located close to the entrance door opening;

- the operation of all switches, outlets and controls should not require the simultaneous use of both hands;
- switched socket outlets should indicate whether they are 'ON' or 'OFF';
- individual switches on panels should be well separated;
- socket outlets should be wall-mounted;
- telephone points and TV sockets should be located 400 mm to 1000 mm above the floor;
- socket outlets should be located no nearer than 350 mm from room corners;
- controls that need close vision should be located between 1200 mm and 1400 mm above the floor;
- **emergency alarm pull cords** should be:

  - coloured red;
  - located as close to a wall as possible;

- red and green should not be used in combination as indicators of 'ON' and 'OFF' for switches and controls.

Aids to communication

- hearing enhancement systems should be installed:

  - in all rooms and spaces designed for meetings, lectures, classes, performances, spectator sport or films;
  - at service or reception counters (especially when they are situated in noisy areas or they are behind glazed screens);

- the availability of an induction loop or infrared hearing enhancement system should be indicated by the standard symbol;
- telephones suitable for hearing aid users should be clearly indicated and incorporate an inductive coupler and volume control;
- artificial lighting should be designed to:

  - be compatible with other electronic and radio frequency installations;
  - give good colour rendering of all surfaces;

- uplighters mounted at low or floor level should be avoided as they can disorientate some visually impaired people.

The problem disabled people face accessing society

There are currently ten million people with physical and sensory impairments in the UK. Only a fraction are out and about, working, shopping, relaxing and taking part in the day-to-day life of wider society. A key reason is the great friction people encounter when attempting to access the shops, services and leisure facilities that the rest of the community takes for granted.

Trying to go out for a meal, visit a dentist, take a train, find a solicitor or even pop to the shops all have to be researched before they are possible. There is no easy way to do it. Should you phone a venue to enquire, the person who answers is unlikely to be able to answer your questions; in fact, people often become the subject of searching personal enquiries such as 'What is wrong with you?' or 'What can you do?' As a result people lose confidence, withdraw and become excluded from family and community life.

In this way society disables people with physical and sensory impairments and limits their opportunities. Businesses miss out on a market of millions of people who, according to government figures, have £80 billion to spend each year.

Bringing disabled people and business together

Accurate and reliable access information is the key to tackling this problem.

Founded by Dr Gregory Burke, a wheelchair user, and launched in January 2003 to mark the European Year of Disabled People, DisabledGo provides detailed and accurate data for people with hearing, vision or mobility impairments about access to shops, pubs, restaurants and offices – tens of thousands of goods and service providers across the UK.

For the first time business and service providers have a means to communicate directly with disabled people and explain their access to them.

People can judge for themselves whether or not a venue will suit their needs. You can check, for example: whether a cinema offers wheelchair access, has an accessible toilet, or a hearing loop; whether a hotel has a vibrating fire alarm; what side of a museum's steps the handrails are located on; if a solicitor will make home visits; if a cafe offers its menu in large print or in Braille or even which shops will welcome an assistance dog!

Service providers often do not realize that simple changes to the way they deliver their service could open up their business to many more customers. DisabledGo can offer advice and consultancy on issues surrounding access audits, the training of staff and the Disability Discrimination Act.

A free service developed by disabled people for disabled people

DisabledGo's innovative service was developed in consultation with over 100 disability groups across the UK. It is free for people to access and is paid for by a partnership of private sector sponsorship, from Marks & Spencer, and sponsorship from local authorities who are committed to opening up their localities to everybody. The guide can be accessed on www.disabledgo.info.

For more information about DisabledGo please contact Chris Sherwood on 01727 739 700 or visit www.disabledgo.info.

Key

The following signs have been used in the access guides to assist disabled people in determining the suitability of a venue they may wish to visit.

 The service provider has pledged to give a little more time to customers with any additional needs, if the customers so wish. A business will only be listed in the guide if it claims to be positive about disabled people.

 The premises/services are fully accessible to a wheelchair user who is travelling unaccompanied.

 The premises/services are accessible to a wheelchair user so long as there is someone to assist them.

 The premises have no more than three steps. If there is more than one step, there must be a handrail.

 A seat is available on the premises for customers, otherwise one may be requested.

 There are toilets on the premises that have been adapted for people with mobility impairments.

 There are level access toilets (excluding adapted ones) on the premises.

 Depending on the nature of the service this means either:
1. there are baby changing facilities within the adapted toilets; or
2. there are adapted changing rooms for people with mobility impairments.

 Bills, menus, programmes or any other written information are available in a larger print size if requested.

 Bills, menus, programmes or any other written information are available in Braille if requested.

The service provider is able to welcome assistance dogs onto the premises.

 A hearing system is available at certain locations within the premises.

 Either the service provider has a mini-com/text phone facility, or they are prepared to conduct any business via fax or email. Please note the contact details of the listing to know which of these facilities are available.

 A home delivery or home visit service is available if requested.

 Staff have received disability-awareness training.

 A personal shopper service is available.

Appendix B

Conservation of fuel and power

(A copy of this Appendix is available on www.routledge.com/books/details/ 9780415809696/)

B.1 The requirement

Part L Conservation of Fuel and Power

Reasonable provision shall be made for the conservation of fuel and power in buildings by:

(a) *limiting heat gains and losses –*
- (i) *through thermal elements and other parts of the building fabric; and*
- (ii) *from pipes, ducts and vessels used for space heating, space cooling and hot water services;*

(b) *providing fixed building services which –*
- (i) *are energy efficient;*
- (ii) *have effective controls; and*
- (iii) *are commissioned by testing and adjusting as necessary to ensure they use no more fuel and power than is reasonable in the circumstances; and*

(c) *providing to the owner sufficient information about the building, the fixed building services and their maintenance requirements so that the building can be operated in such a manner as to use no more fuel and power than is reasonable in the circumstances*

(Approved Document L)

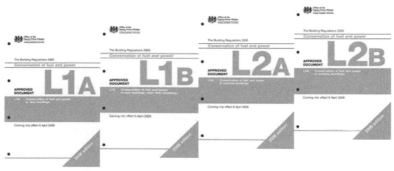

A new Part L came into force on 1 October 2010.

Note: In addition to Part L, some other Approved Documents also have requirements concerning the conservation of fuel and power. In particular these include:

- Part J (Combustion appliances and fuel storage systems);
- Part F (Ventilation);
- Part E (Resistance to the passage of sound);
- Part C (Site preparation and resistance to moisture).

But, as specified in Regulation 9, Part L does **not** apply to:

- buildings which are used primarily as places of worship;
- temporary buildings that are going to be used for less than two years;
- industrial sites;
- workshops;
- non-residential agricultural buildings with a low energy demand (e.g. parts of a building that are used for seed germination and, therefore, only need to be heated for a few days in each year).

B.2 Main changes in the 2010 edition

In 2006, Part L increased (both in size and content) and became a four-part series of Approved Documents that covered the conservation of fuel and power in:

- new dwellings (L1A);
- existing dwellings (L1B);
- new buildings (other than dwellings) (L2A);
- existing buildings (other than dwellings) (L2B).

The 2006 editions represented an in-depth rewrite of Part L, and the main changes (in addition to it now being in four separate volumes) were as follows:

- building control bodies authorized to accept self-certification by competent persons;
- general improvement in the performance standards for work on thermal elements, windows, doors, heating, hot water, ventilation and lighting systems in existing dwellings;
- increased use of technical reference publications to form part of the approved guidance;
- more guidance on building extensions and conservatories;
- more guidance on complying with the new requirements to make cost-effective consequential improvements whenever work is carried out on larger buildings;
- new minimum energy-performance requirements;
- new requirement for the improvement of the energy performance of the buildings if the floor area exceeds $1000\,\text{m}^2$, whenever these buildings are subject to major works;

- new requirements for pressure testing, commissioning and energy calculations;
- significant changes to the definitions of works and exempt works;
- the list of works that need not be notified to building control bodies increased to include minor works on heating, ventilation and lighting systems;
- the scope of the competent persons schemes was widened and more scheme operators were approved.

In 2006 it is estimated that adoption of Part L (and other associated Approved Documents) would result in buildings that are naturally ventilated and heated, achieving an overall improvement of about 23% and for air-conditioned buildings this improvement will be around 28%.

 Note: The 2010 version of Part L now requires that the annual CO_2 emission rate of a new dwelling **must** be 25% better than the previous 2006 requirement!

The contents of the newly formed Part L Approved Documents consist (and basically still do, with some minor exceptions, which are covered later in this appendix) of the items described below.

B.2.1 L1A – New dwellings

- The omission of the Elemental Method and the Target U-value Method in favour of a new approach to show compliance with the energy-efficiency requirements;
- the introduction of this new approach for compliance which addresses five criteria, namely that:
  - *the annual CO_2, emission rate of the completed dwelling must not exceed targets that have been set;*
  - *building fabric and services performance specifications are within reasonable limits;*
  - *solar shading and other measures to limit risks of summer overheating are reasonably acceptable;*
  - *fabric insulation and air tightness are as intended;*
  - *satisfactory information has been provided to the occupier(s);*
- the inclusion of an appendix containing a new checklist for builders and building control bodies to help in assessing compliance;
- the inclusion of an appendix listing the threshold performance values;
- details of new competent persons schemes that have been approved for pressure testing and energy performance calculations.

B.2.2 L1B – Existing dwellings

- A new definition of *thermal element* to address more types of alteration and renovation work;
- a new appendix giving examples of what can be achieved cost-effectively;

- new requirements for providing and/or renovating thermal elements;
- new requirements for commissioning heating, ventilation and lighting systems;
- new section containing guidance on ways of complying with these new requirements concerning the provision and renovation of thermal elements.

B.2.3 L2A – New buildings (other than dwellings)

- The omission of the elemental method and the target U-value method in favour of a new approach to show compliance with the energy efficiency requirements;
- the introduction of this new approach for compliance which addresses five criteria, namely that:
  - the annual CO_2 emission rate of the completed building must not exceed targets that have been set;
  - building fabric and services performance specifications are within reasonable limits;
  - non-air-conditioned buildings do not cause high internal temperatures as a result of excessive solar gains;
  - fabric insulation and airtightness are as intended;
  - satisfactory information has been provided to the occupier(s);
- the inclusion of an appendix containing a compliance checklist for builders and building control bodies.

B.2.4 L2B – Existing buildings (other than dwellings)

- A new definition of 'thermal element' to address more types of alteration and renovation work;
- new requirements for providing and/or renovating thermal elements;
- new requirements for commissioning heating, ventilation and lighting systems;
- new section containing guidance on complying with the new requirements to make cost-effective improvements whenever work is carried out on new buildings.

B.3 Changes and improvements contained in the 2010 edition of Part L

The main changes to the legal requirements and the supporting guidance since the 2006 issue of Approved Document L are as detailed below.

B.3.1 Changes in the legal requirements for the conservation of fuel and power

New dwellings and buildings

- Extensions consisting of a conservatory or porch are only exempted from the energy efficiency provisions of Parts L where the existing walls, windows or doors are retained, or replaced provided that the heating system of the building does not extended into the conservatory or porch;
- a CO_2 emission rate needs to be calculated and given to the building control body before the start of building work on the erection of a new building.

Existing dwellings and buildings

- The installation of thermal insulation in a roof space or loft space need not be notified to the building control body if:
  - this is the only work being carried out; or
  - the work being carried out is not work that needs to comply with any specific requirement of the Building Regulations.

B.3.2 Changes in the technical guidance for the conservation of fuel and power

New dwellings

- The annual CO_2 emission rate of a new dwelling must be 25% better than the previous 2006 requirement;
- secondary heating is now counted as part of the annual CO_2 emission rate;
- credit is allowed if low-energy lighting is installed;
- some performance specifications for building fabrics and services have been strengthened;
- revised guidance is provided for avoiding thermal bridging at construction joints;
- new guidance on how to limit heat loss from a swimming pool basin has been included.

New buildings

- Revised guidance is provided:
  - on shell and core developments and first fit-out work;
  - for avoiding thermal bridging at construction joints;
  - to limit heat loss from a swimming pool basin where this is constructed as part of a new building;
- a revised procedure is provided for demonstrating that reasonable provision has been made to limit the effects of solar gain in summer.

Existing dwellings and buildings

- Guidance is generally based on an elemental approach to demonstrating compliance.
- The main technical changes comprise a general strengthening of energy-efficiency standards for work on thermal elements, controlled fittings and controlled services in existing dwellings.
- Amended guidance is given:
  - for historic and traditional buildings, which may have an exemption from the energy-efficiency requirements (or where special considerations apply);
  - where an extension is a conservatory or porch that is not exempt from the energy-efficiency requirements;
  - for the renovation of a thermal element through the provision of a new layer or the replacement of an existing layer that has been expanded;
  - for swimming pool basins (walls and floor) in existing dwellings and buildings other than dwellings.

B.4 Changes in and improvements due to the 2010 edition of Part F

Some important changes to the legal requirements and the supporting guidance since the 1995 issue of Approved Document F (*Ventilation*) also became effective on 1 October 2010. Those directly linked with the conservation of fuel and power are as detailed below.

B.4.1 Changes in the legal requirements for ventilation

New and existing dwellings and buildings:

- Al fixed mechanical ventilation systems shall be commissioned and a commissioning notice given to the building control body;
- the owner shall be given sufficient information about the ventilation system and its maintenance requirements.

New dwellings

- The air flow rates of mechanical ventilation systems shall be measured on site and a notice given to the building control body.

B.4.2 Changes in the technical guidance for ventilation

- Ventilation provisions have been increased for dwellings with a design air permeability tighter than or equal to $5\,m^3/h\,m^2$) at $50\,Pa$.
- For passive stack ventilators, the stack diameter has been increased to

125 mm for all room types, and use of passive stack ventilation in inner wet rooms has been clarified.
- The guidance on ventilation when a kitchen or bathroom in an existing dwelling is refurbished has been clarified.

B.5 General guidance on the use of the four Part L Approved Documents

B.5.1 Types of work covered by L1A

Approved Document L1A is intended to provide guidance on what, in ordinary circumstances, would be accepted as reasonable conditions for satisfying the requirements for *creating new dwellings by means of new construction works*.

If part of a unit that contains living accommodation also contains space that is going to be used for commercial purposes (e.g. a workshop or office), it should be treated as a dwelling **if** the commercial part could revert to domestic use on a change of ownership. For example, if:

- there was direct access between the commercial space and the living accommodation; and
- both were contained within the same thermal envelope; and
- the living accommodation occupied a substantial proportion of the total area of the building.

When constructing a dwelling as part of a larger building that contains other types of accommodation, L1A should be used for guidance in relation to the individual dwellings, while L2A should be used for the non-dwelling parts of such buildings (e.g. heated common areas, and (in the case of mixed-use developments) commercial or retail space).

In mixed-use developments (i.e. where part of a building is used as a dwelling while another part has a non-domestic use) the requirements for non-domestic use should apply in all shared parts of the building.

If a conservatory is built as part of a new dwelling, the performance of the dwelling should be assessed as if the conservatory were not there.

If a dwelling is being created as the result of a material change of use, Approved Document L1B applies.

Buildings containing rooms for residential purposes (e.g. nursing homes, student accommodation and similar) are **not** considered as dwellings, and in such cases Approved Document L2A applies.

B.5.2 Types of work covered by L1B

L1B contains guidance on building work on existing dwellings, in particular:

- historic buildings;
- the extension of a dwelling;

- conservatories;
- when creating a new dwelling or part of a dwelling through a material change of use;
- material alterations to existing dwellings;
- work on controlled services and/or fittings;
- the provision and/or renovation of a thermal element.

When completing building work in a dwelling that is part of a mixed-use building, L1B should be used for guidance in relation to the dwelling and L2B for those parts of the building that are not a dwelling, including any common areas.

In most instances (but it is best to check with the local authority first) the building control body will need to be notified of the intended work before the work actually commences, either in the form of a deposit of full plans or by a building notice. If, however, the work is being carried out by a person approved under the competent persons self-certification (CP) scheme, no advance notification is required, but that person **must** provide the building owner with a certificate confirming that the installation has been carried out in accordance with the relevant requirements and notify the local authority to that effect.

If the work involves an emergency repair (e.g. a failed boiler or a leaking hot water cylinder), although advance notification is not required, the repair must still comply with the requirements of Part L1B and a completion certificate issued in the normal way.

If the work is of a minor nature (e.g. replacing a damaged cable, adding light fittings and switches to an existing circuit, upgrading supplementary equipotential bonding, work on telephone wiring) the work must still comply with the relevant requirements, but need not be notified to the building control body.

B.5.3 Types of work covered by L2A

L2A contains guidance concerning:

- the construction of new buildings **other** than dwellings;
- fit-out works (i.e. specific requirements made by the incoming occupiers) which are included as part of the construction of the building;
- the construction of extensions to existing buildings (where the total floor area of the extension is greater than $100\,m^2$ and greater than 25 per cent of the total useful floor area of the existing building);
- the construction of parts of a building (e.g. heated common areas in the case of mixed use);
- the construction of boarding houses, hostels and student accommodation blocks.

Note: Buildings used for industrial or commercial purposes (e.g. workshops and/or offices) that also contain living accommodation should be treated as a dwelling **if** the commercial part could revert to domestic use on a change of ownership.

B.5.4 Types of work covered by L2B

L2B contains guidance on building work on existing buildings, in particular:

- historic buildings;
- consequential improvements;
- extensions;
- material change of use;
- material alterations;
- the provision of a controlled service;
- thermal elements.

B.6 Historic and traditional buildings

When undertaking work on any of these, the aim should always be to improve energy efficiency as far as is reasonably practicable without prejudicing the character of the host building or increasing the risk of long-term deterioration of the building fabric or fittings.

Building inspectors normally require the use of 'sympathetic treatment' when restoring the historic character of a building that has been subject to previous inappropriate alteration (e.g. replacement doors).

Many books have been written about the problems related to restoring historic buildings and, before considering any work of this nature, you should probably seek the advice of the local planning authority's conservation officer, particularly if you are contemplating:

- the restoration of a historic building that has been subject to previous inappropriate alteration (e.g. replacement windows, doors and rooflights);
- rebuilding a former building (e.g. following a fire or filling a gap site in a terrace);
- enabling the fabric of historic buildings to 'breathe' so as to control moisture and potential long-term decay problems.

In general, new extensions to historic or traditional buildings should comply with the standards of energy efficiency as set out in Part L:2010. The only exception would be where there is a particular need to match the external appearance or character of the extension to that of the host building

Note: See English Heritage publication *Building Regulations and Historic Buildings* for further advice.

B.7 Material change of use

For the purposes of these Regulations, a material change of use is where there is a change in the purposes for which (or the circumstances in which) a building is used, so that after that change the building:

- is used as a dwelling;
- contains a flat;
- is used as a hotel or a boarding house;
- is used as an institution;
- is used as a public building;
- is not an exempt building (i.e. such as a building not frequented by people, a greenhouse, temporary or ancillary buildings or a small detached building);
- contains a room for residential purposes;
- contains at least one room for residential purposes, or contains a greater or lesser number of such rooms;
- is used as a shop where previously it was not; or
- which contains at least one dwelling, contains a greater or lesser number of dwellings than it did previously.

When a building is subject to a material change of use:

- any thermal element that is being retained should be upgraded;
- any existing window (including roof window or rooflight) or door which separates a conditioned space from an unconditioned space (or the external environment) and which has a U-value that is worse than $3.3\,\text{W/m}^2\text{K}$, should be replaced.

An accredited whole-building calculation model such as SAP 2005 can be used to demonstrate that the CO_2 emissions from the building will be no greater than if the building had been improved.

 The exemption from the energy-efficiency provisions for extensions consisting of a conservatory or porch is amended to grant the exemption only where the existing doors are retained (or replaced if removed).

B.7.1 Material changes of use (dwellings)

The erection of a new dwelling is **not** a material change of use. However, the requirements for the conservation of fuel and power still apply where a dwelling is being created in an existing building as the result of a material change of use of all, or part of, the building.

A material change of use in an existing dwelling is where, after the change, replacement units are provided for an existing door which separates a conditioned space from an unconditioned space (or external environment) which has a U-value worse than $3.3\,\text{W/m}^2\text{K}$.

B.8 Material alterations

Material alterations (i.e. where work, or any part of it, would result in a building or controlled service or fitting not complying with a relevant requirement where previously it did, or making previous compliance more unsatisfactory) should comply with the requirements for conservation of heat and energy as detailed below.

B.8.1 Material alterations (domestic buildings)

If a building is subject to a material alteration by:

- substantially replacing a thermal element;
- renovating a thermal element;
- making an existing element part of the thermal envelope of the building (where previously it was not);
- providing a controlled fitting;
- providing (or extending) a controlled service; then

in addition to the requirements of Part L, all applicable requirements from the following Approved Documents must be taken into account:

- Part A (*Structure*);
- Paragraph B1 (*Means of Warning and Escape*);
- Paragraph B3 (*Internal Fire Spread – Structure*);
- Paragraph B4 (*External Fire Spread*);
- Paragraph B5 (*Access and Facilities for the Fire Service*);
- Part M (*Access to and Use of Buildings*).

B.8.2 Material alterations (buildings other than dwellings)

When an existing element becomes part of the thermal element of a building (where previously it was not) and it has a U-value worse than $3.3\,W/m^2\,K$ it should be replaced (unless the element is a display window or high-usage door).

B.9 Consequential improvements (non-domestic buildings)

Note: *Consequential improvements* means those energy-efficiency improvements required by Regulation 17D.

If work is being undertaken on an existing building which has a total useful floor area of over $1000\,m^2$, then, in addition to the principal works (which must still comply with the energy-efficiency requirements detailed in Part L:2010 in the normal way), consequential improvements (where technically, functionally and economically feasible) will have to be completed.

If a building has a total useful floor area greater than $1000\,m^2$ and the proposed building work includes:

- an extension; or
- the initial provision of any fixed building services; or
- an increase to the installed capacity of any fixed building services;

then *consequential improvements* should be made to improve the energy-efficiency of the whole building. These will include:

- upgrading all thermal units which have a high U-value;
- replacing all existing windows (less display windows), roof windows, rooflights or doors (excluding high-usage entrance doors) within the area served by the fixed building service with an increased capacity;
- replacing any heating system that is more than 15 years old;
- replacing any cooling system that is more than 15 years old;
- replacing any air handling system that is more than 15 years old;
- upgrading any general lighting system that serves an area greater than 100 m² which has an average lamp efficacy of less than 40 lamp-lumens per circuit-watt;
- installing energy metering;
- upgrading existing Low and Zero Carbon (LZC) energy systems if they provide less than 10 per cent of the building's energy demand.

Where the installed capacity per unit area of a heating system is increased, existing doors (but excluding high-usage entrance doors) within the area served and which have U-values worse than 3.3 W/m²K should be replaced.

 Replacing these sorts of doors will normally achieve a simple payback within 15 years.

B.10 Controlled fittings

 Note: A *controlled service or fitting* is defined as a service or fitting in relation to which Part G, H, J, L or P of Schedule 1 imposes a requirement.

Where windows, roof windows, rooflights or doors are to be provided, draught-proofed units with an area-weighted average performance no worse than that shown in Table B1 should be used.

Table B1 Standards for controlled fittings

Fitting	Standard (W/m²K)
Windows, roof windows and glazed rooflights	1.8 for the whole unit
Alternative option for windows in buildings that are essentially domestic in character	Window energy rating of B and C
Plastic rooflight	1.8
Curtain walling	<1.8
Pedestrian doors where the door has more than 50% of its internal face area glazed	1.8 for the whole unit
High-usage entrance doors for people	3.5
Vehicle access and similar large doors	1.5
Other doors	1.8
Roof ventilators (including smoke extract ventilators)	3.5

'Controlled fitting' when applied to a door refers to a whole unit and includes the frame. Consequently, putting a new door in an existing frame is not providing a controlled fitting, and so such work is not notifiable and does not have to meet Building Regulations – although, where practical, it would always be sensible to do so!

B.10.1 Controlled fittings (dwellings)

All doors should be provided with draught-proofed units, the performance of which is no worse than $1.8\,W/m^2\,K$. In addition, insulated cavity closets should be installed where appropriate.

B.10.2 Controlled fittings (buildings other than dwellings)

Where doors are to be provided, reasonable provision would be to install:

- draught-proofed units the area-weighted average performance of which is no worse than that given in Table B2; and
- insulated cavity closers where appropriate.

Table B2 Standards for controlled door fittings

Fitting	Standard ($W/m^2\,K$)
Pedestrian doors where the door has more than 50% of its internal face area glazed	1.8 for the whole unit
High-usage entrance doors for people	3.5
Vehicle access and similar large doors	1.5
Other doors	1.8

 Where the replacement windows are unable to meet the requirements of Table B1 because of the need to maintain the external appearance of the façade or the character of the building, replacement windows should meet a centre pane U-value of $1.2\,W/m^2\,K$, or single glazing should be supplemented with low-e secondary glazing.

If a door is enlarged or a new one created, the area of the pedestrian door, expressed as a percentage of the total floor area of the building, should not exceed the relevant value from Table B3, or should be compensated for in some other way.

Table B3 Opening areas in the extension

Building type	Personnel doors as % of exposed wall	Rooflights as % of area of roof
Residential buildings where people temporarily or permanently reside	30	20
Places of assembly, offices and shops	40	20
Industrial and storage buildings	15	20
Vehicle access doors	As required	N/A

U-values of doors shall be calculated using the methods and conventions set out in BR 443 (*Conventions for U-value Calculations*, available at http://www.bre. co.uk/filelibrary/pdf/rpts/BR_443_(2006_Edition).pdf)) and should be based on the whole unit (i.e. in the case of a window, the combined performance of the glazing and frame).

Note: In certain classes of building with high internal gains, a less demanding U-value for glazing may be an appropriate way of reducing overall CO_2 emissions, in which case the average U-value for doors can be relaxed from the values given in Table B3, but the value should not exceed $2.7\,W/m^2\,K$

B.11 Controlled services (non-domestic buildings)

Where the work involves the provision of a controlled service:

- all new services should be energy efficient;
- all windows, roof windows, rooflights and/or doors should be provided with draught-proofed units;
- new HVAC systems should be provided with controls that are capable of achieving a reasonable standard of energy efficiency;
- fixed building services systems should be subdivided into separate control zones for each area of the building, with a significantly different solar exposure, occupancy period or type of use;
- separate control zones should be capable of independent switching and control set-point.

If both heating and cooling are provided, they should not be capable of operating simultaneously.

The central plant serving zone-based systems should:

- only operate as and when required;
- have a default condition that is 'OFF'.

B.12 Extensions

Under regulation 17D of the Building Regulations, the construction of an extension may trigger the requirement for consequential improvements.

When building an extension to an existing dwelling or building, the following applies.

B.12.1 Dwellings

Regulation 9 of the Building Regulations exempts some conservatory and porch extensions from the energy-efficiency requirements

Extensions to dwellings should either use newly constructed thermal elements (that meet the requirements for the conservation of fuel and power) or use of existing (or new) doors, windows, roof windows and rooflights that meet these standards (Table B4).

Table B4 Standards for control fittings

Fitting	Standard (W/m²K)
Window, roof window or rooflight	1.6
Doors with >50% of internal face glazed	1.8
Other doors	1.8

In most circumstances, the most reasonable provision would be to limit the total area of windows, roof windows and doors in extensions so that it does not exceed the sum of:

- 25 per cent of the floor area of the extension; plus
- the total area of any windows or doors which, as a result of the extension works, no longer exist or are no longer exposed.

One way of complying would be to show that the area-weighted U-value of all the elements in the extension is no greater than that of an extension of the same size and shape.

If upgrades are proposed to the existing dwelling, they should be implemented to a standard that is no worse than that shown in column (b) of Table B5.

Table B5 Upgrading retained thermal elements

Element	(a) Threshold U-value (W/m²K)	(b) Improved U-value (W/m²K)
Wall – cavity insulation	0.70	0.55
Wall – external or cavity insulation	0.70	0.30
Floor	0.70	0.25
Pitched roof – insulation at ceiling level	0.35	0.16
Pitched roof – insulation between rafters	0.35	0.18
Flat roof or roof with integral insulation	0.35	0.18

B.12.2 Large extensions

Where the proposed extension has a total useful floor area that is both:

- greater than 1000 m²; and
- greater than 25 per cent of the total useful floor area of the existing building;

then the work should be regarded as a new building.

B.12.3 Buildings other than dwellings

Reasonable provision would be for the proposed extension to incorporate the following:

*   doors, windows, roof windows, rooflights and smoke vents;
*   new thermal elements (that meet the requirements for the conservation of fuel and power); and
*   existing opaque fabric which becomes a thermal element where previously it was not should be upgraded;

provided that they meet the standards set by the Building Regulations.

Opening areas

The area of windows and rooflights in the extension should generally not exceed the values given in Table B6.

Table B6 Opening areas in the extension

Building type	Windows and personnel doors as % of exposed wall	Rooflights as % of area of roof
Residential buildings where people temporarily or permanently reside	30	20
Places of assembly, offices and shops	40	20
Industrial and storage buildings	15	20
Vehicle access doors and display windows and similar glazing	As required	N/A
Smoke vents	N/A	As required

B.13 Conservatories

Some conservatory and porch extensions are exempt from the energy-efficiency requirements. These consist of conservatories or porches:

*   which are at ground level;
*   where the floor area is less than $30\,m^2$;
*   where the existing walls, doors and windows in the part of the dwelling which separates the conservatory are retained or, if removed, replaced by walls, windows and doors which meet the energy-efficiency requirements; and
*   where the heating system of the dwelling is not extended into the conservatory or porch.

If a new conservatory or porch does **not** meet all these requirements, it is **not** exempt and **must** comply with the relevant energy-efficiency requirements of Part L:2010.

If a building extension is a conservatory or porch that is not exempt from the energy-efficiency requirements, then:

- the effective thermal separation between the heated area in the existing building (i.e. the walls, doors and windows between the building and the extension) should be insulated and draught-proofed to at least the same extent as in the existing building;

Doors with >50% of their internal face glazed should not exceed 1.8 W/m^{2}K

- particular attention should be paid to independent temperature and 'ON/OFF' controls to any heating system installed within the extension; and
- glazed elements should meet the standards set out in Table B7 and opaque elements should meet the standards set out in Table B8.

Table B7 Standards for controlled fittings

Fitting	Standard (W/m^{2}K)
Windows, roof windows and glazed rooflights	1.8 for the whole unit
Alternative option for windows in buildings that are essentially domestic in character	Window energy rating of Band C
Plastic rooflight	1.8
Curtain walling	< 1.8
Pedestrian doors where the door has more than 50% of its internal face area glazed	1.8 for the whole unit
High-usage entrance doors for people	3.5
Vehicle access and similar large doors	1.5
Other doors	1.8
Roof ventilators (including smoke extract ventilators)	3.5

Table B8 Standards for new thermal elements

Element	Standard (W/m^{2}K)
Wall	0.262
Pitched roof – insulation at ceiling level	0.16
Pitched roof – insulation at rafter level	0.18
Flat roof or roof with integral insulation	0.18
Floors	0.224
Swimming pool basin	0.25

Removing and not replacing any of the thermal separation between the building and an existing exempt extension, or extending the building's heating system into the extension, means that the extension ceases to be exempt!

B.14 Thermal elements

- New thermal elements must be reasonably energy efficient.
- Renovated thermal elements shall have improved energy efficiency.
- Retained thermal elements whose U-value is worse than the threshold value shall be upgraded.
- Thermal bridges in the insulation layers around window and door openings shall be avoided.
- Unwanted air leakage through the new envelope parts shall be minimized.
- When a thermal element is being renovated or replaced it should meet the requirements for limiting heat gains and losses.

B.14.1 Fabric standards

Table B9 provides details of the worst acceptable standards for fabric properties.

Table B9 Limiting fabric parameters

Element	Standard (W/m^{2}K)
Roof	0.20
Wall	0.30
Floor	0.25
Party wall	0.20
Windows, roof windows, glazed rooflights, curtain walling and pedestrian doors	2.00
Air permeability	10.00 m^3/h m^2 at 50 Pa

B.14.2 Building fabric

In accordance with Part L:2010 and Regulation 7:

- The building fabric of all dwellings and buildings other than dwellings should be constructed to a reasonable standard so that:
  - the insulation is reasonably continuous over the whole building envelope; and
  - the air permeability is within reasonable limits.
- The building fabric should be constructed so that there are no reasonably avoidable thermal bridges in the insulation layers caused by gaps within the various elements, at the joints between elements and at the edges of elements, such as those around window and door openings

Reductions in thermal performance can occur where the air barrier and the insulation layer are not contiguous and the cavity between them is subject to air movement. To avoid this problem, either the insulation layer should be contiguous with the air barrier at all points in the building envelope, or the space between them should be filled with solid material, such as in a masonry wall

B.14.3 Fabric standards – buildings

Fabric elements and fixed building services should satisfy minimum energy-efficiency standards. Table B10 sets out the worst acceptable standards for fabric properties.

Table B10 Limiting fabric parameters

Element	W/m² K
Roof	0.25
Wall	0.35
Floor	0.25
Windows, roof windows, glazed rooflights, curtain walling and pedestrian doors	2.2
Vehicle access and similar large doors	1.5
High usage entrance doors	3.5
Roof ventilators (including smoke vents)	3.5
Air permeability	10.00 m³/h m² at 50 Pa

B.14.4 Thermal bridges

The heat loss through a party wall can be reduced by fully filling the cavity and/or by providing effective sealing around the perimeter so as to restrict air movement through the cavity.

The building fabric should be constructed so that there are no reasonably avoidable thermal bridges in the insulation layers caused by gaps within the various elements, at the joints between elements, or at the edges of elements such as those around window and door openings.

 Provision should also be made to reduce unwanted air leakage through the new envelope parts.

Reductions in thermal performance can occur where the air barrier and the insulation layer are not contiguous and the cavity between them is subject to air movement. To avoid this problem, either the insulation layer should be contiguous with the air barrier at all points in the building envelope, or the space between them should be filled with solid material, such as in a masonry wall

B.15 Renovation of thermal elements

When a thermal element is renovated it should, in most cases, meet the standard set out in column (b) of Table B11. If such an upgrade is not technically or functionally feasible (or would not achieve a simple payback of 15 years or less) the element should be as per the guidance contained in Appendix A of Approved Document L1B.

Thermal elements whose U-value is worse than the threshold value shown in column (a) of Table B11 should be upgraded to the best standard that is technically and functionally feasible and which delivers a simple payback period of 15 years or less.

Table B11 Upgrading retained thermal elements

Element	(a) Threshold value (W/m²K)	(b) Improved value (W/m²K)
Cavity wall	0.70	0.55
Other wall type	0.70	0.35
Floor	0.70	0.25
Pitched roof – insulation at ceiling level	0.35	0.16
Pitched roof – insulation between rafters	0.35	0.20
Flat roof or roof with integral insulation	0.35	0.25

B.15.1 Conservatories and substantially glazed spaces

If a conservatory is built as part of the new dwelling, the performance of the dwelling should be assessed (i.e. as if the conservatory were not there) and Approved Document L1B should be followed in respect of the construction of the conservatory itself. This means that the thermal separation between dwelling and conservatory **must** be constructed to a standard comparable to the rest of the external envelope of the dwelling.

Substantially glazed spaces that are actually part of the dwelling (i.e. there is no thermal separation and therefore, by definition, the space is not a conservatory) should be included as part of the new dwelling when checking against the five compliance criteria.

B.15.2 Area of windows, roof windows and doors

It is recommended that the total area of windows, roof windows and doors in extensions to existing buildings do not exceed the sum of:

• 25 per cent of the floor area of the extension; plus
• the total area of any windows or doors which, as a result of the extension works, no longer exist or are no longer exposed.

B.16 Demonstrating compliance

Compliance with the requirements and recommendations of Approved Documents L1A and L2A can be achieved by meeting the five criteria shown in Table B12.

Table B12 The five criteria for demonstrating compliance with Approved Document L1A

Criteria	Requirement	In a nutshell
1	Achieving the TER	The predicted rate of CO_2 emissions from the dwelling (i.e. the dwelling emission rate or DER) and/or a building (i.e. the building emission rate, or BER) shall not be greater than the target emission rate (TER)
2	Limits on design flexibility	The performance of the building fabric, the fixed building services and parts of the building provided with comfort cooling should be no worse than the design limits
3	Limiting the effects of solar gains in summer	The dwelling shall have appropriate passive control measures to limit the effect of solar gains on indoor temperatures in summer
4	Building fabric	The performance of the dwelling, as built, shall be the BER or DER (as appropriate)
5	Commissioning of heating and hot water systems	The necessary provisions for energy-efficient operation of the dwelling are put in place

Appendix A to Parts L1A and L2A contain checklists that can be used to confirm that all the provisions in a new dwelling or building have been met satisfactorily. These checklists are included in Appendix 1 to this appendix.

B.16.1 Criterion 1 – achieving the TER

The target CO_2 emission rate (TER) is the minimum energy-performance requirement for all new dwellings and buildings, expressed in terms of the amount of carbon dioxide (CO_2) in units of kg per m^2 of floor area, that is annually emitted by a standardized household through the provision of heating, hot water, ventilation and internal fixed lighting and (in addition) buildings as a result of cooling and lighting.

The requirement in Regulation 17C clearly states that:

New buildings shall meet the target CO_2 emission rate for the building.

In other words, the dwelling (CO_2) emission rate (DER) or the building (CO_2) emission rate (BER) must **not** be worse than the target (CO_2) emission rate (TER), and so the final calculation must be based on the building as constructed, incorporating:

• any changes to the performance specifications that have been made during construction;
• the measured air permeability, ductwork leakage and fan performances as commissioned.

The person carrying out the work shall:

- ensure that pressure testing has been carried out in accordance with approved procedures and the results have been forwarded for approval to the local authority (if the person carrying out the work is registered by the British Institute of Non-destructive Testing, then the local authority will accept a certificate signed by that person as proof of compliance);
- ensure that the local authority is (when necessary) supplied with a notice confirming that the fixed building services have been commissioned in accordance with approved procedures;
- provide the local authority with a notice which specifies:
  - the TER for the building; and
  - the calculated CO_2 emission rate for the building as constructed;
- (if the person carrying out the work is registered by FAERO Limited; or BRE Certification Limited in respect of the calculation of the CO_2 emission rates of buildings, the local authority will accept a certificate signed by that person as proof of compliance);
- air leakage testing of ductwork should be carried out on:
  - all systems served by fans with a design flow rate greater than $1\,\text{m}^3/\text{s}$; and
  - all sections of ductwork where HVCA DW/143 recommends testing.

CO_2 emission rate calculations

Individual dwellings

For individual dwellings, the TER is calculated by using:

- the government's standard assessment procedure (SAP) for individual dwellings with less than $450\,\text{m}^2$ total floor area;
- the simplified building energy model (SBEM) for individual dwellings with more than $450\,\text{m}^2$ total floor area;

as follows:

- first, estimate the amount of CO_2 emissions from a notional dwelling that is the same size and shape as the new dwelling and whose emissions arise from:
  - the provision of heating and hot water (CH) (which includes the energy used by pumps and fans); and
  - the use of internal fixed lighting (CL);
- secondly, determine the TER using the following formula:

$$\text{TER} = (\text{CH} \times \text{fuel factor} + \text{CL} \times (1 - \text{improvement factor}))$$

where the fuel factor is taken from Table B13 and the improvement factor

Note: The specific fuel factor should be used for those appliances that can only burn the particular fuel.

Where an appliance is classed as multi-fuel, the multi-fuel factor should be used, except where the dwelling is in a smoke control area. In such cases the

Table B13 Fuel factor

Heating fuel	Fuel factor
Mains gas	1.00
LPG	1.10
Oil	1.17
Grid electricity (for direct acting, storage and electric heat pump systems)	1.47
Solid mineral fuel	1.28
Renewable energy (including biofuels such as wood pellets)	1.00
Solid multi-fuel	1.00

solid mineral fuel figure should be used, unless the specific appliance type has been approved for use within smoke control areas.

The fuel factor used to calculate the **TER** should be based on the fuel used, as follows:

• Where all the heating appliances are served by the same fuel: the fuel used for the TER calculation is the fuel used in those appliances.
• Where a dwelling has more than one heating appliance and these are served by different fuels, the fuel used for the TER calculation should be:
 – mains gas if any of the heating appliances is fired by mains gas; otherwise
 – the fuel used in the main heating system.
• Where a dwelling is served by a community heating scheme, the fuel used for the TER calculation is the principal fuel used by the community heating system.

Buildings with multiple dwellings

For buildings consisting of multiple dwellings (e.g. terraced houses or blocks of apartments), the floor-area-weighted average of all the individual TERs (using the following formula) should be used:

$$\frac{(\text{TER} \times \text{Floor area}) + (\text{TER}_2 \times \text{Floor area}_2)... + \text{TER}_N \times \text{Floor area}_N}{\text{Floor area} + \text{Floor area}_2... + \text{Floor area}_N}$$

Compliance (i.e. with Regulation 17C) will be achieved if:

• either every individual dwelling has a DER that is no greater than its corresponding TER; or
• the average DER is no greater than the average TER.

Note: The average DER is the floor-area-weighted average of all the individual DERs, and is calculated in the same way as the average TER (see above).

New buildings

The TER is calculated using either the SBEM or another approved software, such as dynamic simulation models (see Annex 1 to ODPM Circular 03/2006).

The TER is calculated in two steps:

- **Step 1** – Use an approved calculation tool to calculate the CO_2 emission rate ($C_{notional}$) from a notional building with the same size, shape, energy performance value and which is subject to the same occupancy levels and environmental conditions as the proposed building.
- **Step 2** – Adjust the CO_2 emissions rate (calculated in Step 1) by an improvement factor using the formula:

$$\text{TER} = C_{notional} \times (1 - \text{improvement factor}) \times (1 - \text{LZC benchmark})$$

where the:

- *improvement factor* is the improvement in energy efficiency shown in column (a) of Table B14 that is appropriate to the classes of building services being provided in the proposed building; and
- *LZC benchmark* is the benchmark provision for low and zero carbon energy sources shown in column (b) of Table B14.

Table B14 Improvement in whole building carbon dioxide emissions

Building services strategy	(a) Improvement factor	(b) LZC benchmark
Heated and naturally ventilated	0.15	0.10
Heated and mechanically ventilated	0.20	0.10
Air conditioned	0.20	0.10

The BER is calculated in a similar manner to the TER (see above), but in this case two calculations are required, as follows:

1. A preliminary calculation as part of the design submission (based on plans and specifications and the CO_2 emissions factors shown in Table B15).

Table B15 CO_2 emission factors

Fuel	CO_2 emission factor (kg CO_2/kWh)
Natural gas	0.194
LPG	0.234
Biogas	0.025
Oil	0.265
Coal	0.291
Anthracite	0.317
Smokeless fuel (including coke)	0.392
Duel fuel appliances (e.g. mineral + wood)	0.187
Biomass	0.025
Grid supplied electricity	0.422
Grid displaced electricity (i.e. building integrated power generation systems)	0.568
Waste heat (e.g. from industrial processes)	0.018

Note: This would normally be provided as part of the full plans submission.

2. A final calculation (based on the dwelling as constructed and incorporating any changes to the performance specifications that have been made during construction) together with the measured air permeability, ductwork leakage and fan performance as commissioned.

In some cases, management features can improve the energy efficiency of a building. When this occurs, the BER can be reduced by the adjustment factor shown in Table B16.

Table B16 Enhanced management and control features

Feature	Adjustment factor
Automatic monitoring and targeting with alarms for out-of-range values	0.050
Power factor correction to achieve a whole-building power factor of at least 0.90	0.010
Power factor correction to achieve a whole-building power factor of at least 0.95	0.025

 Note: The power factor adjustment can only be taken if the whole building power factor is corrected to the level stated.

Technical risk

While all building work must satisfy the technical requirements set out in the Regulations, the inclusion of any particular energy-efficiency measure should not involve excessive technical risk.

Limits on design flexibility (domestic buildings)

*   The performance of the building fabric and the fixed building services should be no worse than the design limits.
*   Reasonable provision should be made to limit heat gains and losses through the fabric of the building.
*   Energy-efficient building services and effective controls should be provided.

Secondary heating

Where a secondary heating appliance is fitted, the efficiency of the actual appliance with its appropriate fuel shall be used in the calculation of the DER.

Where a chimney or flue is provided (but no appliance is actually installed), the presence of the following appliances shall be assumed when calculating the DER:

- If a gas point is located adjacent to the hearth: a decorative fuel-effect fire open to the chimney or flue with an efficiency of 20% shall be assumed.
- If there is no gas point: an open fire in grate with an efficiency of 37% burning multi-fuel (unless the dwelling is in a smoke control area, when the fuel should be taken as smokeless solid mineral fuel) shall be assumed.
- In all other cases: an electric room heater shall be taken as the secondary heating appliance.

Lighting

In all cases the DER should be calculated using a fixed assumption of 30 per cent low-energy lighting.

Fixed internal lighting

Light fittings (including lamp, control gear and an appropriate housing, reflector, shade or diffuser or other device for controlling the output light) should only take lamps with a luminous efficiency greater than 40 lumens per circuit-watt (please note that light fittings in less frequented areas such as cupboards and other storage areas do not count).

Fluorescent and compact fluorescent lighting fittings will meet this requirement, but light fittings for GLS tungsten lamps with bayonet cap or Edison screw bases, or tungsten halogen lamps, will not.

Fixed internal lighting – domestic buildings

Light fittings should only take lamps with a luminous efficiency greater than 40 lumens/circuit-watt.

The occupiers of a dwelling should be provided with efficient electric lighting whenever:

- a dwelling is extended; or
- a new dwelling is created from a material change of use; or
- an existing lighting system is being replaced as part of rewiring works.

All rewiring works **must** comply with Part P. A light fitting may contain one or more lamps.

In some cases, of course, it might be better to install an energy-efficient light fitting in a location that is not part of the building work (e.g. replacing the fitting on the landing when a new bedroom is created via a loft conversion).

Fixed external lighting

- Fixed external lighting, when supplied, should be provided with effective control and/or the use of efficient lamps.
- Rewiring works must comply with Part P.
- Fixed energy-efficient light fittings (one per $25\,m^2$ dwelling floor area (excluding garages) and one per four fixed light fittings) should be installed in the most frequented locations in the dwelling.

- Fixed external lighting (i.e. lighting that is fixed to an external surface of the dwelling and which is powered from the dwelling's electrical system) should either:
  - have a lamp capacity not exceeding 150 W per light fitting that automatically switches off when there is enough daylight and when it is not required at night; or
  - include sockets that can only be used with lamps which have an efficiency greater than 40 lumens/circuit-watt.

 Compact fluorescent lamps would meet this last requirement, but lamps with bayonet cap or Edison screw bases, or tungsten halogen lamps, would not.

Lighting controls (non-domestic buildings)

- Local switches should be:
  - located in easily accessible positions within each working area (or at boundaries between working areas and general circulation routes);
  - operated by the deliberate action of the occupants, either manually or remotely;
  - located within 6 m (or twice the height of the light fitting above the floor if this is greater) of any luminaire it controls.
- Lighting controls should be provided that switch off the lighting during daylight hours and when the area is unoccupied.
- Automatically switched lighting systems should be subject to a risk assessment.
- Manually operated local switches should be in easily accessible positions within each working area, at boundaries between working areas, and at general circulation routes.
- If the space is a naturally lit *space* served by side windows, the perimeter row of luminaires should be separately switched.
- Local (manual) switching can be supplemented by automatic controls which:
  - switch the lighting off when they sense the absence of occupants; or
  - dim (or switch off) the lighting when there is sufficient daylight.

Office, industrial and storage areas (non-domestic buildings)

Classrooms, seminar and conference rooms, etc., shall have an average efficiency of not less than 45 luminaire-lumens/circuit-watt.

Display lighting in all types of space (non-domestic buildings)

- Display lighting should have an average initial (100-hour) efficiency of not less than 15 lamp-lumens/circuit-watt.
- Where possible, display lighting should be connected in dedicated circuits that can be switched off at times when people will not be inspecting exhibits or merchandise or attending entertainment events.

Emergency escape lighting (non-domestic buildings)

Emergency escape lighting, specialist process lighting and vertical transportation systems are not subject to the requirements of Part L.

General lighting efficiency in all other types of space (non-domestic buildings)

Lighting (over the whole of these areas) should have an average initial efficacy of not less than 45 luminaire-lumens/circuit-watt.

Lighting systems serving other types of space may use lower powered and less efficient lamps.

B.16.2 Criterion 2 – limits on design flexibility

Criterion 2 sets out the design limits for the building fabric to meet the requirements for conservation of fuel through thermal elements and other parts of the building fabric, and for the provision and commission of energy-efficient fixed building services and controls.

 Note: In **all** cases the performance of the building fabric and the heating, hot water and fixed lighting systems should be no worse than the design limits.

Design limits for envelope standards

U-values

U-values (i.e. the overall coefficient of heat transmission) indicate the heat flow through materials, and need to be calculated using the methods and conventions set out in BR 443 (*Conventions for U-value Calculations*). Recognized limits for plane element U-values for building fabric elements are shown in Table B17.

Table B17 Limiting U-value standards (W/m²K)

Element	Area-weighted dwelling average	Worst individual subelement
Wall	0.35	0.70
Floor	0.25	0.70
Roof	0.25	0.35
Windows, roof windows, rooflights and doors	2.2	3.3
Pedestrian doors	2.2	3.0
Vehicle access and similar large doors	1.5	4.0
High usage entrance doors	6.0	6.0
Roof ventilators (including smoke vents)	6.0	6.0

 Note: Display windows and similar glazing are **not** required to meet the standard given for *windows and rooflights*.

Air permeability

Air permeability is the physical property used to measure the airtightness of the building fabric, and is defined as *air leakage rate per envelope area at the test reference pressure differential across the building envelope of 50 Pa (50 N/ml)*.

The envelope area of the building is the total area of all floors, walls and ceilings bordering the internal volume subject to the test and including walls and floors below external ground level. Overall internal dimensions are used to calculate this area, and no subtractions are made for the area of the junctions of internal walls, floors and ceilings with exterior walls, floors and ceilings.

A reasonable limit for the design air permeability is $10\,\text{m}^3/(\text{h}\,\text{m}^2)$ at 50 Pa and guidance on some ways of achieving this is given in *Limiting Thermal Bridging and Air Leakage: Robust Construction Details for Dwellings and Similar Buildings* (available from the Energy Saving Trust: http://www.est.org.uk).

Building services

Controls

- Systems should be subdivided into separate control zones.
- Separate control zones should be capable of independent timing, temperature control and (where appropriate) ventilation and air recirculation rate.
- Heating and cooling systems should not operate simultaneously.
- Central plant should only operate when zone systems require it.
- The default condition should be 'OFF'.

Energy meters

- Energy performance monitoring meters should be included in the installation of all new building services equipment.
- Separate meters should be provided to monitor low or zero carbon (LZC) systems.
- Buildings with floor areas greater than $1000\,\text{m}^2$ should include an automatic meter reading and data collection facility.

Heating and hot water system(s)

If a heating or hot water system is being provided or extended, the installed appliance should:

- not be less than that recommended for its type in the *Domestic Heating Compliance Guide*;
- have an efficiency which is not worse than 2 per cent lower than that of the appliance being replaced – if the appliance is the primary heating service;
- be provided with controls that meet the minimum control requirements of the *Domestic Heating Compliance Guide* for the particular type of appliance and heat distribution system;

Appendix B

- be commissioned so that, at completion, the system(s) and their controls are left in working order and can operate efficiently for the purposes of the conservation of fuel and power.

The person carrying out the work shall provide the local authority with a notice (signed by a suitably qualified person) confirming that all fixed building services have been properly commissioned in accordance with the *Domestic Heating Compliance Guide*.

Insulation of pipes, ducts and vessels

Hot and chilled water pipework, storage vessels, refrigerant pipework and ventilation ductwork should be insulated so as to conserve energy and to maintain the temperature of the heating or cooling service.

Air handling plant

- Should have an efficient and effective control system.
- Should be capable of achieving a specific fan power at 25 per cent of design flow rate.
- Fans rated at more than 1100 W should be equipped with variable-speed drives.
- Ventilation ductwork should be reasonably airtight.

Mechanical ventilation

The performance of systems should be better than those described in GPG 2689 (*Energy-efficient Ventilation in Housing*) and their fan powers and heat-recovery efficiency should be no worse than those listed in Table B18.

In dwellings, mechanical ventilation systems **must** satisfy the requirements in Part F.

Table B18 Limits on design flexibility for mechanical ventilation

System type	Performance
Specific fan power (SFP) for continuous supply only and continuous extract only	0.8 l/s W
SFP for balanced systems	2.0 l/s W
Heat-recovery efficiency	66%

Cooling plant

Cooling systems should have a:

- suitably efficient cooling plant; and
- an effective control system.

Mechanical cooling

Fixed air-conditioners in new dwellings should have an energy-efficiency classification equal to or better than Class C in Schedule 3 of the labelling scheme adopted under *The Energy Information (Household Air Conditioners) No. 2 Regulations* (SI 2005/1726).

Building services (non-domestic buildings)

For energy-saving purposes, Approved Document L2A requires that:

- all heating, ventilation and air-conditioning systems should be provided with controls to enable them to achieve reasonable standards of energy efficiency;
- display lighting should be connected in dedicated circuits that can be switched off at times when people will not be inspecting exhibits or merchandise or attending entertainment events;
- heating and hot water service systems (cooling plant and/or air-handling plants) should have an efficiency not less than that recommended for its type in the *Non-domestic Heating, Cooling and Ventilation Compliance Guide*;
- lighting systems should be capable of being automatically (or, in certain circumstances, mechanically) switched off during daylight hours and in unoccupied working spaces, etc.;
- pipes, ducts and vessels should be insulated in accordance with the recommendations contained in the ODPM's *Domestic Heating Compliance Guide*;
- systems should be provided with energy meters to efficiently manage the amount energy used.

Emergency escape lighting, specialist process lighting and vertical transportation systems are **not** subject to the requirements of Part L.

Inspection and commissioning of the building services systems

For non-domestic buildings, when building services systems are commissioned:

- the systems and their controls shall be left in their intended working order and are capable of operating efficiently regarding the conservation of fuel and power;
- the person carrying out the work shall provide the local authority with a notice confirming that all fixed building services have been properly commissioned;
- leakage testing should be carried out in accordance with the procedures set out in DW/143, HVCA, 2000.

B.16.3 Criterion 3 – limiting the effects of solar gains in summer

High internal temperatures caused by solar gains should be minimized by a combination of:

- window size and orientation;
- shading;
- ventilation; and
- high thermal capacity.

Note: For further guidance on how to control overheating, see:

- BR 364 *Solar Shading of Buildings*;
- AM 10 *Natural Ventilation in Non-domestic Buildings*;
- BS 8206: Part 2 Code of *Practice for Daylighting*;
- CE129 *Reducing Overheating – A Designer's Guide*;
- TM37 *Design for Improved Solar Shading Control*;
- Building Bulletins 87 and 101 (concerning school buildings);
- SAP 2005 Appendix P (which contains a procedure enabling designers to check whether solar gains are excessive).

When trying to limit solar gains, consideration should also be given to the provision of adequate levels of daylight. BS 8206: Part 2 (*Code of Practice for Daylighting*) provides some guidance on maintaining adequate levels of daylighting.

Those parts of the building that are provided with comfort cooling should have passive control measures to limit the effect of solar gains.

Dwellings

- Dwellings shall have passive control measures to limit the effect of solar gains on indoor temperatures in summer.
- High internal temperatures caused by solar gains should be minimized by a combination of window size and orientation, shading, ventilation and high thermal capacity.

Buildings other than dwellings

- High internal temperatures caused by solar gains should be minimized by a combination of window size and orientation, shading, ventilation and high thermal capacity.
- In occupied spaces that are not served by a comfort cooling system:
  - the combined solar and internal casual gains (people, lighting and equipment) per unit floor area averaged over the period of daily occupancy should not be greater than $35\,W/m^2$ calculated over a perimeter area not more than 6 m from the window wall and averaged during the period 06.30–16.30 hours GMT;
  - the operative dry resultant temperature does not exceed 28° C for more than a reasonable number of occupied hours per annum.

B.16.4 Criterion 4 – building fabric

The main requirement is that all buildings should be constructed and equipped so that their performance meets the predicted BER (or DER) and the builder can demonstrate (via a formalized site-inspection system) that construction procedures are to the required standard. To meet this requirement, the building fabric should be constructed (depending on whether it is a new building or a new dwelling) so that:

- the insulation is reasonably continuous over the whole building envelope;
- the air permeability is within reasonable limits;
- there are no reasonably avoidable thermal bridges in insulation layers;
- all buildings should be pressure tested.

 Note: This can be achieved by either using accredited design specifications or demonstrating conformance via BRE IP 1/06 and at the edges of windows and door openings.

Continuity of insulation

Dwellings and buildings should be constructed and equipped so that their performance meets the predicted DER and BER.

The building fabric should be constructed so that:

- the insulation is reasonably continuous over the whole building envelope;
- the air permeability is within reasonable limits;
- there are no reasonably avoidable thermal bridges in the insulation layers caused by gaps within the various elements, at the joints between elements, and at the edges of windows and door openings.

Reasonable provision would be to:

- adopt approved design details (see as those shown in *Limiting Thermal Bridging and Air Leakage: Robust Construction Details for Dwellings and Similar Buildings*, available from http://www.est.rg);
- demonstrate that the specified details deliver an equivalent level of performance (using the guidance in BRE Information Paper IP01/06);
- demonstrate that an appropriate system of site inspection is in place to give confidence that the construction procedures achieve the required standards of consistency.

Air permeability and pressure testing

Compliance with the requirements would be demonstrated if:

- the measured air permeability is not worse than $10 \, m^3/(h \, m^2)$ at $50 \, Pa$; and
- the DER calculated using the measured air permeability is not worse than the TER; and
- pressure testing is carried out in accordance with agreed procedures; and
- a notice (record) of the results of the testing is given to the local authority.

Pressure testing

Pressure testing is a requirement of all new buildings and the following shall apply:

- pressure testing shall be completed so as to demonstrate that the specified air permeability of the building has been achieved;
- air pressure testing should be carried out on a unit of each dwelling type selected by the building control body;
- air pressure testing should be carried out so that half of the scheduled tests are carried out during construction of the first 25 per cent of each dwelling type;
- if a dwelling fails to achieve the design air permeability then remedial measures should be carried out.

All buildings should be pressure tested except:

- dwellings;
- buildings with less than $500\,\text{m}^2$ floor area;
- factory-made modular buildings where no site assembly work is required;
- large extensions which cannot be sealed off from the existing building;
- large complex buildings (where due to building size or complexity, it may be impractical to carry out whole building pressure testing);
- buildings compartmentalized into self-contained units;

Note: If a building fails to achieve the design air permeability, remedial measures should be carried out:

- a record of the results of the testing shall be provided to the local authority;
- the person responsible for commissioning the work shall provide the local authority with a notice confirming that the fixed building services have been commissioned in accordance with approved procedures;
- the person responsible for calculating the CO_2 emission rate shall provide to the local authority a notice which specifies the TER, BER and/or DER;
- the person carrying out the work shall provide the local authority with a compliance notice.

Dwellings that have adopted approved construction details

For dwellings that have used an accredited design specification, air pressure testing should be:

- carried out on a unit of each dwelling type selected by the building control body;
- taken from the first completed batch of units of each dwelling type.

Note: For this test, a block of flats is treated as a separate development, irrespective of the number of blocks on the site.

Dwellings that have NOT adopted approved construction details

Air pressure tests for dwellings which have not adopted approved construction details should be carried out according to the dwelling type as specified in Table B19.

Table B19 Number of pressure tests for dwellings that have not adopted accredited construction details

Number of instances of the dwelling type	Number of tests to be carried out on the dwelling
4 or less	One test on each dwelling type
Greater than 4 but equal or less than 40	Two tests on each dwelling type
More than 40	At least 5% of the dwelling type, unless the first 5 units of the types are tested to achieve the design air permeability, when the sampling frequency can be substantially reduced to 2%

An alternative approach to specific pressure testing on development sites with no more than two dwellings is shown in Section 63 of Approved Document L1A.

The building control body (in consultation with the builder) will select dwellings making up the test sample so that about half of the scheduled tests for each dwelling type are carried out during construction of the first 25 per cent of each dwelling type.

If satisfactory performance is not achieved:

- remedial measures should be carried out on the dwelling and new tests completed until the dwelling achieves the criteria; and
- one additional dwelling of the same dwelling type will need to be tested, thereby increasing the overall sample size.

Note: For the purposes of Approved Document L1A, a block of flats should be treated as a separate development, irrespective of the number of blocks on the site.

Commissioning of heating and hot water systems

When heating and hot water systems are commissioned:

- the systems and their controls shall be left in their intended working order and should operate efficiently for the purposes of the conservation of fuel and power;
- the person carrying out the work shall provide the local authority with a notice (signed by a suitably qualified person) confirming that all fixed building services have been properly commissioned in accordance with the *Domestic Heating Compliance Guide.*

B.16.5 Criterion 5 – operating and maintenance instructions

Dwellings

The owner of the building should be provided with sufficient information (including operating and maintenance instructions) about the building, the fixed building services and their maintenance requirements so that the building can be operated and maintained efficiently and without consuming an unreasonable amount of fuel and power, and installed building services, plant and controls can be operated and maintained without consuming an unreasonable amount of fuel and power.

Note: This information will eventually be included in the Building Log Book (see CIBSE TM31 for guidance), as will the data used to calculate the TER and the DER.

In addition, the person carrying out the building work shall affix, as soon as practicable, in a conspicuous place in the dwelling, a notice stating the energy rating of the dwelling.

Guidance on the preparation of the notices is given in DTLR Circular 3/2001.

Buildings other than dwellings

The owner of the building should be provided with operating and maintenance instructions in the form of a building log book.

This log book should include instructions on making adjustments to the timing and temperature control settings and routine maintenance.

The person carrying out the building work shall affix a notice stating the energy rating of the dwelling.

Following completion of the work

On completion of the work, the building log book should be brought up to date with details of any:

- newly provided, renovated or upgraded thermal elements or controlled fittings;
- newly provided fixed building services (including details of their operation and maintenance);
- newly installed energy meters;
- other details that collectively enable energy consumption to be monitored and controlled.

B.17 Further information

B.17.1 Model designs

Some builders may prefer to adopt model design packages (containing details of U-values, boiler seasonal efficiencies, window opening allowances, etc.) instead of completing the design for themselves. The construction industry is in the process of developing these model designs and (according to the DCLG) these will eventually be available at http://www.modeldesigns.info.

B.17.2 Checklists

Appendix A to Approved Documents L1A and L2A contain an extensive checklist for demonstrating compliance to the requirements of Part L, together with an example of a completed checklist. This checklist can be downloaded via the DCLG website, but for your convenience an extract of this checklist has been reproduced in Appendix 1 (see below).

B.17.3 Design features

Appendix B to Approved Document L1A contains details about the important design features available in SAP 2005 (e.g. a checklist of design features that can help in determining whether the input data are correct, and whether compliance with Regulation 17C is in jeopardy). It is hoped that this information will prove very useful to both builders and building control bodies.

Appendix 1 – builders and building control bodies checklist

A1.1 Criterion 1

Dwellings

Check	Evidence	Produced by	Design OK?	As built OK?
TER (kgCO_2/m^{2}a)	Standard output from SAP calculation	SAP assessment	N/A	N/A
DER for dwelling as designed (kg CO_2/m^2.a)	Standard output from SAP calculation	SAP assessment	N/A	N/A
Are emissions from dwelling as designed less than or equal to the target?	Compare TER and DER as designed	SAP assessment		N/A

Buildings other than dwellings

Check	Evidence	Produced by	Design OK?	As built OK?
TER (kgCO_2/m^{2}a)	Standard output from accredited software	Developer		
BER (kgCO_2/m^{2}a)	Standard output from accredited software	Approved competent person or developer		
Are emissions from dwelling as designed less than or equal to the target?	Compare TER and BER as designed	Approved competent person or developer		
Are as built details the same as used in BER calculations?	Declaration	Developer	N/A	

A1.2 Criterion 2

Dwellings

Check	Evidence	Produced by	Design OK?	As built OK?
Fabric values				
Are all U-values better than the design limits?	Schedule U-values produced as standard output from SAP	SAP assessment		
Common areas in buildings with multiple dwellings (where relevant)				
If the common areas are un-heated, are all U-values better than the limits in Table 2? (If heated, use L2A)	Schedule of U-values	Builder's submission		
Heating and hot water systems				
Does the efficiency of the heating systems meet the minimum value set out in the *Domestic Heating Compliance Guide*?	Schedule of appliance efficiencies as standard output from SAP	SAP assessment		
Does the insulation of the hot water cylinder meet the standards set out in the *Domestic Heating Compliance Guide*?	Cylinder insulation specification as output from SAP	SAP assessment		
Do controls meet the minimum controls provision set out in the *Domestic Heating Compliance Guide*?	Controls specification as output from SAP	SAP assessment		
Does the heating and hot water system meet the minimum provisions in the *Domestic Heating Compliance Guide*?	Schedule of compliance provisions	Builder's submission		
Fixed internal and external lighting				
Does fixed internal lighting comply with paragraphs 42 to 44 of L1A?	Schedule of installed fixed internal lighting	Builder's submission		
Does the external lighting comply with paragraph 45 of L1A?	Schedule of installed external lighting	Builder's submission		

Buildings other than dwellings

Check	Evidence	Produced by	Design OK?	As built OK?
Building fabric				
Are all U-values better than the design limits?	Schedule U-values produced as standard output from accredited software	Approved competent person or developer		
Is air permeability no greater than the worse acceptable standard?	Standard output from approved	Approved competent person or developer		
Fixed building services				
Are all building services standards acceptable?	Schedule of system efficiencies produced as standard output from approved software	Approved competent person or developer		
Does fixed internal lighting comply with the requirements of L2A paragraphs 49 to 61	Schedule of installed fixed internal lighting	Developer, builder or electrical contractor who is a Part P approved competent person		
Are energy meters installed in accordance with GIL 65?	Meter strategy document	Developer		

A1.3 Criterion 3

Dwellings

Check	Evidence	Produced by	Design OK?	As built OK?
Does the building have a strong tendency to high summertime temperatures?	Prediction produced as standard output from SAP calculation	SAP assessment		

Buildings other than dwellings

Check	Evidence	Produced by	Design OK?	As built OK?
Is the combined solar and internal casual gains per unit floor area less than 35 W/m^2?	Schedule	Developer's submission		
Is the operative dry resultant temperature less than 28° C?	Schedule	Developer's submission		
Do the spaces served by comfort cooling systems meet the TER?	Schedule	Developer's submission		

A1.4 Criterion 4

Dwellings

Check	Evidence	Produced by	Design OK?	As built OK?
Have the key features of the design been included (or bettered) in practice?	List of key features produced as standard output from SAP to facilitate sample checking by the building control body and to enable the builder to control construction on site	SAP assessment	N/A	
Fabric construction				
Have accredited details been used?	Schedule of details used and their reference codes	Builder's submission		
Have non-accredited details been used?	Evidence that details conform to standards set out in IP 1/06	Builder's submission		
Has satisfactory documentary evidence of site inspection checks been produced?	Completed pro-formas showing checklists that have been completed	Builder's submission	N/A	
Design air permeability				
Design air permeability ($0\,m^3/(h\,m^2)$) at 50 Pa)	Standard output from SAP calculation	SAP assessment	N/A	
Has evidence been provided that demonstrates that the design air permeability has been satisfactorily achieved?	Sample pressure test results in comparison to design value	Builder's submission	N/A	
Commissioning of heating and water systems				
Evidence that the heating and water systems have been commissioned satisfactorily	Commissioning completion certificate	Builder's submission	N/A	

Buildings other than dwellings

Check	Evidence	Produced by	Design OK?	As built OK?
Building fabric				
Have the key features of the design been included (or bettered) in practice?	List of key features produced by accredited software to enable sample checking to be completed by the building control board	Building control body	N/A	
Is the level of thermal bridging acceptable?	Schedule of accredited details used and their reference codes. Evidence that details adopted deliver equivalent performance	Developer's submission		
Has satisfactory documentary evidence of site inspection checks been produced?	Completed pro-formas showing checklists that have been completed	Developer's submission		
Design air permeability				
Has the design air permeability $(m^3/(h\,m^2)$ at 50 Pa been achieved?	Standard output from approved software	Approved competent person or developer		
Commissioning of the fixed building services				
Has evidence been provided that demonstrates that the design air permeability has been satisfactorily achieved?	Pressure test results in comparison to design value or report on agreed program of design development and component testing or modular building type test results	Developer	N/A	
Has commissioning been completed satisfactorily?	Commissioning report submitted in accordance with CIBSE Code M?	Developer	N/A	
Has evidence been provided that demonstrates that the ductwork is sufficiently airtight?	Report confirming that the results of the leakage tests are in line with the leakage specification	Developer	N/A	

A1.5 Criterion 5

Dwellings

Check	Evidence	Produced by	Design OK?	As built OK?
Has all the relevant information been provided?	O&M instructions SAP rating as required by Regulation 16	Builder's submission	N/A	

Buildings other than dwellings

Check	Evidence	Produced by	Design OK?	As built OK?
Has a suitable building log book been provided?	Completed CIBSE TM31 template (or equivalent)	Developer		

Appendix C

Sound insulation

(A copy of this Appendix is available on www.routledge.com/books/details/9780415809696/)

C.1 Requirement E1

C.1.1 The requirement

Dwellings shall be designed so that the noise from domestic activity in an adjoining dwelling (or other parts of the building) is kept to a level that:

- *does not affect the health of the occupants of the dwelling;*
- *will allow them to sleep, rest and engage in their normal activities in satisfactory conditions.*

(Approved Document E1)

C.1.2 Meeting the requirement

Walls, floors and stairs that have a separating function should achieve the sound insulation values for dwelling houses and flats as set out in Table C1.	E0.1
Walls, floors and stairs that have a separating function should achieve the sound insulation values for rooms for residential purposes as set out in Table C1.	E0.1
For walls that separate rooms for residential purposes from adjoining dwelling houses and flats should achieve the sound insulation values for dwelling houses and flats as set out in Table C1.	E0.1

Note: The sound insulation values in these tables include a built-in allowance for 'measurement uncertainty' and so if any of these test values are not met, then that particular test will be considered as failed.

Note: Occasionally a higher standard of sound insulation may be required between spaces used for normal domestic purposes and noise generated in and to an adjoining communal or non-domestic space. In these cases it would be best to seek specialist advice before committing yourself.

Table C1 Dwelling houses and flats – performance standards for separating walls, separating floors and stairs that have a separating function

	Airborne sound insulation $D_{nT,w} + C_{tr}$ dB (maximum values)	Impact sound insulation $L'_{nT,w}$ dB (minimum values)
Purpose built dwelling houses and flats		
Walls	45	–
Floors and stairs	45	62
Dwelling houses and flats formed by material change of use		
Walls	43	–
Floors and stairs	43	64

Figure C1 illustrates the relevant parts of the building that should be protected from airborne and impact sound in order to satisfy Requirement E1.

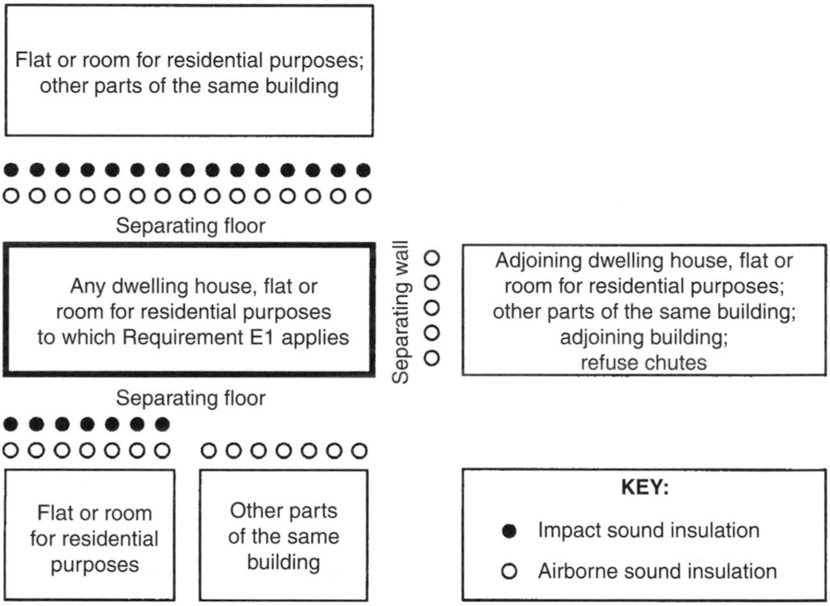

Figure C1 Requirement E1 – resistance to sound

Material change of use

In some circumstances (for example, when a historic building is undergoing a material change of use) it may not be practical to improve the sound insulation to the standards set out in Approved Document E1 particularly if the special characteristics of such a building need to be recognized. In these circumstances the aim should be to improve sound insulation to the 'extent that it is practically possible'.

Note: BS 7913:1998 *The Principles of the Conservation of Historic Buildings* provides guidance on the principles that should be applied when proposing work on historic buildings.

C.2 Requirement E2

> Constructions for new walls and floors within a dwelling-house   E0.9
> (flat or room for residential purposes) – whether purpose built
> or formed by a material change of use – shall meet the labora-
> tory sound insulation values set out in Table C2.

Table C2 Laboratory values for new internal walls within dwelling houses, flats and rooms for residential purposes – whether purpose built or formed by a material change of use

	Airborne sound insulation R_W dB (minimum values)
Purpose built dwelling houses and flats	
Walls	40
Floors	40

Figures C2 and C3 illustrate the relevant parts of the building that should be protected from airborne and impact sound in order to satisfy Requirement E2.

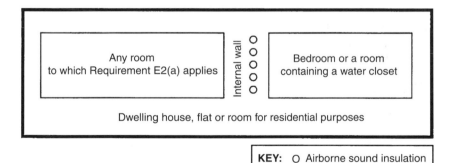

Figure C2 Requirement E2a – internal walls

C.3 Requirement E3

> Sound absorption measures described in Section 7 of   E0.11
> Approved Document N shall be applied.

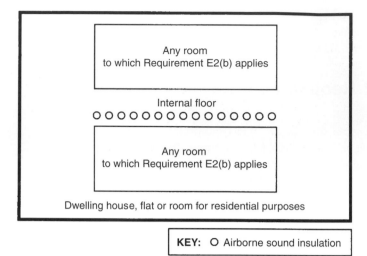

Figure C3 Requirement E2(b) – internal floors

C.4 Requirement E4

> The values for sound insulation, reverberation time and indoor   E0.12
> ambient noise as described in Section 1 of Building Bulletin
> 93 *The Acoustic Design of Schools* (produced by DFES and
> published by the Stationery Office (ISBN 0 11 271105 7))
> shall be satisfied.

C.5 Sound insulation testing

Sound insulation testing has to be completed for:

(a) purpose built dwelling houses and flats;
(b) dwelling houses and flats formed by material change of use;
(c) purpose built rooms for residential purposes;
(d) rooms for residential purposes formed by material change of use.

The person carrying out the building work is responsible for ensuring that sound insulation testing is carried out by a test body with appropriate third-party accreditation (preferably UKAS accredited) for completing field measurements. The person is also responsible for the cost of the testing.

 Note: The procedures for sound insulation testing are described in Annex B to Part E of the Regulations.

Sound insulation testing should be carried out in accordance with the procedure described in Annex B of Approved Document E.	E0.3
The person carrying out the building work should arrange for sound insulation testing to be carried out by a test body with appropriate third-party accreditation.	E0.4
Test bodies conducting testing should preferably have UKAS accreditation (or a European equivalent) for field measurements.	E0.4
Sound insulation testing (to demonstrate compliance with Requirement E1) should be carried out on-site as part of the construction process (i.e. pre-completion testing).	E1.2
Testing should not be carried out between living spaces, corridors, stairwells or hallways.	E1.8
Tests should be carried out between rooms or spaces that share a common area of separating wall or separating floor.	E1.9
Tests should be carried out once the dwelling houses, flats or rooms for residential purposes either side of a separating element are essentially complete, except for decoration.	E1.10
Impact sound insulation tests should be carried out without a soft covering (e.g. carpet, foam backed vinyl etc.) on the floor.	E1.10

 Note: Some properties, for example loft apartments, may be sold before being fitted out with internal walls and other fixtures and fittings. In these cases sound insulation measurements should be made between the available spaces.

Table C3 shows the types of tests that have to be carried out on dwelling houses and flats.

C.5.1 Failed tests

In the event of a failed set of tests, 'appropriate remedial treatment' should be applied to the rooms that failed the test.	E1.33
After a failed set of tests, the rate of testing should be increased until the building control body is satisfied that the problem has been solved.	E1.36

Table C3 Sets of tests

	A test of insulation against airborne sound between one pair of rooms (where possible suitable for use as living rooms) on opposite sides of the separating wall	A test of insulation against airborne sound between another pair of rooms (where possible suitable for use as bedrooms) on opposite sides of the separating wall	Tests of insulation against both airborne and impact sound between one pair of rooms (where possible suitable for use as living rooms) on opposite sides of the bedrooms) separating floor	Tests of insulation against both airborne and impact sound between another pair of rooms (where possible suitable for use as on opposite sides of the separating floor
Sets of tests in dwelling–houses (including bungalows)	Yes	Yes		
Sets of tests in flats with separating floors but without separating walls			Yes	Yes
Sets of tests in flats with a separating floor and a separating wall	Yes	Yes	Yes	Yes

 Note: To conduct a full set of tests, access to at least three flats will be required.

C.5.2 Remedial treatment

Appropriate remedial treatment should be applied following a failed set of tests.

Note: Guidance is available in BRE information paper IP 14/02.   E1.37

Where remedial treatment has been completed, the building control body should be satisfied with its efficacy – normally this will be assessed through additional sound insulation testing.   E1.39

Building control bodies should be satisfied that everything reasonable has been done to improve the sound insulation.   E1.40

Appendix D

Guidance to the requirements of Part P – electrical safety

(A copy of this Appendix in available on www.routledge.com/books/details/9780415809696/)

D.1 Background

For many years, the UK has managed to maintain relatively high electrical safety standards with the support of guidance based on BS 7671, but with a growing number of electrical accidents occurring in the 'home', the government have been forced to consider legal requirement for safety in electrical installation work in dwellings.

As from 1 January 2005, therefore, **all** new electrical wiring or electrical components for domestic premises (or small commercial premises linked to domestic accommodation) have had to be designed and installed in accordance with the Building Regulations, and in particular, Part P which is based on the fundamental principles set out in Chapter 13 of the BS 7671 (i.e. *The IEE Wiring Regulations*). In addition, all fixed electrical installations (i.e. wiring and appliances that are attached to, in or part of a building, such as socket outlets, switches, consumer units and ceiling fittings on the consumer's side of the electricity supply meter) have now to be designed, installed, inspected, tested and certified to BS 7671.

Part P also introduced the requirement for new cable core colours for ac power circuits and with effect 31 March 2006, **all** new installations or alterations to existing installations **must** use these new (harmonized) colour cables. (Further information, concerning cable identification colours for extra-low voltage and dc power circuits, is available from the IET website at www.iee. org/cablecolours.)

For single phase installations in domestic premises, the new colours are the same as those for flexible cables to appliances (namely green-and-yellow, blue and brown for the protective, neutral and phase conductors respectively).

 Note: Part P applies only to fixed electrical installations that are intended to operate at low voltage or extra-low voltage which are not controlled by the Electricity Supply Regulations 1988 as amended, or the Electricity, Safety, Quality and Continuity Regulations, 2002 as amended.

Table D1 Identification of conductors in ac power and lighting circuits

Conductor	Colour
Protective conductor	Green-and-yellow
Neutral	Blue
Phase of single phase circuit	Brown
Phase 1 of 3 phase circuit	Brown
Phase 2 of 3 phase circuit	Black
Phase 3 of 3 phase circuit	Grey

D.2 What is the aim of Approved Document P?

The aim of Part P is to increase the safety of householders by improving the design, installation, inspection and testing of electrical installations in dwellings when they (i.e. the installations) are being newly built, extended or altered.

It is understood that the government is also intending to introduce a scheme whereby domestic installations are checked at regular intervals (as well as when they are sold and/or purchased) to make sure that they comply. This would mean, of course, that if you had an installation which was not correctly certified, then your house insurance might well **not** be valid!

D.3 Who is responsible for electrical safety?

The owner – needs to determine whether the works being carried out are either minor or notifiable work. If the work is notifiable, then the owner needs to make sure that the person(s) carrying out the work is either registered under one of the self-certified schemes (see Figure D1) or is able to certify their work under the local authority building control route.

The designer – needs to ensure that all electrical work is designed, constructed, inspected and tested in accordance with the BS 7671 and either falls under a competent persons scheme or the local authority building control approval route.

The builder/developer – needs to ensure that they have electricians who can self-certify their work or who are qualified/experienced enough to enable them to sign off under the Electrical Installation Certification form.

D.4 What are the statutory requirements?

In future all electrical installations need to:

- be designed and installed to protect against mechanical and thermal damage;
- be designed and installed so that they will not present an electrical shock or fire hazard;

Authorized competent person self-certification schemes for installers who can do all electrical installation work	Authorized competent person self-certification schemes for installers who can do electrical work only if it is necessary when they are carrying out other work
SUPPORTED BY  **BRE Certification Ltd** Phone: 0870 609 6093 www.bre.co.uk	**Benchmark Certification Ltd** Phone: 0844 879 4798 www.benchmark-certification.com
British Standards Institution Phone: 01442 230442 www.bsi-global.com/kitemark	 **ELECSA Limited** Phone: 0845 634 9043 www.elecsa.org.uk
 **ELECSA Limited** Phone: 0845 634 9043 www.elecsa.org.uk	 **NAPIT Certification Ltd** Phone: 0845 643 0330 www.napit.org.uk
NAPIT Certification Ltd Phone: 0845 643 0330 www.napit.org.uk	**NICEIC Certification Services Ltd** Phone: 0870 013 0382 www.niceic.com
NICEIC Certification Services Ltd Phone: 0870 013 0382 www.niceic.com	 **OFTEC (Oil Firing Technical Association)** Phone: 0845 658 5080 www.oftec.co.uk

Figure D1 Authorized competent person self-certification schemes for installers

- be tested and inspected to meet relevant equipment/installation standards;
- provide sufficient information so that persons wishing to operate, maintain or alter an electrical installation can do so with reasonable safety;
- comply with such requirements placed by:

  - Part A (Structure): depth of chases in walls, and size of holes and notches in floor and roof joists;
  - Part B (Fire safety): safety of certain electrical installations; provision of fire alarm and fire detection systems; fire resistance of penetrations through floors and walls;
  - Part C (Site preparation and resistance to moisture): moisture resistance of cable penetrations through external walls;
  - Part E (Resistance to the passage of sound): penetrations through floors and walls;
  - Part F (Ventilation): ventilation rates for dwellings;
  - Part L (Conservation of fuel and power): energy efficient lighting; reduced current carrying capacity of cables in insulation;
  - Part M (Access to and use of buildings): heights of switches and socket outlets.

D.5 What does all this mean?

With a few exceptions any electrical work undertaken in your home which includes the addition of a new electrical circuit, or involves work in your:

- kitchen;
- bathroom;
- garden area;

must from 1 January 2005 be reported to the local authority Building Control for inspection. This includes any work undertaken professionally, by you or another family member, or by a friend.

The **only** exception is when the installer has been approved by a Competent Persons organization such as ELECSA (see Figure D1).

D.6 What types of building does Approved Document P cover?

Part P applies to all electrical installations in (and around) buildings or parts of buildings comprising:

- dwelling houses and flats;
- dwellings and business premises that have a common metered supply;
- in or on land associated with domestic buildings;
- fixed lighting and pond pumps in gardens;
- in shops and public houses with a flat above with a common meter;

- common access areas in blocks of flats such as corridors and stairways;
- shared amenities of blocks of flats such as laundries and gymnasiums.

Table D2 provides the details of works that are notifiable to local authorities and/or must be completed by a company registered as a 'competent firm'.

Table D2 Notifiable work

Locations where work is being completed	Extensions and modifications to circuits	New circuits
Bathrooms	Yes	Yes
Bedrooms	Yes	Yes
Communal area of flats	Yes	Yes
Conservatories		Yes
Dining rooms		Yes
Garages (integral)		Yes
Garages (detached)		Yes*
Greenhouses	Yes	Yes
Halls		Yes
Hot air saunas	Yes	Yes
Kitchen	Yes	Yes
Kitchen diners	Yes	Yes
Landings		Yes
Lounge		Yes
Paddling pools	Yes	Yes
Remote buildings	Yes	Yes
Sheds	Yes	Yes
Shower rooms	Yes	Yes
Stairways		Yes
Studies		Yes
Swimming pools	Yes	Yes
TV rooms		Yes
Workshops (remote)	Yes	Yes

* if the installation requires outdoor wiring.

D.7 What is a competent firm?

For the purposes of Part P, the government has defined 'competent firms' as electrical contractors:

- who work in conformance with the requirements to BS 7671;
- whose standard of electrical work has been assessed by a third party;

- who are registered under the NICEIC Approved Contractor scheme and the Electrotechnical Assessment Scheme.

D.8 What is a competent person responsible for?

When installation work is undertaken by a competent person, that person is responsible for:

- ensuring compliance with BS 7671:2001 and all relevant Building Regulations;
- providing the person ordering the work with a signed Building Regulations compliance certificate;
- providing the relevant building control body with an information copy of the certificate;
- providing the person ordering the work with a completed Electrical Installation Certificate.

D.9 Who is entitled to self-certify an installation?

Part P affects **every** electrical contractor carrying out fixed installation and alteration work in homes. **Only** registered installers are entitled to self-certify the electrical work, however **and** they must be registered as a competent person under one of the following schemes.

Working with industry and consumer organizations, the government has recently developed the TrustMark initiative for builders and specialist firms that work on the home. Schemes that are capable of delivering 'agreed competence and customer-care standards' are approved to use the TrustMark brand by a Board consisting of industry and consumer representatives, with government observers. The brand is owned by DTI, which licences it to the Board. (see Figure D2).

Figure D2 The TrustMark initiative

The TrustMark replaces the Quality Mark scheme which closed on 31 December 2004 because too few firms joined. For more information about TrustMark see the website at: www.trustmark.org.uk .

D.10 What are the consequences of not obtaining approval?

Failure to comply with the requirements of Part P is a criminal offence and local authorities have the power to require the removal or alteration of work that does not comply with the Building Regulations. In addition, a completion certificate for works will not be issued – which could cause severe problems in the future such as:

- the electrical installation may not be safe;
- you will have no record of the work done;
- you may have difficulty in selling your home without records of the installation and the relevant safety certificates.

D.11 Do I have to inform the local authority Building Control body?

All proposals to carry out electrical installation work must be notified to the local authority's building control body before work begins, **unless** the proposed installation work is undertaken by a person registered as a competent person under a government-approved Part P self-certification scheme or the work is agreed non-notifiable work such as:

- connecting an electric gate or garage door to an existing isolator (**but**, be careful, the installation of the circuit up to the isolator **is** notifiable!);
- fitting and replacing cookers and electric showers (unless a new circuit is required);
- installing equipment (e.g. security lighting, air conditioning equipment and radon fans) that is attached to the outside wall of a house (unless there are exposed outdoor connections and/or the installation is a new circuit, or an extension of a circuit in a kitchen, or special location, or is associated with a special installation);
- installing fixed equipment where the final connection is via a 13A plug and socket (unless it involves fixed wiring and the installation of a new circuit or the extension of a circuit in a kitchen or special location);
- installing prefabricated, 'modular' systems such as kitchen lighting systems and armoured garden cabling that are linked by plug and socket connectors (provided that products are CE-marked and that any final connections in kitchens and special locations are made to existing connection units or points – e.g. a 13A socket outlet);

Appendix D

- installing or upgrading main or supplementary equipotential bonding (provided that the work complies with other applicable legislation, such as the Gas Safety (Installation and Use) Regulations);
- installing mechanical protection to existing fixed installations (provided that the circuit's protective measures and current-carrying capacity of conductors are unaffected by increased thermal insulation);
- refixing or replacing the enclosures of existing installation components;
- replacement, repair and maintenance jobs;
- replacing fixed electrical equipment (e.g. socket outlets, control switches and ceiling roses) which do not require the provision of any new fixed cabling;
- replacing the cable of a single circuit cable (where damaged, for example, by fire, rodent or impact – provided that the replacement cable has the same or greater current-carrying capacity, follows the same route and does not serve more than one sub-circuit through a distribution board);
- work that is not in a kitchen or special location, which does not involve a special installation and which only consists of:

  - adding lighting points (light fittings and switches) to an existing circuit;
  - adding socket outlets and fused spurs to an existing ring or radial circuit (provided that the existing circuit protective device is suitable and supplies adequate protection for the modified circuit);

- work that is not in a special location and only concerns:

  - adding telephone, extra-low voltage wiring and equipment for communications, information technology, signalling, control and similar purposes;
  - adding prefabricated equipment sets (and their associated flexible leads) with integral plug and socket connections.

All of this work can be completed by a DIY enthusiast (family member or friend) but still needs to be installed in accordance with manufacturers' instructions and done in such a way that they do not present a safety hazard. This work does **not** need to be notified to a local authority building control body (unless it is installed in an area of high risk such as a kitchen or a bathroom etc.) **but** all DIY electrical work (unless completed by a qualified professional – who is responsible for issuing a Minor Electrical Installation Certificate) *will still need to be checked, certified and tested by a competent electrician* .

Any work that involves adding a new circuit to a dwelling will need to be either notified to the building control body (who will then inspect the work) or needs to be carried out by a competent person who is registered under a government-approved Part P self-certification scheme.

Work involving any of the following will also have to be notified:

- consumer unit replacements;
- electric floor or ceiling heating systems;

- extra-low voltage lighting installations (other than pre-assembled, CE-marked lighting sets);
- garden lighting and/or power installations;
- installation of a socket outlet on an external wall;
- installation of outdoor lighting and/or power installations in the garden or that involves crossing the garden;
- installation of new central heating control wiring;
- solar Photovoltaic (pv) power supply systems;
- small scale generators (such as microCHP units).

Note: Where a person who is **not** registered to self-certify intends to carry out the electrical installation, then a Building Regulation (i.e. a Building Notice or Full Plans) application will need to be submitted together with the appropriate fee, based on the estimated cost of the electrical installation. The building control body will then arrange to have the electrical installation inspected at first fix stage and tested upon completion.

In any event the electrical work will still need to be certified under BS 7671 by a suitably competent person who will be responsible for the design, installation, inspection and testing of the system (on completion) and have the confidence of completing a certificate to say that the work is satisfactory and complies with current codes of practice.

The main things to remember are:

- is the work notifiable or non-notifiable?
- does the person undertaking the work need to be registered as a competent person?
- what records (if any) need to be kept of the installation?

Figure D3 is a quick guide to the requirements.

D.11.1 Work completed by yourself, a friend or a relative

You do **not** need to tell your local authority's Building Control Department about non-notifiable work such as:

- repairs, replacements and maintenance work;
- extra power points or lighting points or other alterations to existing circuits (**unless** that are in a kitchen, bathroom, or is outdoors).

You **do** need to tell them about most other work.

Note: if you are not sure about this, or you have any questions, ask your local authority's building control department.

D.11.2 Work completed by a contractor or an installer

If the work is of a notifiable nature then the installer(s) must be registered with one of the schemes shown in Figure D1 .

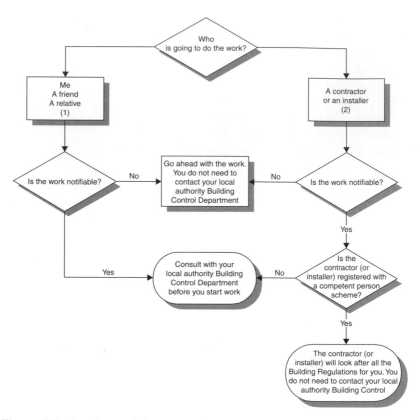

Figure D3 How to meet the new rules

D.12 What inspections and tests will have to be completed and recorded?

As shown in Table D3, there are four types of electrical installation certificates and one Building Regulations compliance certificate that have to be completed.

Figure D4 indicates how to chose what type of inspection is required.

D.13 What should be included in the records of the installation?

All 'original' certificates should be retained in a safe place and be available to any person inspecting or undertaking further work on the electrical installation in the future. If you later vacate the property, this certificate will demonstrate to the new owner that the electrical installation complied with the requirements of British Standard 7671 at the time that the certificate was issued. The Construction (Design and Management) Regulations require that for a project covered by those

Table D3 Types of installation

Type of inspection	When is it used?	What should it contain?	Remarks
Minor Electrical installation Works Certificate (see Appendix 1)	For additions and alterations to an installation that do not extend to the provision of a new circuit	Relevant provisions of Part 7 of BS 7671	An example of a minor electrical installation could be (for example) the addition of a socket outlet or a lighting point to an existing circuit
Electrical installation Certificate (short form) (see Appendix 2)	For use when one person is responsible for the design, construction, inspections and testing of an installation	A schedule of inspections and a schedule of test results as required, by Part 7 (of BS 7671)	
Electrical installation certificate Certificate (full) (see Appendix 3)	For the initial certification of a new installation or for the alteration and/or addition to an existing installation where new circuits have been introduced	A schedule of inspections and test results as required by Part 7 (of BS 7671). A certificate, including guidance for recipients (standard form from Appendix 6 of BS 7671)	An electrical installation certificate is not to be used for a periodic inspection
Periodic Inspection Report (see Appendix 4)	For the inspection of an existing electrical installation	A schedule of inspections and a schedule of test results as required by Part 7 (of BS 7671)	For safety reasons, electrical installations need to be inspected at appropriate intervals by a competent person
Building Regulations Compliance Certificate (see Appendix 5)	Confirmation that the work carried out complies with the Building Regulations	The basic details of the installation, the location, completion date and name of the installer	This document may be requested by a purchaser's solicitor when you sell your property.

**Appendix
D**

Regulations, a copy of this certificate, together with the schedules, is included in the project health and safety documentation.

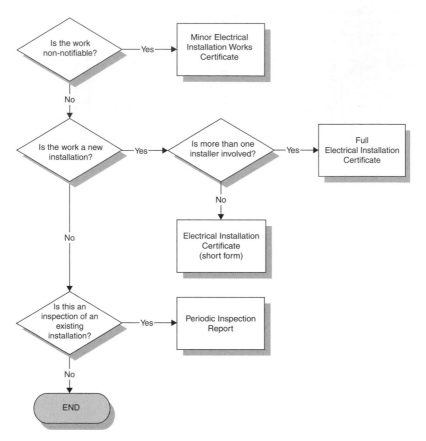

Figure D4 Choosing the correct inspection certificate

D.14 Where can I get more information?

Further guidance concerning the requirements of Part P (Electrical Safety) is available from the:

- IET (Institution of Engineering and Technology) at http://www.iee.org/publish/wireregs/partp.cfm;
- The NICEIC (National Inspection Council for Electrical Installation Contracting) at www.niceic.org.uk;
- the ECA (Electrical Contractors' Association) at www.eca.co.uk.

To download .pdf copies, go to the following:

- for Part P, http://www.planningportal.gov.uk/uploads/br/BR_AD_2006.pdf
- for details of fixed wire colour changes, http://www.niceic.org.uk/downloads/WiringSupp.pdf.

Appendix 1

Minor electrical installation works certificate

(REQUIREMENTS FOR ELECTRICAL INSTALLATIONS - BS 7671 [IEE WIRING REGULATIONS])

To be used only for minor electrical work which does not include the provision of a new circuit

---

PART 1: Description of minor works

1. Description of the minor works: ....

2. Location/Address: ....

3. Date minor works completed: ....

4. Details of departures, if any, from BS 7671

   ....

   ....

   ....

PART 2: Installation details

1. System earthing arrangement: TN-C-S ☐   TN-S ☐   TT ☐

2. Method of protection against indirect contact: ....

3. Protective device for the modified circuit: Type BS ....   Rating ....A

4. Comments on existing installation, including adequacy of earthing and bonding arrangements:
   (see Regulation 130-07) ....

   ....

   ....

   ....

PART 3: Essential Tests

1. Earth continuity: satisfactory   ☐

2. Insulation resistance:

   Phase/neutral ....MΩ

   Phase/earth ....MΩ

   Neutral/earth ....MΩ

3. Earth fault loop impedance: ....Ω

4. Polarity: satisfactory   ☐

5. RCD operation (if applicable): Rated residual operating current $I_{\Delta n}$ ....mA and operating time of ....ms (at $I_{\Delta n}$)

PART 4: Declaration

1. I/We CERTIFY that the said works do not impair the safety of the existing installation, that the said works have been designed, constructed, inspected and tested in accordance with BS 7671: .... (IEE Wiring Regulations), amended to .... and that the said works, to the best of my/our knowledge and belief, at the time of my/our inspection, complied with BS 7671 except as detailed in Part 1.

2. Name: ....   3. Signature: ....

   For and on behalf of: ....   Position: ....

   Address: ....

   ....   Date: ....

   ....

   ....Postcode: ....

Note: check with the latest issue of BS 7671 to make sure that there have been no changes to this form.

Appendix 2

Electrical installation certificate (notes 1 and 2)
(REQUIREMENTS FOR ELECTRICAL INSTALLATIONS - BS 7671 [IEE WIRING REGULATIONS])

DETAILS OF THE CLIENT (note 1)

....

....

....

INSTALLATION ADDRESS

....

....

....Postcode ....

DESCRIPTION AND EXTENT OF THE INSTALLATION Tick boxes as appropriate

Description of installation:	New installation ☐
Extent of installation covered by this Certificate:	
....	Addition to an
....	existing installation ☐
....	
....	Alteration to an
....	existing installation ☐

FOR DESIGN, CONSTRUCTION, INSPECTION & TESTING

I being the person responsible for the Design, Construction, Inspection & Testing of the electrical installation (as indicated by my signature below), particulars of which are described above, having exercised reasonable skill and care when carrying out the Design, Construction, Inspection & Testing, hereby CERTIFY that the said work for which I have been responsible is to the best of my knowledge and belief in accordance with BS 7671 : ...., amended to .... (date) except for the departures, if any, detailed as follows:

> Details of departures from BS 7671 (Regulations 120-01-03, 120-02):

The extent of liability of the signatory is limited to the work described above as the subject of this Certificate.

Name (IN BLOCK LETTERS):...   Position: ....

Signature (note 3): ....   Date:...

For and on behalf of: ....

Address: ....

....

....Postcode ....   Tel No: ....

NEXT INSPECTION

I recommend that this installation is further inspected and tested after an interval of not more than .... years/months (notes 4 and 7)

SUPPLY CHARACTERISTICS AND EARTHING ARRANGEMENTS Tick boxes and enter details, as appropriate

Earthing arrangements	Number and Type of Live Conductors		Nature of Supply Parameters	Supply Protective Device Characteristics
TN-C ☐	a.c. ☐	d.c ☐	Nominal voltage, U/Uo[1] V	
TN-S ☐	1-phase, 2-wire ☐	2-pole ☐	Nominal frequency, f [1] ....Hz	Type:........................
TN-C-S ☐	1-phase, 3-wire ☐	3-pole ☐	Prospective fault current, Ipf [2] (note 6) ....kA	....
TT ☐	2-phase, 3-wire ☐	other ☐	External loop impedance, Ze [2] ....Ω	
IT ☐	3-phase, 3-wire ☐		(Note: (1) by enquiry, (2) by enquiry or by measurement)	Nominal current rating
	3-phase, 4-wire ☐			............................A
Alternative source ☐ of supply (to be detailed on attached schedules)				

Appendix 2   (Continued)

PARTICULARS OF INSTALLATION REFERRED TO IN THE CERTIFICATE Tick boxes and enter details, as appropriate

Means of Earthing	**Maximum Demand**

Distributor's facility ☐   Maximum demand (load)..Amps per phase

Details of Installation Earth Electrode (where applicable)

Installation   Type   Location   Electrode resistance to earth
earth electrode ☐   (e.g. rod(s), tape etc)

....Ω

Main Protective Conductors

Earthing conductor: material .... csa ....mm²   connection verified   ☐

Main equipotential
bonding conductors   material .... csa ....mm²   connection verified   ☐

To incoming water and/or gas service ☐   To other elements ....

Main Switch or Circuit-breaker

BS, Type ....   No. of poles ....   Current rating ....A   Voltage rating V

Location ....   Fuse rating or setting ....A

Rated residual operating current I ∆n =....................... mA, and operating time of............ms (at I ∆n)
(applicable only where an RCD is suitable and is used as a main circuit-breaker)

COMMENTS ON EXISTING INSTALLATION: (In the case of an alteration or additions see Section 743)

....

....

....

....

....

....

....

....

SCHEDULES (note 2)

The attached Schedules are part of this document and this Certificate is valid only when they are attached to it.

.... Schedules of Inspections and .... Schedules of Test Results are attached.

(Enter quantities of schedules attached)

Appendix D

Appendix 3

Form 2   Form No. 12

Electrical installation certificate (notes 1 and 2)

(REQUIREMENTS FOR ELECTRICAL INSTALLATIONS - BS 7671 [IEE WIRING REGULATIONS])

| DETAILS OF THE CLIENT (note 1) .... |
| .... |
| .... |

| INSTALLATION ADDRESS .... |
| .... |
| ....Postcode... |

DESCRIPTION AND EXTENT OF THE INSTALLATION Tick boxes as appropriate (note 1)

Description of installation: ....

Extent of installation covered by this Certificate: ....

....

....

....

New installation   ☐

Addition to an existing installation   ☐

Alteration to an existing installation   ☐

FOR DESIGN

I/We being the person(s) responsible for the design of the electrical installation (as indicated by my/our signatures below), particulars of which are described above, having exercised reasonable skill and care when carrying out the design, hereby CERTIFY that the design work for which I/we have been responsible is to the best of my/our knowledge and belief in accordance with BS 7671: ...., amended to.......... (date) except for the departures, if any, detailed as follows:

> Details of departures from BS 7671 (Regulations 120-01-03, 120-02):

The extent of liability of the signatory or the signatories is limited to the work described above as the subject of this Certificate. For the DESIGN of the installation.   **(Where there is mutual responsibility for the design)

Signature: .... Date .... Name (BLOCK LETTERS): .... Designer No. 1

Signature: .... Date .... Name (BLOCK LETTERS): .... Designer No. 2**

FOR CONSTRUCTION

I/We being the person(s) responsible for the construction of the electrical installation (as indicated by my/our signatures below), particulars of which are described above, having exercised reasonable skill and care when carrying out the construction, hereby CERTIFY that the construction work for which I/we have been responsible is to the best of my/our knowledge and belief in accordance with BS 7671: ....amended to ....(date) except for the departures, If any, detailed as follows:

> Details of departures from BS 7671 (Regulations 120-01-03, 120-02):

The extent of liability of the signatory is limited to the work described above as the subject of this Certificate. For CONSTRUCTION of the installation:

Signature: .... Date ....

Name (BLOCK LETTERS): .... Constructor

FOR INSPECTION & TESTING

I/We being the person(s) responsible for the inspection & testing of the electrical installation (as indicated by my/our signatures below), particulars of which are described above, having exercised reasonable skill and care when carrying out the inspection & testing, hereby CERTIFY that the work for which I/we have been responsible is to the best of my knowledge and belief in accordance with BS 7671:........., amended to .... (date) except for the departures, if any, detailed as follows:

> Details of departures from BS 7671 (Regulations 120-01-03, 120-02):

The extent of liability of the signatory is limited to the work described above as the subject of this Certificate. For INSPECTION & TEST of the installation: **(Where there is mutual responsibility for the design)

Signature: .... Date ....

Name (BLOCK LETTERS): .... Inspector

NEXT INSPECTION (notes 4 and 7)

I/We the designer(s) recommend that this installation is further inspected and tested after an interval of not more than............. years/months

Appendix 3   (Continued)

PARTICULARS OF THE SIGNATORIES TO THE ELECTRICAL INSTALLATION CERTIFICATE (note 3)

Designer (No 1)
Name: .... Company: ....

Address: ....

.... Postcode: .... Tel No: ....

Designer (No 2)
(if applicable)   Name: .... Company: ....

Address: ....

.... Postcode: .... Tel No: ....

Constructor
Name: .... Company: ....

Address: ....

.... Postcode: .... Tel No: ....

Inspector
Name: .... Company: ....

Address: ....

.... Postcode: .... Tel No: ....

SUPPLY CHARACTERISTICS AND EARTHING ARRANGEMENTS Tick boxes and enter details, as appropriate

Earthing arrangements	Number and Type of Live Conductors	Nature of Supply Parameters	Supply Protective Device Characteristics
TN-C ☐	a.c. ☐ d.c ☐	Nominal voltage, U/Uo$^{(1)}$ V	
TN-S ☐	1-phase, 2-wire ☐ 2-pole ☐	Nominal frequency, f $^{(1)}$ ....Hz	Type: ....
TN-C-S ☐	1-phase, 3-wire ☐ 3-pole ☐	Prospective fault current, Ipf $^{(2)}$ (note 6) ....kA	
TT ☐	2-phase, 3-wire ☐ other ☐	External loop impedance, Ze $^{(2)}$ ....Ω	
IT ☐	3-phase, 3-wire ☐	(Note: (1) by enquiry, (2) by enquiry or by measurement)	Nominal current rating
	3-phase, 4-wire ☐		...........................A
Alternative source ☐ of supply (to be detailed on attached schedules)			

PARTICULARS OF INSTALLATION REFERRED TO IN THE CERTIFICATE Tick boxes and enter details, as appropriate

Means of Earthing

Distributor's facility ☐

Maximum Demand

Maximum demand (load) ....Amps per phase

Details of Installation Earth Electrode (where applicable)

Installation earth electrode ☐

Type (e.g. rod(s), tape etc)	Location	Electrode resistance to earth
....	....	....Ω

Main Protective Conductors

Earthing conductor: material .... csa ....mm² connection verified ☐

Main equipotential bonding conductors material .... csa ....mm² connection verified ☐

To incoming water and/or gas service ☐   To other elements ....

Main Switch or Circuit-breaker

BS, Type............................... No. of poles .... Current rating ....A Voltage rating..................................V

Location .... Fuse rating or setting ....A

Rated residual operating current I $_{\Delta n}$ = ... mA, and operating time of ....ms (at I $_{\Delta n}$)

(applicable only where an RCD is suitable and is used as a main circuit-breaker)

COMMENTS ON EXISTING INSTALLATION: (In the case of an alteration or additions see Section 743)

....

....

....

SCHEDULES (note 2)
The attached Schedules are part of this document and this Certificate is valid only when they are attached to it.
.... Schedules of Inspections and .... Schedules of Test Results are attached. (Enter quantities of schedules attached)

Appendix D

Appendix 4

Form 3   Form No. 13

Schedule of inspections

Methods of protection against electric shock

(a) Protection against both direct and indirect contact:

☐ (i) SELV (note 1)

☐ (ii) Limitation of discharge of energy

(b) Protection against direct contact: (note 2)

☐ (i) Insulation of live parts

☐ (ii) Barriers or enclosures

☐ (iii) Obstacles (note 3)

☐ (iv) Placing out of reach (note 4)

☐ (v) PELV

☐ (vi) Presence of RCD for supplementary protection

(c) Protection against indirect contact:

(i) EEBADS including:

☐ Presence of earthing conductor

☐ Presence of circuit protective conductors

☐ Presence of main equipotential bonding conductors

☐ Presence of supplementary equipotential bonding conductors

☐ Presence of earthing arrangements for combined protective and functional purposes

☐ Presence of adequate arrangements for alternative source(s), where applicable

☐ Presence of residual current device(s)

☐ (ii) Use of Class II equipment or equivalent insulation (note 5)

☐ (iii) Non-conducting location: (note 6) Absence of protective conductors

☐ (iv) Earth-free equipotential bonding: (note 7) Presence of earth-free equipotential bonding conductors

☐ (v) Electrical separation (note 8)

Prevention of mutual detrimental influence

☐ (a) Proximity of non-electrical services and other influences

☐ (b) Segregation of band I and band II circuits or band II insulation used

☐ (c) Segregation of safety circuits

Identification

☐ (a) Presence of diagrams, instructions, circuit charts and similar information

☐ (b) Presence of danger notices and other warning notices

☐ (c) Labelling of protective devices, switches and terminals

☐ (d) Identification of conductors

Cables and conductors

☐ (a) Routing of cables in prescribed zones or within mechanical protection

☐ (b) Connection of conductors

☐ (c) Erection methods

☐ (d) Selection of conductors for current-carrying capacity and voltage drop

☐ (e) Presence of fire barriers, suitable seals and protection against thermal effects

General

☐ (a) Presence and correct location of appropriate devices for isolation and switching

☐ (b) Adequacy of access to switchgear and other equipment

☐ (c) Particular protective measures for special installations and locations

☐ (d) Connection of single-pole devices for protection or switching in phase conductors only

☐ (e) Correct connection of accessories and equipment

☐ (f) Presence of undervoltage protective devices

☐ (g) Choice and setting of protective and monitoring devices for protection against indirect contact and/or overcurrent

☐ (h) Selection of equipment and protective measures appropriate to external influences

☐ (i) Selection of appropriate functional switching devices

Inspected by ....   Date ....

Notes:

T   to indicate an inspection has been carried out and the result is satisfactory
C   to indicate an inspection has been carried out and the result was unsatisfactory
N/A   to indicate the inspection is not applicable
LIM   to indicate that, exceptionally, a limitation agreed with the person ordering the work prevented the inspection or test being carried out

1. SELV – an extra-low voltage system which is electrically separated from earth and from other systems. The particular requirements of the Regulations must be checked (see Regulations 411-02 and 471-02)

2. Method of protection against direct contact – will include measurement of distances where appropriate

3. Obstacles – only adopted in special circumstances (see Regulations 412-04 and 471-06)

4. Placing out of reach – only adopted in special circumstances (see Regulations 412-05 and 471-07)

5. Use of Class II equipment – infrequently adopted and only when the installation is to be supervised (see Regulations 413-03 and 471-09)

6. Non-conducting locations – not applicable in domestic premises and requiring special precautions (see Regulations 413-04 and 471-10)

7. Earth-free local equipotential bonding – not applicable in domestic premises, only used in special circumstances (see Regulations 413-05 and 471-11)

8. Electrical separation (see Regulations 413-06 and 471-12)

Appendix 4   (Continued)

Form 4   Form No   /4

SCHEDULE OF TEST RESULTS

Contractor:...   Address/Location of distribution board: Instruments
Test Date: ....   * Type of Supply: TN-S/TN-C-S/TT   loop impedance: ....
                                                          ....   * Ze at origin: ....ohms   continuity: ....
Signature ....   * PFC: ....kA   insulation: ....
Method of protection against indirect contact: ....   RCD tester: ....
Equipment vulnerable to testing: ....

Description of Work: ....

Circuit Description	Overcurrent Device			Test Results										
	*Short-circuit capacity: ....kA	Wiring Conductors		Continuity			Insulation Resistance		Polarity	Earth Loop Impedance	Functional Testing		Remarks	
	type	Rating I_n	live	cpc	(R_1 + R_2)*	R_2*	Ring	Live/Live	Live/Earth		Zs	RCD time	Other	
		A	mm²	mm²	Ω	Ω		MΩ	MΩ		Ω	ms		
1	2	3	4	5	*6	*7	*8	*9	*10	*11	*12	*13	*14	15

Deviations from Wiring Regulations and special notes:

* See notes on schedule of test results

Appendix 5

BUILDING REGULATIONS COMPLIANCE CERTIFICATE

A Customer
1 High Street
New Town
Countyshire
AB1 2CD

Certificate number: 1234

The installation of:

Rewire of existing circuits - Building sharing supply with dwelling

(In any dispute reference should be made to the electrical installation certificate for a full description of the work carried out)

Installed at:

1 High Street, New Town, Countyshire, AB1 2CD

Completed on:

01 January 2005

(This date should be the same as that on the electrical installation certificate for the work carried out)

Is certified by the installer:

A N Electrical Company
123456

To be compliant with sections 4 and 7 of the Building Regulations 2000

This document should be kept in a safe place for future reference

Appendix E

Fire resistance

(A copy of this Appendix is available on www.routledge.com/books/details/9780415809696/)

E.1 Fire resistance

The overall aim of fire safety precautions is to ensure that:

- all corridor doors shall meet the requirements for fire safety as described in Building Regulations Part B – Fire safety (see Part E2);
- all doors shall satisfy the Requirements of Building Regulation Part B – Fire safety (see Part E4);
- a satisfactory means of giving an alarm of fire is available;
- a satisfactory means of escape for persons in the event of fire in a building is available (see Part B1);
- external walls and roofs have adequate resistance to the spread of fire over the external envelope;
- fire stopping should be flexible to prevent a rigid contact between the pipe and the floor (see Parts E3 and E4);
- if a fire stop is required in the cavity between frames, then it should either be flexible or only be fixed to one frame (see Part E2);
- if there is an existing lath and plaster ceiling it should be retained as long as it satisfies Building Regulation Part B – Fire safety (see Part E);
- penetrations through a separating floor by ducts and pipes should have fire protection to satisfy Building Regulation Part B – Fire safety (see Parts E3 and E4);
- that fire spread over the internal linings of buildings is inhibited (see Part B2);
- the ceiling void and roof space detail can only be used where the Requirements of Building Regulation Part B – Fire safety can also be satisfied (see Part E);
- the junction between the separating wall and the roof should be filled with a flexible closer which is also suitable as a fire stop (see Part E2);
- the spread of fire from one building to another is restricted (see Part B4);
- the unseen spread of fire and smoke in concealed spaces in buildings is inhibited (see Part B3);
- the stability of buildings is ensured in the event of fire;
- there are facilities in buildings to assist firefighters in the saving of life of people in and around buildings (see Part B5);
- there is satisfactory access for fire appliances to buildings;

- there is a sufficient degree of fire separation within buildings and between adjoining buildings;
- where a staircase performs a separating function it shall conform to Building Regulation Part B – Fire safety (see Part E4).

Many of the requirements are, of course, closely interlinked. For example, there is a close link between the provisions for means of escape (B1) and those for the control of fire growth (B2), fire containment (B3) and facilities for the fire service (B5). Similarly there are links between B3 and the provisions for controlling external fire spread (B4), and between (B3) and (B5). Interaction between these different requirements should be recognized where variations in the standard of provision are being considered.

Factors that should be taken into account include:

- the ability of a structure to resist the spread of fire and smoke;
- the anticipated probability of a fire occurring;
- the anticipated fire severity;
- the consequential danger to people in and around the building.

Measures that could be incorporated include:

- availability of powers to require staff training in fire safety and fire routines, e.g. under the Fire Precautions Act 1971, the Fire Precautions (Workplace) Regulations 1997, or registration or licensing procedures;
- control of the rate of growth of a fire;
- consideration of the availability of any continuing control under other legislation that could ensure continued maintenance of such systems;
- early fire warning by an automatic detection and warning system;
- facilities to assist the fire service;
- provision of smoke control;
- the adequacy of means to prevent fire;
- the adequacy of the structure to resist the effects of a fire;
- the degree of fire containment;
- fire separation between buildings or parts of buildings;
- the standard of active measures for fire extinguishment or control;
- the standard of means of escape;
- management.

The design of fire safety in hospitals is covered by Health Technical Memorandum (HTM) 81 Fire precautions in new hospitals (revised 1996).

Building Regulations are intended to ensure that a reasonable standard of life safety is provided, in case of fire. The protection of property, including the building itself, may require additional measures, and insurers will in general seek their own higher standards, before accepting the insurance risk. Guidance is given in the LPC Design guide for the fire protection of buildings.

Guidance for assisting protection in Civil and Defence Estates is given in the Crown Fire Standards published by the Property Advisers to the Civil Estate (PACE).

Fire safety engineering

Fire safety engineering can provide an alternative approach to fire safety and in certain circumstances it could be the only practical way to achieve a satisfactory standard of fire safety in some buildings. Fire safety engineering is also suitable for solving problems concerning the design of the building, which although meeting the requirements of the Regulations is still problematic.

Factors that should be taken into account include:

- ability of a structure to resist the spread of fire and smoke;
- anticipated probability of a fire occurring;
- anticipated fire severity;
- consequential danger to people in and around the building.

Many processes are available for consideration such as:

- are facilities available that will assist fire and rescue services?
- are smoke controls provided and are they adequate?
- are there appropriate existing methods for controlling and extinguishing fires?
- are these methods regularly reviewed and where necessary repaired/ replaced? How adequate is the current means of preventing fire?
- how appropriate is the designed means of escape?
- is the fire separation between buildings and/or parts of buildings appropriate?
- is the rate of growth of the fire controllable?
- is the structure capable of resisting the effects of fire?
- is there an automatic early fire warning system in place?
- to what degree can the fire be contained?
- is there an existing process for training staff in fire safety and fire routines?
- do Top Management endorse the requirement for fire safety?

Risk assessment

The assessment and design of means of escape shall take into account: B1.ii

- the nature of the building structure;
- the use of the building;
- the potential of fire spread through the building; and
- the standard of fire safety management proposed.

Fire risk analysis

Part B now includes a requirement for the responsible person (i.e. the person carrying out work to a building) to make available to the owner (other than

Appendix E

houses occupied as single private dwellings) 'fire safety information' concerning the design and construction of the building or extension plus details of the services, fittings and equipment that have been provided in order that they (when required under the new Regulatory Reform (Fire Safety) Order 2005 – Statutory Instrument 2005 No. 1541) may complete a fire risk analysis.

Although these requirements are applicable to premises whilst in operation, it would be useful for the designers of a building to carry out a preliminary fire risk assessment as part of the design process. If a preliminary risk assessment is produced, it can be used as part of the Building Regulations submission and can assist the fire safety enforcing authority in providing advice at an early stage as to what, if any, additional provisions may be necessary when the building is first occupied.

E.1.1 The requirement

Means of escape

- *There shall be an early warning fire alarm system for persons in the building.*
- *There shall be sufficient escape routes that are suitably located to enable persons to evacuate the building in the event of a fire.*
- *Safety routes shall be protected from the effects of fire.*
- *In an emergency, the occupants of any part of the building shall be able to escape without any external assistance.*

(Approved Document B1)

Internal fire spread (linings)

- *The spread of flame over the internal linings of the building shall be restricted.*
- *The heat released from the internal linings shall be restricted.*

(Approved Document B2)

Internal fire spread (structure)

Dependent on the use of the building, its size and the location of the element of construction:

- *loadbearing elements of a building structure shall be capable of withstanding the effects of fire for an appropriate period without loss of stability;*
- *the building shall be subdivided by elements of fire-resisting construction into compartments;*
- *all openings in fire-separating elements shall be suitably protected in order to maintain the integrity of the element (i.e. the continuity of the fire separation);*

- *any hidden voids in the construction shall be sealed and subdivided to inhibit the unseen spread of fire and products of combustion.*

(Approved Document B3)

External fire spread

- *External walls shall be constructed so as to have a low rate of heat release and thereby be capable of reducing the risk of ignition from an external source and the spread of fire over their surfaces.*
- *The amount of unprotected area in the sides of the building shall be restricted so as to limit the amount of thermal radiation that can pass through the wall.*
- *The roof shall be constructed so that the risk of spread of flame and/or fire penetration from an external fire source is restricted.*

(Approved Document B4)

Access facilities for the fire service

- *There shall be sufficient means of external access to enable fire appliances to be brought near to the building for effective use.*
- *There shall be sufficient means of access into and within the building for firefighting personnel to affect search and rescue and fight fire.*
- *The building shall be provided with sufficient internal fire mains and other facilities to assist firefighters in their tasks.*
- *The building shall be provided with adequate means for venting heat and smoke from a fire in a basement.*

(Approved Document B4)

E.1.2 Meeting the requirements

Fire resistance

An element of construction shall provide: B3

- resistance to collapse;
- resistance to fire penetration;
- resistance to the transfer of excessive heat.

The purpose in providing the structure with fire resistance is:

- to minimize the risk to the occupants;
- to reduce the risk to firefighters;
- to reduce the danger to people in the vicinity of the building.

Appendix E

Fire resistance standard

Elements such as structural frames, beams, columns, internal and external loadbearing walls, floor structures and gallery structures should have at least the fire resistance shown in Part B, Appendix A, Table A1.	B3 4.2 (V1) B3 7.2 (V2)
The fire resistance of an element of structure that supports or gives stability to another element of structure shall not be less than the other element.	B3 4.3 (V1) B3 7.3 (V2)

Loft conversions

The floor(s), both old and new, shall have the full 30 minute standard of fire resistance shown in Part B Appendix A, Table A1 unless: • only one storey is being added; • the new storey contains no more than two habitable rooms; and • the total area of the new storey is less than 50 m².	B3 4.7 (V1)
In those places where the floor only separates rooms (and not circulation spaces), a modified 30 minute standard of fire resistance may be applied.	B3 4.7 (V1)
Where the conversion of an existing roof space (such as a loft conversion to a two-storey house) means that a new storey is going to be added, then the stairway will need to be protected with fire-resisting doors and partitions.	B1 2.20b

Flats

If the existing building has timber floors and these are to be retained, the requirements for fire resistance may be difficult to meet. In these cases, provided that the means of escape conforms to Part B Section 3 and are adequately protected. Then:

• those parts of the building that are **no** more than three storeys high shall need to have a 30 minute standard of fire resistance; • if the altered building has four or more storeys, then the full standard of fire resistance (as described in Appendix A to Part B) would be required.	B2 7.10 (V2)

The amount of fire resistance provided by the building structure and other elements of construction is determined by reference to either:

- BS 476 (National classification);
- Commission Decision 2000/367/EC of 3 May 2000 implementing Council Directive 89/1061EEC (European classification);
- BS EN 13501–2:2003;
- BS EN 13501–3:2005;
- BS EN 13501–4:2007.

See also Appendix A to Part B for Tables setting out minimum periods of fire resistance etc.

Escape routes

Planning escape routes

Common corridors that connect two or more storey exits should be subdivided by a self-closing fire door with, if necessary, an associated fire-resisting screen.   B1 2.28 (V2)

Note: Self-closing fire doors should be positioned so that smoke will not affect access to more than one stairway.

Protection of escape routes

Generally, a 30 minute standard is sufficient for the protection of means of escape. (Details of fire resistance test criteria and standards of performance are contained in Appendix A to Part B).   B1 5.2 (V2)

External escape stairs

External escape stairs should meet the following provisions:

where an external escape stair is provided: B1 2.15 (a, b and c)

- all doors giving access to the stair should be fire-resisting;
- any part of the external envelope of the building within 1800 mm of (and 9 m vertically below) the flights and landings of an external escape stair should be of fire-resisting construction (see Figure E1);
- any part of the building (including doors) within 1800 mm of the escape route shall be protected by fire-resisting construction.

External escape stairs greater than 6 m in vertical dextent shall be protected from the effects of adverse weather conditions.   B1 2.15

Note: Glazing in any fire-resisting construction should be fire-resisting and fixed shut.

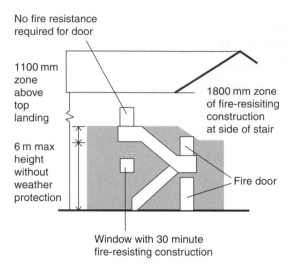

Figure E1 Fire resistance of areas adjacent to external stairs

Dwelling-houses with one floor more than 4.5 m above ground level

The dwelling-house may either have a protected stairway which: B1 2.6 (V1)

- extends to the final exit (see Figure E2(a)); or
- gives access to at least two escape routes at ground level, each delivering to final exits and separated from each other by fire-resisting construction and fire doors (see Figure E2(b)); or

- the top floor can be separated from the lower storeys by fire-resisting construction and be provided with its own alternative escape route leading to its own final exit.

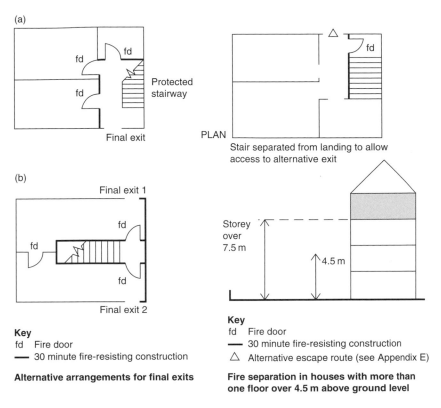

Figure E2 Final exits and fire separation

Flats with a floor more than 4.5 m above ground level

Where any flat has an alternative exit and the
habitable rooms do not have direct access to the
entrance hall (see Figure E2):

B1 2.14 (V2)

- the bedrooms should be separated from the living
  accommodation by fire-resisting construction and
  fire doors); and
- the alternative exit should be located in the part of
  the flat containing the bedroom(s).

Residential care homes

Generally, in care homes for the elderly it is reasonable to assume that at least
a proportion of the residents will need some assistance to evacuate.

> Bedrooms should be enclosed in fire-resisting   B1 3.48 (V2)
> construction with fire resisting doors and every
> corridor serving bedrooms should be a protected
> corridor.

Shopping complexes

Part B is primarily intended to cover shops that are contained in a single separate building. If a shop forms part of a complex, such as a covered mall then the requirements for fire resistance, walls separating shop units, surfaces and boundary distances will probably be different (see Sections 5 and 6 of BS 5588–10:1991 for further guidance).

Ancillary accommodation

> Ancillary accommodation such as: B1 3.50 (V2)
>
> • day rooms;
> • chemical stores;
> • cleaners' rooms;
> • clothes storage;
> • disposal rooms;
> • kitchens;
> • laundry rooms;
> • linen stores;
> • plant rooms;
> • smoking rooms (now non-existent in view of
>   government legislation);
> • staff changing and locker rooms; and
> • store rooms;
>
> should be enclosed by fire-resisting construction.

Portal frames

> Where a portal framed building is near a relevant   B4 12.4 (V2)
> boundary, the external wall near the boundary may
> need fire resistance to restrict the spread of fire
> between buildings.

Access lobbies and corridors

Any such protected exit passageway should have the same standard of fire resistance and lobby protection as the stairway it serves.

Balconies and flat roofs

Any flat roof that forms part of a means of escape should:	B1 2.10 (V1)
• be part of the same building from which escape is being made;	B1 2.7 (V2)
• lead to a storey exit or external escape route; and	
• provide 30 minutes fire resistance (see Appendix A, Table A1 of Approved Document B for fire resistance figures for elements of structure etc.).	
✎ **Note:** If a balcony or flat roof is provided for escape purposes guarding may be required (see Approved Document K Protection).	B1 2.11 (V1)

Ceilings

The need for cavity barriers in some concealed floor or roof spaces can be reduced by using a fire resisting ceiling below the cavity.	B2 3.6 (V1) B2 6.6 (V2)

Suspended ceilings

A suspended, fire-protected ceiling should meet the requirements in Table E1: for further details see Part B, Appendix A, Table A3.

Table E1 Limitations on fire-protecting suspended ceilings

Height of building or separated part (m)	Type of floor	Provision for fire resistance of floor (minutes)
Less than 18	Not compartment	60 or less
	Compartment	Less than 60
18 or more	Any	60 or less
No limit	Any	More than 60

Appendix E

Cavity barriers

Cavity barriers should be provided at the junction between an internal cavity wall and any assembly which forms a fire-resisting barrier.	B3 9.3 (V2)
Every cavity barrier should be constructed to provide at least 30 minutes fire resistance.	B3 6.5 (V1) B3 9.13 (V2)
Openings in a cavity barrier should be limited to those for:	B3 6.8 (V1) B3 9.13 (V2)

- doors which have at least 30 minutes fire resistance;
- the passage of pipes which meet the provisions in Part P Section 7;
- the passage of cables or conduits containing one or more cables;
- openings fitted with a suitably mounted automatic fire damper;
- ducts which are fire-resisting or are fitted with a suitably mounted automatic fire damper where they pass through the cavity barrier.

Compartment walls and compartment floors

Every compartment wall and compartment floor should:

- form a complete barrier to fire between the compartments they separate; and — B2 8.20a (V2) / B3 5.6 (V1)
- have the appropriate fire resistance as indicated in Appendix A, Tables A1 and A2. — B2 8.20b (V2) / B3 5.6 (V1)

Junctions between a compartment floor and an external wall that has no fire resistance (such as a curtain wall) should be restrained at floor level to reduce the movement of the wall away from the floor when exposed to fire. — B2 8.26 (V2)

Junction of compartment wall or compartment floor with other walls

At the junction of a compartment floor with an external wall that has no fire resistance (such as a curtain wall) the external wall should be restrained at floor level to reduce the movement of the wall away from the floor when exposed to fire.	B3 5.10 (V1)

Junction of compartment wall with roof

A compartment wall should be:

- taken up to meet the underside of the roof covering   B3 5.11 (V1)
  or deck, with fire-stopping (where necessary) at
  the wall/roof junction to maintain the continuity of
  fire resistance;
- continued across any eaves.   B3 8.28 (V2)

Openings in compartment walls separating buildings or occupancies

Any openings in a compartment wall which is   B3 5.13 (V1)
common to two or more buildings should be limited to   B3 8.32 (V2)
those for:

- a door which is providing 'means of access' in
  case of fire (and which has the same fire resistance
  as that required for the wall);
- the passage of a pipe.

All other openings in compartment walls or
compartment floors should be limited to those for:

- doors which have the appropriate fire resistance;   B3 8.34 (V2)
- the passage of pipes, ventilation ducts, service
  cables, chimneys, appliance ventilation ducts or
  ducts encasing one or more flue pipes;
- refuse chutes of non-combustible construction;
- atria designed in accordance with BS 5588–
  7:1997; and
- protected shafts (see B3 8.35 V2 for details of the
  relevant requirements).

Appendix
E

External walls

External walls of the building should have sufficient fire resistance to prevent fire spread across the relevant boundary.	B4 8.1 (V1) B4 12.1 (V2)
The external surfaces of walls of dwellings within 1000 mm of the relevant boundary should meet Class 0 (National Class) or Class B-s3,d2 or better (European class). For all buildings other than dwellings, they should meet the relevant European requirements as shown in Diagram 40 (page 95) of Part B Volume 2.	B4 8.4 (V1) B4 12.6 (V2)

Note: Any part of an external wall which has less fire resistance than that shown in Part B, Volume 1, Appendix A, Table A2, is considered to be an unprotected area.

Fire doors

Two fire doors may be fitted in the same opening so that the total fire resistance is the sum of their individual fire resistances, provided that each door is capable of closing the opening.	App B 4 (V2)

Table E2 Provision for fire doors (dwellings)

Position of door	Minimum fire resistance of door in terms of integrity (minutes) when tested to BS 476–22:1987	Minimum fire resistance of door in terms of integrity (minutes) when tested to the relevant BSEN 1634 European standard
Any door:		
with a cavity barrier	FD 30	E30
between a dwelling-house and a garage	FD 30s	E30Sa
forming part of the enclosure to a protected stairway in a single family dwelling-house	FD 20	E20
within any other fire resisting construction in a dwelling-house not described elsewhere in the table	FD 20	E20

Notes:

(1)   Minimum fire resistance for doors in buildings other than dwellings is given in Table B1 (page 134) of Part B Volume 2.

(2)   BS 8214:1990 gives recommendations for the specification, design, construction, installation and maintenance of fire doors constructed with non-metallic door leaves.

Guidance on timber fire-resisting doorsets may be found in *Timber fire-resisting doorsets: maintaining performance under the new European test standard* published by TRADA.

Galleries

All galleries shall be provided with an alternative exit or, where the gallery floor is not more than 4.5 m above ground level, an emergency egress window (which complies with paragraph 2.8).
 	
If the gallery floor is not provided with an alternative exit or escape window:
• the gallery should overlook at least 50% of the room below (see Figure E3);
• the distance between the foot of the access stair to the gallery and the door to the room containing the gallery should not exceed 3 m;
• the distance from the head of the access stair to any point on the gallery should not exceed 7.5 m; and
• any cooking facilities within a room containing a gallery should either:
  – be enclosed with fire-resisting construction; or
  – be remote from the stair to the gallery.
	B1 2.12 (V1)
	B1 2.8 (V2)

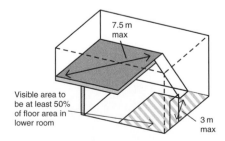

Notes:
1   This diagram does not apply where the gallery is
    i. provided with an alternative escape route; or
    ii. provided with an emergency egress window
        (where the gallery floor is not more than 4.5 m
        above gound level).
2   Any cooking facilities within a room containing a
    gallery should either:
    i. be enclosed with fire-resisting construction; or
    ii. be remote from the stair to the gallery and
        positioned such that they do not prejudice the
        escape from the gallery.

Visible area to be at least 50% of floor area in lower room

7.5 m max

3 m max

Figure E3 Gallery floors with no alternative exit

Inner rooms

> Store rooms should be enclosed with fire-resisting construction.   B1 3.35 (V2)

Openings and fire-stopping

 Note: Detailed guidance on door openings and fire doors is given in Part B Appendix B.

> Every joint, imperfection or opening of a fire-separating element should be protected by sealing or fire-stopping so that the fire resistance of the element is not weakened.   B3 7.2 (V1)
> B3 10.2 (V2)

Openings for pipes

Unless the pipe is in a protected shaft, all pipes which pass through fire-separating elements should conform to one the three alternatives shown below.

	Type	Requirements
Alternative A	Proprietary seals (any pipe diameter)	Provide a proprietary sealing system which has been shown by test to maintain the fire resistance of the wall, floor or cavity barrier
Alternative B	Pipes with a restricted diameter	Where a proprietary sealing system is not used, fire stopping may be used around the pipe, keeping the opening as small as possible (see Table 3 of Part B Volume 1 for dimension details)
Alternative C	Sleeving	A pipe of lead, aluminium, aluminium alloy, fibre-cement or uPVC, with a maximum nominal internal diameter of 160 mm, may be used with a sleeving of non-combustible pipe as shown in Figure E4

Ventilation ducts and flues etc.

> If a flue or duct passes through a compartment wall or compartment floor, or is built into a compartment wall, each wall of the flue or duct should have a fire resistance of at least half that of the wall or floor (see Figure E5).   B1 7.11 (V1)

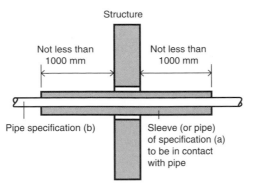

Notes:
1 Make the opening in the structure as small
   as possible and provide fire-stopping
   between pipe and structure.

Figure E4 Pipes penetrating structure

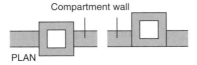

In each case flue walls should have a fire resistance at
least one half of that required for the compartment wall
and be of non-combustible construction.

Figure E5 Flues penetrating compartment walls or floors

If a flue (or a duct containing flues and/or ventilation   B3 10.16 (V2)
duct(s)), passes through a compartment wall or
compartment floor, or is built into a compartment
wall, each wall of the flue or duct should have a fire
resistance of at least half that of the wall or floor.

Passenger lifts

Passenger lifts in dwelling houses which serve any   B1 2.18
floor more than 4.5 m above ground level should
either be located in the enclosure to the protected
stairway or be contained in a fire-resisting lift shaft.

Protected shafts

Protected shafts (see Figure E6) should: B2 8.37 (V2)

- form a complete barrier to fire between the different compartments which the shaft connects;
- have the appropriate fire resistance given in Appendix A, Table A1; and
- meet the requirements for ventilation and the treatment of openings (see Part B Sections 8.41 and 8.42).

Protected shafts provide for the movement of people (e.g. stairs, lifts), or for passage of goods, air or services such as pipes or cables between different compartments. The elements enclosing the shaft (unless formed by adjacent external walls) are compartment walls and floors. Figure E6 shows three common examples which illustrate the principles.

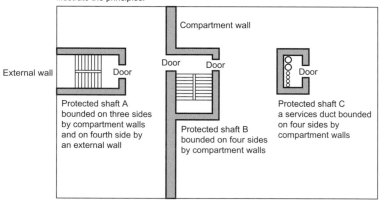

Figure E6 Protected shafts

An uninsulated glazed screen may be incorporated in B2 8.38 (V2)
the enclosure to a protected shaft between a stair and a lobby or corridor which is entered from the stair provided that:

- the fire resistance for the stair enclosure is not more than 60 minutes;
- the glazed screen has at least 30 minutes fire resistance; and
- the lobby or corridor is enclosed to at least a 30 minute standard (see Figure E7).

Generally speaking, an external wall of a protected shaft does not need to have fire resistance (but see BS 5588–5:2004 for fire resistance of external walls of firefighting shafts).

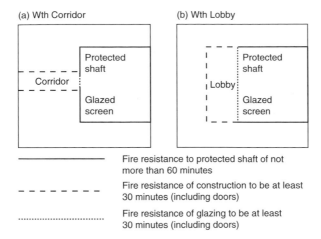

Figure E7 Uninsulated glazed screen separating protected shaft from lobby or corridor

Roof coverings

In thatched roofs:	B4 10.9 (V1)
	B4 14.9 (V2)
• the rafters should be overdrawn with construction having not less than 30 minutes fire resistance;	
• a smoke alarm should be installed in the roof space.	

 Note: See Part B, Volume 1, Table 5 for limitations on roof coverings and Tables 6 and 7 for the limitations on using plastic rooflights and thermoplastic materials.

Stairs

Where an external escape stair is provided in accordance with paragraph 4.44, it should meet the following provisions:	B1 5.25 (V2)
• all doors giving access to the stair should be fire-resisting and self-closing;	
• any part of the external envelope of the building within 1800 mm of (and 9 m vertically below) the flights and landings of an external escape stair should be fire resisting;	
• there is protection by fire-resisting construction for any part of the building within 1800 mm of the escape route from the stair to a place of safety;	
• glazing should also be fire resistant and fixed shut.	

Common stairs

Any stair used as a firefighting stair should be at least 1100 mm wide (see Part B V2, Appendix C for measurement of width).	B1 2.33 (V2)
All common stairs should be situated within a fire-resisting enclosure (i.e. it should be a protected stairway), to reduce the risk of smoke and heat making use of the stair hazardous.	B1 2.36 (V2)
In single stair buildings, meters located within the stairway should be enclosed within a secure cupboard which is separated from the escape route with fire-resisting construction.	B1 2.40 (V2)

Fire detection and fire alarm systems

General

All new dwelling-houses should be provided with a BS 5839–6:2004 Grade D Category LD3 fire detection and fire alarm system.	B1 1.3
There should be at least one smoke alarm on every storey of a dwelling house.	B1 1.12
An installation and commissioning certificate should be provided when a fire alarm system is installed.	B1 1.23
Occupants should be provided with information on the use of the equipment and on its maintenance (or guidance on suitable maintenance contractors).	B1 1.24
The rapid spread of smoke and fumes shall be limited.	B1.iv
The design and installation of fire detection and fire alarm systems in dwelling houses shall be in accordance with BS 5839–6:2004.	B1 1.10

Large houses

Large dwelling houses of two storeys (excluding basement storeys) should be provided with a BS 5839–6:2004 Grade B Category LD3 fire detection and fire alarm system.	B1 1.6

Large dwelling houses of three or more storeys (excluding basement storeys) should be provided with a BS 5839–6:2004 Grade A Category LD3 fire detection and fire alarm system in accordance with BS 5839–6:2004 Grade D Category LD3.

B1 1.7

Fire detectors used in large dwelling houses of three or more storeys (excluding basement storeys) should be sited in accordance with BS 5839–6:2004 Category L2.

B1 1.7

Material alteration

New habitable rooms that are the result of a material alteration and which are above ground floor level (or at ground floor level where no final exit has been provided) shall be equipped with:

B1 1.8

- a fire detection and fire alarm system;
- smoke alarms (in accordance with paragraphs 1.10 to 1.18) in the circulation spaces.

Sheltered housing

Fire detection equipment used in sheltered housing which are overseen by a warden or supervisor, shall be linked to a central monitoring point or alarm receiving centre.

B1 1.9

Smoke alarms

The provision of smoke alarms shall be based on an assessment of the risk to the occupants in the event of fire.

B1.ii

Smoke alarms should:

B1 1.4

- be mains-operated and conform to BS 5446–1:2000;
- have a standby power supply such as a rechargeable (or non-rechargeable) battery;
- be positioned in the circulation spaces between sleeping spaces and places where fires are most like to start (e.g. in kitchens and living rooms).

B1 1.11

The design and installation of smoke alarms shall be in accordance with BS 5839–6:2004.

B1 1.10

Appendix E

There should be at least one smoke alarm on every storey of a dwelling-house.	B1 1.12
Where the kitchen area is not separated from the stairway or circulation space by a door, there should be a compatible interlinked heat detector or heat alarm in the kitchen, in addition to whatever smoke alarms are needed in the circulation space(s).	B1 1.13
If more than one alarm is installed they should be linked so that the detection of smoke or heat by one unit operates the alarm signal in all of them.	B1 1.14

Smoke alarms/detectors should be sited so that:	B1 1.15a
• there is a smoke alarm in the circulation space within 7.5 m of the door to every habitable room;	B1 1.15b
• they are ceiling-mounted and at least 300 mm from walls and light fittings;	B1 1.15c
• the sensor in ceiling-mounted devices is between 25 mm and 600 mm below the ceiling (25–150 mm in the case of heat detectors or heat alarms).	

Smoke alarms should **not** be fixed:	B1 1.16
• over a stair or any other opening between floors;	
• next to or directly above heaters or air-conditioning outlets;	B1 1.17
• in bathrooms, showers, cooking areas or garages;	B1 1.17
• in any place where steam, condensation or fumes could give false alarms;	B1 1.17
• in places that get very hot (such as a boiler room);	B1 1.18
• in places that get very cold (such as an unheated porch);	B1 1.18
• to surfaces which are normally much warmer or colder than the rest of the space.	B1 1.18

Power supplies

The power supply for a smoke alarm system:	B1 1.19
• should be derived from the dwelling-house's mains electricity supply;	
• should comprise a single independent circuit at the dwelling-house's main distribution board (consumer unit or a single regularly used local lighting circuit.	

It should be possible to isolate the power to the smoke alarms without isolating the lighting.

B1 1.19

The electrical installation should comply with Approved Document P (Electrical safety).

B1 1.20

Any cable suitable for domestic wiring may be used for the power supply and interconnection to smoke alarm systems (except in large buildings where the cable needs to be fire resistant (see BS 5839–6:2004).

B1 1.21

Conductors used to interconnect alarms (e.g. signalling) should be colour coded so as to distinguish them from those supplying mains power.

B1 1.21

Mains powered smoke alarms may be interconnected using radio-links, provided that this does not reduce the lifetime or duration of any standby power supply below 72 hours.

B1 1.21

Heat alarms

Heat alarms should:

B1 1.4

- be mains-operated and conform to BS 5446–2:2003;
- have a standby power supply such as a rechargeable (or non-rechargeable) battery;
- be designed and installed in accordance with BS 5839–6:2004.

B1 1.10

Heat detectors and heat alarms should be sited so that the sensor in ceiling-mounted devices is between 25 mm and 150 mm below the ceiling.

B1 1.15c

Appendix E

Appendix F

Means of escape

(A copy of this Appendix is available on www.routledge.com/books/details/9780415809696/)

F.1 Means of escape

F.1.1 The requirement

Subject to Section 30(3) of the Fire Precautions Act 1971, if a building (or proposed building) exceeds two storeys in height and the floor of any upper storey is more than 20ft above the surface of the street or ground on any side of the building and is:

* *let out as flats or tenement dwellings;*
* *used as an inn, hotel, boarding house, hospital, nursing home, boarding school, children's home or similar institution; or is*
* *used as a restaurant, shop, store or warehouse and has on an upper floor sleeping accommodation for persons employed on the premises;*

then it must be equipped with adequate means of escape in case of fire, from each storey.

(Building Act 1984 Section 72)

The building shall be designed and constructed so that there are appropriate provisions for the early warning of fire, and appropriate means of escape in case of fire from the building to a place of safety outside the building capable of being safely and effectively used at all material times.

(Approved Document B1)

For a typical one- or two-storey dwelling, the requirement is limited to the provision of smoke alarms and to the provision of openable windows for emergency exit (see B1.i).

There shall be sufficient escape routes that are suitably located to enable persons to evacuate the building in the event of a fire.	B1
Safety routes shall be protected from the effects of fire.	B1
In an emergency, the occupants of any part of the building shall be able to escape without any external assistance.	B1 i

There should be alternative means of escape from 'most situations'.   B1 v (a)

The design of means of escape shall be based on an assessment of the risk to the occupants in the event of fire.   B1 ii

If direct escape to a place of safety is impracticable, it should be possible to reach a place of relative safety such as a protected stairway within a reasonable travel distance.   B1 v (b)

Unprotected escape routes should not require people to have to travel excessive distances while exposed to the immediate danger of fire and smoke.   B1 vii

People should be able to turn their backs on a fire wherever it occurs and travel away from it to a final exit or protected escape route leading to a place of safety.   B1 vii

The following are **not** considered acceptable as a means of escape:

- lifts (except suitably designed and installed evacuation lifts);   B1
- portable ladders;
- throw-out ladders;
- fold-down ladders and chutes;
- escalators (although it is recognized that they are likely to be used by people who are escaping).

These facilities may, however, be used as an additional feature.

 Note: Mechanized walkways could be accepted and their capacity assessed on the basis of their use as a walking route, while in the static mode.

Risk assessment

A risk assessment shall be carried out and the design of means of escape shall take into account: B1.ii

- the nature of the building structure;
- the use of the building;
- the potential of fire spread through the building; and
- the proposed standard of fire safety management.

Dwelling houses

One- or two-storey dwelling houses shall be provided with:	B1 2.1 (V1)

• an early warning system in the event of fire;
• suitable means for emergency egress from each storey via windows or doors.

Floors more than 7.5 m above ground shall be provided with an alternative escape route.	B1 2.1 (V1)

Ground floor dwelling houses and flats

Except for kitchens, all habitable rooms on the ground floor should:	B1 2.3 (V1)

• either open directly onto a hall leading to the   B1 2.11 (V2)
  entrance or other suitable exit; or
• be provided with an emergency window (or door).

Any inner room that is a kitchen, laundry or utility room, dressing room, bathroom, WC or shower or situated not more than 4.5 m above ground level and whose only escape route is through another room, shall be provided with an emergency egress window.	B1 2.9 (V1) B1 2.5 (V2)

 Note: The means of escape from a flat with a floor not more than 4.5 m above ground level is relatively simple to provide. Few provisions are specified in the 2006 edition of Part B beyond ensuring that means shall be provided for giving early warning in the event of fire and suitable means are provided for emergency egress from these storeys. With increasing height, however, the situation becomes more complex because emergency egress through upper windows will become increasingly hazardous.

Upper floors not more than 4.5 m above ground level

Except for kitchens, all habitable rooms in the upper storey(s) of a dwelling-house that are served by only one stair, should be provided with:	B1 2.4 (V1) B1 2.12 (V2)

• a window (or external door); or have
• direct access to a protected escape route.

Dwelling houses with one floor more than 4.5 m above ground level

The dwelling-house may either have a protected stairway which: **B1 2.6 (V1)**

- extends to the final exit (see Figure E7); or
- gives access to at least two escape routes at ground level, each delivering to final exits and separated from each other by fire-resisting construction and fire doors (see Figure E7);

or the top floor can be separated from the lower storeys by fire-resisting construction and be provided with its own alternative escape route leading to its own final exit (see Figure E7).

Dwelling houses with more than one floor more than 4.5 m above ground level

Dwelling houses with floors more than 4.5 m above ground level shall (in addition to meeting requirement B1 2.6) have: **B12.7 (V1)**

- an alternative escape route for each story or level that is more than 7.5 m above ground level; or
- a sprinkler system designed and installed in accordance with BS 9251:2005.

Note: The access to the alternative escape route should either be:

- via a protected stairway to an upper storey; or
- a landing within the protected stairway enclosure to an alternative escape route on the same storey; or
- the protected stairway that is at (or about) 7.5 m above ground level and which is separated from the lower storeys or levels by fire-resisting construction.

Buildings other than flats

Where the means of escape is based on phased evacuation, then a staged alarm system should be used.   **B1 1.25 (V2)**

Where the means of escape is based on simultaneous evacuation, operation of a manual call point or fire detector should give an almost instantaneous warning from all the fire alarm sounders.   **B1 1.25 (V2)**

Note: Automatic sprinkler systems can be used to operate a fire alarm system.

Sheltered housing

Whilst many of the provisions made in Part B 2007 for means of escape from flats are applicable to sheltered housing, the nature of the occupancy may necessitate some additional fire protection measures.

Institutional buildings

Special considerations may apply to some institutional buildings if residents need the assistance of staff to evacuate the building.

Basements

Owing to the risk that a single stairway may be blocked by smoke from a fire in the basement or ground storey:

basement storeys in a dwelling-house that contain a habitable room shall be provided with either:	B1 2.13 (V1)
• an external door or window suitable for egress from the basement: or	B1 2.6 (V2)
• a protected stairway leading from the basement to a final exit.	

Galleries

All galleries shall be provided with an alternative exit or, where the gallery floor is not more than 4.5 m above ground level, an emergency egress window.	B1 2.12 (V1) B1 2.8 (V2)

If the gallery floor is not provided with an alternative exit or escape window:

- the gallery should overlook at least 50% of the room below (see Figure F1);
- the distance between the foot of the access stair to the gallery and the door to the room containing the gallery should not exceed 3 m;
- the distance from the head of the access stair to any point on the gallery should not exceed 7.5 m; and
- any cooking facilities within a room containing a gallery should either:

  – be enclosed with fire-resisting construction; or
  – be remote from the stair to the gallery.

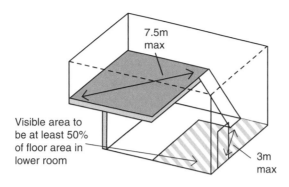

Figure F1 Gallery floors with no alternative exit

Balconies and flat roofs

Any flat roof that forms part of a means of escape should:	B1 2.10 (V1)
• be part of the same building from which escape is being made;	B1 2.7 (V2)
• lead to a storey exit or external escape route: and	
• provide 30 minutes fire resistance (see Appendix A, Table Al of Approved Document B for fire resistance figures for elements of structure etc.)	
Note: If a balcony or flat roof is provided for escape purposes, guarding may be required (see Approved Document K – Protection from falling, collision and impact).	B1 2.11 (V1)

Fire protected stairways

Fire protected stairways that, as far as reasonably possible:	B1 1.viii
• exclude all flames, smoke and gases shall be designed to provide effective 'fire sterile' areas that lead to places of safety outside the building;	
• consist of fire-resistant material and fire-resistant doors and have an appropriate form of smoke control system.	

External escape stairs

Where an external escape stair is provided: B1 2.15 (a, b and c)

- all doors giving access to the stair should be fire-resisting;
- any part of the external envelope of the building within 1800 mm of (and 9 m vertically below) the flights and landings of an external escape stair should be of fire resisting construction (see Figure F1);
- any part of the building (including doors) within 1800 mm of the escape route shall be protected by fire resisting construction.

External escape stairs greater than 6 m in   B1 2.15d
vertical extent shall be protected from the
effects of adverse weather conditions.

 Note: Glazing in any fire-resisting construction should be fire resisting and fixed shut.

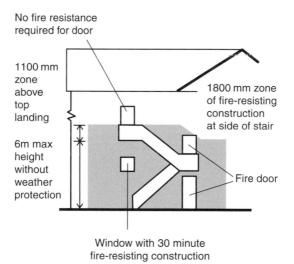

No fire resistance required for door

1100 mm zone above top landing

1800 mm zone of fire-resisting construction at side of stair

6m max height without weather protection

Fire door

Window with 30 minute fire-resisting construction

Figure F2 Fire resistance of areas adjacent to external stairs

Emergency egress windows and external doors

The window should be at least 450 mm high and   B1 2.8 (V1)
450 mm wide and have an unobstructed openable area   B1 2.9 (V2)
of at least 0.33 m^2.

The bottom of the openable area should be not more than 1100 m above the floor.

The window or door should enable the person escaping to reach a place free from danger of fire (e.g. a courtyard or back garden which is at least as deep as the dwelling-house is high – see Figure F3).

Notes:

1. Approved Document K (Protection from falling, collision and impact) specifies a minimum guarding height of 800 mm, except in the case of a window in a roof where the bottom of the opening may be 600 mm above the floor.
2. Locks (with or without removable keys) and stays may be fitted to egress windows, provided that the stay is fitted with a child-resistant release catch.
3. Windows should be designed so that they remain in the open position without needing to be held open by the person making their escape.

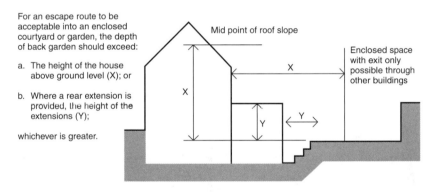

For an escape route to be acceptable into an enclosed courtyard or garden, the depth of back garden should exceed:

a. The height of the house above ground level (X); or

b. Where a rear extension is provided, the height of the extensions (Y);

whichever is greater.

Mid point of roof slope

Enclosed space with exit only possible through other buildings

Figure F3 Ground or basement storey exit into an enclosed space

Means of escape from the common parts of flats

The following requirements are primarily concerned with means of escape from the entrance doors of flats to a final exit.

Every flat should have access to alternative escape routes (but see Part B V2 paragraphs 2.20 to 2.22 for variations to this rule).	B1 2.20 (V2)
Escape routes in the common areas should comply with the limitations on travel distance shown in Table F1.	B1 2.23 (V2)

Table F1 Limitations on distance of travel in common areas of blocks of flats

Maximum distance of travel (m) from flat entrance door to common stair, or to stair lobby

Escape in one direction only	Escape in more than one direction
7.5 m	30 m

Escape routes should be planned so that people do not have to pass through one stairway enclosure to reach another.	B1 2.23 (V2)
Common corridors should be protected corridors.	B1 2.24 (V2)
The wall between each flat and the corridor should be a compartment wall (see Section 8).	B1 2.24 (V2)
Means of ventilating common corridors/lobbies (i.e. to control smoke and so protect the common stairs) should be available.	B1 2.25 (V2)
In large buildings, the corridor or lobby adjoining the stair should be provided with a vent that is located as high as practicable and with its top edge at least as high as the top of the door to the stair.	B1 2.26 (V2)
There should also be a vent, with a free area of at least 1.0 m^2 from the top storey of the stairway to the outside.	B1 2.26 (V2)
In single stair buildings the smoke vents on the fire floor and at the head of the stair should be actuated by means of smoke detectors in the common access space providing access to the flats.	B1 2.26 (V2)
In buildings with more than one stair the smoke vents may be actuated manually.	B1 2.26 (V2)
Vents should either:	B1 2.26 (V2)
• be located on an external wall with minimum free area of 1.5 m^2; • discharge into a vertical smoke shaft that is closed at the base (and meets the criteria listed in Part b V2 Paragraph 2.26b).	
Smoke control of common escape routes by mechanical ventilation is permitted provided that it meets the requirements of BS EN 12101-6:2005.	B1 2.27 (V2)
Common corridors that connect two or more storey exits should be subdivided by a self-closing fire door with, if necessary, an associated fire-resisting screen (see Figure F4).	B1 2.28 (V2)

Note: Self-closing fire doors should be positioned so that smoke will not affect access to more than one stairway.

The dead-end portion of any common corridor should be separated from the rest of the corridor by a self-closing fire door (see Figure F4).

B1 2.29 (V2)

Stores and other ancillary accommodation should not be located within, or entered from, any protected lobby or protected corridor forming part of the only common escape route from a flat on the same storey as that ancillary accommodation.

B1 2.30 (V2)

If more than one escape route is available from a storey, or part of a building, one of those routes may be by way of a flat roof.

B1 2.31 (V2)

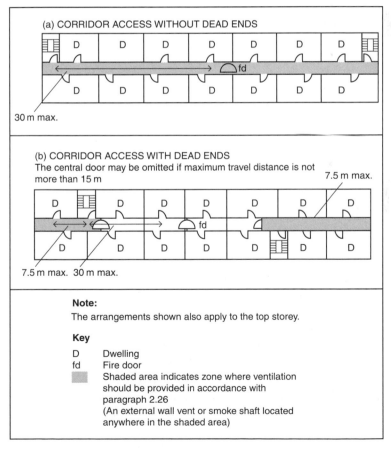

(a) CORRIDOR ACCESS WITHOUT DEAD ENDS

30 m max.

(b) CORRIDOR ACCESS WITH DEAD ENDS
The central door may be omitted if maximum travel distance is not more than 15 m

7.5 m max.

7.5 m max. 30 m max.

Note:
The arrangements shown also apply to the top storey.

Key

D   Dwelling
fd   Fire door
      Shaded area indicates zone where ventilation should be provided in accordance with paragraph 2.26
(An external wall vent or smoke shaft located anywhere in the shaded area)

Figure F4 Flats served by more than one common stair

Appendix F

Stairs

The flights and landings of every escape stair should be constructed using materials of limited combustibility particularly if it is:	B1 5.19 (V2)

* the only stair serving the building;
* within a basement storey;
* serves any storey having a floor level more than 18 m above ground or access level;
* external;
* a firefighting stair.

Single steps may only be used on escape routes where they are prominently marked.	B1 5.21 (V2)

Helical and spiral stairs forming part of an escape route should be:	B1 5.22a (V2)

* designed in accordance with BS 5395-2:1984;
* type B (Public stair) if they are intended to serve members of the public.

Fixed ladders should not be used as a means of escape for members of the public.	B1 5.22b (V2)

 Note: See Part K for guidance on the design of helical and spiral stairs and fixed ladders.

If a protected stairway projects beyond, or is recessed from, or is in an internal angle adjoining external wall of the building, then the distance between any unprotected area in the external enclosures to the building and any unprotected area in the enclosure to the stairway should be at least 1800 mm (see Figure F5).	B1 5.24 (V2)

Where an external escape stair is provided in addition to another type of escape route (see paragraph 4.44) it should meet the following provisions:	B1 5.25 (V2)

* all doors giving access to the stair should be fire-resisting and self-closing;
* any part of the external envelope of the building within 1800 mm of (and 9 m vertically below) the flights and landings of an external escape stair should be fire resisting;
* there is protection by fire-resisting construction for any part of the building within 1800 mm of the escape route from the stair to a place of safety;
* glazing should also be fire resistant and fixed shut.

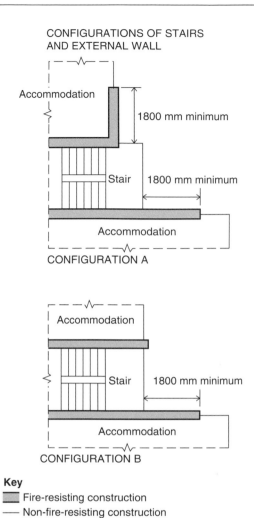

CONFIGURATIONS OF STAIRS
AND EXTERNAL WALL

Key

Fire-resisting construction

Non-fire-resisting construction

Figure F5 External protection to protected stairways

Width of escape stairs

The width of escape stairs should: B1 4.15 (V2)

- not be less than the width of any exit(s);
- not be less than the minimum widths given in Table F2;
- not exceed 1400 mm if their vertical extent is more than 30 m, **unless** it is provided with a central handrail;
- not reduce in width at any point on the way to a final exit.

> In public buildings, if the width of the stair is more than 1800 mm, the stair should have a central handrail.   B1 4.16 (V2)
>
> Every escape stair should be wide enough to accommodate the number of persons needing to use it in an emergency.   B1 4.18 (V2)

 Note: For further guidance and worked examples see Appendix C to Part B and Sections 4.18 (V2) to 4.25 (V2).

Table F2 Minimum widths of escape stairs

	Stair situation	Maximum number of people served	Minimum stair width
1a	In an Institutional building (unless the stair is only used by staff)	150	1000 mm
1b	In an assembly building and serving an area used for assembly purposes (unless the area is less than 100 m²)	220	1100 mm
1c	In any other building and serving an area with an occupancy of more than 50 people	Over 2200	1000–1800 mm*
2	Any stair not described above	50	800 mm

Protection of escape stairs

> Escape stairs need to have a satisfactory standard of fire protection.   B1 4.31 (V2)
>
> Internal escape stairs should be a protected stairway within a fire-resisting enclosure.   B1 4.32 (V2)
>
> Except for bars and restaurants, stairs may be open provided that: B1 4.33 (V2)
>
> • it does not connect more than two storeys and reaches the ground storey not more than 3 m from the final exit; and
> • the storey is also served by a protected stairway; or
> • it is a single stair in a small premises with the floor area in any storey not exceeding 90 m².

Basement stairs

Because of their situation, basement stairways are more likely to be filled with smoke and heat than stairs in ground and upper storeys. Special measures are therefore needed in order to prevent a basement fire endangering upper storeys.

If an escape stair forms part of the only escape route from an upper storey of a building it should **not** be continued down to serve any basement storey (i.e. the basement should be served by a separate stair).	B1 4.42 (V2)
If there is more than one escape stair from an upper storey of a building only one of the stairs serving the upper storeys of the building need be terminated at ground level.	B1 4.43 (V2)

External escape stairs

An external escape stair may be used, provided that: • there is at least one internal escape stair from every part of each storey (excluding plant areas); • in the case of an assembly and recreation building, the route is not intended for use by members of the public; or • in the case of an institutional building, the route serves only office or residential staff accommodation.	B1 4.44 (V2)

 Note: External escape stairs should meet the following provisions:

- all doors giving access to the stair should be fire-resisting and self-closing;
- any part of the external envelope of the building within 1800 mm of (and 9 m vertically below) the flights and landings of an external escape stair should be fire resisting;
- there is protection by fire-resisting construction for any part of the building within 1800 mm of the escape route from the stair to a place of safety;
- glazing should also be fire resistant and fixed shut.

Common stairs

Normally a single common stair can be acceptable. In some cases, however, there should be access to more than one common stair for escape purposes.

Any stair used as a firefighting stair should be at least 1100 mm wide (see Part B V2, Appendix C for measurement of width).	B1 2.33 (V2)
All common stairs should be situated within a fire-resisting enclosure (i.e. it should be a protected stairway), to reduce the risk of smoke and heat making use of the stair hazardous.	B1 2.36 (V2)

Protected stairways should discharge: B1 2.38 (V2)

- directly to a final exit; or
- via a protected exit passageway to a final exit.

Where two protected stairways (or exit passageways leading to different final exits) are adjacent, they should be separated by an imperforate enclosure.	B1 2.39 (V2)
A protected stairway needs to be relatively free of potential sources of fire.	B1 2.40 (V2)
If an escape stair forms part of the only escape route from an upper storey of a large building it should not be continued down to serve any basement storey.	B1 2.44 (V2)

The basement should be served by a separate stair.

If there is more than one escape stair from an upper storey of a building, only one of the stairs serving the upper storeys of the building need be terminated at ground level.	B1 2.45 (V2)

Note: Other stairs may connect with the basement storey(s) if there is a protected lobby or a protected corridor between the stair(s) and accommodation at each basement level.

Where a common stair forms part of the only escape route from a flat, it should **not** also serve any covered car park, boiler room, fuel storage space or other ancillary accommodation of similar fire risk.	B1 2.46 (V2)
Common stairs which do not form part of the only escape route from a flat may also serve ancillary accommodation if they are separated from the ancillary accommodation by a protected lobby or a protected corridor.	B1 2.47 (V2)
If the stair serves an enclosed (non-open-sided) car park, or place of special fire hazard, the lobby or corridor should have not less than 0.4 m² permanent ventilation or be protected by a mechanical smoke control system.	B1 2.47 (V2)

In single stair buildings, meters located within the stairway should be enclosed within a secure cupboard which is separated from the escape route with fire-resisting construction. | B1 2.40 (V2)

Gas service and installation pipes or associated meters should not be incorporated within a protected stairway unless the gas installation is in accordance with the requirements for installation and connection set out in the Pipelines Safety Regulations 1996, SI 1996 No 825 and the Gas Safety (installation and use) Regulations 1998 SI 1998 No 2451. | B1 2.42 (V2)

External escape stairs

If the building (or part of the building) is served by a single access stair, that stair may be external if it serves a floor not more than 6 m above the ground level. | B1 2.48 (V2)

If there is more than one escape route available from a storey (or part of a building) an external escape stair may be used, provided that there is at least one internal escape stair from every part of each storey (excluding plant areas) and the external stair(s). | B1 2.49 (V2)

Flats in mixed-use buildings

The stairs of buildings which are **no more than** three storeys above the ground storey, may serve both flats and other occupancies, **provided** that the stairs are separated from each occupancy by protected lobbies at all levels. | B1 2.50 (V2)

The stairs of buildings which are more than three storeys above the ground storey, may serve both flats and other occupancies, **provided** that: | B1 2.51 (V2)

- the flat is ancillary to the main use of the building and is provided with an independent alternative escape route;
- the stair is separated from any other occupancies on the lower storeys by protected lobbies (at those storey levels);
- any automatic fire detection and alarm system with which the main part of the building is fitted also covers the flat;
- any security measures should not prevent escape at all material times.

Appendix
F

Live/work units

If a flat is used as a workplace, the following **additional** fire precautions will be necessary: • the maximum travel distance to the flat entrance door or an alternative means of escape (not a window) from any part of the working area should not exceed 18 m; and • all windowless accommodation should have escape lighting (in accordance with BS 526, 6–1:2005) which illuminates the route.	B1 2.52 (V2)

Design for horizontal escape buildings other than flats

Exits in the central core of a building should be remote from one another.	B1 3.11 (V2)
An escape route should not be within 4.5 m of an opening between floors (i.e. such as an escalator) unless: • the direction of travel is away from the opening; or • there is an alternative escape route which does not pass within 4.5 m of the open connection.	B1 3.12 (V2)
Any storey which has more than one escape stair should be planned so that it is **not** necessary to pass through one stairway to reach another.	B1 3.13 (V2)
Storeys containing areas for the consumption of food and/or drink (and which are in addition to the main use of the building) shall have not less than two escape routes from each such area which lead directly to a storey exit without entering any kitchen or similar area of high fire hazard.	B1 3.15 (V2)
The means of escape from storeys that are divided into separate occupancies: • shall ensure that each occupancy does not have to pass through other occupancy; and (if the means of escape includes a common corridor or circulation space) • should either be a protected corridor, or be equipped with an automatic fire detection and alarm system throughout the storey.	B1 3.16 (V2)
Except doorways, all escape routes should have a clear headroom of not less than 2 m.	B1 3.17 (V2)

Although the width of escape routes and exits depends on the number of persons needing to use them, they should not be less than that shown in Table F3.

B1 3.18 (V2)

Table F3 Widths of escape routes and exits

Maximum number of persons	Minimum width
60	750 mm
110	850 mm
220	1050 mm
More than 220	5 per person

Protected corridors shall be installed for:

B1 3.24 (V2)

- corridors serving bedrooms;
- dead-end corridors (excluding recesses not exceeding 2 m deep);
- any corridor that is common to more than one different occupancies.

If, instead of a protected corridor, the means of escape is enclosed by partitions, those partitions shall:

B1 3.25 (V2)

- be carried up to the main soffit of the floor above or to a suspended ceiling:
- be fitted with doors into all openings into rooms off that corridor.

Every corridor more than 12 m long and which connects two or more storey exits, should be subdivided by self-closing fire doors positioned approximately midway between the two storey exits.

B1 3.26 (V2)

Unless escape stairways and corridors are protected by a pressurization system complying with BS EN 12101-6:2005, every dead-end corridor exceeding 4.5 m in length should be separated by self-closing fire doors (together with any necessary associated screens) from any part of the corridor which:

B1 3.27 (V2)

- provides two directions of escape;
- continues past one storey exit to another.

Appendix F

If an external escape route is beside an external wall of the building, that part of the external wall that is within 1800 mm of the escape route should be fire resistant up to a height of 1100 mm above the paving level of the route. B1 3.30 (V2)

An escape over flat roofs is permissible if:

- the route does not serve an institutional building:
- is not part of a building intended for use by members of the public. B1 3.31 (V2)

In small premises:

- floor areas should be generally undivided (except for kitchens, ancillary offices and stores) to ensure that exits are clearly visible from all parts of the floor areas; B1 3.34 (V2)
- store rooms should be enclosed with fire-resisting construction; B1 3.35 (V2)
- clear glazed areas should be provided in any partitioning separating a kitchen or ancillary office from the open floor area to enable any person within the kitchen or office to obtain early visual warning of an outbreak of fire. B1 3.36 (V2)

Escape routes

The number of escape routes and exits to be provided will depend on the number of occupants in the room, tier or storey in question and the travel distance to the nearest exit (see Table 2 of B1 (V2)). B1 3.2 (V2)

In mixed-use buildings, separate means of escape should be provided from any storeys (or parts of storeys) used for residential or assembly and recreation purposes. B1 3.4 (V2)

There should be alternative escape routes from all parts of the building unless the travel distance is within set limits (see Table F2 of B1 (V2)). B1 3.5 (V2)

Access control measures incorporated into the design of a building should not adversely affect fire safety provisions. B1 3.7 (V2)

The minimum number of escape routes and exits from a room or storey shall be in accordance with the number of occupants (see Table F3). B1 3.2 (V2)

 Note: Further guidance concerning the number of occupants and exits is contained in Appendix C of Part B.

Inner rooms

If the only escape route from an inner room is through another room then:

B1 3.10 (V2)

* the occupant capacity of the inner room should not exceed 60;
* the inner room should not be a bedroom;
* the inner room should be entered directly off the access room (but not via a corridor);
* the escape route from the inner room should not pass through more than one access room;
* the travel distance from any point in the inner room to the exit(s) from the access room should not exceed the distances given in Table F2 of B1 (V2));
* the access room should not be a place of (i.e. potentially with) a special fire hazard;
* the access room should be in the control of the same occupier; and
* one of the following arrangements should be made;
  - the enclosures (walls or partitions) of the inner room should be stopped at least 500 mm below the ceiling; or
  - a vision panel not less $0.1\,m^2$ should be located in the door or walls of the inner room;
  - the access room should be fitted with an automatic fire detection and alarm system.

Residential care homes

 Note: Generally speaking, in care homes for the elderly it is reasonable to assume that at least a proportion of the residents will need some assistance to evacuate.

Buildings should be designed for Progressive Horizontal Evacuation (PHE).

B1 3.39 (V2)

Areas used for the care of residents shall be subdivided into protected areas separated by compartment walls and compartment floors.

B1 3.41 (V2)

Note: This is to allow horizontal escape into an adjoining protected area.

Appendix F

Each storey used for the care of residents should be:	B1 3.42 (V2)
• divided into at least three protected areas by compartment wall; and • all floors should be compartment floors.	
Protected areas should be provided with at least two exits to adjoining, but separate, protected areas.	B1 3.43 (V2)
The maximum travel distances within a protected area to these exits should:	
• not exceed those given in Table F2;	B1 3.43 (V2)
• not be more than 64 m to a storey exit or a final exit.	B1 3.43 (V2)
A fire in a protected area should not prevent the occupants of any other area from reaching a final exit.	B1 3.44 (V2)
Escape routes should not pass through ancillary accommodation (also see section 3.50 of Part B).	B1 3.44 (V2)
The number of residents beds in protected areas should not exceed ten.	B1 3.45 (V2)
A fire detection and alarm system should be provided to an Ll standard in accordance with BS 5839-1:2002.	B1 3.47 (V2)
Bedrooms should be enclosed in fire-resisting construction with fire-resisting doors and every corridor serving bedrooms should be a protected corridor.	B1 3.48 (V2)
Bedrooms should not contain more than one bed (this includes a double bed).	B1 3.49 (V2)

Design for vertical escape

An important aspect of means of escape in multi-storey buildings is the availability of a sufficient number of adequately sized and protected escape stairs.

The number of escape stairs needed in a building (or part of a building) will be determined by:	B1 4.2 (V2)
• the constraints imposed by the design of horizontal escape routes; • whether independent stairs are required in mixed occupancy buildings; • whether a single stair is acceptable; and • the width for escape and the possibility that a stair may have to be discounted because of fire or smoke.	

Provided that independent escape routes are not necessary   B1 4.6 (V2)
from areas in different purpose groups, single escape stairs
may be used from:

- small premises (other than bars or restaurants);
- office buildings comprising not more than five storeys
  above the ground storey;
- factories comprising not more than one storey above
  the ground storey if the building is of normal risk (two
  storeys if the building is of low risk); or
- process plant buildings with an occupant capacity of not
  more than ten people.

Note: In mixed-use buildings (i.e. where a building contains storeys (or parts
of storeys) in different purpose groups) it is important to consider the effect of
one risk on another – for example, a fire in a shop, or unattended office – could
have serious consequences on a residential use in the same building. It is, therefore, important to consider whether completely separate routes of escape should
be provided from each different use within the building or whether other effective means to protect common escape routes can be provided.

General requirements

All escape routes should have a clear headroom of not   B1 5.26 (V2)
less than 2 m with no projection below this height (except
for door frames).

The floors of all escape routes (including the treads of   B1 5.27 (V2)
steps and surfaces of ramps and landings) should be
chosen to minimize their slipperiness when wet.

Any sloping floor or tier should be constructed with a   B1 5.28 (V2)
pitch of not more than 35° to the horizontal.

Where a ramp forms part of an escape route it shall   B1 5.28 (V2)
meet the requirements of Part M Access to and Use of
Buildings (see also Part K).

Final exits should:

- not be less in width than the minimum width required   B1 5.30 (V2)
  for the escape route(s) they serve;
- be sited to ensure rapid dispersal of persons from the   B1 5.31 (V2)
  vicinity of the building;

• not present an obstacle to wheelchair users and other people with disabilities;	B1 5.32 (V2)
• be immediately apparent to persons who may need to use them;	B1 5.33 (V2)
• be sited so that they are clear of any risk from fire or smoke in a basement, or from openings to transformer chambers.	B1 5.34 (V2)

If an escape route is over a flat roof: B1 5.35 (V2)

* the roof should be part of the same building from which escape is being made;
* the route across the roof should lead to a storey exit or external escape route;
* the part of the roof forming the escape route and its supporting structure, together with any opening within 3 m of the escape route, should be fire-resisting;
* the route should be adequately defined and guarded by walls and/or protective barriers (which meet the provisions in Approved Document K).

All escape routes should have adequate artificial lighting.	B1 5.36 (V2)
Routes and areas listed in Table 9 of B1 (V2) should also have escape lighting which illuminates the route if the main supply fails.	B1 5.35 (V2)
Lighting to escape stairs should be on a separate circuit from that supplying any other part of the escape route.	B1 5.36 (V2)

The installation of an escape lighting system shall be in accordance with BS 5266-1:2005.

Escape routes (other than those in ordinary use and/or within a flat) should be marked by emergency exit sign(s) in accordance with BS 5499-1:2002.	B1 5.37 (V2)
Note: Suitable signs should also be provided for refuges (see paragraph 4.10).	
Where it is critical for electrical circuits to be able to continue to function during a fire, protected circuits (meeting the requirements of BS EN 50200:2006) are needed.	B1 5.38 (V2)

Access lobbies and corridors

Escape stairs shall have a protected lobby or protected corridor at all levels (except the top storey, all basement levels and when the stair is a firefighting stair) if: • the stair is the only one serving a building which has more than one storey above or below the ground storey; • where the stair serves any storey at a height greater than 18 m; or • where the building is designed for phased evacuation.	B1 4.34 (V2)
Protected lobbies (with not more than $0.4\,m^2$ permanent ventilation) should be provided between an escape stairway and a place of special fire hazard.	B1 4.35 (V2)
Protected stairways should discharge: • directly to a final exit; or • by way of a protected exit passageway to a final exit.	B1 4.36 (V2)

 Note: Any such protected exit passageway should have the same standard of fire resistance and lobby protection as the stairway it serves.

If two protected stairways are adjacent, they (and any protected exit passageways linking them to final exits) shall be separated by an imperforate enclosure.	B1 4.37 (V2)
Protected stairways shall be free of potential sources of fire.	B1 4.38 (V2)

Cavity barriers

Cavity barriers should be provided above the enclosures to a protected stairway in a dwelling-house with a floor more than 4.5 m above ground level (see Figure F6).	B1 2.14 (V1)

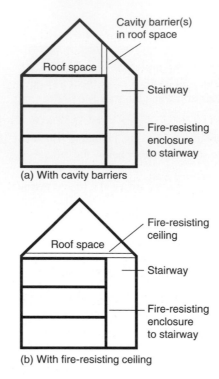

(a) With cavity barriers

(b) With fire-resisting ceiling

Figure F6 Alternative cavity barrier arrangements in roof space over protected stairway in a house with a floor more than 4.5 m above

Conversion to flats

If the existing building has timber floors and these are to be retained, the requirements for fire resistance may be difficult to meet. In these cases, provided that the means of escape conforms to Part B Section 3 and are adequately protected:

• doors on escape routes (both within and from the building) should be readily openable;	B1 5.10 (V2)
• doors on escape routes (whether or not the doors are fire doors), should either not be fitted with lock, latch or bolt fastenings, or they should only be fitted with simple fastenings that can be readily operated from the side approached by people making an escape, without the use of a key and without having to manipulate more than one mechanism.	B1 5.11 (V2)

Where a secure door is operated by a code, combination, swipe or proximity card, biometric data or similar means, it should also be capable of being overridden from the side approached by people making their escape.	B1 5.11 (V2)
Electrically powered locks should return to the unlocked position:	B1 5.11 (V2)

* on operation of the fire alarm system;
* on loss of power or system error;
* on activation of a manual door release unit.

In assembly places, shops and commercial buildings, doors on escape routes from rooms with an occupant capacity of more than 60 should either not be fitted with lock, latch or bolt fastenings (or be fitted with panic fastenings in accordance with BS EN 1125:1997).	B1 5.12 (V2)
See also Appendix B for guidance about door closing and 'hold open' devices for fire doors.	
The door of any doorway or exit should be hung to open in the direction of escape.	B1 5.14 (V2)
All doors on escape routes should be hung to open not less than 90°.	B1 5.15 (V2)
A door that opens towards a corridor or a stairway should be sufficiently recessed to prevent its swing from encroaching on the effective width of the stairway or corridor.	B1 5.16 (V2)
Vision panels shall be provided where doors on escape routes subdivide corridors, or where any doors are hung to swing both ways.	B1 5.17 (V2)
See also Parts M and N.	
Revolving doors, automatic doors and turnstiles should not be placed across escape routes.	B1 5.18 (V2)

Fire doors

Roller shutters across a means of escape should only be released by a heat sensor, such as a fusible link or electric heat detector, in the immediate vicinity of the door.	App B 6 (V2)

Appendix F

Protection of escape routes

Generally, a 30 minute standard is sufficient for the protection of means of escape. (Details of fire resistance test criteria and standards of performance are contained in Appendix A to Part B.)	B1 5.2 (V2)
Walls, partitions and other enclosures that need to be fire-resisting (including roofs that form part of a means of escape), should have the appropriate performance given in Tables A1 and A2 of Appendix A to Part B.	B1 5.3 (V2)
All doors that need to be fire-resisting should meet the requirements given in Table B1 of Appendix B to Part B.	B1 5.6 (V2)
The use of glazed elements in fire-resisting enclosures and/or doors depends on whether that element forms part of a protected shaft (see Appendix A, Table A4 and also Part N).	B1 5.7 (V2)

Raised storage areas

Raised free-standing floors in single storey industrial and storage buildings which are effectively galleries or a floor forming an additional storey, in certain circumstances, might not be able to meet the requirements of Appendix A, Table A1. For the purposes of fire safety, they are, however, deemed acceptable provided that:

• the structure has only one tier and is used for storage purposes only; • the number of persons likely to be on the floor at any one time is low and does not include members of the public; • the floor is not more than 10 m wide or long and does not exceed one half of the floor area of the space in which it is situated; • the floor is open above and below to the room or space in which it is situated; and • the means of escape from the floor meets the relevant requirements of Part B (particularly Sections 3).	B2 7.7 (V2)

Table F4 Limitations on the use of uninsulated glazed elements on escape routes

Position of glazed element	Maximum total glazed area in parts of the building with access to			
	A single stairway		More than one stairway	
	Walls	Door leaf	Walls	Door leaf
Single family dwelling houses				
1 (a) Within the enclosures of (i) protected stairway (ii) existing stair (b) Within fire resisting separation (c) Existing window between an attached/integral garage and the house	Fixed fanlights only Unlimited 100 mm from floor Unlimited	Unlimited Unlimited 100 mm from floor N/A	Fixed fanlights only Unlimited 100 mm from floor Unlimited	Unlimited Unlimited 100 mm N/A
Flats and maisonettes				
2 Within the enclosures of a protected entrance hall or protected landing	Fixed fanlights only	1100 mm from floor	Fixed fanlights only	1100 mm from floor

Appendix G

Entrance and access

(A copy of this Appendix is available on www.routledge.com/books/details/ 9780415809696/)

G.1 Entrance and access

G.1.1 The requirement (Building Act 1984 Sections 24 and 71)

The Building Act is very specific about exits, passageways and gangways and local authorities are required to consult with the fire authority to ensure that proposed methods of ingress and egress are deemed satisfactory (depending on the type of building). The purpose for which the building is going to be used needs to be considered as each case can be different. In particular, is the building going to be:

- a theatre, hall or other public building that is used as a place of public resort;
- a restaurant, shop, store or warehouse to which members of the public are likely to be admitted;
- a club;
- a school;
- a church, chapel or other place of worship (erected or used after the Public Health Acts Amendment Acts 1890 came into force).

At all times, the means of ingress and egress and the passages and gangways, while persons are assembled in the building, are to be kept free and unobstructed.

You are required by the Building Act 1984 to ensure that all courts, yards and passageways giving access to a house, industrial or commercial building (not maintained at the public expense) are capable of allowing satisfactory drainage of its surface or subsoil to a proper outfall.

The local authority can require the owner of any of the buildings to complete such works as may be necessary to remedy the defect.

All entrances to courts and yards must allow the free circulation of air and are not allowed to be closed, narrowed, reduced in height or in any other way altered so that it can impede the free circulation of air through the entrance.

(Building Act 1984 Section 85)

Private houses are not restricted by the actual Building Act of 1984 provided that members of the public are only admitted occasionally or exceptionally.

Fire safety

- *There shall be sufficient means of external access to enable fire appliances to be brought near to the building for effective use.*
- *There shall be sufficient means of access into, and within, the building for firefighting personnel to affect search and rescue and fight fire.*
- *The building shall be provided with sufficient internal fire mains and other facilities to assist firefighters in their tasks.*
- *The building shall be provided with adequate means for venting heat and smoke from a fire in a basement.*

(Approved Document B5)

Vehicle barriers and loading bays

- *Vehicle barriers should be provided that are capable of resisting or deflecting the impact of vehicles.*
- *Loading bays shall be provided with an adequate number of exits (or refuges) to enable people to avoid being crushed by vehicles.*

(Approved Document K3)

Disabled people

During 2002/3, Approved Document M was thoroughly overhauled and restructured in order to meet the changed requirements of the Disability Discrimination Act 1995 (which are enforced by the *Discrimination Act 1995 (Amendment) Regulations 2003* (SI 2003/1673)). Part M now covers:

- the use of a building to disabled people (redefined to include parents with children, elderly people and people with all types of disabilities – such as mobility, sight and hearing etc.) whether as residents, visitors, spectators, customers or employees, or participants in sports events, performances and conferences.

 Note: See Annex A for further guidance on access and facilities for disabled people.

Access and facilities for disabled people

In addition to the requirements of the Disability Discrimination Act 1995 precautions need to be taken to ensure that:

- *new non-domestic buildings and/or dwellings (e.g. houses and flats used for student living accommodation etc.);*
- *extensions to existing non-domestic buildings;*

- *non-domestic buildings that have been subject to a material change of use (e.g. so that they become a hotel, boarding house, institution, public building or shop);*

are capable of allowing people, regardless of their disability, age or gender, to:

- *gain access to buildings;*
- *gain access within buildings;*
- *be able to use the facilities of the buildings (both as visitors and as people who live or work in them);*
- *use sanitary conveniences in the principal storey of any new dwelling.*

(Approved Document M)

 Note: See Annex A for guidance on access and facilities for disabled people.

Access Statements

To assist building control bodies it is recommended that an 'Access Statement' is also provided when plans are deposited, a building notice is given, or details of a project are provided to an approved inspector.

 Note: A building control file should also be prepared for all new buildings, changes of use and where extensive alterations are being made to existing buildings.

In its simplest form, an Access Statement should show where an applicant wishes to deviate from the guidance in Approved Document M, either to:

- make use of new technologies (e.g. infrared activated controls);
- provide a more convenient solution; or
- address the constraints of an existing building.

The Access Statement should include:

- the reasons for departing from the guidance;
- the rationale for the design approach adopted;
- constraints imposed by the existing structure and its immediate environment (why it is not practicable to adjust the existing entrance or provide a suitable new entrance);
- convincing arguments that an alternative solution will achieve the same, a better, or a more convenient outcome (e.g. why a fully compliant independent access is considered impracticable);
- evidence (e.g. current validated research) to support the design approach;
- the identification of buildings (or particular parts of buildings) where access needs to be restricted (e.g. processes that are carried out which might create hazards for children, disabled people or frail, elderly people).

 Note: Further guidance on Access Statements is available on the Disability Rights Commission's website at http://www.drc-gb.org.

G.1.2 Meeting the requirement

Access and facilities for the fire service

Guidance

- There should be sufficient means of external access to enable fire appliances to be brought near to the building for effective use.
- There should be sufficient means of access into, and within, the building for firefighting personnel to effect search and rescue and fight fire.
- The building should be provided with sufficient internal fire mains and other facilities to assist firefighters in their tasks.
- The building should be provided with adequate means for venting heat and smoke from a fire in a basement.

 Note: For dwelling houses and small buildings, it is usually only necessary to ensure that the building is sufficiently close to a point accessible to fire and rescue service vehicles.

Vehicle access

To enable high reach appliances, such as turntable ladders and hydraulic platforms, to be used and to enable pumping appliances to supply water and equipment for firefighting, search and rescue activities, access to the building is required.

There should be vehicle access for a pump appliance to within 45 m of all points within the dwelling-house.	B5 11.2 (V2)
Every elevation to which vehicle access is provided should have a suitable door(s), not less than 750 mm wide, giving access to the interior of the building.	B5 11.3 (V2)

Design of access routes and hard standings (dwelling houses)

A vehicle access route may be a road or other route which, including any inspection covers and the like, meets the standards in Table G1.	B5 11.4 (V2) B5 16.8 (V2)
Turning facilities should be provided in any dead-end access route that is more than 20 m long (see Figure G1).	B5 11.5 (V2) B5 16.9 (V2)

Table G1 Typical fire and rescue service vehicle access route specification

Appliance type	Minimum width of road between kerbs (m)	Minimum width of gateways (m)	Minimum turning circle between kerbs (m)	Minimum turning circle between walls (m)	Minimume clearance height (m)	Minimum carrying capacity (tonnes)
Pump	3.7	3.1	16.8	19.2	3.7	12.5
High reach	3.7	3.1	26.0	29.0	4.0	17.0

Fire and rescue service vehicles should not have to reverse more than 20 m from the end of an access road

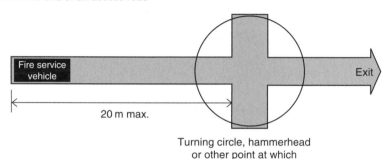

Turning circle, hammerhead or other point at which vehicle can turn

Figure G1 Turning facilities

Buildings not fitted with fire mains

There should be vehicle access for a pump appliance to small buildings (those of up to 2000 m² with a top storey up to 11 m above ground level) to either:

B5 16.2 (V2)

- 15% of the perimeter; or
- within 45 m of every point on the projected plan area.

There should be vehicle access for a pump appliance to blocks of flats to within 45 m of all points within each dwelling.

B5 16.3 (V2)

Every elevation to which vehicle access is provided should have a suitable door(s), not less than 750 mm wide, giving access to the interior of the building.

B5 16.5 (V2)

Door(s) should be provided such that there is no more than 60 m between each door and/or the end of that elevation (e.g. a 150 m elevation would need at least two doors). B5 16.5 (V2)

Buildings fitted with fire mains

If a building is fitted with dry fire mains:

- there should be access for a pumping appliance to within 18 m of each fire main inlet connection point (typically on the face of the building); B5 16.6 (V2)
- the inlet should be visible from the appliance.

If a building is fitted with wet mains, the pumping appliance access:

- should be to within 18 m and within sight of a suitable entrance; and
- in sight of the inlet for the emergency replenishment of the suction tank for the main. B5 16.7 (V2)

Vehicular access

If vehicles have access to a floor or roof edge, barriers of at least 375 mm should be provided to any edges that are level with (or above) the adjoining floor or ground.	K3 (3.7)
If vehicles have access to a ramp edge, barriers of at least 610 mm should be provided to any edges that are level with (or above) the adjoining floor or ground.	K3 (3.7)
Any wall, parapet, balustrade or similar obstruction may serve as a barrier.	K3 (3.8)
All barriers should be capable of resisting forces set out in BS 6399.	K3 (3.8)
Loading bays should be provided with at least one exit point from the lower level (preferably near the centre of the rear wall).	K3 (3.9)

Access and facilities for disabled people

Access (i.e. approach, entry or exit) to a building is frequently a problem for wheelchair users, people who need to use walking aids, people with impaired sight and parents with prams etc. In designing the approach to a building (and routes between buildings within a complex) the following should, therefore, always be taken into consideration:

- changes in level between the entrance storey and the site entry point should be minimized;
- access routes should be wide enough to let people pass each other;
- potential hazards (e.g. windows from adjacent buildings opening onto access routes) should be avoided.

 Note: See also *Mobility: A Guide to Best Practice on Access to Pedestrian and Transport Infrastructure.*

General

The primary aim should be to make it reasonably possible for a disabled person to approach and gain access to the dwelling from the entrance point at the boundary of the site (and from any car parking that is provided on the site) to the building. It is also important that routes between buildings within a complex are also accessible.

Approach to a building

Access from the boundary of the site (and/or from car parking designated for disabled people) to the principal entrance should be level.	M (1.2, 1.4, 1.6 and 1.13) M (6.2)
If a difference in level is unavoidable (i.e. due to site constraints) the approach can have a gentle gradient (provided that it is over a long distance) or can include a number of shorter parts (at steeper gradients) as long as level landings are provided as rest points.	M (1.7)
The principal entrance (entrances, main staff entrance and any lobbies) should be accessible to disabled people and mothers pushing prams etc.	M (2.1)
If this is not possible, an alternative accessible entrance should be provided.	M (2.2)
Risks to people when entering the building should be minimal.	M (2.3)
Access routes should be wide enough to let people pass each other.	M (1.11)

Note: A surface width of 1800 mm is ideal but this can be reduced on restricted sites to 1200 mm, provided that a case is made in the Access Statement.

The route to the principal entrance (or alternative accessible entrance) should be clearly identified and well lit.   M (1.13 g)

A separate pedestrian route should be provided.   M (1.13 h)

Uncontrolled vehicular crossing points should be identified by a buff coloured blister surface (see Figure G2).   M (1.13 h)

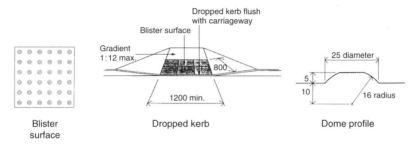

Blister surface

Dropped kerb

Dome profile

Figure G2 An example of tactile paving used at an uncontrolled crossing

Gradients

Approach gradients:

- should ideally be no steeper than 1:60 along their whole length;   M (1.13c)
- if steeper than 1:20, should be designed as a ramped access;   M (1.8)
- if less than 1:20, should be provided with level landings for each 500 mm rise of the access.   M (1.13c)
- Cross-fall gradients should be no steeper than 1:40.   M (1.13c)

Surface

The surface of all access routes should:

- allow people to travel along them easily, without excessive effort and without the risk of tripping or falling;   M (1.9)
- be at least 1.5 m wide;   M (1.13a)
- be firm, durable and slip resistant;   M (1.13d)
- have undulations not exceeding 3 mm (under a 1 m straight edge);   M (1.13d)

Appendix G

• be made of the same material and similar frictional characteristics (loose sand or gravel should **not** be used).	M (1.13d and e)

Building perimeters

The perimeter of the building should be well lit.	M (1.12)

Passing places

Passing places should be: • free of obstructions to a height of 2.1 m; • at least 1.8 m wide and at least 2 m long.	M (1.13a) M (1.13b)

Joints

Joints should be: • filled flush or (if recessed) no deeper than 5 mm; • no wider than 10 mm or (if unfilled) no wider than 5 mm.	M (1.13f) M (1.13f)
The difference in level at joints between paving units should be no greater than 5 mm.	M (1.13f)

On-site car parking and setting down – parking bays

At least one parking bay designated for disabled people should be provided as close as possible to the principal entrance of the building.	M (1.18a)
The dimensions of the designated parking bays should be as per Figure G3 (with a 1200 mm accessibility zone between and a 1200 mm safety zone on the vehicular side of the parking bays and with a dropped kerb when there is a pedestrian route at the other side of the parking bay).	M (1.18b)

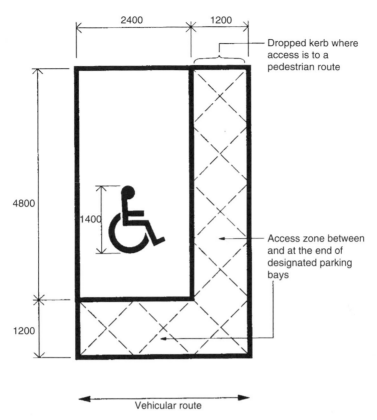

Figure G3 Parking bay designated for disabled people

A clearly signposted setting down point should be located on firm, level ground as close as possible to the principal (or alternative) entrance.	M (1.18e)
The surface of the accessibility zone should be firm, durable and slip resistant, with undulations not exceeding 3 mm (under a 1 m straight edge).	M (1.18c)
The surface of a parking bay designated for disabled people (in particular the area surrounding the bay) should allow the safe transfer of a passenger or driver to a wheelchair and transfer from the parking bay to the access route to the building without undue effort.	M (1.15)
Ticket machines should: • have their controls between 750 mm and 1200 mm above ground level;	M (1.16 and 1.18d)

- be located near parking bays designated for disabled people;
- be located so that a person in a wheelchair (or a person of short stature) is able to reach the controls.

The plinth of the ticket machine should not project in front   M (1.18d)
of the face of the machine.

Note: See also BS 8300 for guidance on:

- provision of parking bays designated for disabled people;
- ticket dispensing machines;
- vehicular control barriers; and
- multi-storey car parks.

Ramped access

If the constraints of the site mean that there is an approach gradient of 1 in 20 or steeper, then a ramped access should be provided as they are beneficial not only for wheelchair users but also people pushing prams and bicycles.

Where a ramped access is provided:
- the approach should be clearly signposted;   M (1.26a)
- the going should be no greater than 10 m;   M (1.26c)
- the rise should be no more than 500 mm;   M (1.26c)
- if the total rise is greater than 2 m then an alternative   M (1.26d)
  means of access (e.g. a lift) should be provided for
  wheelchair users;
- the surface width should be at least 1.5 m;   M (1.26e)
- gradients should be as shallow as practicable;   M (1.20)
- the ramp surface should be slip resistant;   M (1.26f)
- the ramp surface should be of a contrasting colour to   M (1.26f)
  that of the landings;
- frictional characteristics of ramp and landing surfaces   M (1.26g)
  should be similar;
- landings at the foot and head of a ramp should be at   M (1.26h)
  least 1.2 m long and clear of any obstructions;
- intermediate landings should be at least 1.5 m long and   M (1.26i)
  clear of obstructions;
- intermediate landings (at least 1800 mm wide and   M (1.26j)
  1800 mm long) should be provided at passing places;

- all landings should:
  - be level;
  - have a maximum gradient of 1:60 along their length;
  - have a maximum cross fall gradient of 1:40;

  M (1.26k)

- there should be a handrail on both sides;
- in addition to the guarding requirements of Part K, there should be a visually contrasting kerb on the open side of the ramp (or landing) at least 100 mm high;

  M (1.26l)
  M (1.26m)

- when the rise of the ramp is greater than two 150 mm steps, signposted steps should be provided;

  M (1.26n)

- the gradient of a ramp flight and its going between landings should be in accordance with Table G2 and Figure G4.

  M (1.26b)

Table G2 Limits for ramp gradients

Going of a flight (m)	Maximum gradient	Maximum rise (mm)
10	1 : 20	500
5	1 : 15	333
2	1 : 12	166

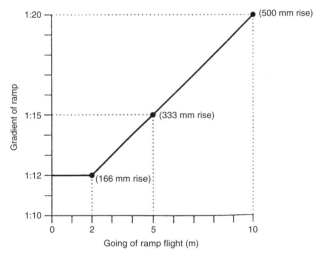

Figure G4 Relationship for ramp gradients to the going of a flight

Note: Approved Document K (*Protection from falling, collision and impact*) contains general guidance on stair and ramp design. The guidance in Approved Document M reflects more recent ergonomic research conducted to support BS 8300 and takes precedence over Approved Document K in conflicting areas.

Appendix G

Possible future amendment	Further research on stairs is currently being undertaken and will be reflected in future revisions of Approved Document K.

Stepped access

People with impaired sight risk tripping or losing their balance if there is no warning that there are steps that provide a change in level. The risk is most hazardous at the head of a flight of stairs when a person is descending.

A stepped access should:

- have a level landing at the top and bottom of each flight; M (1.33a)
- be 1200 mm long each landing and unobstructed; M (1.33b)
- have a corduroy hazard warning surface at top and bottom landings of a series of flights so as to give advance warning of a change in level (see Figure G5); M (1.33c)

Note: Approved Document K (*Protection from falling, collision and impact*) contains general guidance on stair and ramp design. The guidance in Approved Document M reflects more recent ergonomic research conducted to support BS 8300 and takes precedence over Approved Document K.

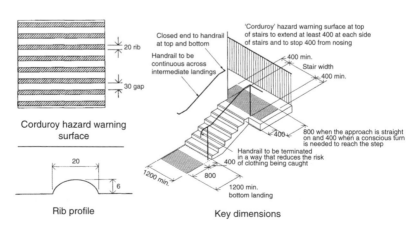

Figure G5 Stepped access – key dimensions and use of hazard warning surfaces

In addition:

> - side accesses onto intermediate landings should have   M (1.33d)
>   a 400 mm deep corduroy hazard warning surface;
> - doors should not swing across landings;   M (1.33e)
> - the surface width of flights between enclosing walls,   M (1.33f)
>   strings or upstands, should not be less than 1.2 m;
> - there should be no single steps;   M (1.33g)
> - the rise of a flight between landings should contain no   M (1.33h)
>   more than:
>   - 12 risers for a going of **less** than 350 mm;
>   - 18 risers for a going of 350 mm or **greater**
>   (see Figure G6).

Note: For school buildings, the preferred dimensions are a rise of 150 mm and a going of 280 mm.

Figure G6 External steps and stairs – key dimensions

Nosings for the tread and the riser should be 55 mm wide and of a contrasting material.	M (1.33i)
Step nosings should not project over the tread below by more than 25 mm (see Figure G7).	M (1.33j)
The rise and going of each step should be consistent throughout a flight.	M (1.33k)
The rise of each step should be between 150 mm and 170 mm.	M (1.33l)
Rises should not be open.	M (1.33n)
The going of each step should be between 280 mm and 425 mm.	M (1.33m)
There should be a continuous handrail on each side of a flight and landings.	M (1.33o)
If additional handrails are used to divide the flight into channels, then they should not be less than 1 m wide or more than 1.8 m wide.	M (1.33p)

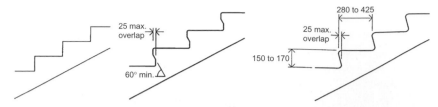

The rise and going dimensions apply to all step profiles

Figure G7 Examples of step profiles and key dimensions for external stairs

Warnings should be placed sufficiently in advance of a hazard to allow time to stop.	M (1.28)
Warnings should not be so narrow that they could be missed in a single stride.	M (1.28)
Materials for treads should not present a slip hazard.	M (1.29)

Handrails to external stepped and for ramped access

Handrails to external stepped or ramped access should be positioned as per Figure G8.	M (1.37a)
Handrails to external stepped or ramped access should:	
• be continuous across flights and landings;	M (1.37c)
• extend at least 300 mm horizontally beyond the top and bottom of a ramped access;	M (1.37d)
• not project into an access route;	M (1.37d)
• contrast visually with the background;	M (1.37e)

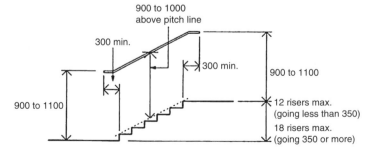

Figure G8 Handrails to external stepped and ramped access – key dimensions

- have a slip resistant surface which is **not** cold to the touch;   M (1.37f)
- terminate in such a way that reduces the risk of clothing being caught;   M (1.37g)
- either be circular (with a diameter of between 40 and 45 mm) or oval with a width of 50 mm (see Figure G9).   M (1.37h)

In addition, handrails should:

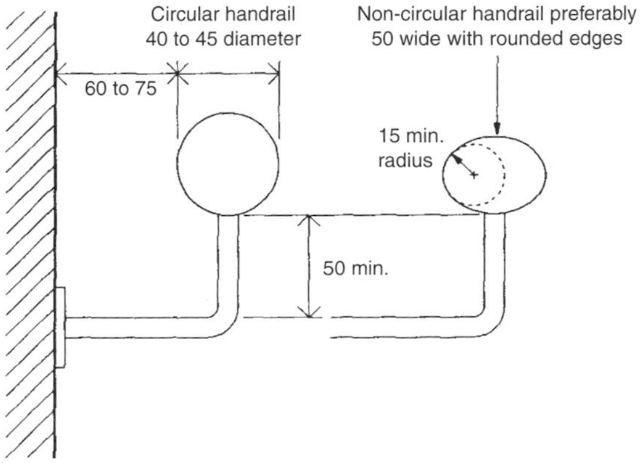

Figure G9 Handrail designs

- not protrude more than 100 mm into the surface width of the ramped or stepped access where this would impinge on the stair width requirement of Part B1;   M (1.37i)
- have a clearance of between 60 and 75 mm between the handrail and any adjacent wall surface;   M (1.37j)
- have a clearance of at least 50 mm between a cranked support and the underside of the handrail;   M (1.37k)
- ensure that its inner face is located no more than 50 mm beyond the surface width of the ramped or stepped access;   M (1.37l)
- be spaced away from the wall and rigidly supported in a way that avoids impeding finger grip;   M (1.35)
- be set at heights that are convenient for all users of the building.   M (1.36)

Appendix G

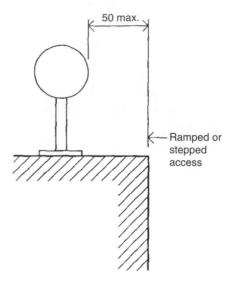

Figure G10 Handrail location

Hazards on access routes

Features of a building (e.g. windows and doors) that can occasionally obstruct an access route should not present a hazard.	M (1.38)
Areas below stairs or ramps with a soffit less than 2.1 m above ground level should be protected by guarding and low-level cane detection.	M (1.39b)
Any feature projecting more than 100 mm an access route should be protected by guarding that includes a kerb (or other solid barrier) that can be detected using a cane (see Figure G11).	M (1.39b)

Accessible entrances

Accessible entrances should be clearly signposted (e.g. with the International Symbol of Access) and easily recognized.	M (2.5 and 2.7a)
Accessible entrances should also:	
• be easily identifiable (e.g. by lighting and/or visual contrast);	M (2.7b)

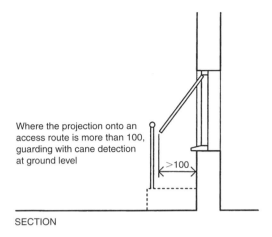

Where the projection onto an access route is more than 100, guarding with cane detection at ground level

>100

SECTION

Figure G11 Avoiding hazards on access routes

- have a level landing at least 1500 × 1500 mm,   M (2.7d)
  clear of any door swings, immediately in front of
  the entrance;
- avoid raised thresholds (if unavoidable,   M (2.7e)
  then the total height should not be more
  than 15 mm, with a minimum number of
  upstands and slopes);
- ensure that all door entry systems are accessible to   M (2.7f)
  deaf and hard of hearing people, plus people who
  cannot speak;
- not have internal floor surface material   M (2.7h)
  (e.g. coir matting) adjacent to the threshold
  that could impede the movement of
  wheelchairs;
- not have changes in floor materials that could   M (2.7h)
  create a potential trip hazard;
- if mat wells are provided, have the surface of the   M (2.7i)
  mat level with the surface of the adjacent floor
  finish;
- have the route from the exterior across the   M (2.6 and 2.7a)
  threshold weather protected;
- not present a hazard for visually impaired people   M (2.5 and 2.7c)
  (e.g. have structural elements such as canopy
  supports).

 Note: See BS 8300 for further guidance on signposting.

Doors to accessible entrances

Doors to the principal (or alternative accessible) entrance should be accessible to all, particularly for wheelchair users and people with limited physical dexterity.	M (2.8)
Entrance doors should be capable of being held closed when not in use.	M (2.8)
A power operated door opening and closing system should be used if a force greater than 20 N is required to open or shut a door.	M (2.13a)

The premises have no more than three steps

The premises are fully accessible for wheelchair users

Wheelchair assistance required

Figure G12 Typical access signs for disabled people

Once open, all doors to accessible entrances should be wide enough to allow unrestricted passage for a variety of users, including wheelchair users, people carrying luggage, people with assistance dogs, and parents with pushchairs and small children.	
People should be able to see other people approaching from the opposite direction.	M (2.12)
The effective clear width through a single leaf door (or one leaf of a double leaf door) should be in accordance with Table G3.	M (2.13b)
Door leaves and side panels wider than 450 mm should incorporate vision panels: • towards the leading edge of the door; • between 500 mm and 1500 mm from the floor (see Figure G13).	M (2.13c)

Table G3 Minimum effective clear widths of doors

Direction and width of approach	New buildings (mm)	Existing buildings (mm)
Straight-on (without a turn or oblique approach)	800	750
At right angles to an access route at least 1500 mm wide	800	750
At right angles to an access route at least 1200 mm wide	825	775
External doors to buildings used by the general public	1000	775

Manually operated non-powered entrance doors

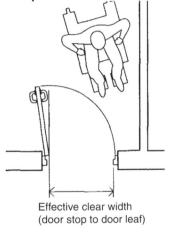

Effective clear width
(door stop to door leaf)

Figure G13 Effective clear width and visibility requirements of doors

Self-closing devices on manually operated (i.e. non-powered) doors can be a great disadvantage to people who have limited upper body strength, or people who are pushing prams or carrying heavy objects. To rectify this matter:

The opening force at the leading edge of the door should be no greater than 20 N.	M (2.17a)
A space alongside the leading edge of a door should be provided to enable a wheelchair user to reach and grip the door handle.	M (2.15)

Door opening furniture should:

- be easy to operate by people with limited manual dexterity;   M (2.16)
- be capable of being operated with one hand using a
  closed fist (e.g. a lever handle);   M (2.17c)
- contrast visually with the surface of the door and not be
  cold to the touch.   M (2.17d)

There should be an unobstructed space of at least 300 mm   M (2.17b)
on the pull side of the door between the leading edge of
the door and any return wall (unless the door is a powered
entrance door – see Figure G14).

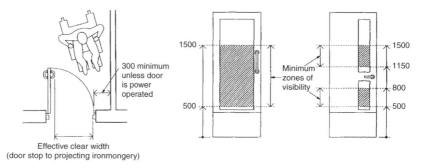

Effective clear width
(door stop to projecting ironmongery)

Figure G14 Effective clear width and visibility requirements of doors

Power-operated entrance doors

Power-operated entrance doors should have a sliding, swinging or folding
action controlled manually (by a push pad, card swipe, coded entry, or remote
control) or be automatically controlled by a motion sensor or proximity sensor such as a contact mat.

Power-operated entrance doors should:

- open towards people approaching the doors;   M (2.21a)
- provide visual and audible warnings that they are   M (2.21c)
  operating (or about to operate);
- incorporate automatic sensors to ensure that they open   M (2.21c)
  early enough (and stay open long enough) to permit safe
  entry and exit;
- incorporate a safety stop that is activated if the doors   M (2.21b)
  begin to close when a person is passing through;

- revert to manual control (or fail-safe) in the open position in the event of a power failure;   M (2.21d)
- when open, not project into any adjacent access route;   M (2.21e)
- ensure that its manual controls: M (2.21f)
  - are located between 750 mm and 1000 mm above floor level;
  - are operable with a closed fist;
- be set back 1400 mm from the leading edge of the door when fully open;   M (2.21g)
- be clearly distinguishable against the background;   M (2.21g)
- contrast visually with the background.   M (2.19 and 2.21g)

 Note: Revolving doors are not considered 'accessible' as they create particular difficulties (and possible injury) for people who are visually impaired, people with assistance dogs or mobility problems and for parents with children and/or pushchairs.

Glass entrance doors and glazed screens

The presence of the door should be apparent not only when it is shut but also when it is open.   M (2.23)

Glass entrance doors and glazed screens should:

- be clearly marked (i.e. with a logo or sign) on the glass at two levels, 850 to 1000 mm and 1400 to 1600 mm above the floor;   M (2.24a)

  Note: The logo or sign should be at least 150 mm high (repeated if on a glazed screen), or a decorative feature such as broken lines or continuous bands, at least 50 mm high.

- when adjacent to, or forming part of, a glazed screen, be provided with a high contrast strip at the top, and on both sides;   M (2.24c)
- ensure that glass entrance doors (if capable of being held open) are protected by guarding to prevent the leading edge from becoming a possible hazard.   M (2.24d)

For dwellings:

> An external door providing access for disabled
> people should have a minimum clear opening width
> (taken from the face of the door stop on the latch side
> to the face of the door when open at 90°) of
> 775 mm.
>
> The door opening width should be sufficient to enable a
> wheelchair user to manoeuvre into the dwelling.

M (6.23)

M (6.22)

Possible *future* *amendment*	Approved Document N (Glazing – safety in relation to impact, opening and cleaning) contains guidance on the use of symbols and markings on glazed doors and screens. The guidance now given in Approved Document M is as a result of more recent experience of 'door manifestation' and takes precedence over the guidance currently provided in Approved Document N in conflicting areas until such time as Approved Document N is revised.

Entrance (and internal) lobbies

> Lobbies should be:
> - large enough to allow a wheelchair user or a
>   person pushing a pram to move clear of one
>   door before opening the second door;
> - capable of accommodating a companion
>   helping a wheelchair user to open doors and
>   guide the wheelchair through.
>
> The minimum length of the lobby is related to
> the chosen door size, the swing of each door, the
> projection of the door into the lobby and the size
> of an occupied wheelchair with a companion
> pushing.
>
> Within the lobby:
> - glazing should not create distracting
>   reflections;

M (2.27 and 3.15)

M (2.29d and
3.16d)

- floor surface materials should not impede the movement of wheelchairs etc.;   M (2.29e)
- changes in floor materials should not create a and potential trip hazard;   M (2.29e 3.16e)
- the floor surface should assist in removing rainwater from shoes and wheelchairs;   M (2.29f)
- any columns and ducting etc. that project into the lobby by more than 100 mm should be protected by a visually contrasting guard rail.   M (2.29h and 3.16a–c)

The length and width of an entrance and/or an internal lobby should be as per Table G4.   M (2.29a, b and c)

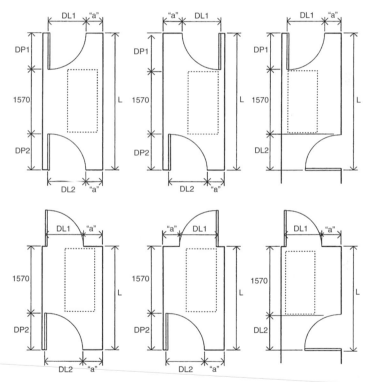

DL1 and DL2 = door leaf dimensions of the doors to the lobby
DP1 and DP2 = door projection into the lobby (normally door leaf size)
L = minimum length of lobby, or length up to door leaf for side entry lobby
"a" = at least 300 mm wheelchair access space (can be increased to reduce L)
1570 = length of occupied wheelchair with a companion pushing (or a large scooter)
NB: For every 100 mm increase above 300 mm in the dimension "a" (which gives a greater overlap of the wheelchair footprint over the door swing), there can be a corresponding reduction of 100 mm in the dimension L, up to a maximum of 600 mm reduction.

Figure G15 Key dimensions – lobbies and entrance doors

Table G4 Entrance lobbies – dimensions

	Length	Width
Entrance/internal lobby with single swing door	as per Figure 6.202	at least 1200 mm (or the width of the two doors plus 300 mm whichever is the greater)
Entrance/internal lobby with double swing doors	at least the size (i.e. width) of the two doors plus 1570 mm	at least 1800 mm

Entrance hall and reception area

If there is a reception point:

- it should be easily accessible and convenient to use;   M (3.2)
- it should be located away from the principal entrance;   M (3.6a)
- it should be easily identifiable from the entrance doors or lobby;   M (3.6b)
- relevant information about the building should be clearly available from noticeboards and signs;   M (3.5)
- the floor surface should be slip resistant;   M (3.6)
- the approach to it should be direct and free from obstructions;   M (3.6b)
- the design of the approach should allow space for wheelchair users to gain access;   M (3.6c)
- there should be a clear manoeuvring space in front of any reception desk at least 1200 mm deep and 1800 mm wide.   M (3.6d)

If there is a reception desk or counter:

- it should be designed to accommodate both standing and seated visitors;   M (3.6e)
- if there is a knee recess, then this should be at least 500 mm deep (or 1400 mm deep and 2200 mm wide if there is no knee recess);   M (3.6d)
- at least one section of the reception desk should be no less than 1500 mm wide, its surface no higher than 760 mm and a knee recess not less than 700 mm, above floor level;   M (3.6e)
- it should be provided with a hearing enhancement system (e.g. an induction loop).   M (3.6f)

Note: See BS 8300 for guidance on aids to communication.

Domestic buildings

If there is a reception point:

- it should be easily accessible and convenient to use;   M (3.2 to 3.5)
- information about the building should be clearly available from noticeboards and signs;
- the floor surface should be slip resistant.

Approach to a dwelling

On plots which are reasonably level, wheelchair users   M (6.2)
should normally be able to approach the principal entrance.

Wheelchair users (having approached the entrance)   M (6.3)
should be able to gain access into the dwelling-house
and/or entrance level of flats.

A suitable approach should be provided from the point of   M (6.11)
access to the entrance of the dwelling.

The whole, or part, of the approach may be a driveway.   M (6.12)

The approach should:

- not have crossfalls greater than 1 in 40;   M (6.11)
- be safe and convenient for disabled people as is   M (6.6)
  reasonable possible;   M (6.8)
- ideally be level or ramped.

Note: On steeply sloping plots, a stepped approach
is permissible.

If a stepped approach to the dwelling is unavoidable,   M (6.7)
the aim should be for the steps to be designed to suit
the needs of ambulant disabled people (see
paragraph 6.19).

The surface of the wheelchair user's approach   M (6.9)
should:

- be firm enough to support the weight of the user and
  their wheelchair;
- be smooth enough to permit easy manoeuvre;
- not be made up of loose-laid materials (such as gravel
  and shingle);
- take account of the needs of stick and crutch users.

For steeply sloping plots, it would be reasonable to
provide for stick or crutch users.

Level approach

A 'level' approach should: M (6.13)

- be no steeper than 1 in 20;
- have a firm and even surface;
- have a width not less than 900 mm.

Note: The width of the approach, excluding space
for parked vehicle, should take account of the needs of a
wheelchair user, or a stick or crutch user.

Ramped approach

If a plot gradient exceeds 1 in 20, a ramped approach may   M (6.14)
be provided in which case:

- the surface should be firm and even;   M (6.15a)
- the flights should have unobstructed widths of at least   M (6.15b)
  900 mm;
- individual flights should be no longer than 10 m (for   M (6.15c)
  gradients not steeper than 1 in 15) or 5 m for gradients
  not steeper than 1 in 12;
- it should have top and bottom landings (and   M (6.15c)
  intermediate landings if necessary) not less than 1.2 m
  in length – exclusive of the swing of any door or gate
  which opens onto it.

Stepped approach

A stepped approach should be used if the plot gradient is   M (6.16)
greater than 1 in 15.

A stepped approach should:

- have flights with an unobstructed width of at least 900 mm; M (6.17a)
- have a flight rise of not more than 1.8 m;   M (6.17b)
- have a top and bottom (and if necessary intermediate)
  landing not less than 900 mm in length;   M (6.17c)
- have steps: M (6.17d)
  - with suitable tread nosing profiles (see Figure G16);
  - with a uniform rise between 75 mm and 150 mm;   M (6.17e)
- ensure that the going of each step is not less than   M (6.17f)
  280 mm;

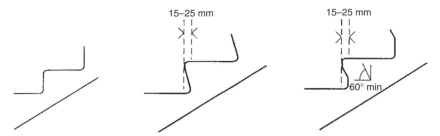

Figure G16 External step profiles

- comprise three or more risers;
- have a suitable continuous handrail on one side of the flight between 850 mm and 1000 mm above the pitch line of the flight; and extend 300 mm beyond the top and bottom nosings.

Approach using a driveway

Where a driveway provides a means of approach towards the entrance, the approach past any parked cars must be in accordance with requirements M6.11–6.27.	M (6.18)

Access into the dwelling

The point of access should be reasonably level.	M (6.11)
Where the approach to the entrance consists of a level or ramped approach, an accessible threshold at the entrance should be provided.	M (6.19)

Note: An accessible threshold into entrance level flats should also be provided.

If a stepped approach is provided into the dwelling, the rise should be no more than 150 mm.	M (6.20)
An accessible threshold should be provided into the entrance.	M (6.21)

 Note: The design of an accessible threshold should also satisfy the requirements of Part C2: 'Dangerous and offensive substances' and Part C4: 'Resistance to weather and ground moisture'.

Appendix G

Appendix H_____
Acronyms

ABBE	Awarding Body for the Built Environment;
ACE	Amalgamated Chimney Engineers
ach	air changes per hour
AD	Approved Document
ADAHIPP	Association of Home Information Pack Providers
ATTMA	Air Tightness Testing and Measurement Association
BAFSA	British Automatic Fire Sprinkler Association
BBA	British Board of Agrément
BCO	Building Control Officer
BER	Building (CO_2) Emission Rate
BFCMA	British Flue and Chimney Manufacturers Association
BHIF	British Hardware Industry Federation
BRE	Building Research Establishment
BREEAM	BRE Environmental Assessment Method
BS	British Standard
BSI	British Standards Institution
BSRIA	Building Services Research and Information Association
CCL	Climate Change Levy
CEA	Commercial Energy Assessor
CIBSE	Chartered Institution of Building Services Engineers
CIRIA	Construction Industry Research and Information Association
CO_2	Carbon Dioxide
CoPSO	Council of Property Search Organisations
CP	Competent Persons Self-certification Scheme
CRed	Carbon Reduction
CSH	Code for Sustainable Homes
DCER	Dwellings Carbon Emission Rate
DCLG	Department of Communities and Local Government
DEA	Domestic Energy Assessors
DEC	Display Energy Certificates
DECC	Department of Energy and Climate Change

DER	Dwelling (CO_2) Emission Rate
DfES	Department for Education and Skills
DH	Department of Health
DIAG	Directive Implementation Advisory Group
DQRA	Detailed Quantitative Risk Assessment
DSMA	Door and Shutter Manufacturers Association
EEA	European Economic Area
EFBD	Energy Performance of Buildings Directive
EPC	Energy Performance Certificate
EST	Energy Saving Trust
EU	European Union
FPA	Fire Protection Association
GGF	Glass and Glazing Federation
GQRA	Generic Quantitative Risk Assessment
H&S	Health and Safety
HBF	House Builders Federation
HETAS	Heating Equipment Testing and Approval Scheme
HIP	Home Information Pack
HSE	Health and Safety Executive
HVAC	Heating Ventilation And Cooling
IACSC	International Association of Cold Storage Contractors
LABC	Local Authority Building Control
LZC	Low and Zero Carbon
MEV	Mechanical Extract Ventilation
MOU	Memorandum of Understanding
MVHR	Continuous Mechanical Supply and Extract with Heat Recovery
NACE	National Association of Chimney Engineers
NACS	National Association of Chimney Sweeps
NBS	National Building Specification
NCM	National Calculation Methodology
NDEPC	Non-Domestic Energy Performance Certificate
NEF	National Energy Foundation
NES	National Energy Services
NFA	National Fireplace Association
NHER	National Home Energy Rating
NIMBY	Not In My Back Yard
ODPM	Office of the Deputy Prime Minister
OFTEC	Oil Firing Technical Association Ltd
PCCB	Property Codes Compliance Board
PCST	Pre-Completion Sound Testing
PSA	Property Services Agency
PSV	Passive Stack Ventilation
RCEP	Royal Commission on Environmental Pollution
RdSAP	Reduced data Standard Assessment Procedure
RVA	Residential Ventilation Association
SAP	Standard Assessment Procedure

SBEM	Simplified Building Energy Model
SCI	Steel Construction Institute
SFA	Solid Fuel Association
SRHRV	Single Room Heat Recovery Ventilator
TEHVA	The Electric Heating and Ventilation Association
TER	Target Emissions Rate
TRADA	Timber Research and Development Association
TVOC	Total Volatile Organic Compound
VOC	Volatile Organic Compounds

Appendix I

Useful contact names and addresses

The following professional body is willing to provide general and informal advice about the Act. **However**, any advice given should **not** be seen as being endorsed by the UK government.

The Royal Institution of Chartered Surveyors (RICS)
Parliament Square
London SW1P 3AD
Tel: 0870 333 1600
Fax: 020 7334 3811

The following bodies hold lists of their members who may be willing to provide professional advice or act as a 'surveyor' under the Act – again with the proviso that any advice given should **not** be seen as being endorsed by the UK government.

Architecture and Surveying Institute
Register of Party Wall Surveyors
St Mary House
15 St Mary Street
Chippenham
Wiltshire SN15 3WD
Tel: 01249 444505
Fax: 01249 443602

The Association of Building Engineers (ABE)
Private Practice Register
Lutyens House
Billing Brook Road
Weston Favell

Northampton NN3 8NW
Tel: 0845 126 1058
Fax: 01604 784220

The Pyramus & Thisbe Club
Florence House
53 Acton Lane
London NW10 8UX
Tel: 020 8961 3311
Fax: 020 8963 1689

The Royal Institute of British Architects (RIBA)
Clients Advisory Service
66 Portland Place
London W1B 1AD

Tel: 020 580 5533
Fax: 020 255 1541

The Royal Institution of Chartered Surveyors (RICS)
Parliament Square
London SW1P 3AD
Tel: 0870 333 1600
Fax: 020 7334 3811

Professional contacts

Asbestos specialist

Asbestos Information Centre Ltd
5a The Maltings
Stowupland Road
Stowmarket
Suffolk 1P14 SAG
Tel: 0904 517 0156

Concrete specialist

British Ready Mixed Concrete Association
The Bury
Church Street
Chesham
Buckinghamshire HP5 1JE
Tel: 01494 791050

Damp, rot, infestation

British Wood Preserving & Damp Proofing Association
Building No. 6
The Office Village
4 Romford Road
Stratford
London E15 4EA
Tel: 020 8519 2588

English Nature
Northminster House
Northminster Road

Peterborough PE1 1VA
Tel: 01733 340345

Countryside Council for Wales
Maps y Ffynnon
Penrhosgarnedd
Bangor
Gwynedd LL57 2DW
Tel: 01248 355782

Scottish Natural Heritage
12 Hope Terrace
Edinburgh EH9 2AS
Tel: 0131 447 4784

Local Department of Environmental Health
Refer to your local directory.

Decorators

British Decorators Association
32 Coton Road
Nuneaton
Warwickshire CV11 5TW
Tel: 0247 635 3776

Scottish Decorators Federation
Castlecraig
Business Park
Players Road
Stirling FK7 7SH
Tel: 01786 448838

Electricians

National Inspection Council for Electrical Installation Contracting
Warwick House
Houghton Hall Park
Houghton Regis
Dunstable LU5 5ZX
Tel: 0870 013 0382

Fencing erectors

Fencing Contractors Association
Warren Rd
Trellech
Monmouthshire NP25 4PQ
Tel: 07000 560722

Glazing specialists

Glass and Glazing Federation
44–48 Borough High Street
London SE1 1XB
Tel: 020 7403 7177

Heating installers

British Gas Regional Office
Refer to your local directory.

Electricity Supply Company
Refer to your local directory.

British Coal Corporation
Hobart House
Grosvenor Place
London SW1X 7AE
Tel: 020 7235 2020

**Heating and Ventilating
Contractors' Association**
Esca House
34 Palace Court
London W2 4JG
Tel: 020 7229 2488

**National Association of Plumbing,
Heating and Mechanical Services
Contractors**
Ensign House
Ensign Business Centre
Westwood Way
Coventry CV4 8JA
Tel: 01203 470626

Home security

Local Crime Prevention Officer
Refer to your local directory.

Local Fire Prevention Officer
Refer to your local directory.

**National Approval Council for
Security Systems**
Queensgate House
14 Cookham Road
Maidenhead S16 8AJ
Tel: 01628 37512

Master Locksmiths Association
5d Great Central Way
Woodford Halse
Daventry
Northants NN1 3PZ
Tel: 01327 262 255

**British Security Industry
Association**
Kirkham House
John Comyn Drive
Worcester WR3 7NS
Tel: 0845 389 3889

Insulation installers

**Draught Proofing Advisory
Association Ltd**
PO Box 12
Haslemere
Surrey GU27 3AH
Tel: 01428 654011

**External Wall Insulation
Association**
PO Box 12
Haslemere
Surrey GU27 3AH
Tel: 01428 654011

National Insulation Association
2 Vimy Court
Vimy Road
Leighton Buzzard
Beds LU7 1FG
Tel: 08451 636363

**National Association of Loft
Insulation Contractors**
PO Box 12
Haslemere
Surrey GU27 3AH
Tel: 01428 654011

Plasterers

Federation of Master Builders
Gordon Fisher House
14–15 Great James Street
London WC1N 3DP
Tel: 020 7242 7583

Plumbers

**National Association of Plumbing,
Heating and
Mechanical Services Contractors**
Ensign House
Ensign Business Centre
Westwood Way
Coventry CV4 8JA
Tel: 01203 470626

Robust Detail

Robust Details Ltd
Davy Avenue
Knowlhill
Milton Keynes MK5 8NB
Business line: 0870 240 8210
Technical support line: 0870 240 8209
Fax: 0870 240 8203
Technical email support: technical@robustdetails.com

Administrative email support:
administration@robustdetails.com
Other support: customerservice@robustdetails.com

Roofers

Builders' Merchants Federation
15 Soho Square
London W1V 3HL
Tel: 020 7439 1753

**National Federation of Roofing
Contractors**
24 Worship Street
London EC2A 2DY
Tel: 020 7638 7663

Ventilation

**Heating and Ventilating
Contractors' Association**
Esca House
34 Palace Court
London W2 4JG
Tel: 020 7229 2488

Other useful contacts

British Board of Agrément (BBA)
Bucknalls Lane
Garston
Watford WD5 9BA
Tel: 01923 665300
Fax: 01923 665301
Email: contact@bba.star.co.uk
Internet: http://www.bbacerts.co.uk

British Standards Institution (BSI)
389 Chiswick High Road
London W4 4AL
Tel: 020 8996 9001
Fax: 020 8996 7001
Email: csenices@bsigroup.com
Internet: http://www.bsigroup.com

Fenestration Self-Assessment Scheme
Fensa Ltd
54 Agres Street
London SE1 1EU
Tel: 020 7645 3700
Fax: 020 7407 8307

HETAS Ltd
Orchard Business Centre
Stoke Orchard
Cheltenham
Gloucestershire GL52 7RZ
Tel: 0845 634 5626
Note: Registration scheme for companies and engineers involved in the installation and maintenance of domestic solid-fuel-fired equipment.

HMSO
The Stationery Office
PO Box 29
Norwich NR3 1GN
Tel: orders/general enquiries 0870 600 5522
Fax: orders 0870 600 5533
Internet: http://www.tsoshop.co.uk

Institute of Plumbing
64 Station Lane
Hornchurch
Essex RM12 6NH
Tel: 01708 472791
Fax: 01708 448987
Note: Approved Contractor Person Scheme (Building Regulations).

OFTEC
Oil Firing Registration Scheme
Foxwood House
Dobbs Lane
Kesgrave
Ipswich IPS 2QQ
Tel: 0845 658 5080
Fax: 0845 6585181

Robust Details Limited
Davy Avenue
Knowlhill
Milton Keynes MK5 8NB
Tel: customer service 0870 240 8210
Tel: technical support 0870 240 8209
Fax: 0870 240 8203

UKAS
United Kingdom Accreditation Service
21–47 High Street
Feltham
Middlesex TW3 4UN
Tel: 0208 917 8400

WIMLAS
WIMLAS Limited
St Peter's House
6–8 High Street
Aver
Buckinghamshire SL0 9NG
Tel: 01753 737744
Fax: 01753 792321
Email: wimlas@compuserve.com

Useful websites

The Building Act and Building Regulations: http://www.communities.gov.uk (the new DCLG site)

Approved Documents: http://www.planningportal.gov.uk/buildingregulations/approveddocuments

Air Tightness Testing and Measurement Association (ATTMA)	http://www.attma.org
Building Research Establishment Ltd (BRE)	http://www.bre.co.uk
British Flue and Chimney Manufacturers Association (BFCMA)	http://www.feta.co.uk/bfcma
British Standards Institution (BSI)	http://www.bsigroup.com
Building Services Research and Information Association (BSRIA)	http://www.bsria.co.uk
Centre for Window and Cladding Technology (CWCT)	http://www.cwct.co.uk
Chartered Institution of Building Services Engineers (CIBSE)	http://www.cibse.org
Department for Education	http://www.education.gov.uk
Department for Business, Innovation and Skills	http://www.bis.gov.uk
Department of Energy and Climate Change (DECC)	http://www.decc.gov.uk
Department for Transport (DFT)	http://www.dft.gov.uk
Energy Saving Trust (EST)	http://www.energysavingtrust.org.uk
English Heritage	http://www.english-heritage.org.uk
Environment Agency	http://www.environment-agency.gov.uk
Gas Safe Register	http://www.gassaferegister.co.uk
Heating Equipment Testing and Approval Scheme (HETAS)	http://www.hetas.co.uk
Health and Safety Executive (HSE)	http://www.hse.gov.uk
Heating and Ventilating Contractors' Association (HVCA)	http://www.hvca.org.uk
Institution of Gas Engineers & Managers	http://www.igem.org.uk
Metal Cladding and Roofing Manufacturers Association (MCRMA)	http://www.mcrma.co.uk
Modular and Portable Buildings Association (MPBA)	http://www.mpba.biz
National Association of Rooflight Manufacturers (NARM)	http://www.narm.org.uk

Local councils (A–Z list)	http://www.idea.gov.uk/idk/org/la-data.do
Oil Firing Technical Association Ltd (OFTEC)	http://www.oftec.org
Solid Fuel Association (SFA)	http://www.solidfuel.co.uk
Thermal Insulation Manufacturers and Suppliers Association (TIMSA)	http://www.timsa.org.uk
TrustMark	http://www.trustmark.org.uk

Assessable thresholds	http://www.tso.co.uk
British Standards	http://www.bsonline.techindex.co.uk
Building near trees	http://www.nhbc.co.uk
Building research	http://www.bre.co.uk
Carbon dioxide from natural sources and mining areas	http://www.bgs.ac.uk http://www.tso.co.uk http://www.defra.gov.uk http://www.ciria.org.uk
Cladding	http://www.mcrma.co.uk
Concrete in aggressive ground	http://www.bre.co.uk
Contaminated land	http://www.defra.gov.uk http://www.ciria.org.uk http://www.hse.gov.uk
Contamination in disused coal mines	http://www.tso.co.uk
Demolition	http://www.ciria.org.uk
Electrical safety	http://www.theiet.org
Environmental aspects	http://www.arup.com http://www.environment-agency.gov.uk
Excavation and disposal	http://www.defra.gov.uk http://www.ciria.org.uk
Flood protection	http://www.ciria.org.uk http://www.environment-agency.gov.uk
Flooding from sewers	http://www.defra.gov.uk http://www.ciria.org.uk
Foundations	http://www.bre.co.uk
Gas contaminated land	http://www.defra.gov.uk http://www.bre.co.uk http://www.ciria.org.uk

Geoenvironmental and geotechnical investigations	http://www.ags.org.uk
Glass and glazing	http://www.ggf.org.uk
Hardcore	http://www.bre.co.uk
Health and safety	http://www.hse.gov.uk http://www.defra.gov.uk
Land quality	http://www.environment-agency.gov.uk
Landfill gas	http://www.ciwm.co.uk http://www.defra.gov.uk http://www.gassim.co.uk
Laying water pipelines in contaminated ground	http://www.fwr.org
Low-rise buildings	http://www.bre.co.uk
Materials and workmanship	http://www.tso.co.uk
Methane	http://www.bgs.ac.uk http://www.tso.co.uk http://www.defra.gov.uk http://www.ciria.org.uk
Oil seeps from natural sources and mining areas	http://www.bgs.ac.uk http://www.tso.co.uk http://www.defra.gov.uk http://www.ciria.org.uk
Petroleum retail sites	http://www.petroleum.co.uk
Pollution control	http://www.communities.gov.uk
Protection of ancient buildings	http://www.spab.org.uk
Radon	http://www.bre.co.uk
Robust construction details	http://www.tso.co.uk
Roofing design	http://www.mcrma.co.uk
Shrinkable clay soils	http://www.bre.co.uk
Soil sampling	http://www.defra.gov.uk
Soils, sludge and sediment	http://www.ciria.org.uk
Subsidence	http://www.bre.co.uk
Thermal bridging	http://www.bre.co.uk http://www.tso.co.uk
Thermal insulation	http://www.bre.co.uk
Timbers	http://www.bre.co.uk

Legislation

SI 1991/1620	Construction Products Regulations 1991
SI 1994/3051	Construction Products (Amendment) Regulations 1994
SI 1994/3260	Electrical Equipment (Safety) Regulations 1994
SI 2000/2531	The Building (Approved Inspectors etc.) Regulations 2010
SI 2000/2532	The Building (Approved Inspectors etc.) Regulations 2000
SI 2006/3418	Electromagnetic Compatibility Regulations 2006
SI 2007/991	Energy Performance of Buildings (Certificates and Inspections) (England and Wales) Regulations 2007
Decision No 1/95	of the EC–Turkey Association Council of 22 December 1995

Bibliography

Standards referred to

British Standards

Compliance with a British, European or International Standard does not in itself confer immunity from legal obligations. British Standards can, however, provide a useful source of information that could be used to supplement or provide an alternative to the guidance given in an Approved Document.

When an Approved Document makes reference to a named standard, the relevant version of the standard is the one listed at the end of the publication. However, if this version of the standard has been revised or updated by the issuing standards body, the new version may be used as a source of guidance, provided it continues to address the relevant requirements of the Regulations.

Drafts for Development (DDs) are not British Standards. They are issued in the DD series of publications and are provisional in nature. They are intended to be applied on a provisional basis so that information and experience of their practical application may be obtained and the document developed. Where the recommendations of a DD are adopted, care should be taken to ensure that the requirements of the Building Regulations are adequately met. Any observations that a user may have in relation to any aspect of a DD should be passed onto the British Standards Institution.

Title	Standard
Investigation of potentially contaminated land. Code of practice	BS 10175:2001
Specification for fibre boards	BS 1142:1989
Specification for clay flue linings and flue terminals	BS 1181:1999
Specification for Portland cements	BS 12:1989
Specification for metal ties for cavity wall construction	BS 1243:1978
Specification for open fireplace components	BS 1251:1987
Wood preservatives. Guidance on choice, use and application	BS 1282:1999
Flue blocks and masonry terminals for gas appliances Part 1:1986 *Specification for precast concrete flue blocks and terminals* Part 2:1989 *Specification for clay flue blocks and terminals*	BS 1289–1:1986

Title	Standard
Specification for tongued and grooved softwood flooring	BS 1297:1987
Steel plate, sheet and strip Part 2:1983 Specification for stainless and heat-resisting steel plate, sheet and strip	BS 1449
Steel plate, sheet and strip. Carbon and carbon manganese plate, sheet and strip. General specifications	BS 1449–1:1991
Specification for stainless and heat-resisting steel plate sheet and strip. AMD 4807 AMD 6646 and AMD 8832	BS 1449–2:1983
Copper indirect cylinders for domestic purposes. Open vented copper cylinders. Requirements and test methods	BS 1566–1:2002
Glossary of terms relating to solid fuel burning equipment Part 1:1994 Domestic appliances	BS 1846
Glossary of terms relating to solid fuel burning equipment Part 1:1994 Domestic appliances	BS 1846–1:1994
Specification for calcium silicate (sandlime and flintlime) bricks	BS 187:1978

Part 4:1980 Field measurement of airborne sound insulation
    between rooms
Part 6:1980 Laboratory measurement of impact sound
    insulation of floors
Part 7:1980 Field measurements of impact sound insulation
    of floors
	BS 2750

Methods of testing plastics
Part 1: Thermal properties: Methods 120A to 12OE: 1990
    Determination of the Vicat softening temperature of
    thermoplastics
	BS 2782

Methods of testing. Plastics. Introduction	BS 2782–0:2004

Fuel oils for non-marine use
Part 2:1988 Specification for fuel oil for agricultural and
    industrial engines and burners (classes A2, C1, C2, D,
    E, F, G and H)
	BS 2869:1998

Fuel oils for agricultural domestic and industrial engines and boilers. Specification	BS 2869:2006
Fuel oils for non-marine use. Specification for fuel oil for agricultural and industrial engines and burners (Classes AZ C1 C2 0 E F G and H). AMD 6505	BS 2869–2:1998
Specification for copper and copper alloys. Tubes Part 1:1971 Copper tubes for water, gas and sanitation	BS 2871
Vitrified clay pipes and fittings and pipe joints for drains and sewers Part 1:1991 Test requirements Part 2:1991 Quality control and sampling Part 3:1991 Test methods	BS 295
Specification for copper hot water storage combination units for domestic purposes	BS 3196:1981
Specification for copper hot water storage combination units for domestic purposes	BS 3198:1981
Specification. Indicator plates for fire hydrants and emergency water supplies	BS 3251:1976
Specification for quality of vitreous china sanitary appliances	BS 3402:1969

Title	Standard
Stairs, ladders and walkways. Code of practice for the design, construction and maintenance of straight stairs and winders	BS 5395–1:2000
Specification for asbestos-cement pipes, joints and fittings for sewerage and drainage	BS 3656:1981 (1990)
Specification for prefabricated drainage stack units in galvanized steel	BS 3868:1995
Specification for topsoil	BS 3882:1994
Specification for clay bricks	BS 3921:1985
Specification for plastics waste traps	BS 3943:1979 (1988)
Specification for electrical controls for household and similar general purposes	BS 3955:1986
Specification for cast iron spigot and socket flue or smoke pipes and fittings	BS 41:1973 (1998)
Discharge and ventilating pipes and fittings, sand-cast or spun in cast iron Part 1:1990 Specification for spigot and socket systems Part 2:1990 Specification for socketless systems	BS 416
Specification for galvanized low carbon steel cisterns, cistern lids, tanks and cylinders. Metric units	BS 417–2:1987
Specification for thermostats for gas-burning appliances	BS 4201:1979 (1984)
Specification for ladders for permanent access to chimneys, other high structures, silos and bins	BS 4211:1987
Cisterns for domestic use. Cold water storage and combined feed and expansion (thermoplastic) cisterns up to 500 l	BS 4213:2004
Specification for cast iron spigot and socket drain pipes and fittings	BS 437:1978
Specification for safety aspects in the design, construction and installation of refrigerating appliances and systems	BS 4434:1989
Specification for sizes of sawn and processed softwood	BS 4471:1987
Specification for the use of structural steel in building Part 2:1969 Metric units	BS 449
Unplasticized PVC soil and ventilating pipes of 82.4 mm minimum mean outside diameter, and fittings and accessories of 82.4 mm and of other sizes	BS 4514:2001
Specification for domestic appliances. Specification for installation design	BS 4543
Factory-made insulated chimneys. Methods of test. AMD 8379	BS 4543–1:1990
Factory-made insulated chimneys. Specification for chimneys with stainless steel flue linings for use with solid fuel fired appliances. AMD 8380	BS 4543–2:1990
Factory-made insulated chimneys. Specification for chimneys with stainless steel flue lining for use with oil fired appliances. AMD 8381	BS 4543–3:1990
Specification for un-plasticized polyvinyl chloride (PVC-U) pipes and plastics fittings of nominal sizes 110 and 160 for below ground drainage and sewerage	BS 4660:1989

Title	Standard
Fire tests on building materials and structures Part 3:1958 *External fire exposure roof tests* Part 4:1970 *(1984) Non combustibility test for materials* Part 6:1981 *Method of test for fire propagation for products* Part 6:1989 *Method of test for fire propagation for products* Part 7:1971 *Surface spread of flame tests for materials* Part 7:1987 *Method for classification of the surface spread of flame of products* Part 7:1997 *Method of test to determine the classification of the surface spread of flame of products* Part 8:1972 *Test methods and criteria for the fire resistance of elements of building construction* Part 11:1982 (1988) *Method for assessing the heat emission from building materials* Part 20:1987 *Method for determination of the fire resistance of elements of construction (general principles)* Part 21:1987 *Methods for determination of the fire resistance of load bearing elements of construction* Part 22:1987 *Methods for determination of the fire resistance of non-load bearing elements of construction* Part 23:1987 *Methods for determination of the contribution of components to the fire resistance of a structure* Part 24:1987 *Method for determination of the fire resistance of ventilation ducts* Part 31: *Methods for measuring smoke penetration through doorsets and shutter assemblies.* Section 31.1:1983 *Measurement under ambient temperature conditions*	BS 476
Fire tests on building materials and structures. Method for assessing the heat emission from building materials	BS 476–11:1982
Fire tests on building materials and structures. Method for determination of the fire resistance of elements of construction (general principles)	BS 476–20:1987
Fire tests on building materials and structures. Methods for determination of the fire resistance of loadbearing elements of construction	BS 476–21:1987
Fire tests on building materials and structures. Methods for determination of the fire resistance of non-loadbearing elements of construction	BS 476–22:1987
Fire tests on building materials and structures. Methods for determination of the fire resistance of non-load bearing elements of construction	BS 476–22:1987
Fire tests on building materials and structures. Methods for determination of the contribution of components to the fire resistance of a structure	BS 476–23:1987
Fire tests on building materials and structures. Method for determination of the fire resistance of ventilation ducts	BS 476–24:1987
Fire tests on building materials and structures. Classification and method of test for external fire exposure to roofs	BS 476–3:2004
Fire tests on building materials and structures. Non-combustibility test for materials. AMD 2483 and MAD 4390	BS 476–4:1970 (2007)
Fire tests on building materials and structures. Method of test for fire propagation for products	BS 476–6:1989

Title	Standard
Fire tests on building materials and structures. Method of test to determine the classification of the surface spread of flame of products	BS 476–7:1997
Fire tests on building materials and structures. Test methods and criteria for the fire resistance of elements of building construction (withdrawn)	BS 476–8:1972
Internal and external wood doorsets, door leaves and frames Part 1:1980 (1985) *Specification for dimensional requirements*	BS 4787
Specification for performance requirements for domestic flued oil burning appliances (including test procedures)	BS 4876:1984
Specification for softwood grades for structural use	BS 4978:1988
Structural fixings in concrete and masonry Part 1:1993 *Method of test for tensile loading*	BS 5080:1993
Code of practice for sheet roof and wall coverings Part 14:1975 *Corrugated asbestos-cement*	BS 5247
Code of practice for control of condensation in buildings	BS 5250:2002
Code of practice for the control of condensation in buildings	BS 5250:2002
Specification for thermoplastics waste pipe and fittings	BS 5255:1989
Code of practice for external renderings	BS 5262:1991
Emergency lighting Part 1:1988 *Code of practice for the emergency lighting of premises other than cinemas and certain other specified premises used for entertainment*	BS 5266
Emergency lighting code of practice for the emergency lighting of premises	BS 5266–1:2005
Structural use of timber Part 2:2002 *Code of practice for permissible stress design, materials and workmanship* Part 3:1998 *Code of practice for trussed rafter roofs* Part 6 *Code of practice for timber framed walls* Part 6.1:1988 *Dwellings not exceeding three storeys*	BS 5268
Fire extinguishing installations and equipment on premises Part 1:1976 (1988) *Hydrant systems, hose reels and foam inlets* Part 2:1990 *Specification for sprinkler systems*	BS 5306
Concrete Part 1:1990 *Guide to specifying concrete* Part 2:1990 *Method for specifying concrete mixes* Part 3:1990 *Specification for the procedures to be used in producing and transporting concrete* Part 4:1990 *Specification for the procedures to be used in sampling, testing and assessing compliance of concrete*	BS 5328
Code of practice for stone masonry	BS 5390:1976 (1984)
Stairs, ladders and walkways Part 1:1977 *Code of practice for stairs* Part 2:1984 *Code of practice for the design of helical and spiral stairs* Part 3:1985 *Code of practice for the design of industrial type stairs, permanent ladders and walkways*	BS 5395

Title	Standard
Code of practice for oil firing. Installations up to 44 kW output capacity for space heating and hot water supply purposes. AMD 3637	BS 5410–1:1997
Code of practice for oil firing. Installations of 45 kW and above output capacity for space heating hot water and steam supply services	BS 5410–2:1978
Method for specifying thermal insulating materials for pipes, tanks, vessels, ductwork and equipment operating within the temperature range 240° C to 170° C	BS 5422:2001
Methods of test for flammability of textile fabrics when subjected to a small igniting flame applied to the face or bottom edge of vertically oriented specimens, Test 2	BS 5438:1989
Installation and maintenance of flues and ventilation for gas appliances of rated input not exceeding 70 kW net (1st, 2nd and 3rd family gases). Specification for installation and maintenance of flues	BS 5440-1:2008
Installation and maintenance of flues and ventilation for gas appliances of rated input not exceeding 70 kW net (1st, 2nd and 3rd family gases). Specification for installation and maintenance of ventilation for gas appliances	BS 5440–2:2000
Fire detection and fire alarm devices for dwellings. Specification for heat alarms	BS 5446–2:2003
Fire safety signs, notices and graphic symbols Part 1:1990 Specification for fire safety signs	BS 5449
Recommendations for the storage and exhibition of archival documents	BS 5454:2000
Specification for unplasticized PVC pipe and fittings for gravity sewers	BS 5481:1977 (1989)
Code of practice for domestic butane- and propane-gas-burning installations. Installations at permanent dwellings residential park homes and commercial premises with installation pipework sizes not exceeding DN 25 for steel and ON 28 for corrugated stainless steel or copper	BS 5482–1:2005
Graphical symbols and signs. Safety signs, including fire safety signs. Specification for geometric shapes, colours and layout	BS 5499–1:2002
Buildings and Structures for Agriculture. Various relevant parts including Part 33:1991 Guide to the control of odour pollution Part 52:1991 Code of practice for design of alarm systems, emergency ventilation and smoke ventilation for livestock housing	BS 5502
Buildings and structures for agriculture. Various relevant parts including Part 33:1991 Guide to the control of odour pollution, AMD 10014 1998 Part 52:1991 Code of practice for design of alarm systems, emergency ventilation and smoke ventilation for livestock housing, AMD 10014 1998	BS 5502:2003
Vitreous china washdown WC pans with horizontal outlet. Specification for WC pans with horizontal outlet for use with 7.5 L maximum flush capacity cisterns	BS 5503–3:1990

Title	Standard
Wall hung WC pan, specification for weight hung WC pans for Specification for WC pans with horizontal outlet for use with 7.5 L maximum flush capacity cisterns	BS 5504–4:1990
Specification for installation of hot water supplies for domestic purposes using gas fired appliances of rated input not exceeding 70 kW	BS 5546:2000
Code of practice for sanitary pipework	BS 5572:1978
Fire precautions in the design, construction and use of buildings Part 0:1996 *Guide to fire safety codes of practice for particular premises* Part 1:1990 *Code of practice for residential buildings* Part 4:1998 *Code of practice for smoke control using pressure differentials* Part 5:1991 *Code of practice for fire fighting in stairs and lifts* Part 6:1991 *Code of practice for places of assembly* Part 7:1997 *Code of practice for the incorporation of atria in buildings* Part 8:1999 *Code of practice for means of escape for disabled people* Part 9:1989 *Code of practice for ventilation and air conditioning ductwork* Part 10:1991 *Code of practice for shopping complexes* Part 11:1997 *Code of practice for shops, offices, industrial, storage and other similar buildings* Part 12:2004 *Code of practice for construction and use of buildings. Managing fire safety*	BS 5588
Specification for urea–formaldehyde (UF) foam systems suitable for thermal insulation of cavity walls with masonry or concrete inner and outer leaves	BS 5617:1985
Code of practice for thermal insulation of cavity walls (with masonry or concrete inner and outer leaves) by filling with urea–formaldehyde (UF) foam systems	BS 5618:1985
Code of practice for use of masonry Part 1:1992 *Structural use of un-reinforced masonry* Part 2:2000 *Structural use of reinforced and prestressed masonry* Part 3:2001 *Materials and components, design and workmanship*	BS 5628
Lifts and service lifts: Part 1:1986 *Safety rules for the construction and installation of electric lifts* (Part 1 to be replaced by BS EN 81-1, when published) Part 2:1988 *Safety rules for the construction and installation of hydraulic lifts* (Part 2 to be replaced by BS EN 81-2, when published) Part 5:1989 *Specifications for dimensions for standard lift arrangements* Part 7:1983 *Specification for manual control devices, indicators and additional fittings. Amendment slip*	BS 5655
Code of practice for mechanical ventilation and air-conditioning in buildings	BS 5720:1979

Title	Standard
Specification for powered stairlifts	BS 5776:1996
Thermal insulation for use in pitched roof spaces in dwellings Part 5: Specification for installation of man-made mineral fibre and cellulose fibre insulation	BS 5803–5:1985 (as amended 1999)
Methods for rating the sound insulation in building elements. Part 1:1984 Method for rating the airborne sound insulation in buildings and interior building elements Part 2:1984 Method for rating the impact sound insulation	BS 5821
Fire detection and alarm systems for buildings Part 1:2002 Code of practice for system design, installation and servicing Part 2:1983 Specification for manual call points Part 6:1995 Code of practice for the design and installation of fire detection and alarm systems in dwellings. Part 6:2004 Code of practice for the design, installation and main- tenance of fire detection and fire alarm systems in dwellings Part 8:1998 Code of practice for the design, installation and servicing of voice alarm systems Part 9 Code of practice for the design, installation, commissioning and maintenance of emergency voice communication systems	BS 5839
Wood stairs Part 1:1989 Specification for stairs with closed risers for domestic use, including straight and winder flights and quarter and half landings	BS 585
Code of practice for flues and flue structures in buildings	BS 5854:1980 (1996)
Specification for installation in domestic premises of gas-fired ducted-air heaters of rated input not exceeding 60 kW	BS 5864:2004
Specification for fabrics for curtains and drapes Part 2:1980 Flammability requirements	BS 5867
Specification for installation of gas fires convector heaters fire/back boilers and decorative fuel effect gas appliances. Gas fires convector heaters and fire/back boilers and heating stoves (1st, 2nd and 3rd family gases)	BS 5871–1:2005
Specification for installation of gas fires convector heaters fire/back boilers and decorative fuel effect gas appliances. Inset live fuel effect gas fires of heat input not exceeding 15 kW (2nd and 3rd family gases)	BS 5871–2:2005
Specification for installation of gas fires convector heaters fire/back boilers and decorative fuel effect gas appliances. Decorative fuel effect gas appliances of heat input not exceeding 20 kW (2nd and 3rd family gases)	BS 5871–3:2005
Code of practice for the storage and on-site treatment of solid waste from buildings	BS 5906:1980 (1987)
Precast concrete pipes, fittings and ancillary products Part 2:1982 Specification for inspection chambers and street gullies Part 100:1988 Specification for un-reinforced and reinforced pipes and fittings with flexible joints	BS 5911

Title	Standard
Part 101:1988 *Specification for glass composite concrete (GCC) pipes and fittings with flexible joints* Part 120:1989 *Specification for reinforced jacking pipes with flexible joints* Part 200:1989 *Specification for un-reinforced and reinforced manholes and soakaways of circular cross section*	
Code of practice for solar heating systems for domestic hot water	BS 5918:1989
Code of practice for ventilation principles and designing for natural ventilation. AMD 8930 1995	BS 5925:1991
Code of practice for site investigations	BS 5930:1999
Structural use of steelwork in building Part 1:2000 *Code of practice for design* Part 2:2001 *Specification for materials, fabrication and erection* Part 3:1990 *Design in composite construction* Part 4:1994 *Code of practice for design of composite slabs with profiled steel sheeting* Part 5:1998 *Code of practice for design of cold formed thin gauge sections*	BS 5950
Precast concrete masonry units Part 1:1981 *Specification for precast concrete masonry units*	BS 6073
Specification for flexible joints for grey or ductile cast iron drain pipes and fittings (BS 437) and for discharge and ventilating pipes and fittings (BS 416)	BS 6087:1990
Specification for installation of domestic gas cooking appliances (1st, 2nd and 3rd family gases).	BS 6172:2004
Specification for installation of gas fired catering appliances for use in all types of catering establishments (1st, 2nd and 3rd family gases)	BS 6173:2001
Code of practice for protective barriers in and about buildings	BS 6180:1995
Specification for impact performance requirements for flat safety glass and safety plastics for use in buildings	BS 6206:1981
Flat roofs with continuously supported coverings. Code of practice	BS 6229:2003
Flat roofs with continuously supported coverings. Code of practice	BS 6229:2003
Thermal insulation of cavity walls by filling with blown man-made mineral fibre Part 1:1982 *Specification for the performance of installation systems* Part 2:1982 *Code of practice for installation of blown man-made mineral fibre in cavity walls with masonry and/or concrete leaves*	BS 6232
Safety and control devices for use in hot water systems Part 2:1991 *Specification for temperature relief valves for pressures from 1 bar to 10 bar* Part 3:1991 *Specification for combined temperature and pressure relief valves for pressures from 1 bar to 10 bar*	BS 6283
Safety and control devices for use in hot water systems. Specifications for temperature relief valves for pressures from 1 bar to 10 bar	BS 6283–2:1991

Title	Standard
Safety and control devices for use in hot water systems. Specification for combined temperature and pressure relief valves for pressures from 1 bar to 10 bar	BS 6283–3:1991
Code of practice for design and installation of small sewage treatment works and cesspools	BS 6297:1983
Guide to development and presentation of fire tests and their use in hazard assessment	BS 6336:1998
Code of practice for drainage of roofs and paved areas	BS 6367:1983
Performance of windows and doors. Classification for weathertightness and guidance on selection and specification	BS 6375–1:2009
Specification for performance requirements for cables required to maintain circuit integrity under fire conditions	BS 6387:1994
Loading for buildings Part 1:1996 *Code of practice for dead and imposed loads* Part 2:1997 *Code of practice for wind loads* Part 3:1988 *Code of practice for imposed roof loads*	BS 6399
Code of practice for powered lifting platforms for use by disabled persons (Amendment due 1999)	BS 6440:1983
Installation of chimneys and flues for domestic appliances burning solid fuel (including wood and peat) Part 1:1984 (1998) *Code of practice for masonry chimneys and flue pipes*	BS 6461
Sanitary installations Part 1:1984 *Code of practice for scale of provision, selection and installation of sanitary appliances*	BS 6465
Sanitary installations. *Code of practice for space requirements for sanitary appliances*	BS 6465–2:1996
Sanitary installations. *Code of practice for the selection, installation and maintenance of sanitary and associated appliances*	BS 6465–3:2006
Sanitary installations. *Code of practice for the design of sanitary facilities and scales of provision of sanitary and associated appliances*	BS 6485–1:2006 1 A1:2009
Specification for vitrified clay pipes, fittings and ducts, also flexible mechanical joints for use solely with surface water pipes and fittings	BS 65:1991
Specification for clay and calcium silicate modular bricks	BS 6649:1985
Guide for design, construction and maintenance of single-skin air supported structures	BS 6661:1986
Specification for design, installation, testing and maintenance of services supplying water for domestic use within buildings and their curtilages	BS 6700:1987
Design, installation, testing and maintenance of services supplying water for domestic use within buildings and their curtilages	BS 6700:2006 1 Al :2009
Specification for modular co-ordination in building	BS 6750:1986
Specification for installation of gas-fired boilers of rated input not exceeding 70 kW	BS 6798:2009

Title	Standard
Specification for design and construction of fully supported lead sheet roof and wall coverings	BS 6915:2001
Specification for copper direct cylinders for domestic purposes	BS 699:1984
Specification for vitreous-enamelled low-carbon-steel fluepipes, other components and accessories for solid-fuel-burning appliances with a maximum rated output of 45 kW	BS 6999. 1989 (1996)
Specification for metal flue pipes fittings – terminals and accessories for gas-fired appliances with a rated input not exceeding 60 kW. AMD 8413	BS 715:2005
Method of test for ignitability of fabrics used in the construction of large tented structures	BS 7157:1989
Specification for plastics inspection chambers for drains	BS 7158:2001
Specification for unvented hot water storage units and packages	BS 7206:1990
Code of practice for the operation of fire protection measures. Electrical actuation of gaseous total flooding extinguishing systems	BS 7273–1:2006
Code of practice for the operation of fire protection measures. Mechanical actuation of gaseous total flooding and local application extinguishing systems	BS 7273–2:1992
Code of practice for the operation of fire protection measures. Electrical actuation of pre-action sprinkler systems	BS 7273–3:2000
Thermoplastics pipes and associated fittings for hot and cold water for domestic purposes and heating installations in buildings. General requirements	BS 7291–1:2006
Thermoplastics pipes and associated fittings for hot and cold water for domestic purposes and heating installations in buildings. Specification for polybutylene (PR) pipes and associated fittings	BS 7291–2:2006
Thermoplastics pipes and associated fittings for hot and cold water for domestic purposes and heating installations in buildings. Specification for cross-linked polyethylene (PE-X) pipes and associated fittings	BS 7291–3:2006
Specification for direct surfaced wood chipboard based on thermosetting resins	BS 7331:1990
Components for smoke and heat control systems Part 2:1990 *Specification for powered smoke and heat exhaust ventilators* Part 6:2005 *Components for smoke and heat control systems. Specifications for cable systems* Part 7:2006 *Components for smoke and heat control systems. Code of practice on functional recommendations and calculation methods for smoke and heat control systems for covered car parks*	BS 7346
Fibre cement flue pipes fittings and terminals. Specification for light quality fibre cement flue pipes fittings and terminals	BS 7435–1:1991 (1998)
Fibre cement flue pipes fittings and terminals. Specifications for heavy quality cement flue pipes fittings and terminals	BS 7435–2:1991
Installation of factory-made chimneys to BS 4543 for domestic appliances	BS 7566

Title	Standard
Part 1:1992 (1998) *Method of specifying installation design information* Part 2:1992 *(1998) Specification for installation design* Part 3:1992 (1998) *Specification for site installation* Part 4:1992 (1998) *Recommendations for installation design and installation*	
Code of practice for the audio-frequency induction-loop systems (AFILS)	BS 7594:1993
Requirements for electrical installations (IEE wiring regulations, 16th edition)	BS 7671:2001 (2004)
Requirements for electrical installations (IET wiring regulations, 17th edition)	BS 7671:2008
Guide to the principles of the conservation of historic buildings	BS 7913:1998
Principles of the conservation of historic buildings	BS 7913:1998
Application of fire safety engineering principles to the design of buildings. Code of practice	BS 7974:2001
Oil burning equipment. Specification for oil storage tanks	BS 799–5:1987
Workmanship on building sites Part 6:1990 *Code of practice for slating and tiling of roofs and claddings* Part 13:1989 *Code of practice for above ground drainage and sanitary appliances* Part 14:1989 *Code of practice for below ground drainage*	BS 8000
Workmanship on building sites. Code of practice for hot and cold water services (domestic scale)	BS 8000–15:1990
Code of practice for earth retaining structures	BS 8002:1994
Code of practice for foundations	BS 8004:1986
Code of practice for protection of structures against water from the ground	BS 8102:1990
Structural design of low-rise buildings Part 1:1995 *Code of practice for stability, site investigation, foundations and ground floor slabs for housing* Part 2:1996 *Code of practice for masonry walls for housing* Part 3:1996 *Code of practice for timber floors and roofs for housing* Part 4:1995 *Code of practice for suspended concrete floors for housing*	BS 8103
Code of practice for assessing exposure of walls to wind-driven rain	BS 8104:1992
Structural use of concrete Part 1:1997 *Code of practice for design and construction* Part 2:1985 *Code of practice for special circumstances* Part 3:1995 *Design charts for single reinforced beams, doubly reinforced beams and rectangular columns*	BS 8110
Structural use of aluminium Part 1:1991 *Code of practice for design. Amendment slip* Part 2:1991 *Specification for materials, workmanship and protection*	BS 8118

Title	Standard
Code of practice for design of non-load bearing external vertical enclosures of buildings	BS 8200:1985
Code of practice for daylighting	BS 8206: Part 2
Lighting for buildings. Code of practice for daylighting	BS 8206–2:2008
Guide to assessment of suitability of external cavity walls for filling with thermal insulants Part 1:1985 Existing traditional cavity construction	BS 8208
Windows, doors and rooflights Part 1 Code of practice for safety in use and during cleaning of windows and doors (including guidance on cleaning materials and methods)	BS 8213: Part 1:1991
Code of practice for fire door assemblies with non-metallic leaves	BS 8214:1990
Code of practice for design and installation of damp-proof courses in masonry construction	BS 8215:1991
Profiled fibre cement. Code of practice	BS 8219:2001
Sound Insulation and Noise Reduction for Buildings. Code of practice	BS 8233:1999
Code of practice for design and installation of non-loadbearing precast concrete cladding	BS 8297:2000
Code of practice for design and installation of natural stone cladding and lining	BS 8298:1994
Design of buildings and their approaches to meet the needs of disabled people. Code of practice	BS 8300:2001
Code of practice for building drainage	BS 8301:1985
Installation of domestic heating and cooking appliances burning solid mineral fuels specification for the design of installations	BS 8303–1:1994
Installation of domestic heating and cooking appliances burning solid mineral fuels specification for installing and commissioning on site	BS 8303–2:1994
Installation of domestic heating and cooking appliances burning solid mineral fuels recommendations for design and on site installation	BS 8303–3:1994
Code of practice for accommodation of building services in ducts	BS 8313:1997
Fire performance of external cladding systems. Test methods for non-loadbearing external cladding systems applied to the face of a building	BS 8414–1:2002
Fire performance of external cladding systems. Test method for non-loadbearing external cladding systems fixed to and supported by a structural steel frame	BS 8414–2:2005
Concrete Part 1:2002 Method of specifying and guidance for the specifier Part 2:2002 Specification for constituents materials and concrete	BS 8500
Rainwater harvesting systems. Code of practice	BS 8515:2009

Title	Standard
Specification for vessels for use in heating systems. Calorifiers and storage vessels for central heating and hot water supply	BS 853–1:1996
Capillary and compression tube fittings of copper and copper alloy Part 2:1983 Specification for capillary and compression fittings for copper tubes	BS 864
Specification for aggregates from natural sources for concrete	BS 882:1983
Sprinkler systems for residential and domestic occupancies. Code of practice	BS 9251:2005
Code of practice for non-automatic fire-fighting systems in buildings	BS 9990:2006
Protection of buildings against water from the ground	BS CP 102:1973
Code of practice for sheet roof and wall coverings	BS CP 143
Stainless steels. List of stainless steels	BS EN 10088–1:2005
Cold rolled steel flat products with high yield strength for cold forming. Technical delivery conditions	BS EN 10268:2006
Copper and copper alloys. Seamless, round copper tubes for water and gas in sanitary and heating applications	BS EN 1057:1996
Vacuum sewerage systems outside buildings	BS EN 1091:1997
Sanitary tapware. Thermostatic mixing valves (PN 10). General technical specification	BS EN 1111:1999
Building hardware. Panic exit devices operated by a horizontal bar. Requirements and test methods	BS EN 1125:1997
Wastewater lifting plants for buildings and sites – principles of construction and testing Part 1 Lifting plants for wastewater containing faecal matter Part 2 Lifting plants for faecal-free wastewater Part 3 Lifting plants for wastewater containing faecal matter for limited application	BS EN 12050:2001
Wastewater lifting plants for buildings and sites. Principles of construction and testing. Lifting plants for wastewater containing faecal matter for limited applications	BS EN 12050–3:2001
Gravity drainage systems inside buildings Part 1 Scope, definitions, general and performance requirements Part 2 Wastewater systems, layout and calculation Part 3 Roof drainage layout and calculation Part 4 Effluent lifting plants, layout and calculation Part 5 Installation, maintenance and user instructions	BS EN 12056:2000
Smoke and heat control systems. Specification for powered smoke and heat exhaust ventilators	BS EN 12101–3:2002
Smoke and heat control systems. Specification for pressure differential systems. Kits	BS EN 12101–6:2005
Vacuum drainage systems inside buildings	BS EN 12109:1999
Building materials and products. Hygrothermal properties. Tabulated design values	BS EN 12524:2000
Copper and copper alloys. Plumbing fittings Part 1 Fittings with ends for capillary soldering or capillary brazing to copper tubes	BS EN 1254:1998

Title	Standard
Part 2 *Fittings with compression ends for use with copper tubes*	
Part 3 *Fittings with compression ends for use with plastic pipes*	
Part 4 *Fittings combining other end connections with capillary or compression ends*	
Part 5 *Fittings with short ends for capillary brazing to copper tubes*	
Small wastewater treatment plants less than 50 PE	BS EN 12566–1:2000
Aggregates for concrete	BS EN 12620:2002
Thermal performance of building materials and products. Determination of thermal resistance by means of guarded hot plate and heat flow meter methods. Dry and moist products of low and medium thermal resistance	BS EN 12664:2001
Thermal performance of building materials and products. Determination of thermal resistance by means of guarded hot plate and heat flow meter methods. Products of high and medium thermal resistance	BS EN 12667:2000
Fixed firefighting systems. Automatic sprinkler systems. Design, installation and maintenance	BS EN 12845:2004
Sanitary tapware. Low pressure thermostatic mixing valves. General technical specifications	BS EN 1287:1999
Water supply. Specification for indirectly heated unvented (closed) storage water heaters	BS EN 12897:2006
Thermal performance of building materials and products. Determination of thermal resistance by means of guarded hot plate and heat flow meter methods. Thick products of high and medium thermal resistance	BS EN 12939:2001
Structural design of buried pipelines under various conditions of loading. General requirements	BS EN 1295–1:1998
Thermal solar systems and components. Factory made systems. General requirements	BS EN 12976–1:2006
Ventilation for buildings. Performance testing of components/ products for residential ventilation. Externally and internally mounted air transfer devices	BS EN 13141–1:2004
Ventilation for buildings. Performance testing of components/ products for residential ventilation. Range hoods for residential use	BS EN 13141–3:2004
Ventilation for buildings. Performance testing of components/ products for residential ventilation. Fans used in residential ventilation systems	BS EN 13141–4:2004
Ventilation for buildings. Performance testing of components/ products for residential ventilation. Exhaust ventilation system packages used in a single dwelling	BS EN 13141–6:2004
Ventilation for buildings. Performance testing of components/ products for residential ventilation. Performance testing of a mechanical supply and exhaust ventilation units (including heat recovery) for mechanical ventilation systems intended for single family dwellings	BS EN 13141–7:2004

Title	Standard
Ventilation for buildings. Performance testing of components/ products for residential ventilation. Performance testing of unducted mechanical supply and exhaust ventilation units [including heat recovery] for mechanical ventilation systems intended for a single room	BS EN 13141–8:2006
Reaction to fire tests for building products. Conditioning procedures and general rules for selection of substrates	BS EN 13238:2001
Plastics piping systems for soil and waste discharge (low and high temperature) within the building structure. Unplasticized polyvinyl chloride (PVC-U). Specifications for pipes, fittings and the system	BS EN 1329–1:2000
Chimneys. Thermal and fluid dynamic calculation methods. Chimneys serving one appliance	BS EN 13384–1:2002 + A22008
Plastics piping systems for non-pressure underground drainage and sewerage. Structured-wall piping systems of unplasticized poly(vinyl chloride) (PVC-U), polypropylene (PP) and polyethylene (PE). General requirements and performance characteristics	BS EN 13476–1:2007
Fire classification of construction products and building elements Part 1:2002 *Classification using test data from reaction to fire tests* Part 2:2003 *Classification using data from fire resistance tests, excluding ventilation services* Part 3:2005 *Classification using data from fire resistance tests on products and elements used in building set-vice installations: fire resisting ducts and fire dampers* Part 4:2005 *Classification using data from fire resistance tests on smoke control systems* Part 5 :2005 *Classification using data from external fire exposure to roof tests*	BS EN 13501
Fire resistance tests for non-loadbearing elements Part 1:1999 *Walls* Part 2:1999 *Ceilings* Part 3:2006 *Curtain walling. Full configuration (complete assembly)*	BS EN 1364
Fire resistance tests for loadbearing elements Part 1:1999 *Walls* Part 2:2000 *Floors and roofs* Part 3:2000 *Beams* Part 4:1999 *Columns*	BS EN 1365
Fire resistance tests for service installations Part 1:1992 *Ducts* Part 2:1999 *Fire dampers* Part 3:2004 *Penetration seals* Part 4:2006 *Linear joint seals* Part 5:2003 *Service ducts and shafts* Part 6:2004 *Raised access and hollow core floors*	BS EN 1366
Reaction to fire tests for building products. Building products excluding footings exposed to thermal attack by a single burning item	BS EN 13823:2002

Title	Standard
Wood-based panels for use in construction. Characteristics, evaluation of conformity and marking	BS EN 13986:2004
Plastics piping systems for non-pressure underground drainage and sewerage. Unplasticized polyvinylchloride (PVC-U). Specifications for pipes, fittings and the system	BS EN 1401–1:1998
Heating fuels. Fatty acid methyl esters (FAME). Requirements and test methods	BS EN 14213:2003
Windows and doors. Product standard, performance characteristics. Windows and external pedestrian doorsets without resistance to fire and/or smoke leakage characteristics	BS EN 14351–1:2006
Windows and doors. Product standard, performance characteristics. Windows and external pedestrian doorsets without resistance to fire and/or smoke leakage characteristics	BS EN 14351–1:2006
Chimneys. General requirements	BS EN 1443:2003
Plastics piping systems for soil and waste discharge (low and high temperature) within the building structure. Polypropylene (PP). Specifications for pipes, fittings and the system	BS EN 1451–1:2000
Plastics piping systems for soil and waste (low and high temperature) within the building structure. Acrylonitrilebutadiene-styrene (ABS). Specifications for pipes, fittings and the system	BS EN 1455–1:2000
Chimneys. Clay/ceramic flue liners. Requirements and test methods	BS EN 1457:2009
Building valves. Combined temperature and pressure relief valves. Tests and requirements	BS EN 1490:2000
Reaction to fire tests for building products. Non-combustibility test	BS EN 150 1182:2002
Plastics. Thermoplastic materials. Determination of Vicat softening temperature (VST)	BS EN 150 306:2004
Building valves. Inline hot water supply tempering valves. Tests and requirements	BS EN 15092:2008
Plastics piping systems for soil and waste discharge (low and high temperature) within the building structure. Polyethylene (PE). Specifications for pipes, fittings and the system	BS EN 1519–1:2000
Chimneys. Design installation and commissioning of chimneys	BS EN 15287–1:2007
Plastics piping systems for soil and waste discharge (low and high temperature) within the building structure	BS EN 1565–1:2000
Plastics piping systems for soil and waste discharge (low and high temperature) within the building structure	BS EN 1566–1:2000
Plastics piping systems for soil and waste discharge (low and high temperature) within the building structure. Chlorinated polyvinyl chloride) (PVC-C). Specification for pipes, fittings and the system	BS EN 1566–1:2000
Construction and testing of drains and sewers	BS EN 1610:1998
Fire resistance tests for door and shutter assemblies. Fire doors and shutters	BS EN 1634–1:2000
Fire resistance tests for door and shutter assemblies Part 2 Fire door hardware	BS EN 1634–2:2008

Title	Standard
Fire resistance tests for door and shutter assemblies. Smoke control doors and shutters	BS EN 1634–3:2001
Pressure sewerage systems outside buildings	BS EN 1671:1997
Chimneys. Clay/ceramic flue blocks for single wall chimneys. Requirements and test methods	BS EN 1806:2006
Chimneys. Requirements for metal chimneys. System chimney products	BS EN 1856–1:2003
Chimneys. Requirements for metal chimneys. Metal liners and connecting flue pipes	BS EN 1856–2:2004
Chimneys. Components. Concrete flue liners	BS EN 1857:2003 + A1:2008
Chimneys. Components. Concrete flue blocks	BS EN 1858:2003
Chimneys. Metal chimneys. Test methods	BS EN 1859:2009
Cement Part 1:2000 Composition, specifications and conformity criteria for common elements Part 2:2000 Conformity evaluation	BS EN 197
Acoustics. Measurement of sound absorption in a reverberation room	BS EN 20354:1993
Mechanical thermostats for gas-burning appliances	BS EN 257:1992
Sanitary tapware. Waste fittings for basins, bidets and baths. General technical specifications	BS EN 274:1993
Acoustics. Method for the determination of dynamic stiffness Part 1 Materials used under floating floors in dwellings	BS EN 29052–1:1992
Vitrified clay pipes and fittings and pipe joints for drains and sewers Part 1:1991 Test requirements Part 2:1991 Quality control and sampling Part 3:1991 Test methods Part 6:1996 Requirements for vitrified clay manholes	BS EN 295
Heating boilers. Heating boilers with forced draught burners. Terminology general requirements – testing and marking	BS EN 303–1:1999
Particleboards. Specifications. Requirements for load-bearing boards for use in humid conditions	BS EN 312–5:1997
Refrigerating systems and heat pumps. Safety and environmental requirements. Installation site and personal protection.	BS EN 378–3:2008
Glass in building. Determination of luminous and solar characteristics of glazing	BS EN 410:1998
Specification for dedicated liquid petroleum gas appliances. Domestic flueless space heaters (including diffusive catalytic combustion heaters)	BS EN 449:2002 + AI :2007
Method of test for resistance to fire of unprotected small cables for use in emergency circuits	BS EN 50200:2006
Fire detection and fire alarm systems. Manual call points	BS EN 54–11:2001
Ductile iron pipes, fittings, accessories and their joints for sewerage applications. Requirements and test methods	BS EN 598:1995

Title	Standard
Household and similar electrical appliances. Safety. Particular requirements for storage water heaters	BS EN 60335–2–21:2003
Specification for safety of household and similar electrical appliances	BS EN 60335–2–35:2002
Specification for safety of household and similar electrical appliances. Particular requirements for fixed immersion heaters	BS EN 60335–2–73:2003
Specification for low-voltage switchgear and controlgear assemblies. Particular requirements or low-voltage switchgear and control assemblies intended to be installed in places where unskilled persons have access to their use	BS EN 60439–3:1991
Automatic electrical controls for household and similar use. Particular requirements for temperature sensing controls	BS EN 60730–2–9:2002
Drain and sewer systems outside buildings Part 1:1996 Generalities and definitions Part 2:1997 Performance requirements Part 3:1997 Planning Part 4:1997 Hydraulic design and environmental aspects Part 5:1997 Rehabilitation Part 6:1998 Pumping installations Part 7:1998 Maintenance and operations	BS EN 752
Specification for masonry units Part 1:2003 Clay masonry units Part 2:2001 Calcium silicate masonry units Part 3 Aggregate concrete masonry units Part 4:2001 Autoclaved aerated concrete masonry units Part 5 Manufactured stone masonry units Part 6:2001 Natural stone masonry units	BS EN 771
Specification for masonry units. Clay masonry units	BS EN 771–1:2003
Safety rules for the construction and installation of lifts. Electric lifts	BS EN 81–1:1998
Safety rules for the construction and installation of lifts. Hydraulic lifts	BS EN 81–2:1998
Safety rules for the construction and installation of lifts. Particular applications for passenger and goods passenger lifts. Accessibility to lifts for persons including persons with disability	BS EN 81–70:2003
Safety rules for the construction and installation of lifts. Particular applications for passenger and goods passenger lifts. Fire fighters lifts	BS EN 81–72:2003
Specification for ancillary components for masonry Part 1 :2001 Ties, tension straps, hangers and brackets Part 2 :2001 Lintels Part 3 :2001 Bed joint reinforcement of steel meshwork	BS EN 845
Installations for separation of light liquids (e.g. petrol or oil) Part 1 Principles of design, performance and testing, marking and quality control	BS EN 858:2001
Cast iron pipes and fittings, their joints and accessories for the evacuation of water from buildings. Requirements, test methods and quality assurance	BS EN 877:1999

Title	Standard
Specification for mortar for masonry Part 2:2002 *Masonry mortar*	BS EN 998
Thermal performance of windows, doors and shutters. Calculation of thermal transmittance Part 1 *Simplified methods*	BS EN ISO 10077–1:2000
Thermal bridges in building construction. Calculation of heat flows and surface temperatures Part 1 *General methods*	BS EN ISO 10211–1:1996
Thermal bridges in building construction. Calculation of heat flows and surface temperatures Part 2 *Linear thermal bridges*	BS EN ISO 10211–2:2001
Plastics. Symbols and abbreviated terms. Basic polymers and their special characteristics	BS EN ISO 1043–1:2002
Acoustics. Sound absorbers for use in buildings. Rating of sound absorption	BS EN ISO 11654:1997
Reaction to fire tests. ignitability of building products subjected to direct impingement of flame. Single-flame source test	BS EN ISO 11925–2:2002
Thermal performance of windows and doors. Determination of thermal transmittance by hot box method Part 1 *Complete windows and doors*	BS EN ISO 12567–1:2000
Thermal performance of buildings. Heat transfer via the ground. Calculation methods	BS EN ISO 13370:2007
Hygrothermal performance of building components and building elements	BS EN ISO 13788:2002

Acoustics. Measurement of sound insulation in buildings and
of building elements
Part 1:1980 *Recommendations for laboratories*
Part 3:1995 *Laboratory measurement of airborne sound
    insulation of building elements*
Part 4:1998 *Field measurements of airborne sound insulation
    between rooms*
Part 6:1998 *Laboratory measurements of impact sound
    insulation of floors*
Part 7:1998 *Field measurements of impact sound insulation
    of floors*
Part 8:1998 *Laboratory measurements of the reduction of
    transmitted impact noise by floor coverings on a
    heavyweight standard floor*
	BS EN ISO 140

Reaction to fire tests for building products. Determination of the heat of combustion	BS EN ISO 1716:2002
Plastics. Thermoplastic materials. Determination of Vicat softening temperature (VST)	BS EN ISO 306:2004
Building components and building elements. Thermal resistance and thermal transmittance. Calculation method	BS EN ISO 6946:1997
Acoustics. Rating of sound insulation in buildings and of building elements Part 1:1997 *Airborne sound insulation* Part 2:1997 *Impact sound insulation*	BS EN ISO 717
Thermal insulation. Determination of steady-state thermal transmission properties. Calibrated and guarded hot box	BS EN ISO 8990:1996

Title	Standard
Fire-resistance tests. Fire dampers for air distribution systems. Classification, criteria and field of application of test results	BS ISO 10294–2:1999
Fire-resistance tests. Fire dampers for air distribution systems. Intumescent fire dampers	BS ISO 10294–5:2005
Ventilation for buildings – design criteria for the indoor environment	BSI PD CR 1752:1999
Components for residential sprinkler systems. Specification and test methods for residential sprinklers	DD 252:2002
Test methods for external fire exposure to roofs	DD ENV 1187:2002
Test methods for external fire exposure to roofs	ENV 1187:2002, test 4
Anti-flooding valves	Pr En 13564
Thermal solar systems and components. Custom built systems. General requirements	prCEN/TS 12977–1:2008
Ventilation for buildings. Performance testing of components/products for residential ventilation Part 8:2004 Performance testing of unducted mechanical supply and exhaust ventilation units (including heat recovery) for mechanical ventilation systems intended for a single room Part 9:2004 Humidity controlled external air inlet Part 10:2004 Performance testing of unducted mechanical supply and exhaust ventilation units [including heat recovery] for mechanical ventilation systems intended for a single room	prEN 13141

 Note: Copies of all British Standards are available from: BSI, PO Box 16206, Chiswick, London W4 4ZL (http://www.bsonline.techindex.co.uk).

Other publications

Air Tightness Testing and Measurement Association (ATTMA)
http://www.attma.org

• Measuring Air Permeability of Building Envelopes

Association for Specialist Fire Protection (ASFP)
http://www.asfp.org.uk

• ASFP Red Book – *Fire stopping and penetration seals for the construction industry*
• ASFP Yellow Book – *Fire protection for structural steel in buildings*
• ASFP Grey Book – *Fire and smoke resisting dampers*
• ASFP Blue Book – *Fire resisting ductwork*

Building Research Establishment Ltd (BRE)
http://www.bre.co.uk

• BR 128 *Guidelines for the construction of fire resisting structural elements*
• BR 135 *Fire performance of external thermal insulation for walls of multi-storey buildings*

- BR 187 *External fire spread. Building separation and boundary distances*
- BR 208 *Increasing the fire resistance of existing timber floors*
- BR 262 *Thermal insulation: avoiding risks*
- BR 274 *Fire safety of PFTE based materials used in buildings*
- BR 364 *Solar shading of buildings*
- BR 369 *Design methodologies for smoke and exhaust ventilation*
- BR 443 *Conventions for U-value calculations*
- BR 498 *Selecting lighting controls*
- BR 454 *Multi-storey timber frame buildings – a design guide*
- Information Paper 1P1/06 *Assessing the effects of thermal bridging at junctions and around openings in the external elements of buildings*
- Information paper 1P14103 *Preventing hot water scalding in bathrooms: using TMVs*

The British Automatic Fire Sprinkler Association (BAFSA)
http://www.bafsa.org.uk

- *Sprinklers for safety. Use and benefits of incorporating sprinklers in buildings and structures*

Builders Hardware Industry Federation

- *Hardware for fire and escape doors*

Centre for Window and Cladding Technology
http://www.cwct.co.uk

- *Thermal assessment of window assemblies, curtain walling and non-traditional building envelopes*

CIBSE
http://www.cibse.org

- CIBSE Commissioning Code M *Commissioning management*
- CIBSE Guide A *Environmental design*
- AM 10 *Natural ventilation in non-domestic buildings*
- *Solar heating design* and *installation guide*
- TM 31 *Building log book toolkit*
- TM 36 *Climate change and the indoor environment.* 28 750 2 *impacts and adaptation*
- TM 37 *Design for improved solar shading control*
- TM 39 *Building energy metering*

Construction (Design and Management) Regulations 2007
http://www.hse.gov.uk/construction/cdm.htm

Department for Communities and Local Government (DCLG)
http://www.communities.gov.uk

- *Fire safety in adult placements: a code of practice*

Department for Education (DFE)
http://www.education.gov.uk

- Building Bulletin 10*1 Ventilation of school buildings, School Building and Design Unit*

Department for Environment, Food and Rural Affairs (Defra)
http://www.defra.gov.uk

- The government's Standard Assessment Procedure for energy rating of dwellings, SAP 2005. (available at: http://www.bre.co.uk/sap2005)

Department of Health (DH)
http://www.dh.gov.uk

- Health Technical Memorandum 05-02 *Guidance to support of functional provisions in healthcare premises*

Door and Shutter Manufacturers' Association (DSMA)
http://www.dhfonline.org.uk

- Code of practice for fire-resisting metal doorsets

Electrical Contractors' Association (ECA) and National Inspection Council for Electrical Installation Contracting (NICEIC)
http://www.eca.co.uk and http://www.niceic.org.uk

- *ECA comprehensive guide to harmonised cable colours*
- *Electrical installers' guide to the Building Regulations*
- *New fixed wiring colours – A practical guide*

Energy Saving Trust (EST)
http://www.est.org.uk

- CE66 *Windows for new and existing housing*
- CE129 *Reducing overheating – a designer's guide*
- GPG268 *Energy efficient ventilation in dwellings – A guide for specifiers*, 2006
- GIL20 *Low energy domestic lighting*

English Heritage
http://www.english-heritage.org.uk

- *Building Regulations and historic buildings*

Environment Agency
http://www.environment-agency.gov.uk

- *Pollution prevention guidelines* (PPG18) – *Managing fire water and major spillages*

Fire Protection Association (FPA)
http://www.thefpa.co.uk

- Design guide

Food Standards Agency
http://www.food.gov.uk

- *Code of practice. Food hygiene – a guide for businesses*

Football Licensing Authority
http://www.flaweb.org.uk

- Concourses

Glass and Glazing Federation (GGF)
http://www.ggf.org.uk

- *A guide to best practice in the specification and use of fire resistant glazed systems*

Health and Safety Executive (HSE)
http://www.hse.gov.uk

- *Workplace (Health, Safety and Welfare) Regulations*
- *Legionnaires' disease: control of Legionella bacteria in water systems*

Heating and Ventilating Contractors Association
http://www.hvca.org.uk

- DW/143 *A practical guide to ductwork leakage testing*
- DW/144 *Specification for sheet metal ductwork*

Institution of Engineering and Technology
http://www.theiet.org

Electrician's guide to the Building Regulations

- *IEE Guidance Note 1: Selection and erection of equipment,* 4th edition
- *IEE Guidance Note 2: Isolation and switching,* 4th edition
- *IEE Guidance Note 3: Inspection and testing,* 4th edition
- *IEE Guidance Note 4: Protection against fire,* 4th edition
- *IEE Guidance Note 5: Protection against electric shock,* 4th edition
- *IEE Guidance Note 6. Protection against overcurrent,* 4th edition
- *IEE Guidance Note 7: Special locations,* 2nd edition
- *IEE On-site guide* (BS 7671 IEE Wiring Regulations, 16th edition)
- *New wiring colours,* 2004

International Association of Cold Storage Contractors (IACSC)
http://www.iarw.org/iacsc

- *Design, construction, specification and fire management of insulated envelopes for temperature controlled environments*

Market Transformation Programme
http://efficient-products.defra.gov.uk/

- *Rainwater and greywater: technical and economic feasibility*
- *Rainwater and greywater: a guide for specifiers*
- *Rainwater and greywater: review of water quality standards and recommendations for the UK*

Metal Cladding and Roofing Manufacturers Association
http://www.mcrma.co.uk

- *Guidance for design of metal cladding and roofing to comply with Approved Document L2*

Modular & Portable Building Association (MPBA)
http://www.mpba.biz

- *Energy performance standards for modular and portable buildings*

National Association of Rooflight Manufacturers
http://www.narm.org.uk

- *Use of rooflights to satisfy the 2002 Building Regulations for the conservation of fuel and power*

NHS

- *National Health Service Model Engineering* Specifications D 08. *Thermostatic Mixing valves (healthcare premises).* Revision 1 (Archived historical document – no longer current), 1997 (http://www.thenbs.com/PublicationIndex/DocumentSummary.aspx?PubID=412&DocID=275997)

Passive Fire Protection Federation (PFPF)
http://www.pfpf.org

- *Ensuring best practice for passive fire protection in buildings*

Planning Portal
http://www.planningportal.gov.uk

- Copies of Approved Documents
- *New rules for electrical safety in the home* (http://www.planningportal.gov.uk/england/professionals/buildingregs/technicalguidance/bcelectricalsafetypartp/bcassociateddocuments2).

Steel Construction Institute (SCI)
http://www.steel-sci.org

- SCI P!97 *Designing for structural safety: a handbook for architects and engineers*
- SCI Publication 288 *Fire safe design. a new approach to multi-storey steel-framed buildings (2nd edition)*
- SCI Publication P313 *Single storey steel framed buildings in fire boundary conditions*

Thermal Insulation Manufacturers and Suppliers Association (TIMSA)
http://www.timsa.org.uk

- *HVAC guidance for achieving compliance with Part L of the Building Regulations*

Timber Research and Development Associations (TRADA)
http://www.trada.co.uk

- *Timber fire-resisting Doorsets; maintaining performance under the new European test standard*

Water Regulations Advisory Service
http://www.wras.co.uk

- *Water Regulations Advisory Scheme. Water Regulations Guide*
- *WRAS information & guidance Note No. 9-02-05 Marking and identification of pipework for reclaimed (greywater) systems*

Legislation

Standard	Title
SI 1991/1620	Construction Products Regulations 1991
SI 1992/2372	Electromagnetic Compatibility Regulations 1992
SI 1992/3004	The Workplace (Health, Safety and Welfare) Regulations 1992
SI 1994/1886	The Gas Safety (Installation and Use) Regulations 1994
SI 1994/3051	Construction Products (Amendment) Regulations 1994
SI 1994/3080	Electromagnetic Compatibility (Amendment) Regulations 1994
SI 1994/3260	Electrical Equipment (Safety) Regulations 1994.
SI 1999/1148	The Water Supply (Water Fittings) Regulations 1999
SI 2000/3184	The Water Supply (Water Quality) Regulations 2000
SI 2001/3335	Building (Amendment) Regulations 2001
SI 2005/1726	Energy Information (Household Air Conditioners) (No. 2) Regulations 2005
SI 2006/14	The Food Hygiene (England) Regulations 2006
SI 2006/31	The Food Hygiene (Wales) Regulations 2006
SI 2006/652	Building And Approved Inspectors (Amendment) Regulations 2006
SI 2009/3101	The Private Water Supplies Regulations 2009
SI 2010/66	The Private Water Supplies (Wales) Regulations 2010

Other

- Council Directive 89/1 06, IEEC on construction products
- Council Directive 89/106/EC, as regards the classification of the external fire performance of roofs and roof coverings
- Commission Decision 2005/823iEC of 22 November 2005 amending Decision 2001/671/EC regarding the classification of the external fire performance of roofs and roof coverings
- 89/106/EEC, *The Construction Products Directive*
- 93/68/EEC, *The CE Marking Directive*
- Disability Discrimination Act 1995 Education Act 1996
- Electrical Equipment (Safety) Regulations 1994 (SI 1994 No. 3260)
- Electromagnetic Compatibility (Amendment) Regulations 1994 (.SI 1994 No. 3080)
- Electromagnetic Compatibility Regulations 1992 (ISI 1992 No. 2372)
- The European Communities Act 1972
- Health and Safety (Safety Signs and Signals) Regulations 1996

- The Health and Safety at Work etc. Act 1974
- Pipelines Safety Regulations 1996, SI 1996 No. 825; and Gas Safety (Installation and Use) Regulations 1998 SI 1998 No. 2451
- The Water Industry Act 1991
- The Workplace (Health, Safety and Welfare) Regulations 1992

Water efficiency

(a) A record of the sanitary appliances and relevant white goods   G2
     (washing machines and dishwashers) used in the water consumption
     calculation and installed in the dwelling shall be provided;
(b) a record of the alternative sources of water used in the water
     consumption calculation and supplied to the dwelling shall be
     provided.

Note: For all new dwellings, the estimated consumption of wholesome water resulting from the design of cold and hot water systems should not be greater than 125 litres/head/day of wholesome water.

Notification of Water Efficiency Calculation to the BCB

A notice specifying the calculated potential consumption of wholesome water   G2
per person per day relating to the dwelling as constructed should be given to
the appropriate BCB (Building Control Body)%, not later than five days after the
completion of the building work

Books by the same authors

Title	Extracts from Book Reviews	ISBN
Wiring Regulations in Brief (2nd edition)	A complete guide to the requirements of the 17th edition of the IEE Wiring Regulations, BS 7671 and Part P of the Building Regulations. • Tired of trawling through the Wiring Regulations? • Perplexed by Part P? • Confused by cables, conductors and circuits? Then look no further! This handy guide provides an on-the-job reference source for electricians, designers, service engineers, inspectors, builders, students, DIY enthusiasts.	Routledge ISBN-13: 978-0750689731

Title	Extracts from Book Reviews	ISBN
Water Regulations in Brief 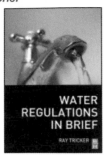	*Water Regulations in Brief* is a unique reference book, providing all the information needed to comply with the regulations, in an easy to use, **full colour** format. Crucially (unlike other titles on this subject) this book **does not** just cover the Water Regulations, it also clearly shows how they link in with the Building Regulations **and** the Wiring Regulations.	Routledge ISBN-13: 978–1–85617–628–6
Scottish Building Standards in Brief 	*Scottish Building Standards in Brief* takes the highly successful formula of Ray Tricker's previous '*In Brief*' series and applies it to the requirements of the Building (Scotland) Regulations 2004. With the same no-nonsense and simple to follow guidance – but written specifically for the Scottish Building Standards – it is the ideal book for builders, architects, designers and DIY enthusiasts working in Scotland.	Routledge ISBN-13: 978-0750685580
ISO 9001:2008 for Small Businesses (4th edition) 	The new edition of this top-selling quality management handbook. Contains a full description of ISO 9001:2008 plus detailed information on quality control and quality assurance. Fully updated following 11 years of practical field experience of the standard. Includes a sample Quality Manual (that can be customised to suit individual requirements) and on-line assistance on self-certification, etc.	Routledge ISBN-13: 978–1856178617

Title	Extracts from Book Reviews	ISBN
ISO 9001:2008 Quality Manual & Audit Checksheets (Second Edition) 	A CD containing a soft copy of the generic Quality Management System featured in *ISO 9001 for Small Businesses* (4th edition) plus a soft copy of all the checksheets and example audit forms contained in *ISO 9001 Audit Procedure* (2nd edition). A comprehensive CD containing all the vital documentation, information and guidance to develop a full quality management system.	Herne European Consultancy Ltd ISBN-13: 978–0954864798
ISO 9001:2008 Audit Procedures (2nd edition) 	This book usefully describes methods for completing management reviews and quality audits. It contains a complete set of audit checksheets and explanations to assist quality managers and auditors in completing internal, external and third-party audits of ISO 9001 Quality Management Systems.	Routledge ISBN-13: 978–0750666152
Quality Management System for ISO 9001: 2008 (2nd Edition) 	This book and the accompanying CD is probably the most comprehensive set of ISO 9001:2008 compliant documents available world-wide. Fully customisable, it can be used as a basic template for any organisation wishing to work in compliance with, or gain registration to, ISO 9001.	Herne European Consultancy Ltd ISBN-13: 978–0954864743

Title	Extracts from Book Reviews	ISBN
ISO 9001:2008 in Brief (2nd edition) 	Revised and expanded, this new edition of an easy to understand guide provides practical information on how to set up a cost-effective ISO 9001 compliant quality management system.	Routledge ISBN-13: 978–0750666169
Auditing Quality Management Systems (2nd edition) 	This publication is the result of over a decade's experience of all major international standards for integrated quality management systems. It is a comprehensive CD containing all the major audit checksheets and forms that are required to conduct either a simple internal audit or an external assessment of an organisation against the formal requirements of ISO 9001:2008. Now fully updated to include checklists for: • project management • health and safety in the workplace. Note: Also includes 'background notes for auditors'.	Herne European Consultancy Ltd ISBN-13: 978–0954864774
MDD Compliance using Quality Management Techniques 	The Medical Device Directive (MDD) is difficult to understand and interpret, but this book covers the subject superlatively. In summary, the book is a good reference for understanding the requirements of the MDD and would aid companies of all sizes in adding these requirements to an existing quality management system.	Routledge ISBN-13: 978–0750644419

Title	Extracts from Book Reviews	ISBN
Quality and Standards in Electronics 	A manufacturer or supplier of electronic equipment or components needs to know the precise requirements for component certification and quality conformance to meet the demands of the customer. This book ensures that the professional is aware of all the UK, European and international necessities, knows the current status of these regulations and standards, and where to obtain them.	Newnes ISBN-13: 978–0750625319
Environmental Requirements for Electromechanical and Electronic Equipment 	This is the definitive reference containing all the background guidance, typical ranges, details of recommended test specifications, case studies and regulations covering the environmental requirements on designers and manufacturers of electrical and electromechanical equipment worldwide.	Routledge ISBN-13: 978–0750639026
CE Conformity Marking 	CE marking can be regarded as a product's trade passport for Europe. It is a mandatory European marking for certain product groups to indicate conformity with the essential health and safety requirements set out in the European Directive. This practical and easy to understand book contains all the essential information for any manufacturer or distributor wishing to trade in the European Union.	Routledge ISBN-13: 978–0750648134

Title	Extracts from Book Reviews	ISBN
ISO 9001:2008 in Brief (2nd edition) 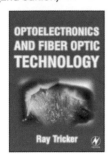	An introduction to the fascinating technology of fibre optics. Students, technicians and professional readers could benefit from this publication. Simply written in an easily accessible style which does not put the reader off and covers all the basic topics in an appropriate and logical order. Topical areas such as optoelectronics in LANs and WANs, cable TV systems, and the global fibre-optic highway make this book essential reading for anyone who needs to keep up with the technology of modern data communications.	Routledge ISBN-13: 978–0750653701

And for those who would like to relax with some cooking recipes – based on cyder and apples!

| *The Cyder Book* | A unique combination of an historical overview of cider making through the ages, the cider making process and a collection of recipes using cider and cider apples. | Herne European Consultancy Ltd ISBN-13: 978–0954864767 |

Index

Contents